W9-CEG-872

Electrical construction wiring

WALTER N. ALERICH

Journeyman Wireman; Coordinator of Instruction,
Electrical-Mechanical Department, Los Angeles Trade-Technical College;
Consultant, California State Department
of Education; B. V. E.; M.A.

AMERICAN TECHNICAL SOCIETY
Chicago, Ill. 60637

Library of Congress Catalog Number: 75-124938
ISBN: 0-8269-1420-9

1st Printing 1971
2nd Printing 1971
3rd Printing 1973
4th Printing 1974
5th Printing 1975
6th Printing 1977
7th Printing 1978

PRINTED IN THE UNITED STATES OF AMERICA

Illustrations on the cover are by courtesy of Minnesota Mining and Mfg. Co., Boron Electric, Inc., Greenlee Tool Co., and Los Angeles Trade Technical College. Illustration on the title page is by courtesy of Minnesota Mining and Mfg. Co.

Preface

Recent technological changes and developments have modified many of the materials and procedures used for wiring. At the same time, changes in the *National Electric Code* have caused many of the old ways to become obsolete.

Electrical Construction Wiring includes the latest accepted changes in wiring; it also covers the more traditional wiring procedures that are still recognized by the NEC. This revised text gives particular emphasis to safe wiring procedures.

This text not only discusses and illustrates what is *safe* but also what is *good* electrical construction. The book details the acceptable practices of quality electrical construction. The student is given accurate, well organized information and is shown "how to" wire the basic systems.

Electrical Construction Wiring is intended for use in giving related instruction for apprenticeship classes, and for journeymen electrician training for wiring and code classes. It gives detailed information necessary to the contractor. The text is also suitable for self study.

Questions at the end of each chapter are for personal check-up, classroom discussion, or assignment. They are designed to reinforce the information given in the chapter and to aid in giving direction for the review of the material in the chapter. Where pertinent, questions are also given on the *National Electric Code*. These may be answered directly from the Code, for minimum standards, and are designed to give practical experience in working electrical problems, and job installations.

The author, Mr. Walter N. Alerich, is a graduated apprentice, journeyman wireman, electrical instructor, administrator, and technical writer. He has attempted to develop a more modern standard for electrical systems, using the instruction methods developed in his many years of work experience and teaching.

The author would like to thank officers and members of the National Electrical Contractors Association, Los Angeles County Chapter, who were most helpful in constructive guidance, as were officers and members of the International Brotherhood of Electrical Workers, Local Number 11, Los Angeles. (The author is a member of the I.B.E.W. Local 11, inside wiremen's unit.) Grateful appreciation is also expressed to electrical instructors and students in pre-employment and apprenticeship classes and journeymen training to whom this text is directed.

The preparation of this instructional material for publication was carried out under the direct supervision of Robert E. Putnam, Senior Technical Editor, American Technical Society.

THE PUBLISHERS

Contents

Chapter

Basic Electrical Theory

Chapter

1

Current (amperes) flows through an electrical conductor because of electrical pressure (volts) in much the same way that water flows through a pipe because of water pressure. Fig. 1-1 shows a simple electrical circuit which contains a source of electricity, lamp or other loads which draws electricity, and connecting wires.

The source of electricity, which may be an electromagnetic unit, a battery, or other similar device, supplies the electrical pressure. This pressure, or voltage, sends an electrical current from the negative terminal of the generator through a connecting wire, the lamp, and the second wire, back to the positive terminal. The path formed by the conductors and the load is termed a *circuit*. Current will not flow unless the circuit is complete. Thus, if the wire is broken at point X, the flow will cease.

Should a wire be connected around the load, or the positive and negative wires connected together, a danger-

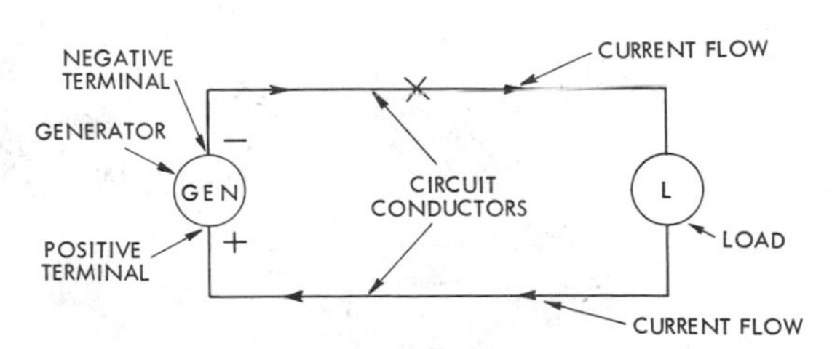

Fig. 1-1. Simple electrical circuit.

ous short circuit would occur because the current limiting resistance would be reduced to zero.

Fig. 1-1 illustrates a generator producing "direct current." This means the current flows in one direction and does not reverse. "Alternating current" is produced by a generator constructed differently. "A.C." reverses in direction of current flow are called "cycles" or "frequency" of reversals. The common A.C. frequency in the United States is 60 *cycles per second*, called a *hertz* (Hz).

Useful Electrical Terms

When working with electricity, and to be able to understand manufacturer's manuals of electrical equipment, it is necessary to become familiar with the various electrical terms. Terms most commonly used, together with definitions, follow:

Ampere (amp). The rate at which a given quantity of electricity flows through a conductor or circuit.

Ampacity. Current carrying capacity expressed in amps.

Ohm. The unit of electrical resistance. The current in an electric circuit is equal to the pressure divided by the resistance. This is known as Ohm's Law.

Volt. The unit used in measuring the electrical pressure causing the current to flow. Voltage is electrical pressure. It is normally supplied to farms and homes in dual voltage 115-230 and 120-240 volts.

Watt. The unit of electrical power. Amps × Volts = Watts. For example, 10 amps × 120 volts equals 1,200 watts.

Kilowatt. One thousand watts.

Kilowatt hours (k.w.h.). A kilowatt of power used for one hour. Electrical energy is metered and sold by the k.w.h.

Conductor. Anything that will carry electric current. For example, the copper wire in a circuit is a conductor.

Insulation. Material used to isolate a charged conductor.

Circuit. Two or more conductors (wires) making a continuous path for the current to travel, from the source of supply to the point where it is used and back to the source.

Fuse. A device used in the circuit to limit the rate of flow (current). It contains a piece of soft metal which melts and opens the circuit when it is overloaded. Most fuses cannot be reused if burned out (blown). Some are renewable, however.

Fusetron and Fusetrat. Devices similar to fuses. They permit an overload for a limited time. Sometimes called delayed action fuses, they cannot be reused.

Circuit Breaker. A switch which stops the flow of current by opening the circuit automatically when more electricity flows through the circuit than the circuit is capable of carrying. Resetting may be either automatic or manual.

Switch. A device to start or stop the flow of electricity.

Horsepower (h.p.). One horsepower equals 746 watts.

Direct Current (D.C.). Current in which the electricity (electrons) flows in one direction only. Examples, a dry cell; an automobile battery.

Alternating Current (A.C.). Current in which the flow of electricity (electrons) is reversed in direction at regular intervals as it passes through the circuit. This is the type of current usually supplied by power companies for use in the home and on the farm and in industry. Generally, a 60-cycle (hertz) current is supplied.

Phase. The type of power available, either single-phase or three-phase. Only single-phase is available in most residential and rural areas, three-phase is used for heavy power consumers, like factories.

Cycle. The flow of alternating current, first in one direction and then in the opposite direction, in one cycle. This occurs 60 times every second in a 60-cycle circuit (also referred to as *hertz*).

Transformer. An apparatus used to increase or decrease the voltage. There are types, *step-up* and *step-down.*

National Electrical Code (NEC). A set of standards covering the design and manufacture of electrical devices and materials, and the manner of their installation. All wiring should conform with the NEC and local codes as well. The validity of most fire insurance policies is based on electrical work that meets NEC approval. Thus, all electrical wiring should be done only by those familiar with the NEC.

Three Primary Circuit Elements

The unit of electrical pressure is the *volt*. The unit of electrical current is the *ampere*. In order to find how many amperes will flow under a pressure of a given number of volts, it is necessary to know the resistance offered to passage of current by the material of which the conductors are made. The unit of electrical resistance is the *ohm*.

These units are so related that current in amperes is equal to pressure in volts divided by resistance in ohms. This relationship is known as *Ohm's law*, which may be written:

$$\text{Current} = \frac{\text{Pressure}}{\text{Resistance}}$$

$$\text{or Amperes} = \frac{\text{Volts}}{\text{Ohms}}$$

The following symbols are commonly employed to represent these three quantities:

Pressure (volts) = E
Current (amperes) = I
Resistance (ohms) = R

Ohm's law may be expressed, therefore, as:

$$I = \frac{E}{R}$$

Ohms are sometimes expressed with the symbol Ω (Greek letter, *Omega*).

If any two of the three quantities are known, the third may be determined by use of the formula, the unknown element being placed at

Fig. 1-2. Voltmeter.

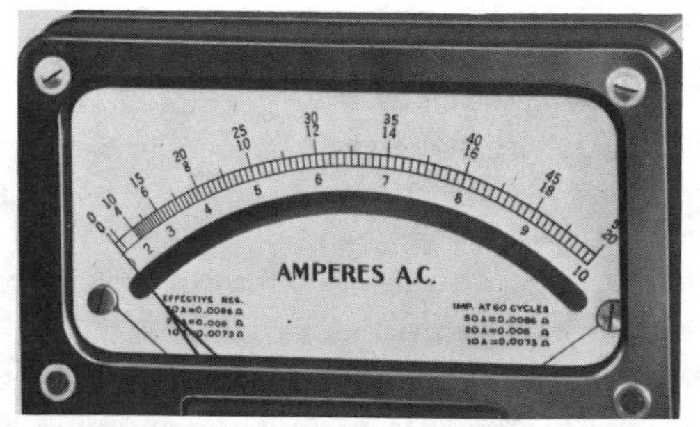

Fig. 1-3. Ammeter. (Weston Electrical Instruments Corp.)

the left of the equality sign, the other two at the right. Thus, if current and resistance are known, and it is desired to find what voltage is acting in the circuit, the formula may be written: $E = I \times R$. If current and voltage are known, the resistance of the circuit is found by writing the formula:

$$R = \frac{E}{I}$$

A voltmeter for measuring voltage and an ammeter for measuring current are illustrated in Figs. 1-2 and 1-3.

Learning Ohm's Law

Since Ohm's law embodies one of the fundamental principles of electricity, it is essential that it be memorized. A simple way of doing so is given in Figs. 1-4 to 1-7. If any one part is removed or covered, the relative position of the other two gives the value of the one covered in terms of the other two.

Thus, if I of Fig. 1-4 is covered, $E \div R$ is left, Fig. 1-5. Therefore, the value of I in terms of E and R is E divided by R. If R is covered, $E \div I$ remains, Fig. 1-6, giving the value of R in terms of E and I, which is E divided by I. In the same way, if E is covered, its value remains in terms of I and R, namely, I times R, Fig. 1-7.

A great amount of practice is required to learn how to apply Ohm's law. Once the principle is firmly grasped, it is possible to handle a wide range of electrical problems. A few examples are given as follows:

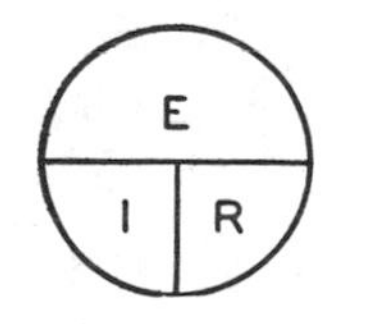

Fig. 1-4. Ohms law.

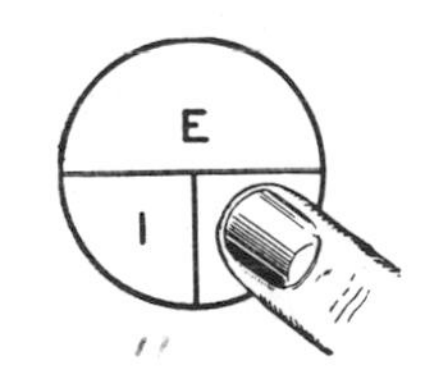

Fig. 1-6. Resistance.

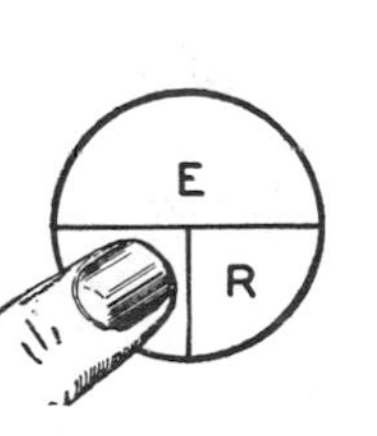

Fig. 1-5. Current.

Fig. 1-7. Voltage.

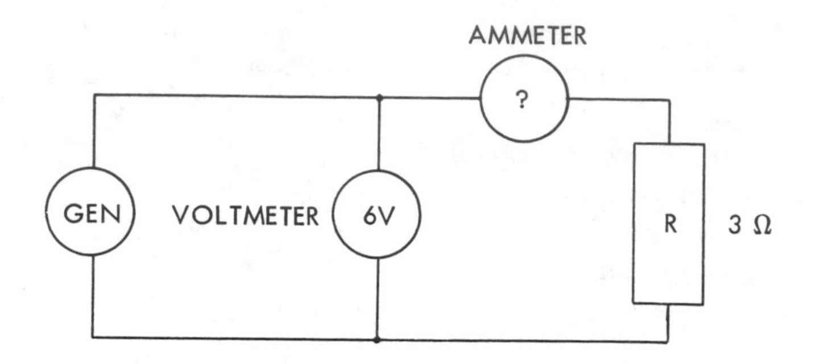

Example 1

Example 1

A voltage of 6 volts is used to force a current through a resistance of 3 ohms. What is the current?

Solution

The voltage (E) is 6 volts and the resistance (R) is 3 ohms. We wish to find the current (I). Using the first statement of Ohm's law we find that

$$I = \frac{E}{R} = \frac{6}{3} = 2 \text{ amperes}$$

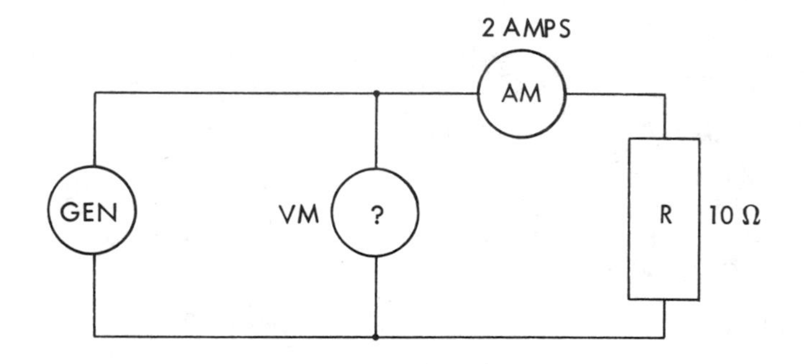

Example 2

Example 2

What voltage is required to force a current of 2 amperes through a resistance of 10 ohms?

Solution

The current (I) is 2 amperes and the resistance (R) is 10 ohms. We want to find the voltage (E).

$$E = I \times R =$$
$$2 \text{ amperes} \times 10 \text{ ohms} = 20 \text{ volts}$$

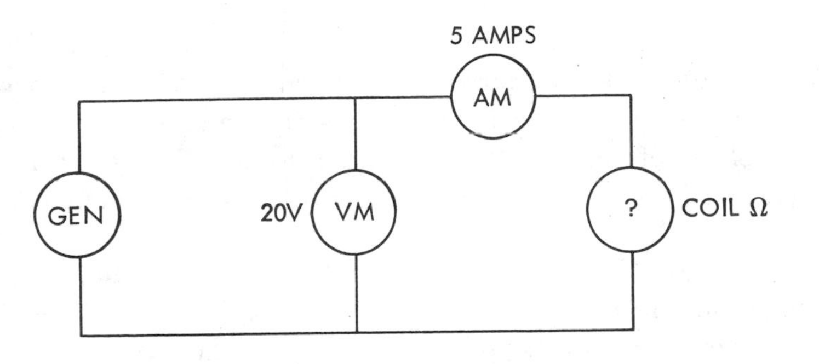

Example 3

Example 3

A voltage of 20 volts is required to force a current of 5 amperes through a coil. What is the resistance of the coil?

Solution

Voltage (E) = 20 volts.

Current (I) = 5 amperes

$$R = \frac{E}{I} = \frac{20 \text{ volts}}{5 \text{ amperes}} = 4 \text{ ohms}$$

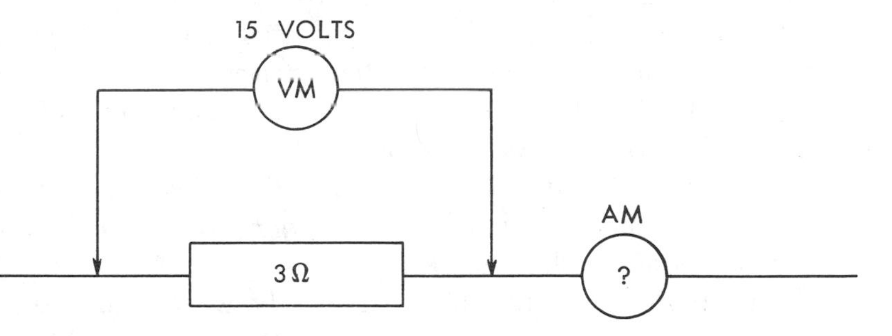

Example 4

Example 4

The voltage across the leads of a resistor measures 15 volts and its resistance is 3 ohms. What current will flow through it?

Solution

Covering the symbol I in the diagram, Fig. 1-5, there remains $E \div R$. Substituting the values of voltage and resistance given,

$$15 \div 3 = 5 \text{ amperes}$$

Example 5

A current of 10 amperes is forced through a resistive conductor by a pressure or voltage of 30 volts. What is the resistance of the conductor?

Solution

Covering R in the diagram, Fig. 1-6, $E \div I$ remains. Substituting for E and I their values from the conditions as stated, $30 \div 10 = 3$ ohms.

Example 6

A current of 10 amperes flows through a resistance of 2 ohms. What is the voltage that is forcing the current through the resistance?

Solution

Covering E, Fig. 1-7, we have left I times R. Substituting their values as before, we have $10 \times 2 = 20$ volts.

Electrical Power and Energy

The rate at which electric energy is delivered and consumed is called *power*. It is measured in *watts*. The *kilowatt*, which is 1000 watts, is used for the larger amounts of power. Thus, an electric-light bulb is rated at 120 volts and uses 50 watts of power. A larger bulb is rated at 120 volts and 200 watts. A 200-watt bulb requires four times as much power as the 50-watt bulb, even though both operate on the same voltage.

An electric generator supplying many lamps could be rated as 120 volts and 50 kilowatts. It is easier to read and write 50 kilowatts than to use 50,000 watts. A kilowatt is abbreviated *kw*.

When the amount of power in watts ends in three or more ciphers such as: 1000 watts, 2000 watts, 3000 watts, or 5000 watts, it is referred to as 1 kw, 2 kw, 3 kw, or 5 kw. If the power in watts is 1050 or 2780 watts or 5875 watts, the term watt is used instead of kilowatt.

The *power equation* is a rule for determining the amount of power in a circuit.

RULE: *Power equals the amperes flowing in a circuit times volts of the circuit.*

It can be written in simpler form by using the letters for abbreviation. Thus,

$$P \text{ is equal to } I \times E, \text{ or } P = IE$$

In the formula, P stands for power in watts. The letter I stands for current. It is the same letter used in Ohm's law. The letter E stands for volts or voltage of the circuit. In formulas, the multiplication sign is often omitted as shown at the right.

The power equation can be arranged as shown in Fig. 1-8. Here a square is used instead of a circle so it would not be confused with Ohm's law. The method of applying it is the same.

To find the power of a circuit, place a finger on the letter P, as shown in Fig. 1-9. Then $I \times E$ gives power. It is read, *current in amperes times voltage gives power*.

In like manner, to find the current represented by the letter I, place a finger on that letter, Fig. 1-10. Then, $I = P/E$, which is the same as $I = P \div E$. Thus, current equals power in watts divided by voltage.

When power and current are known, the voltage rating is found by placing a finger on the letter E, Fig. 1-11, so that $E = P/I$, which means E is equal to $P \div I$.

Following are several different voltages and current which will produce the same amount of power in watts:

6 volts × 200 amps = 1200 watts
12 volts × 100 amps = 1200 watts
24 volts × 50 amps = 1200 watts
120 volts × 10 amps = 1200 watts
240 volts × 5 amps = 1200 watts
600 volts × 2 amps = 1200 watts

As the voltage increases, the amount of current required for the same power is reduced. In an automobile, a 12-volt storage battery, located a few feet from the engine, may supply 100 amperes. In residential wiring where the distance from the power company's transformer to the customer's home is a hundred feet or more, the voltage is 120 for a two-wire service, or 120-240 for a three-wire service. (Note that voltages of

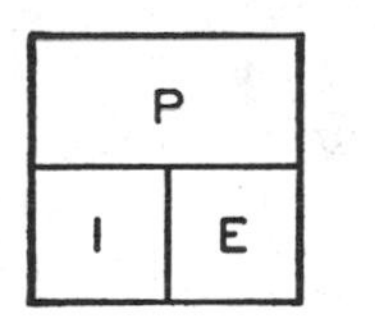

Fig. 1-8. Power equation.

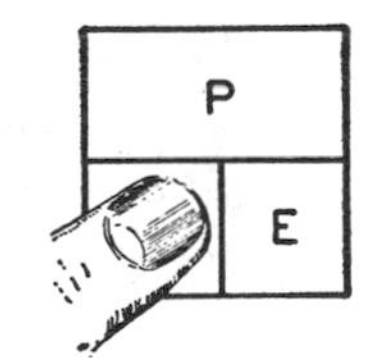

Fig. 1-10. Current.

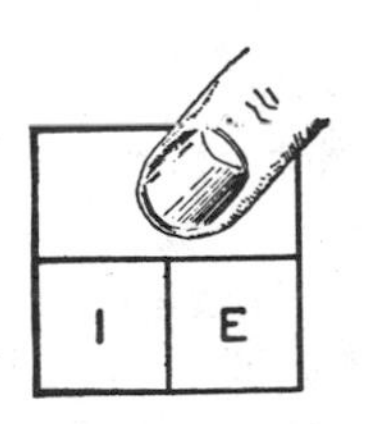

Fig. 1-9. Power.

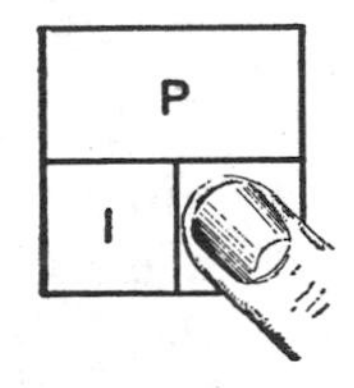

Fig. 1-11. Voltage.

115-230 are used interchangeably throughout the text with 120-240.) The current flowing to the residence at 120 volts would be 10 amps for a 1200 watt load.

Voltage loss or voltage drop due to resistance of the wires is not desirable because it reduces voltage to the lamps or other appliances in the circuit. "Copper losses" are often referred to as "I^2R Losses." This formula is sometimes used when the current and the resistance are known. For example, if the current were 2 amps and the resistance 3Ω:

$$2 \text{ amps squared} \times 3\Omega = 2 \times 2 \times 3 = 12 \text{ watts}$$

The formula $E^2/R = P$ is available but not too often used.

Two Common Types of Circuits

There are two common types of circuits: *series* and *parallel.* In *series,* current is the same, individual voltages add to equal the total. In *parallel,* voltage is the same, individual currents add to equal the total.

Series Circuit

The series form is illustrated in Fig. 1-12. Part (A) shows a string of Christmas tree lights which are connected in series. Part (B) is a line diagram of the circuit. Current flows through each lamp, one after the other, before returning to the supply wires or to the generator.

Certain facts should be observed in regard to the series circuit:

1. The current flowing in all parts of a series circuit is the same. This fact can be proved by connecting an ammeter anywhere in the circuit and noting that it reads the same at all points.

2. Removing any lamp or electrical appliance from a series circuit interrupts the flow of current.

3. The amount of current is inversely proportional to the resistance of the units connected in that circuit (as units of greater resistance are substituted, the current will be reduced).

4. The total voltage applied equals the sum of the individual voltage drops measured at each lamp or resistance.

If a voltmeter with insulated flexible leads is used to take successive readings around the circuit, the total of those readings will equal the applied voltage.

5. There is only one current path in a series circuit and only one conductor or wire connected to a terminal as shown in Fig. 1-12.

If ammeter No. *1* reads 1 ampere, ammeter No. *2* will also read 1 am-

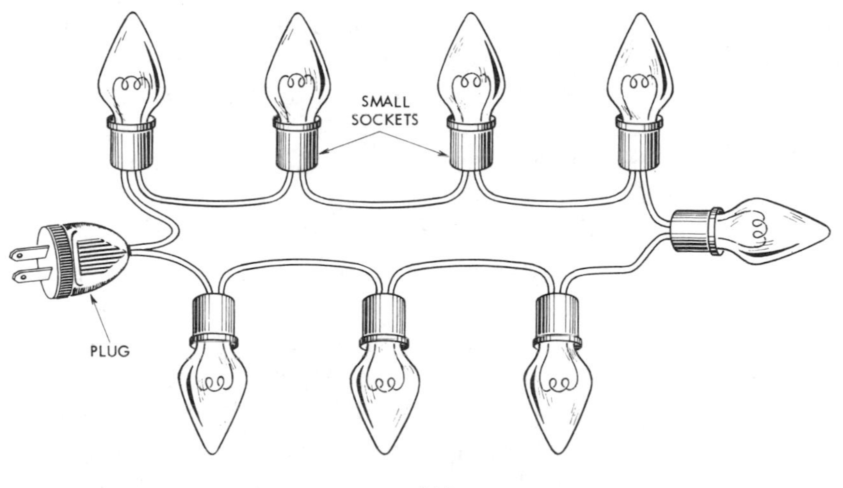

(A)

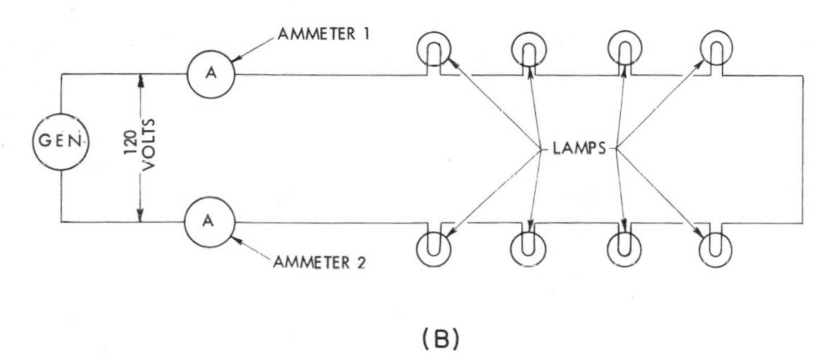

(B)

Fig. 1-12. Series circuit.

pere because all of the current (amperes) *leaving the generator returns to it.* This is an important point to remember in all circuit work no matter how complicated. The total resistance of a series circuit is equal to the sum total of all the individual resistors.

Note that the ammeters in Fig. 1-12 are part of the series circuit. An ammeter is always connected in this manner, so that it is necessary to interrupt the flow of current, at least momentarily, while the instrument is "cut into" one or the other of the circuit conductors.

Parallel Circuit

In the parallel arrangement, Fig. 1-13, each lamp is connected in a sub-circuit of its own, a complete path being formed through the generator and connecting wires without involving the other lamps.

Important facts associated with the parallel circuit are:

1. The current flowing in the main lines of a parallel circuit is the sum of the currents in the various parallel paths of that circuit. If each lamp in Fig. 1-13 takes one ampere, the total current will be 5 amperes, as indicated. The total current equals the sum of the individual currents. If any one of the lamps were removed, the total generated current would reduce by one ampere.

2. Removing any lamp or electrical appliance from a parallel circuit does not break the flow of current to other units. For example, removing lamp L_1 would not prohibit the other lamps from operating. There is more than one current path and more than one conductor connected to a terminal. The currents are divided.

3. The amount of current flowing in a parallel circuit is directly proportional to the number of units connected. (As more lamps are connected, more current flows through the wires.)

4. The total voltage is involved in doing work around any closed path from the generator. If a voltmeter with insulated flexible leads is used for successive readings around any closed loop, the voltage drops will equal the applied voltage. The voltages are all the same.

If three voltmeters are used, as shown in Fig. 1-13, the reading of voltmeter No. *1* will equal the voltage lost in the section of line across which it is connected, namely, 2 volts. In like manner, voltmeter No. 2 will read the voltage lost in lamp

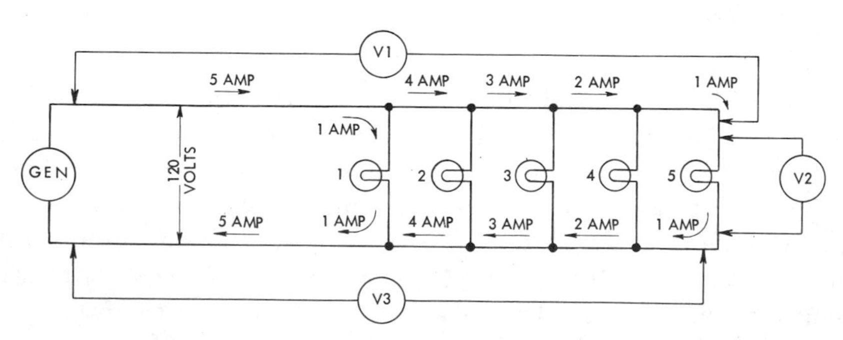

Fig. 1-13. Parallel circuit.

No. *5*, 116 volts. Voltmeter No. *3* will read the voltage lost in the return line to the generators, 2 volts. Adding the voltage drops, 2 + 116 + 2 equals 120 volts, which is generator voltage.

The voltage lost in the two wires feeding lamp *5* depends on the length of the line wire.

If this same test is made around any other closed circuit, it will be found that the total of the voltage losses equals the generator voltage.

Note that the voltmeter is connected in parallel with all or part of the circuit, as desired. Thus, the current is not interrupted when using the voltmeter to take readings.

Voltage Drop

When current flows in a conductor, part of the voltage is lost in overcoming resistance. If the loss is excessive, that is, more than a few percent, the lamps or other devices will not operate satisfactorily. Lamps decrease in brilliancy, heating devices deliver a reduced output, and motors have difficulty in starting their loads. The National Electrical Code should be consulted regarding specific tolerances of voltage drop.

According to Ohm's law, voltage $E = I \times R$. This formula may be used to determine voltage drop. If the resistance of the circuit wires is 0.5 ohm, and the current 20 amperes (hereafter abbreviated "amps"), the voltage drop will be 20 amps × 0.5 ohm, or 10 volts.

In most cases, the resistance of the wires is not given (and tables may not be readily available) but their size is known. A simple formula can be employed to find resistance. For example, in Fig. 1-14, a circuit of two

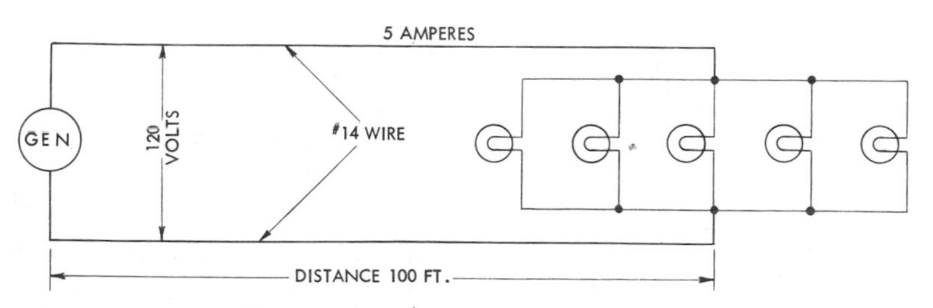

Fig. 1-14. Generator supplying lighting load.

No. 14 AWG (American Wire Gage) copper wires supplies 5 lamps, each of which consumes 1 amp, at a distance of 100 ft from the source of supply. It is desired to calculate the voltage drop in the circuit wires. In order to find voltage, current in amps and resistance in ohms must be known. Here, only the current is known.

The resistance formula reads as follows: $R = \dfrac{K \times L}{d^2}$

R = resistance in ohms (when found).

K = a constant whose value depends on the conductor; for example, 10.8 for copper wire, or 17 for aluminum wire. See Table I.

L = length in feet (both wires).

D = diameter of the wire in thousandths of an inch or mils (1/1000-inch is known as a "mil").

D^2 = diameter squared (multiplied by itself), this diameter being expressed in mils, and the squared result in "circular mils." Electrical tables give the circular areas of wires.

The length, here is 2 × 100 ft, or 200 ft.

The diameter of No. 14 AWG wire is 64.1 mils.

The circular mil area (diameter squared) is equal to 4109.

Substituting these values in the formula:

$$R = \frac{10.8 \times 200}{4109} = \frac{2160}{4109}$$
$$= .526 \text{ ohm.}$$

Voltage drop equals:

$E = I \times R = 5$ amps × .526 ohm = 2.63 volts.

If the voltage at the source is 120, that at the lamps must be equal to:

TABLE I K VALUES FOR OHMS RESISTANCE PER MIL-FOOT (AT 70°F)

Aluminum	17
Brass	42
Copper	10.8
Iron	60
Silver	9.6
Steel	75
Tungsten	33

page 587 in code book size of wire gives resistance for 500 ft wire

120 volts − 2.63 volts, or 117.37 volts.

This drop would not be considered too far out of line, but if the current were 10 amps in Fig. 1-14 it would amount to twice 2.63 volts, or 5.26 volts, which would be somewhat high. In order to cut down the loss of voltage, wires of a larger diameter must be employed.

The next larger size is No. 12 AWG, which has a diameter of 81 mils, and a circular mil area of 6561. Substituting this value in the formula for Fig. 1-14:

$$R = \frac{10.8 \times 200}{6561} = \frac{2160}{6561}$$

$= .330$ ohm.

Voltage drop now equals: 10 amp × .331 ohm = 3.31 volts. This drop is still high, so that a larger diameter wire must be used. The next larger size is No. 10 AWG, with a diameter of about 102 mils and a circular mil area of 10,380. The resistance of 200 ft of this wire is equal to 2160/10,-380, or .208 ohm. Voltage drop becomes: 10 amps × .208 ohm, or 2.08 volts, which is acceptable.

Here again the National Electrical Code, or prevailing local codes, should be consulted for exact tolerances. Different conductor conductivity can be compared in Table I. Three rules to remember:

1. Resistance depends upon the length of wire. For example, 200 feet will have twice as much resistance as 100 feet; that is, if the resistance for 100 feet was 2 ohms, then the resistance for 200 feet of the same wire would be 4 ohms.
2. Resistance also depends upon the size of the material. The larger the diameter of the material, the less the resistance.
3. Resistance also depends upon the kind of material. Table I gives K factors for some common conductors.

Power Loss

Voltage drop results in power loss, which simply heats up the wires and serves no useful purpose. This power loss is registered on the meter, and is paid for by the customer. According to the power formula, Power = $E \times I$. Power Loss = E (line drop) $\times I$ (or $I^2R = P =$ copper losses).

If the current in a long feeder wire is 50 amps and the resistance is .35 ohm, the voltage drop = 50 amps × .35 ohm, or 17.5 volts. The power loss = 17.5 volts × 50 amps, or 875 watts, almost enough to keep nine 100-watt lamps burning continuously.

Three-Phase Current

Only single-phase current has been considered up to this point. As will be seen in industrial work, motor loads are frequently supplied by three-phase current. A three-phase supply, Fig. 1-15, consists of three wires, the phase voltages (that be-

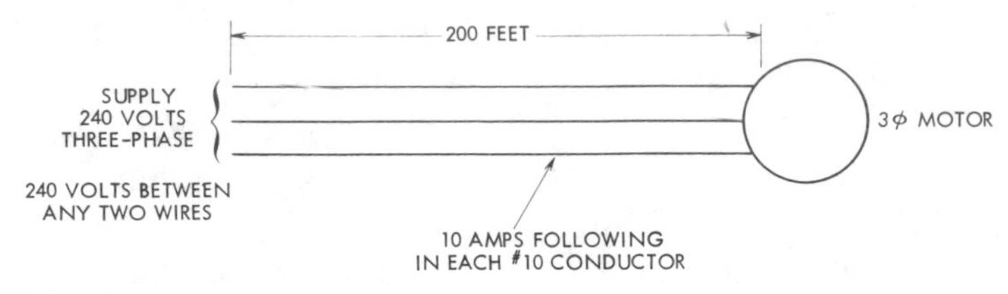

Fig. 1-15. Three-phase system supplying a motor.

tween any two wires) being exactly alike. Here, the voltage is 230 volts. Drop in any one of the supply wires is determined in the same way as for single-phase current. In order to determine the drop per phase, this value is multiplied by the number 1.73.

For example, the three No. 10 conductors in Fig. 1-15 carry a 10-amp, three-phase current to a motor which is 200 ft from the supply point. The voltage drop in a single #10 wire 200 ft long (100 ft × 2 wires, both ways) has already been calculated as 2.08 volts. The voltage drop per phase equals 1.73 × 2.08 volts, or approximately 3.6 volts. (The 1.73 is a constant used with 3 phase calculations. It is the resultant voltage developed by internal connections of a 3 phase generator and is equal to $\sqrt{3}$ or 1.73 times the voltage developed in one generator coil or phase.) The voltage delivered at motor terminals, therefore, is 240 volts minus 3.6 volts, or 236.4 volts.

For single phase, the return wire is considered with the distance. For 3 phase, 1.73 is considered instead of return wire lengths. It is important to avoid voltage drop to a degree. Voltage drop results in power loss and lowers operating efficiency of electrical machinery.

To review finding voltage drop, Ohm's law method:

1. Find current in the line.
2. Find total conductor length: Length equals distance times two.

   Example: 500′ × 2 = 1000′ length

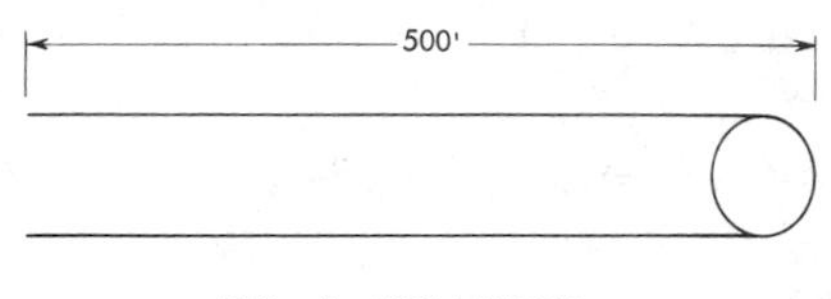

3. Determine allowable voltage drop: Percent × line voltage = voltage drop ($\% \times E_L = E_D$)
4. Select from table or calculate resistance of random wire size. (Consider current, its regard to wire capacity and length.)
5. Find voltage drop of random selection: $E_D = IR$.

6. If selection is not within voltage drop tolerances, choose larger wire size and repeat calculation.

Other Useful Voltage Drop Methods

Circular Mils =

$$\frac{\text{Distance 1 way in feet} \times 2 \times 10.4 \times \text{amperes}}{\text{Line voltage} \times \text{percent drop}}$$

$$CM = \frac{D \times 21 \times I}{E_L \times \%D}$$

Voltage Drop =

$$\frac{\text{Distance 1 way in feet} \times 2 \times 10.4 \times \text{amperes}}{\text{Circular Mils}}$$

$$E_D = \frac{D \times 21 \times I}{CM}$$

Distance =

$$\frac{\text{Voltage Drop} \times \text{Circular Mils}}{\text{Amps} \times 2 \times 10.4}$$

$$D = \frac{E_D \times CM}{I \times 21}$$

Network System

Both lighting and power loads may be supplied by the network system of three-phase distribution illustrated in Fig. 1-16. Three-phase motors are connected to the main or phase wires, lighting circuits between any one of the main wires and the neutral conductor. This is called "3 phase-4 wire." See Fig. 1-16. The voltage between any two of the phase wires is 208 volts, while that between any phase wire and the neutral conductor is 120 volts. In a perfectly balanced system, where only lighting

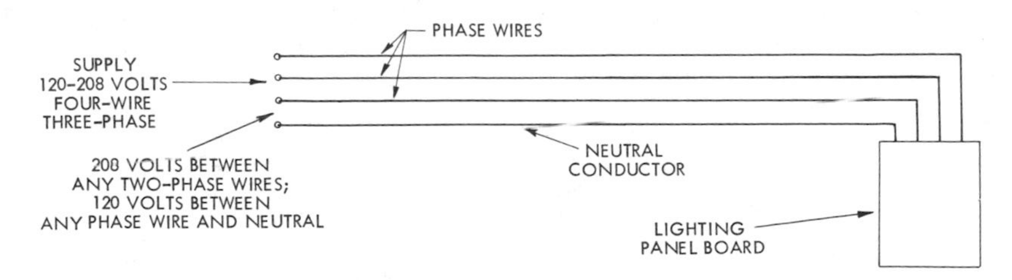

Fig. 1-16. Four-wire, 120-208 volts, three-phase, network system.

circuits are supplied, voltage drop per 120 volt circuit and power loss per 120 volt circuit are equal only to that of a single 3 phase, 208 volt conductor. Total loss in all three phases, under this condition, is equal to three times that in a single conductor.

Using values given in the previous calculation, where the distance from load to center of distribution is 200 ft, the current 10 amps, and the size of conductor No. 10, the voltage drop per circuit is equal to 2.08 volts, and the voltage supplied to the lamps equals 120 volts minus 2.08 volts, or approximately 118 volts. The power loss in each wire is $I \times E$, which is 10 amps $\times$ 2.08 volts, or 20.8 watts. The power loss in all three wires is three times this amount, or 62.4 watts.

QUESTIONS FOR DISCUSSION OR SELF STUDY

1. What term describes the path formed by the conductors and the load?
2. Name the three primary circuit elements.
3. State Ohm's law when voltage and resistance are known.
4. State Ohm's law when voltage and current are known.
5. State Ohm's law when current and resistance are known.
6. What current flows when the circuit voltage is 120 and its resistance is 10 ohms?
7. What circuit voltage is required to force 5 amps through a 20-ohms resistance?
8. How many amps will flow if circuit voltage is 100 and resistance is 40 ohms?
9. What happens to the circuit if one of ten *series* lamps burns out?
10. What happens to the circuit if one of ten *parallel* lamps burns out?
11. Draw a sketch of an ammeter measuring current of a lamp.
12. Draw a sketch of a voltmeter measuring voltage of a lamp.
13. What effect does resistance have on voltage delivered to the load?
14. What is meant by *power loss?*
15. State the power equation.
16. State the formula for determining resistance of a given size of copper wire.
17. Would the resistance increase or decrease if aluminum were substituted for copper in the wire?
18. What is the voltage drop when resistance is 10 ohms and the current 12.5 amps?
19. What is the resistance of 100 ft of No. 14 copper wire if its circular mil area is 4107?
20. What power loss results if 10 amps flow through the above wire?

Introduction to Electrical Codes and Standards

Chapter

2

When electricity was first applied to the lighting of homes several decades ago, one 25-watt lamp in each room was considered enough to supply average needs. The current taken by these few lamps amounted to not more than two, three, or perhaps five amperes. Number 14 AWG (American Wire Gage) rubber-covered copper wire, which had been generally accepted as the smallest size capable of withstanding the mechanical strain of installation, was able to carry this low current without overheating.

After the turn of the century, however, electrical consumption multiplied rapidly. Not only were stronger light bulbs introduced, but also such household appliances as irons and fans were made available to the general public. Installations previously considered adequate were then no longer able to carry the loads. As a result, overheating often destroyed wire insulation and caused fires, accidents, and even deaths. Fig. 2-1 illustrates one result of overheating.

Municipal authorities became interested; insurance companies and building industries began to take notice. Groups of interested parties clamored for laws to govern electrical wiring standards. Finally, the National Fire Protection Association was formed in Boston for the express purpose of preventing fires of an electrical origin. The set of regulations drawn up by this organization is known as the National Electrical Code, abbreviated NEC. This code is revised frequently, a new manual being issued every three years.

Although Congress may not make the NEC a national electrical law since states are independent of the

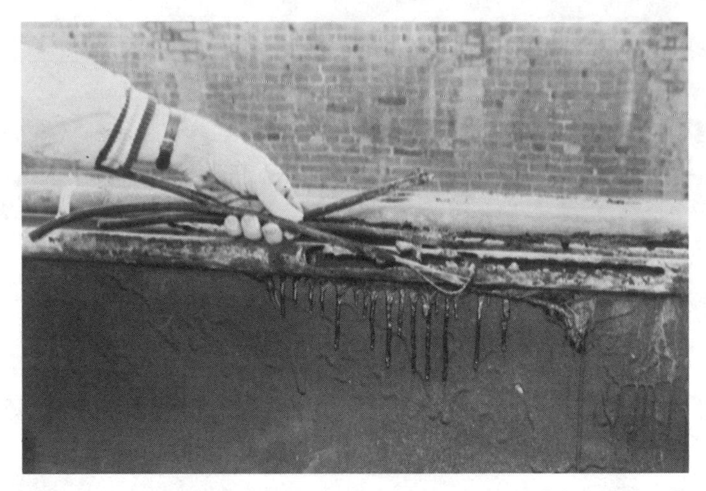

Fig. 2-1. Insulation failed on an illegal wire splice in service conduit with no fuse protection. Arc burned through conduit but building did not burn because of a rainstorm.

Federal government in regard to such matters, ordinances based upon this code, or taken from it word for word, have been enacted by states and cities. For this reason, the electrical worker should familiarize himself with NEC rules, and with such minor variations of these rules as may be recognized in his community. In Canada, electrical construction work is governed by the Canadian Electrical Code, the contents of which are almost identical with those of NEC.

Contents of the Code

The National Electrical Code includes an introduction setting forth its purpose, scope, fundamental rules and recommendations, general application, and applications to installations that involve special occupancies, special equipment, or other special conditions. A chapter governs installation of communications systems and is independent of the preceding chapters except as they may be referred to specifically.

Fig. 2-2. Fire and safety hazard due to tampered and deteriorated wiring system.

Various tables and examples are included.

Methods of wiring are thoroughly treated in the NEC. For instance, a table sets forth the size and number of wires permitted in conduits without danger of overheating or damaging insulation. One section requires that electrical equipment shall be installed in a neat, workmanlike manner. A companion section requires that equipment shall be securely mounted. This provision of orderly rules suggests another favorable result, essentially an economic one. When equipment is neatly and securely placed, it is likely to provide better service, not only when new, but also after the deteriorating effects of time have been imposed on it. Fig. 2-2 illustrates the results of poor workmanship and planning.

Although the NEC states only minimum requirements, most local electrical codes are based on it and provide for any unusual local conditions by incorporating more restrictive measures. Its effect is far reaching: it defines rules of procedures for the apprentice electrician and the electrical inspector; it regulates the work of the electrical contractor and the products of the electrical manufacturer; it demands intelligent layout from the engineer or architect. It does not provide for future load requirements, but it safeguards the public from dangers inherent in poor installations. However, it must be remembered that the NEC has no direct authority until it has been adopted by an agency of local government and properly enforced.

In order to provide a uniform interpretation of the literal text or the actual intent of the code, the NEC provides a formal authorized procedure, stated at the end of the code, that anyone may follow to secure a ruling when a question or problem of interpretation arises.

Local Codes

Generally, municipalities adopt the NEC by reference as the local electrical code. This legislative adoption may embrace the NEC as written or with some modification, sometimes more restrictive when special conditions warrant more stringent measures.

An example of a local condition not specifically covered by the NEC is "hot soils." Here we have a condition where the chemical pattern or contamination of the soil is such that buried metallic conduit would be chemically attacked, causing accelerated corrosion of the conduits. This condition could prove to be both expensive and hazardous if, for example, live conductors became exposed to gasoline-soaked ground following complete deterioration of the electrical conduits. Under these circumstances, a local code would prohibit the placing of galvanized conduit in the ground without protection against chemical corrosion.

A code book of local jurisdiction should be always carried in the electrical workers tool box for ready reference. It is practically impossible to remember all laws pertaining to all phases of electrical construction. The latest electrical code book is an excellent supplement to this text in learning to construct electrical systems.

Standards for Electrical Materials

Even though electrical wiring is carefully installed, the materials and devices themselves can be sources of hazard unless properly designed for specific purposes. This fact caused the National Board of Fire Underwriters to create the Underwriters' Laboratories Inc. Manufacturers who wish to obtain U-L approval submit samples to one of the testing laboratories—at New York, Chicago, or Santa Clara, California. Many large cities require U-L approval on all electrical materials used or sold within their jurisdiction. See Figs. 2-3, 2-4 and 2-5 for U-L Seals.

Fig. 2-3. Underwriters' Laboratories seal.

Fig. 2-4. U-L seal for power supply cords.

Fig. 2-5. U-L seal for ladder.

Some large cities maintain electrical test laboratory facilities similar to U-L. These labs are in the Department of Building and Safety and require that all electrical equipment and devices installed must have either a U-L sticker or the approval of the city in which it is installed. This is strictly enforced by the electrical inspectors within the department. This is all additional protection for life and property. Approved electrical devices, materials and equipment must be installed in an approved manner according to the prevailing code.

QUESTIONS FOR DISCUSSION OR SELF STUDY

1. What size of copper wire was first used in wiring homes?
2. Who writes the National Electrical Code?
3. How often is the NEC revised?
4. Name one of the early electrical appliances.
5. Who established the Underwriters' Laboratories, Inc.?
6. Basically what does the NEC include?
7. Where can a uniform interpretation of the intent of the NEC be found?
8. How are electrical codes enforced?
9. When do codes prevail?
10. Why is it recommended that an electrical worker carry a code book in his tool box?

Residential Blueprint Reading

Chapter

3

Before starting electrical construction for a specific job it is necessary to obtain a wiring permit from the city, town, or county having local jurisdiction. On large housing installations an architect delivers plans to his electrical engineer, who proceeds to mark in the outlets, conduit runs, panelboards, and other items. He also draws up specifications which set forth detailed standards for material and workmanship.

On smaller jobs, the engineer's services are seldom needed. The architect or draftsman furnishes the home builder with blueprints on which a comparatively small amount of electrical data is supplied. If electrical outlets are not indicated, their locations may be determined by a conference between homeowner, builder, and electrical contractor. The contractor is usually asked to sign a statement to the effect that he will use only approved materials, and that he will guarantee the completed installation to pass local inspection.

Blueprints

Blueprints are a set of instructions on how to do some particular piece of work. They convey certain information that could not be given by words alone.

"Blueprints" got their name originally by having blue paper with white lines and symbols. This process has generally been reversed now, with blue lines (or black or brown lines) on white paper. "Blueprints" are also called "plans" and "drawings." These blueprints or plans show by means of lines and symbols the shape and size (by reduction to scale) of a structure, the materials required, and the location of fixtures and outlets. The several sheets of drawings used should, in conjunction with the specifications, contain sufficient instruction for the completion of the structure.

When an architect prepares a set of drawings containing all of the information and dimensions necessary to carry a job through to successful completion, he has made a set of *working drawings*. Reproductions of these working drawings are called a set of blueprints.

Correct reading of drawings comes with experience. Skill in reading and working with architectural plans comes through practice and through a knowledge of symbols and their application.

Symbols

The location of equipment, fixtures, outlets, etc., is shown on a drawing by means of pictorial symbols. In most instances these symbols have been standardized, for ease of understanding, by those who design buildings and by those who construct buildings.

The American National Standards Institute (ANSI), formerly known as the United States of America Standards Institute (USASI) or the American Standards Association (ASA), standardizes commonly used symbols. The electrical construction student or apprentice should make every effort to learn material identification in order to properly associate

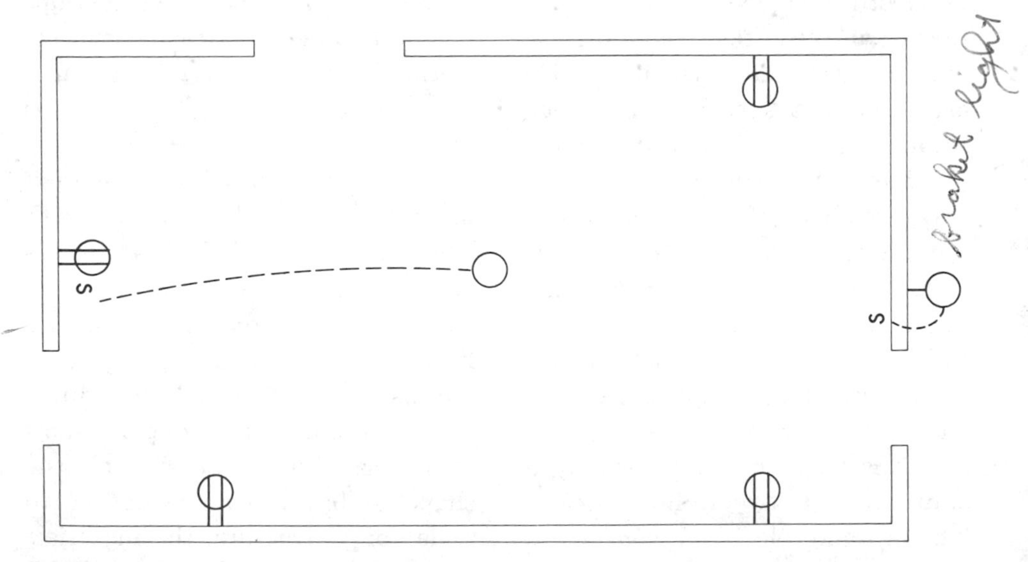

Fig. 3-1. Simplified one room electrical floor plan.

symbols with wiring devices and materials. Use of manufacturer's catalogs with material descriptions are very helpful in this respect.

A complete set of plans consists of many different drawings. To install an electrical wiring system, an electrician must be aware of the information to be found on all drawings. Probably the drawing most often used by the wireman is the floor plan because most wiring requirements are found on this sheet. Fig. 3-1 shows a simplified floor plan of one room in which typical electrical outlets and switches are designated by standard electrical symbols. The curved broken lines in the figure indicate which outlet the switches control. See Fig. 3-2 which identifies the symbols used in Fig. 3-1.

The electrician must wire these outlets so that they will operate as desired and so they will satisfy the code requirements. (Switching circuits and methods are explained in Chapter 6, "Switching Circuits".) The methods of routing the cable or conduit to be installed is shown in Fig. 3-3 by lines drawn connecting switches and outlets. The line symbols added to Fig. 3-1 represent conduit or cable used to connect outlets. Referring to Fig. 3-3 note that the end of the run is indicated to be at the outside wall lighting outlet (on the right side of the figure). The run then drops down to the switch con-

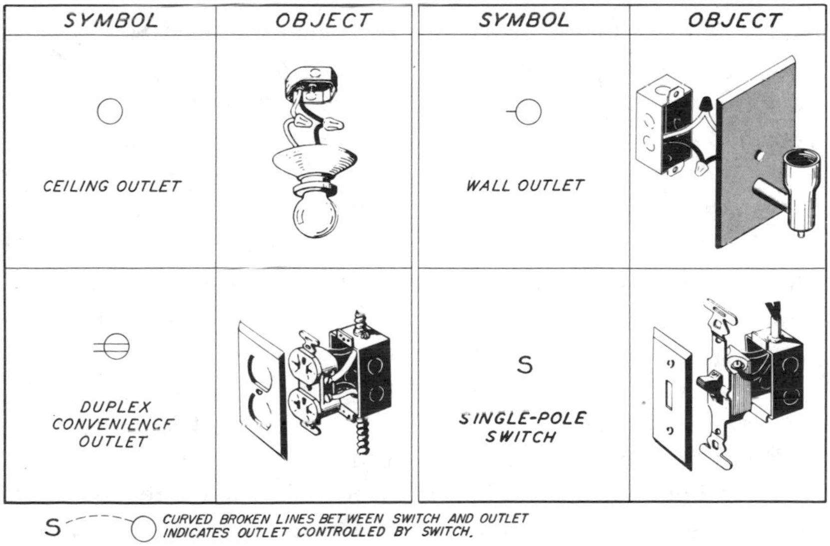

SYMBOL	OBJECT	SYMBOL	OBJECT
CEILING OUTLET		WALL OUTLET	
DUPLEX CONVENIENCE OUTLET		S SINGLE-POLE SWITCH	

S CURVED BROKEN LINES BETWEEN SWITCH AND OUTLET INDICATES OUTLET CONTROLLED BY SWITCH.

Fig. 3-2. Symbol identification of Fig. 3-1.

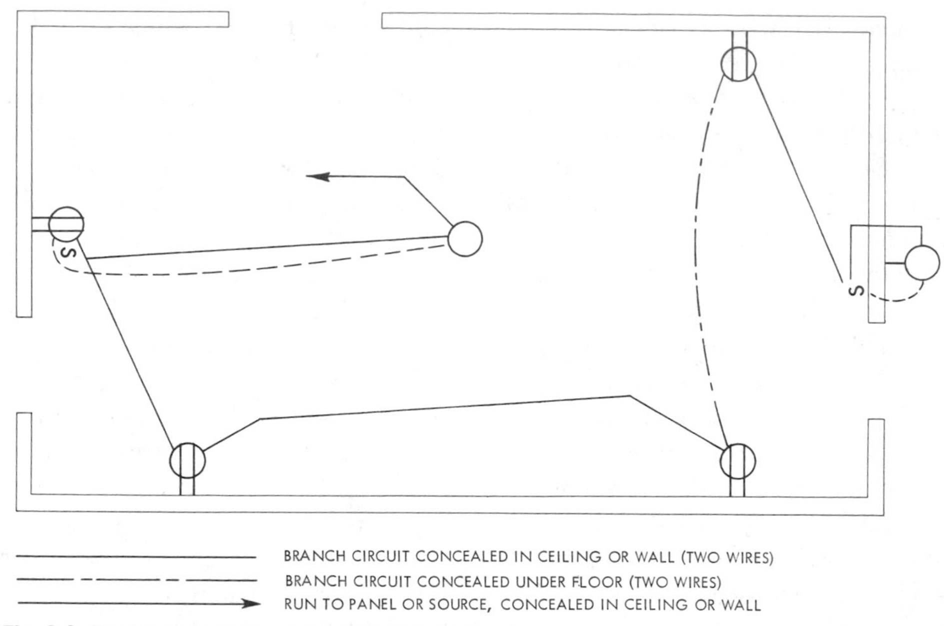

Fig. 3-3. Routing of cables or conduit for outlets.

trolling it. The route may be traced back from the switch. There is a ceiling or wall wiring run from the switch down to a convenience outlet. There is an under floor run from this outlet to another duplex convenient outlet on the other side of the room. The solid line from this outlet indicates a wall or ceiling concealed run, in this case probably through the studs, to another duplex convenient outlet. From there an overhead run to the other outlet, then to a switch, then to the ceiling lighting outlets, and then "home" to the branch-circuit distribution panel. The run to the panel is indicated by an arrow in an "overhead" run.

Local codes should be investigated before an attempt is made to mix lights and plugs on the same circuit. Outlets may be numbered to show on which circuit they are connected.

Fig. 3-4 shows additional symbols. (These circuits should be coordinated with Chapter 6, *Switching Circuits*, for operational definition.) Convenience outlets have been omitted here for simplicity. Unless switched, these are normally connected in parallel on a circuit. Fig. 3-4 shows the ceiling light controlled

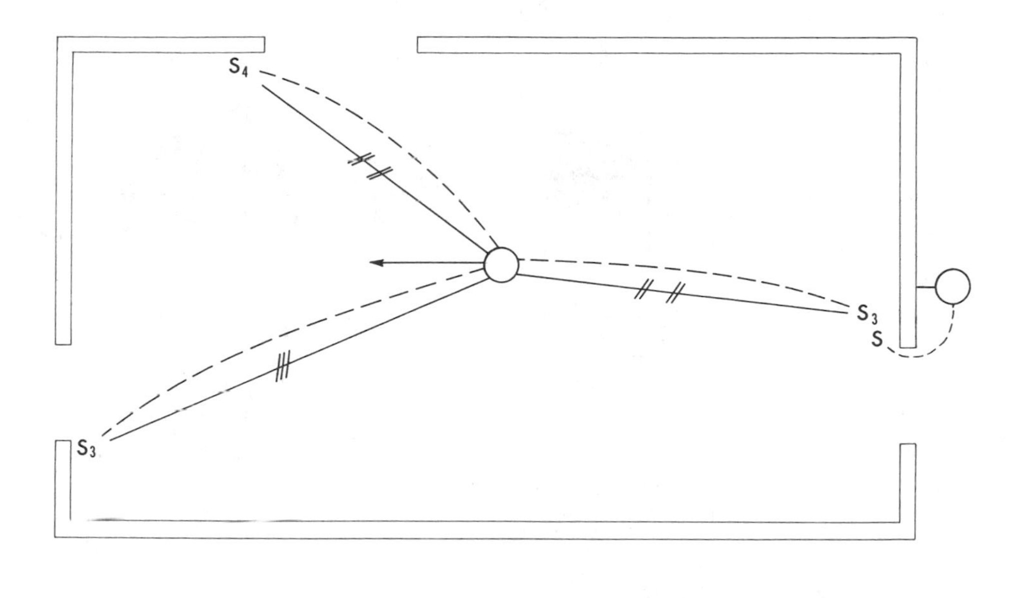

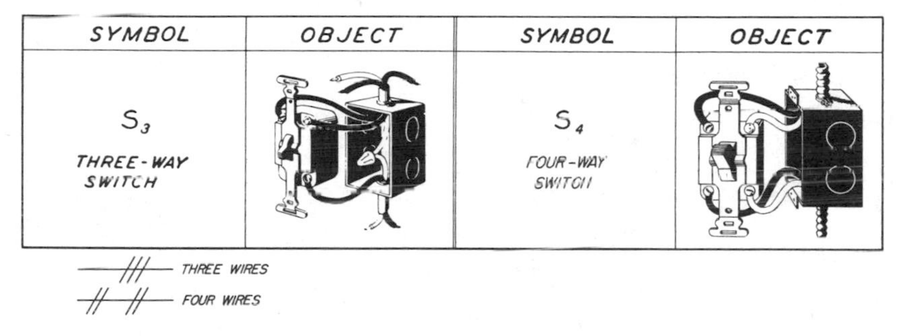

Fig. 3-4. Top: Symbols showing control of outlets; bottom: switches used in top floor plan.

by a switch at each door and the type of switches necessary to do this job. Compare this switching to that shown in Fig. 3-3 where only one switch could turn on the overhead light. In Fig. 3-4 three switches can turn it on. The solid line between a switch and the ceiling outlet shows cable or conduit routing. The three and four small cross lines indicate the number of wires in the conduit. Sometimes the size of conduit or tubing is indicated in addition to the number of conductors.

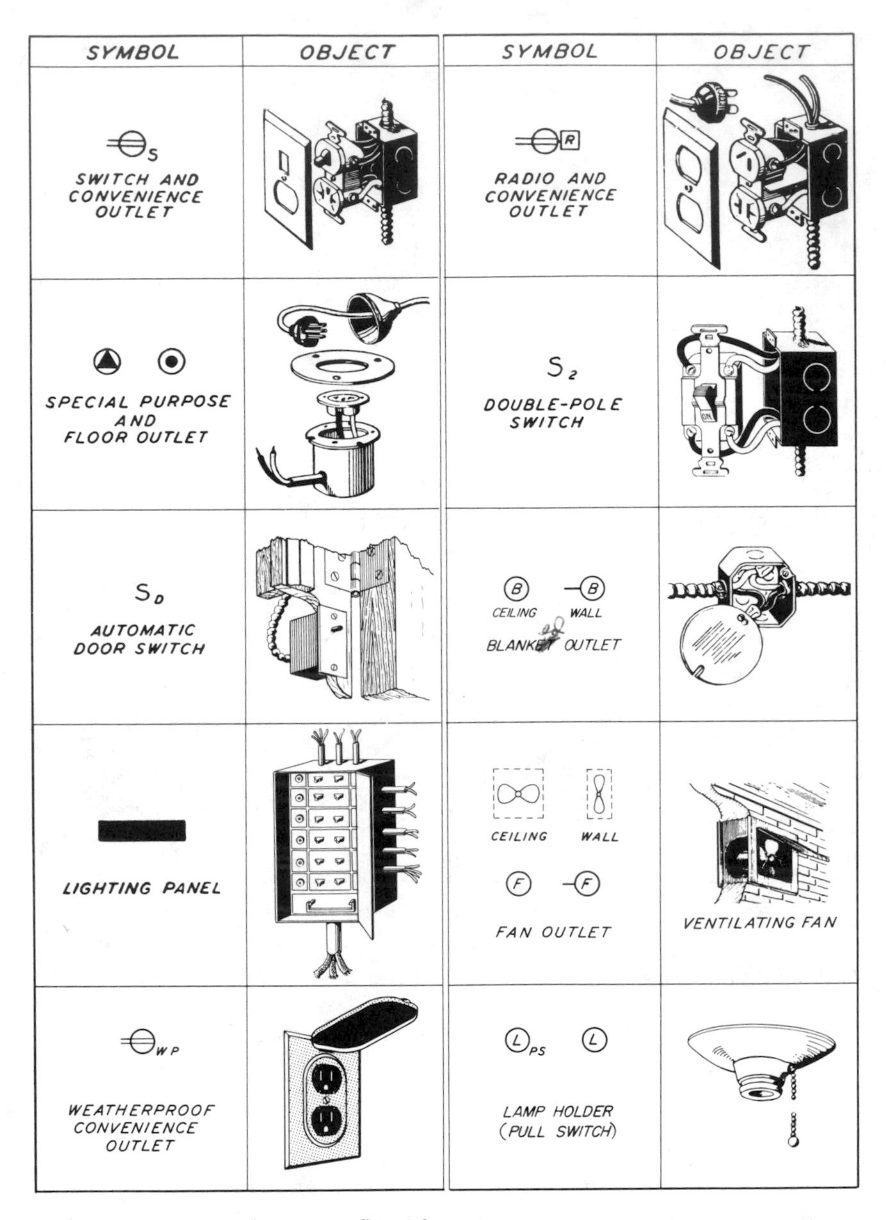

Fig. 3-5. Symbols commonly used on floor plans.

Other commonly used symbols are shown in Fig. 3-5, along with the object they represent. The student can readily understand the importance of electrical material identification; without this the symbols will be meaningless.

The combination switch and con-

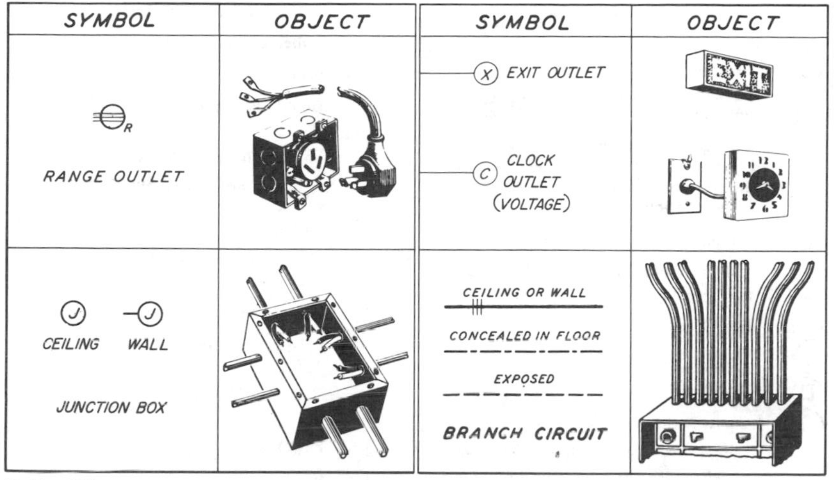

Fig. 3-5. Cont'd.

venience outlets shown in Fig. 3-5, top left, is often installed. If the switch controls another outlet a broken line indicating this will be shown on the plan. (See Figs. 3-1, 3-2, 3-3 and 3-4, for example.) Special purpose outlets (second from top Fig. 3-5, left) are generally designated with letters telling what kind of special purpose: "WH" for example means "water heater"; "DW" means dishwasher"; "H" means "heater"; and "GD" means "garbage disposal". These symbols are for outlets that do not require receptacles and are to be permanently attached to and connected with the wiring system. These may be compared with the conventional symbols for outlets such as the weatherproof outlet, range, clock, exit, lampholders, fan outlets, blanked outlets and junction boxes, as shown in Fig. 3-5. Switches with double pole applications (second from top Fig. 3-5, right) are sometimes used for disconnecting both conductors of a circuit. Automatic door switches (third from top Fig. 3-5, left) are used in closets, walk-in pantries and rooms that require such convenience lighting control. The lighting panel symbol in Fig. 3-5 (fourth from the top, left) is where branch circuits originate.

Electrical Symbols for architectural plans are shown in Fig. 3-6. All drawings would be simpler to read if every draftsman, architect, engineer contractor and draftsman used the same standard symbol. Unfortunately this is not always done. However, with the ability to read standard symbols, slight deviations from standards are recognizable.

General Outlets

Lighting Outlet

Ceiling Lighting Outlet for recessed fixture (Outline shows shape of fixture.)

Continuous Wireway for Fluorescent Lighting on ceiling, in coves, cornices, etc. (Extend rectangle to show length of installation.)

L — Lighting Outlet with Lamp Holder

L PS — Lighting Outlet with Lamp Holder and Pull Switch

F — Fan Outlet

J — Junction Box

D — Drop-Cord Equipped Outlet

C — Clock Outlet

To indicate wall installation of above outlets, place circle near wall and connect with line as shown for clock outlet.

Convenience Outlets

Duplex Convenience Outlet

3 — Triplex Convenience Outlet (Substitute other numbers for other variations in number of plug positions.)

Duplex Convenience Outlet — Split Wired

GR — Duplex Convenience Outlet for Grounding-Type Plugs

WP — Weatherproof Convenience Outlet

X" — Multi-Outlet Assembly (Extend arrows to limits of installation. Use appropriate symbol to indicate type of outlet. Also indicate spacing of outlets as X inches.)

S — Combination Switch and Convenience Outlet

R — Combination Radio and Convenience Outlet

Floor Outlet

R — Range Outlet

DW — Special-Purpose Outlet. Use subscript letters to indicate function. DW-Dishwasher, CD-Clothes Dryer, etc.

Switch Outlets

S — Single-Pole Switch

S_3 — Three-Way Switch

S_4 — Four-Way Switch

S_D — Automatic Door Switch

S_P — Switch and Pilot Light

S_{WP} — Weatherproof Switch

S_2 — Double-Pole Switch

Low-Voltage and Remote-Control Switching Systems

S — Switch for Low-Voltage Relay Systems

MS — Master Switch for Low-Voltage Relay Systems

R — Relay—Equipped Lighting Outlet

Low-Voltage Relay System Wiring

Auxiliary Systems

Push Button

Buzzer

Bell

Combination Bell-Buzzer

CH — Chime

Annunciator

D — Electric Door Opener

M — Maid's Signal Plug

Interconnection Box

T — Bell-Ringing Transformer

Outside Telephone

Interconnecting Telephone

R — Radio Outlet

TV — Television Outlet

Miscellaneous

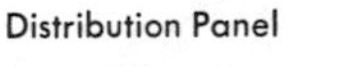

Service Panel

Distribution Panel

Switch Leg Indication. Connects outlets with control points.

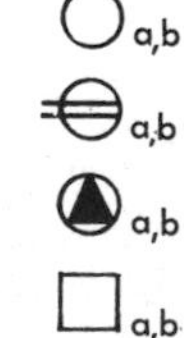

Special Outlets. Any standard symbol given above may be used with the addition of subscript letters to designate some special variation of standard equipment for a particular architectural plan. When so used, the variation should be explained in the Key of Symbols and, if necessary, in the specifications.

Fig. 3-6. Electrical symbols for architectural plans. (American National Standards)

—— Branch Circuit; Concealed in Ceiling or Wall.
–-– Branch Circuit; Concealed in Floor.
----- Branch Circuit; Exposed.
—•—• Home Run to Panel Board. Indicate number of Circuits by number of arrows.
Note: Any circuit without further designation indicates a two-wire circuit. For a greater number of wires indicate as follows —///— (3 wires) —////— (4 wires), etc.
—— Feeders. Note: Use heavy lines and designate by number corresponding to listing in Feeder Schedule.

Fig. 3-6. Cont'd.

Architect Scales

Working drawings are proportional reductions of the final proposed structure. The degree of reduction depends on the actual size of the object and the desired size of the drawing. For architectural drawing it is customary to reduce the original dimensions in feet to measurements in parts of an inch. For example, one foot may be reduced to ¼ of an inch or ⅛ inch. This reduction from feet to inches is called the *scale* of the drawing.

Common Scales. The most common small scales are the ⅛ inch equal to one foot which is written ⅛″ = 1′-0″, and ¼ inch to one foot, or ¼″ = 1′-0″. Also used are ⅜″ = 1′-0″; ½″ = 1′-0″; ¾″ = 1′-0″; and 1″ = 1′-0″. The scale that an architect selects depends upon the size of the building for which he is drawing the plans. For ordinary houses the ¼″ = 1′-0″ scale is suitable. For large buildings ⅛″ = 1′-0″ is preferable in order to keep down the size of blueprints. In many cases the ⅜″ = 1′-0″ scale is employed for elevation views because it is more suitable for complicated symbols, such as windows, than the ¼″ scale.

Although most scale drawings can be read with an ordinary ruler, special architectural scales make reading easier.

Architects often use what is called a three-sided, or triangular scale. This device, Fig. 3-7, is similar to a ruler except that instead of one edge, there are six edges, each of which has different graduations.

It is advisable for the student of construction wiring to obtain and familiarize himself with an architect's flat scale, such as the one in Fig. 3-8. The scale may be used for the following purposes:

1. Spotting fixtures and receptacles.

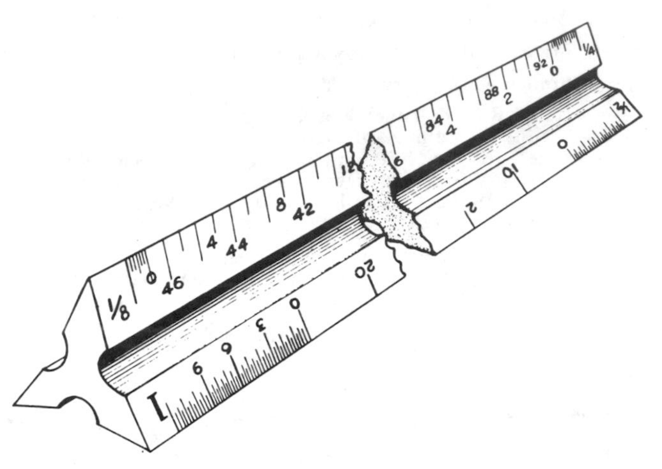

Fig. 3-7. Architects scale.

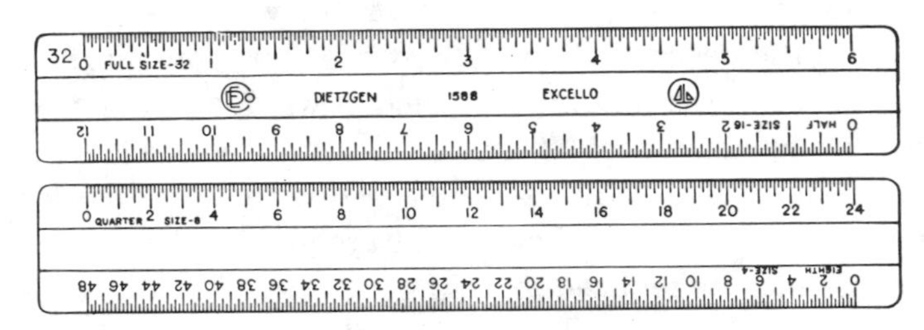

Fig. 3-8. Flat or pocket scale. (Eugene Dietzgen Co.)

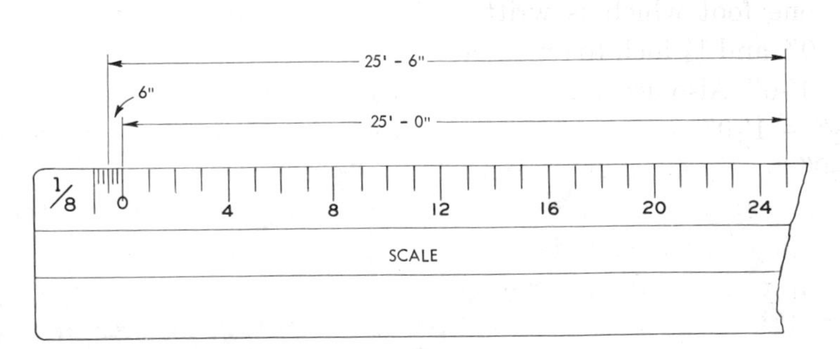

Fig. 3-9. Measuring with the flat scale.

2. Determining locations for appliances.
3. Estimating wire lengths.
4. Drawing up circuits layouts that require minimum use of wire.

In order to draw a line 25′ long according to the 1/8″ = 1′-0″, with the flat scale, Fig. 3-9, an architect actually draws it in 25 eighth-inch divisions. The eighths are numbered along the edge so they can be measured easily. Now suppose the line distance were 25′-6″. To draw this line, it is necessary to measure off 25 eighths plus one half of another eighth-inch. A 1/8″ space at one end has 6 divisions, each of which represents two inches. Therefore, 3 of these lengths must be added to the 25 eighth-inches in order to mark off the whole distance.

Blueprints

The Plot Plan

The plot plan is the starting point for building construction and is an important source of data for the crafts and on certain phases of construction. The plot plan, Fig. 3-10, shows the shape and dimensions of the property on which the building is to be constructed. When the plot is bounded by streets, such information is included. The plot plan in Fig. 3-10 shows that 72 feet of the property faces Lake Shore Drive.

A legal description of the property is also included, listing the lot number, block number, tract number and other data that accurately describes, the property.

The plot plan locates the building or buildings on the property. See *A* in Fig. 3-10. Additional information includes the front setback (the distance the building is located from the street), *B;* the rear set back, *C;* and the side yard clearances (the distance from the side and property lines to the building), *D.* The lines of the roof are usually shown. These include the overhang and covered area outside of the building proper, *E,* and the direction of the roof slope. Note Arrows indicating roof slope. Also included in the plot plan is the location of concrete driveways, walks and other paved areas.

The plot plan aids the electricians in locating the point of entry of electrical service, in determining the height of the service head and in establishing safe clearances of the power companies' service drop wires to the building. The power company should be consulted regarding power pole location. It also assists in estab-

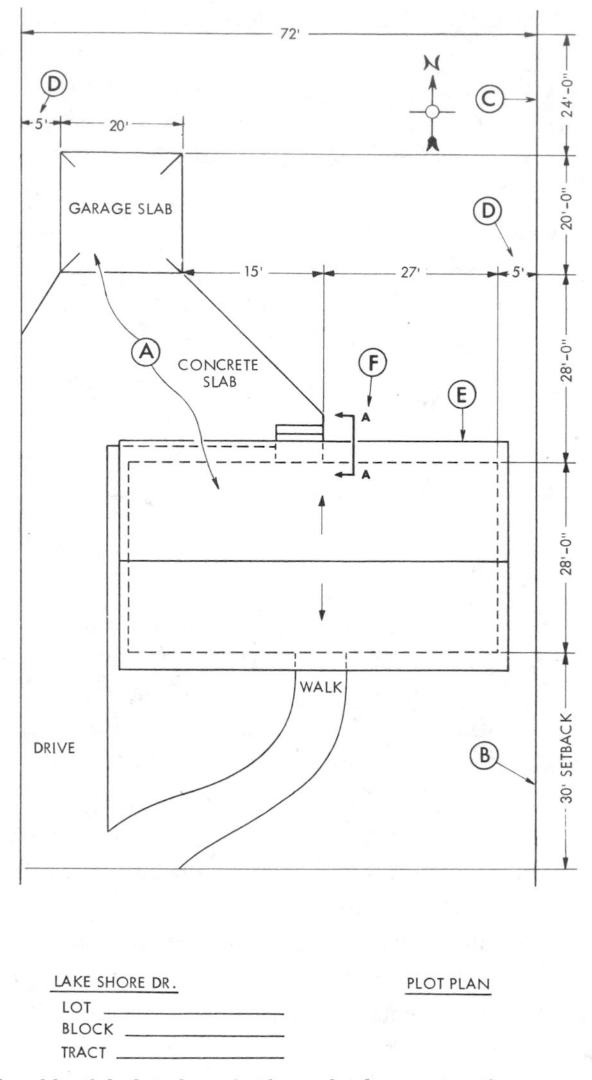

Fig. 3-10. Typical residential plot plan, starting point for construction.

lishing underground wiring for post lights and other landscape lighting, and assists the wireman in routing underground branch-circuit or feeder conduits under the drive or walk slabs or around them. The level the ground will be around the building when it is complete (finished grade) is helpful in determining depths of underground runs.

Foundation Plan

The Foundation Plan shows the location of the foundation of a building. Girder details and the direction, size and spacing of the floor joists are included. The location and size of access openings to the space under the house is also shown. Lettered section notes tie the foundation to the footing details.

A Typical Electrical Floor Plan

The electrical floor plan for a small frame bungalow is shown in Fig. 3-11. The house is 55′ long and 28′ wide. Although blueprints are drawn customarily to an "even" scale such as 1/8″ or 1/4″ to a foot, as mentioned earlier, the illustration here is drawn to an "uneven" scale in order to fit the page.

This one-story dwelling consists of a living room, dining room, kitchen, three bedrooms, bath, laundry, entryway, and central hall. Symbols for the most common types of electrical outlets are shown beneath the drawing. A ceiling outlet for an incandescent lamp is indicated by a small circle, a wall bracket by a circle with a supporting line or foot projecting from one side, a convenience outlet or plug receptacle by a circle with two parallel lines drawn through it, and a switch by a capital S.

As mentioned, there are different kinds of switches, the most common being single-pole, three-way, and four-way switches. A single-pole switch is represented by a plain S, a three-way by an S with the number 3 beside it, namely, S_3, and a four-way switch by the symbol S_4. Three-way switches enable one to turn a light on or off from widely separated points, for example, the two ends of a long hallway. A combination of two three-way switches and the required number of four-way switches permits a light to be operated from three or more points should it be so desired. Methods for connecting these special switches will be explained later on in the chapter on switching, Chapter 6.

In addition to electrical outlets, Fig. 3-11 shows doors, windows, partitions, closets, and bathroom and kitchen fixtures. Swinging doors are indicated by a straight line and the arc of a circle, and sliding doors by short, straight lines which do not meet. Partitions are marked by parallel lines, the distance apart depending on the drawing scale. Bathroom fixtures are shown by descriptive outlines: a bathtub, toilet and lavatory are shown on Fig. 3-11. A sink is located against the north wall of the kitchen.

Accordion doors are shown between kitchen and dining room, and between dining room and living room. Plaster arches are noted in two locations, between hall and living room and between kitchen and laundry. A brick fireplace occupies a space along the east living room wall.

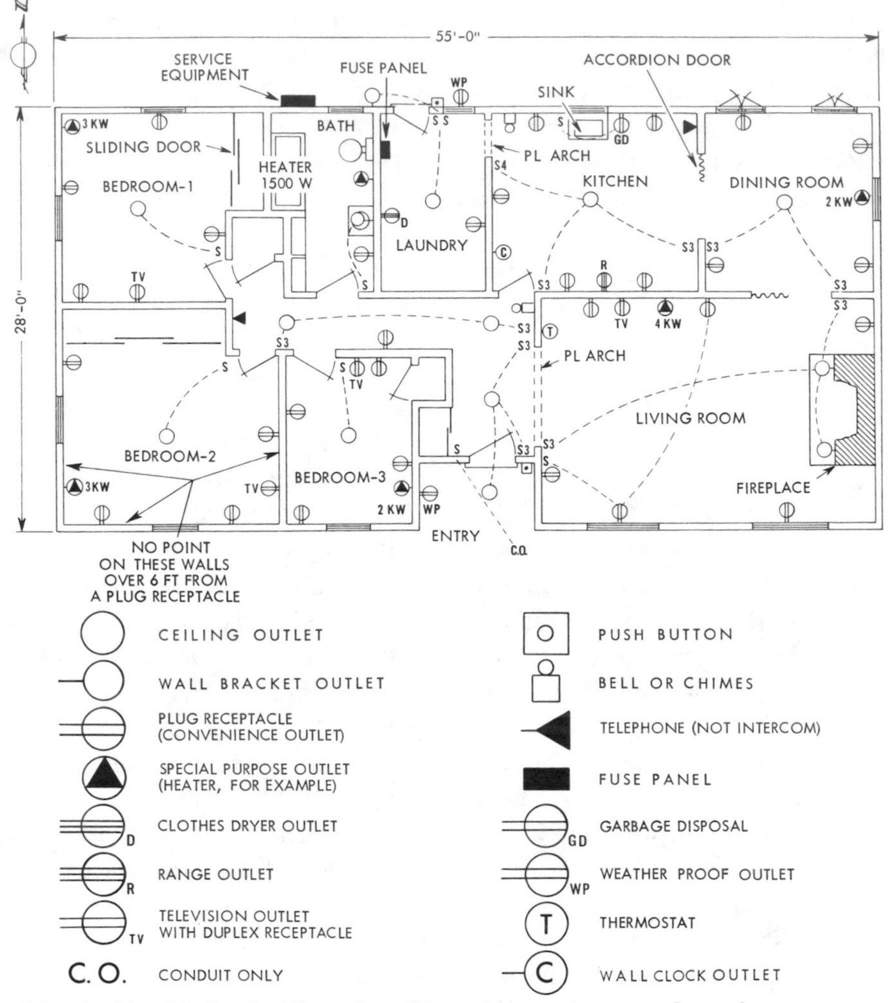

Fig. 3-11. Residential electrical floor plan with symbols most commonly used.

Windows are shown in all four walls, those in the living room being single pane, and all the other double-hung, with the exception of two casement windows in the dining room.

Electrical Outlets: Floor Plan

Starting at the entry in Fig. 3-11, a circle marks the overhead lighting outlet. A weatherproof convenience outlet in the left wall can be used for connection to hedge trimmer, seasonal decorative lights, or other electric devices. A dashed line connects the outside light to another ceiling outlet inside the house. From this point, a dashed line connects with a three-way switch to the right of the door. A second one connects to another three-way beyond the living room arch. These dashed lines are not intended to show paths of wires between outlets or switches but merely to indicate which outlets are controlled by the various switches. Branch circuitry layout according to code is the responsibility of the experienced journeyman wireman or electrical contractor.

At this point, it should be mentioned that circuit wires, conduit, or cable runs, are not usually indicated in drawings for small, medium, or even large residences. To aid in explaining estimating procedure, however, circuit runs for a similar house are illustrated in Chapter 17 on *Estimating Electrical Wiring.* Branch circuits can be laid out by the student here using the examples in the back of the N.E.C. (which tell how to calculate branch circuits) or by following requirements from a local code.

Continuing with the survey of Fig. 3-11, two ceiling lights, one in the entry and the other inside the front door, are controlled by a pair of three-way switches, as noted earlier. Another pair of three-way switches control two lights in the central hall, the first switch being located at the living room arch, the second on the wall between doors of bedroom 2 and 3.

Entering the living room, bracket lights on either side of the fireplace are operated by two three-way switches, one at the arch on the west wall, the other at the side of the dining room door. A single-pole switch at the hall archway connects with two plug receptacles across the room from each other. Switching circuits will be explained later. For the present, it may be noted that codes call for convenience outlets to be placed so that at no point along the floor line (in any usable wall space) is more than 6 feet from an outlet. In Fig. 3-11 the 6 feet specification is observed with regard to convenience outlets in living room, dining room, kitchen, and the three bedrooms.

The dining room ceiling light in Fig. 3-11, is controlled by two three-way switches at the sides of the accordion doors. The kitchen light is operated from three points. A three-way switch is located near the dining-room door, a second three-way is placed at the side of the hall door, and a four-way near the laundry arch. Also note the electric range outlet in the kitchen. The laundry ceiling outlet is connected with a single-pole switch near the outside door. Another switch controls an outside light. All convenience outlets in the kitchen and the laundry room should be on special utility circuits, as required by most codes.

The bathroom has a bracket light, a convenience outlet, and a single-pole switch. There is also a 1500 watt wall heater, which is designated by a triangle inside a circle. Others are found in bedrooms, living room and dining room. Reference to specifics would be found in the specification. The heater in the living room has a remote thermostat temperature control.

A fuse panel on the west wall of the laundry contains protective devices for circuits which radiate from it to the various outlets. This panel is supplied with current taken from service equipment, which is mounted on the north wall of the house adjacent to the bathroom wall. Power companies and codes discourage installations of meters on bedroom walls because of the possibility of a noisy meter. The service installation includes a meter and a disconnect switch. Its details will be explained later in Chapter 11, *Installing Service Equipment.*

Two bell push buttons are shown, one near the entry door and the other beside the laundry door. A pair of two-tone chimes is provided, one in the central hallway, the other in the kitchen. Details here are normally given in the written specifications, a part of the building plans. One tone informs that the back, or north, push button has been used, the other that the south button has been pressed. (Wiring details for chimes and push buttons are given in Chapter 6, *Switching Circuits.*) A telephone outlet occupies a space at the end of the central hallway, between bedrooms 1 and 2 and another in the kitchen area. A weatherproof outlet is shown outside the kitchen. Also an outside light is controlled from the laundry room. (See Figs. 3-12 and 3-13 for outdoor outlets and lights.) Also in the laundry room the plan shows a three-wire 115/230 volt receptacle for clothes dryer.

Fig. 3-12. Weatherproof electrical outlets. (Sierra Electric Corp.)

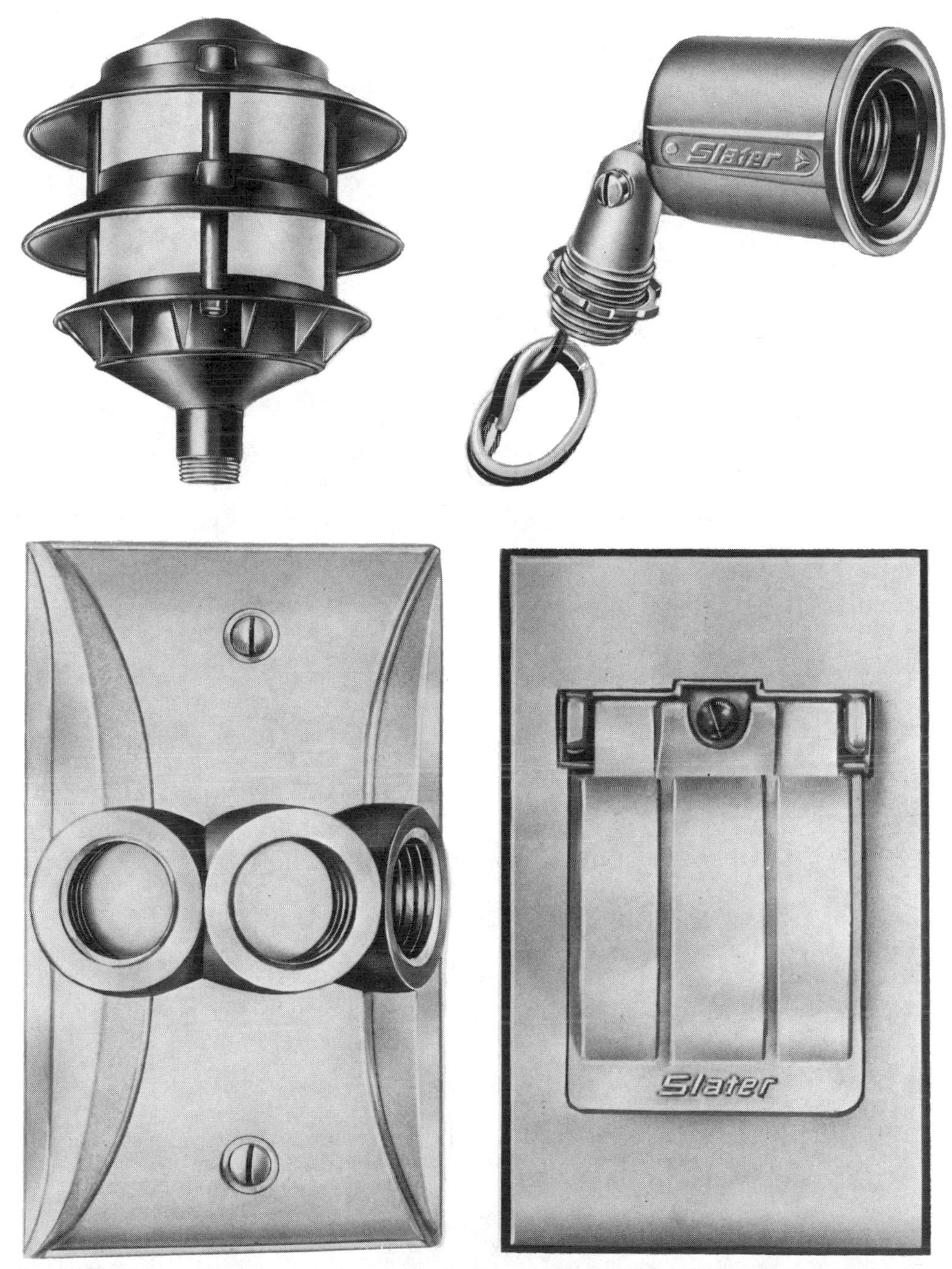

Fig. 3-13. Top left: Weatherproof louver light for walkways or landscape lighting. Top right: Weatherproof lamp holder for outdoor floodlighting. Bottom left: Weatherproof lamp cluster box cover. Bottom right: Weatherproof, hinged receptacle outlet cover. (Slater Electric Inc.)

A switch by the kitchen sink controls a garbage disposal. A switch behind the front door is connected to an underground circuit leading outside to in front of the house. This is marked on the print "C.O." for conduit only." This assembly is for future landscape lighting or weatherproof outlets such as shown in Fig. 3-12.

Fig. 3-13, top left, shows a louver light used in outdoor lighting. This

Fig. 3-14. Television outlets for power and antenna. (Sierra Electric Corp.)

Fig. 3-15. Left: Combination electrical outlet and antenna outlet; center: 2 gang TV or FM outlet; right: single gang TV or FM antenna outlet. (Sierra Electric Corp.)

fixture has ½ inch conduit threads for conduit or fitting mounting. Fig. 3-13, top right, is a commonly used reflector floodlight holder. These are used singularly or in clusters mounted on box covers as shown in Fig. 3-13, bottom left. Fig. 3-13, bottom right, shows weatherproof receptacle box cover.

Fig. 3-14 illustrates the convenience of combination electrical and television outlets combined. This is also shown in Fig. 3-15. Fig. 3-15, left, shows a combination unit commonly used. Fig. 3-15, center, illustrates a wall plate used for efficient transmission of television or FM radio signals from the antenna to receiver set. The most commonly used is that shown in Fig. 3-15, right.

Specifications with building plans give specific information regarding details of selection. Antenna lead-in wires are generally "stubbed" (brought) into the attic space and then down to location, where they are nailed into place behind a plaster ring on rough construction. Be sure to nail between the conductors. See Fig. 3-16. They are then easily located when cover plates are connected and installed.

Note the symbol for the wall clock outlet in the kitchen area of Fig. 3-11. Details of this is shown in Fig. 3-17. Other uses of clock outlets, besides hanging clocks, are shown in Fig. 3-18: they may be used for lighting paintings and pictures, and for electric fans.

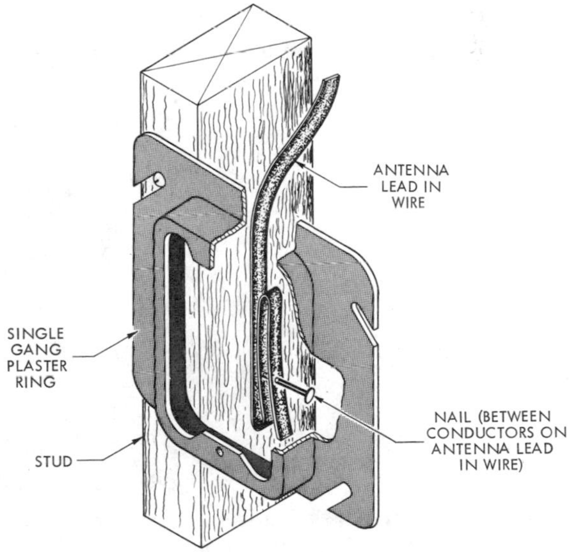

Fig. 3-16. Method of securing antenna wire behind plaster ring. The nail should be removed when the wire is connected to the trim.

ANTENNA LEAD IN WIRE

SINGLE GANG PLASTER RING

STUD

NAIL (BETWEEN CONDUCTORS ON ANTENNA LEAD IN WIRE)

Fig. 3-17. Details of a 120 volt recessed clock outlet. (Sierra Electric Corp.)

Fig. 3-18. Uses of wall clock outlets. (Sierra Electric Corp.)

Fig. 3-19. Top: Uses of electro-luminescent, soft glow night lights; bottom: uses of incandescent night lights. (Sierra Electric Corp.)

Fig. 3-19 illustrates night lights that may be designated on plans by special purpose outlet symbols which are explained in the specifications.

Fig. 3-19, top, shows an electro-luminescent, long-lived light source that provides a soft safety glow in the dark. They have many uses, two

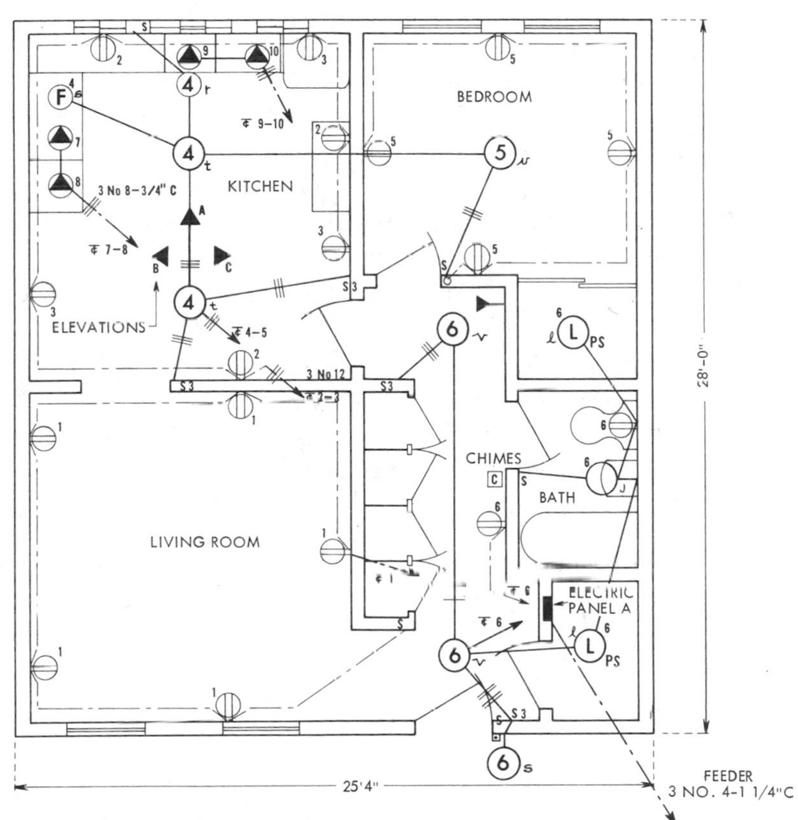

NOTE 1: CROSS-LINES ON CONDUIT RUNS INDICATE NUMBER OF WIRES IN RACEWAY WHERE NO CROSS-LINES ARE SHOWN, TWO CONDUCTORS ARE PRESENT.

NOTE 2: NOTATION "⊄" SOMETIMES USED AS SYMBOL FOR "CIRCUIT".

FIXTURE LIST

DES.	QUAN.	RATING	MFGR	CAT.	TYPE
r	5	100W	MORBRITE	612	RECESSED
s	6	50	ALLRAY	1117	BRACKET
t	12	150	ALLRAY	1129	DRUM
ʋ	5	100	MORBRITE	700	DRUM
v	12	75	ALLRAY	1119	CEILING
J	6	75	MORBRITE	804	BRACKET
ℓ	12	50	TITAN	322	PORC P.C. SOC
h	3	50	TITAN	323	PARC SOCKET
P	3	200	FRANKLIN	1791	W.P. FLOOD

Fig. 3-20. Floor plan showing electrical layout keyed to fixture list.

of which are shown. Fig. 3-19, bottom, shows the incandescent night-light type designated for use with louvered plates and different voltages.

Wiring Layout

As mentioned, the wiring layout is the responsibility of the journeyman wireman or electrical contractor. Fig. 3-20 shows a method of laying out circuits on a floor plan. Circuit No. 1 serves all outlets in the living room. The switch controls one outlet. Note the number by each of the receptacles and lights. This number refers to the circuit number. The number 1 by each of the receptacles in the living room means that they are all on No. 1 circuit and connected to No. 1 circuit breaker or fuse at the panel. Ceiling outlets may have their circuit numbers inside the circle symbol or beside it.

Arrows indicating "home runs" to the panel are marked with circuit numbers. The living room shows circuit No. 1. The kitchen has two circuits each per branch circuit feed. The kitchen probably has the most complicated wiring in the house. The special purpose outlets are explained by the use of interior elevations, Fig. 3-22. (The arrows with A, B and C in the kitchen of Fig. 3-20 indicate the direction of view in the elevations in Fig. 3-22.) A lighting fixture list is also shown in Fig. 3-20. This shows which fixture is to be hung where.

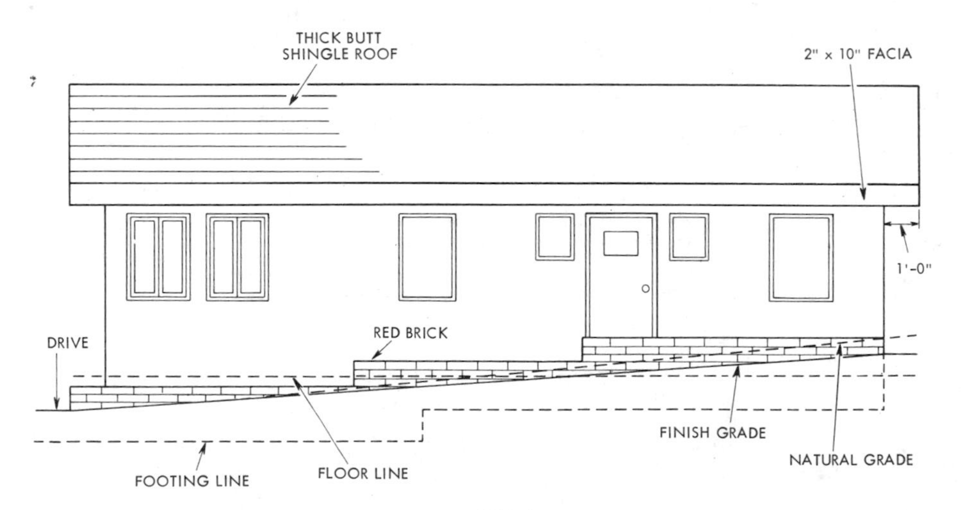

Fig. 3-21. An exterior elevation. See Fig. 3-11 for the plan view.

Exterior Elevations

Views of the finished exterior sides of the building are included in the building plans and are called *exterior elevations*. These elevations show the exterior finish, trim, window and door openings, roofing material, steps, and brick work. Finished grade and floor lines are also usually given. The electrician may find these elevations helpful in locating wall bracket lights, weatherproof convenience outlets and locations for landscape lighting. See Fig. 3-21.

Interior Elevations

Views of interior walls containing counters, sinks, cupboards and other special features are called *interior elevations*. These elevations are helpful in determining the height of convenience outlets and switches in bathrooms and kitchens. An interior elevation of a kitchen is shown in Fig. 3-22. (The floor plan is shown in Fig. 3-20; the letters A, B and C refer to views on the floor plan.) The material to be used in this area affects the depth to which the outlet boxes are to be mounted. Special attention may be required for the kitchen hood with light and exhaust fan in Fig. 3-22. Outlets in furred down ceilings also require special attention. It is worth the time to study these elevations in order to avoid electrical outlet omissions.

Sectional Drawings

Sections are often used on different plans. Section A-A shown in Fig. 3-10, *F*, for example, refers to an enlarged view looking in the direction of the arrows and encompassing the span by the distance between the arrows in "A-A". Fig. 3-23 is the detail referred to by *F*, "A-A" in Fig. 3-10. Detail drawings, such as the ones shown in Fig. 3-23, indicate the various shapes and dimensions of each type of building support.

The electrician may want to place conduits in or under the foundation or footing. Using the footing detail he can determine wiring and crawling clearance under the structure and the boring problems which will be encountered with respect to partitions, joists and girders.

The draftsman or architect also furnishes other drawings of the structure. See Fig. 3-24 for example. Fig. 3-25 shows story height and ceiling height of each floor in the building. *Story height* is the distance from the top of the finished floor of one story to the top of the finished floor in the next upper or lower story. *Ceiling height* is the distance from the top of the finished floor of one story to the finish line of the ceiling of the same story.

Section drawings show what floors are composed of; for example, whether they have wooden joists, Fig. 3-24, left, or some other type of construction, Fig. 3-24, right. This fact influences the method of doing electrical work and the kind of ma-

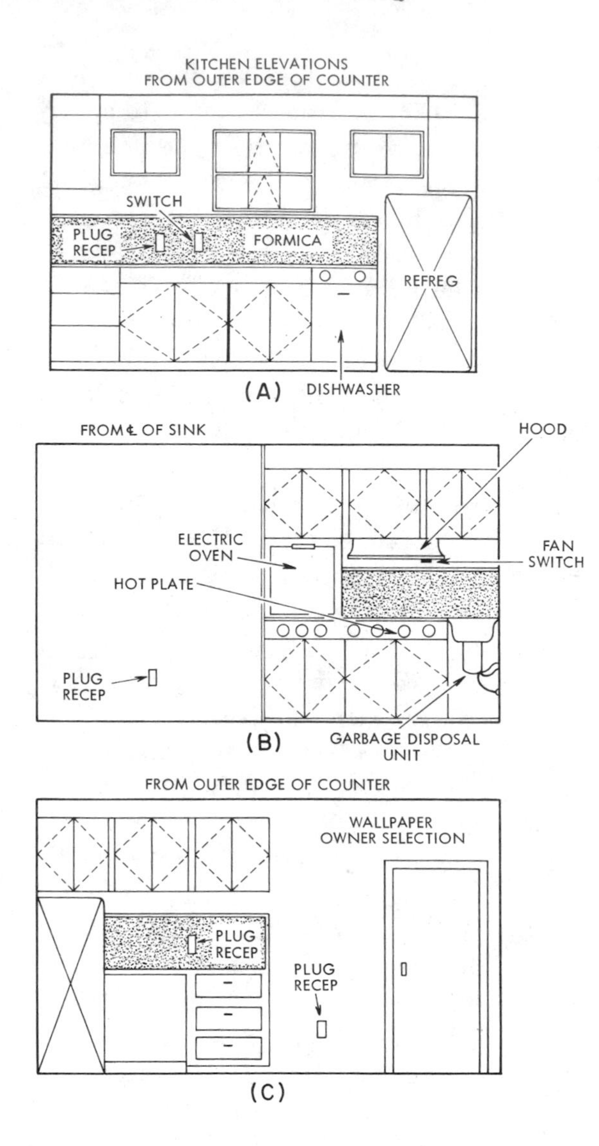

Fig. 3-22. An interior elevation. See Fig. 3-20 for the floor plan; the direction of viewing for the A, B and C elevations are noted on the floor plan.

terials needed. Thus, if floors and partitions are wood, concealed wiring may be done with flexible conduit, nonmetallic sheathed cable, or armored cable, provided local rules allow. If the structure is concrete, either rigid conduit or electrical metallic tubing must be employed for this work.

Ceiling Finish. The sectional drawing may also show the amount of space there is to be between the

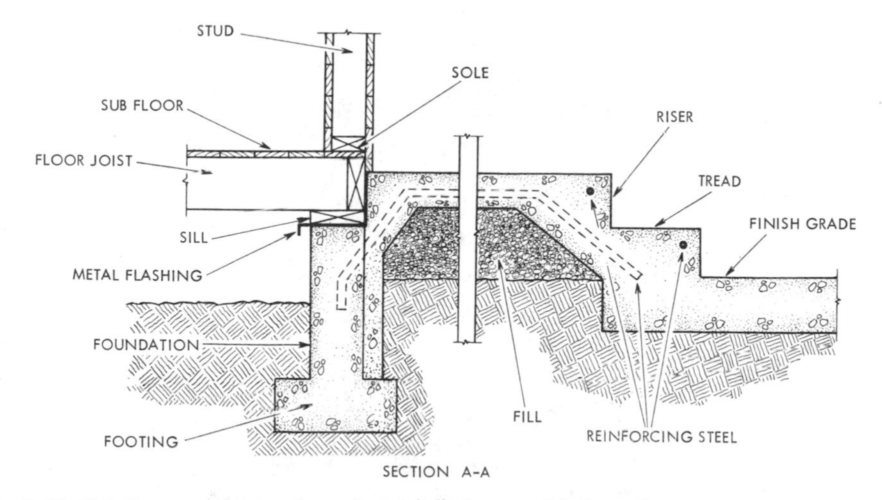

Fig. 3-23. Detail or section drawings give detailed support information.

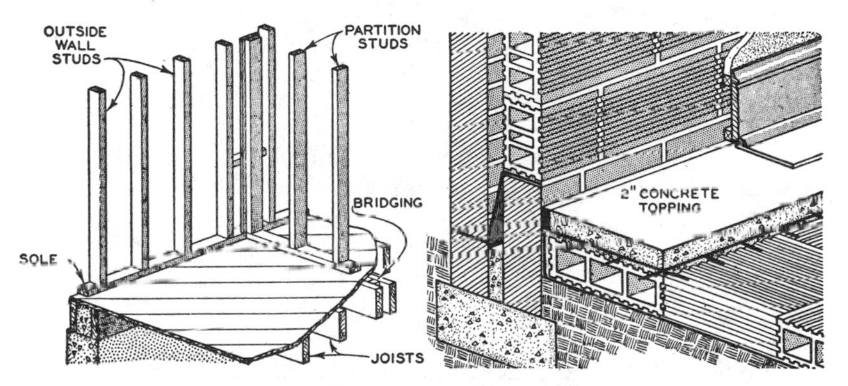

Fig. 3-24. Section drawings of a frame and a concrete structure.

bottom of the joists and the finish line of the plaster or other ceiling material, called *ceiling finish* for short. They may also show the space between the top of the floor joists and the top surface of the finished floor, called *floor finish* for short. These spaces at times are considerable, and must be allowed for in the selection of the outlet box and the plaster ring or cover as well as in the placement of the outlet box. Otherwise the edge will either project beyond the ceiling finish or be too far withdrawn from it.

In buildings of wood joist con-

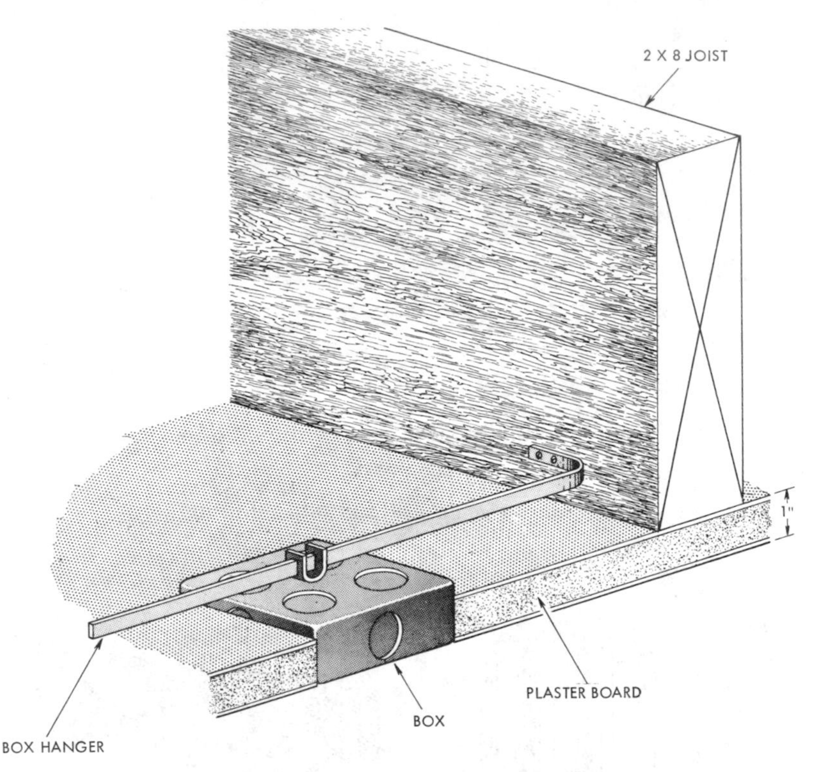

Fig. 3-25. Mounted ceiling outlet box flush on the ceiling finish line.

struction, the plaster board is nailed directly to the bottom of the ceiling joists, Fig. 3-25. In some cases, wallboard, sheetrock, or wood may be used for ceiling or wall finish. The materials vary in thickness; hence it will be necessary to check on this point if the wiring is to be concealed. With a proper hanger it is possible to use a 1½″ × 4″ box without plaster ring, Fig. 3-26, top left. If a 4 inch opening is too large for the fixture canopy, a plaster ring may be installed on the box, thus reducing the size of the opening to 3½ inches, Fig. 3-26, top right.

There are times when the ceiling is furred down several inches. When the amount of furring is greater than can be taken care of with an extension box, Fig. 3-26, bottom, it is necessary to choose between (1) running the conduit on a two story building down to the outlet box or (2) running the conduit in the space between floor and suspended ceiling.

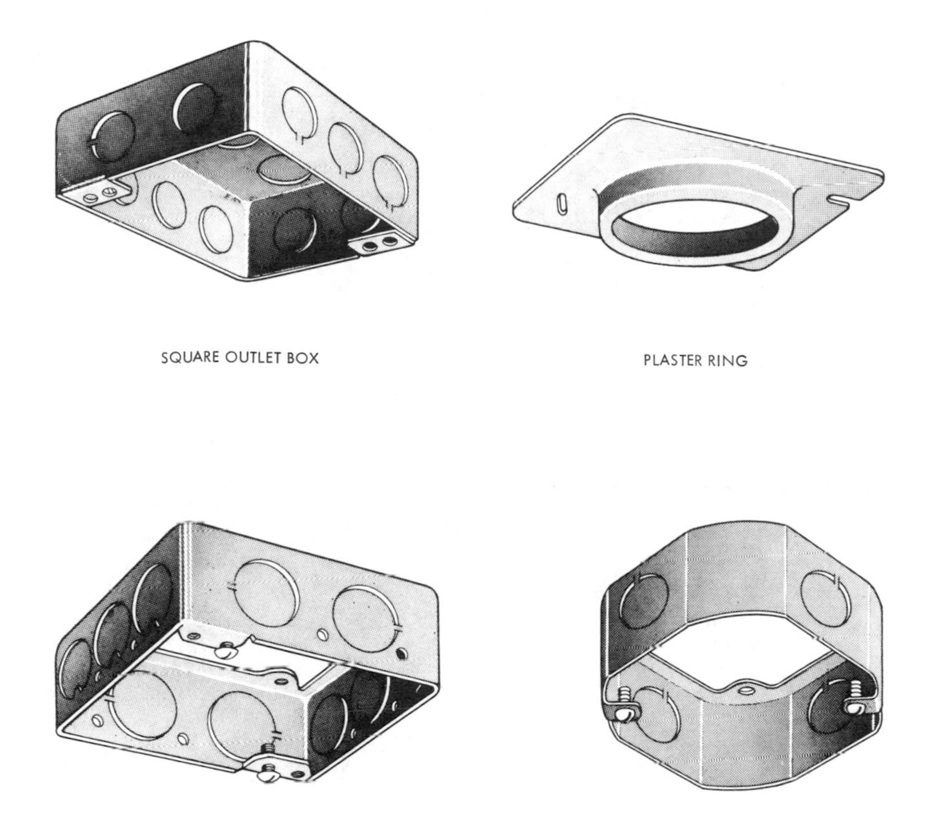

Fig. 3-26. Top, left: 1½" x 4" square outlet box; top, right, 4" square, raised ½" plaster ring; bottom, left, 1½" deep, 4" square outlet box extension ring; bottom, right, octagon outlet box extension ring. (Top right and bottom: Appleton Electric Co.)

Structural Problems

Single-story dwelling wiring problems may be suggested by Fig. 3-27. Problems for cutting, notching and boring or for pulling cable or flexible conduit may be lessened by understanding wood frame construction methods. This is shown in Figs. 3-27 and 3-28. A study of frame construction will assist in understanding references in the code also. The electrician will have to seek a wiring path when drilling through the sole. Floor joists and girders could be obstructions which must be drilled through.

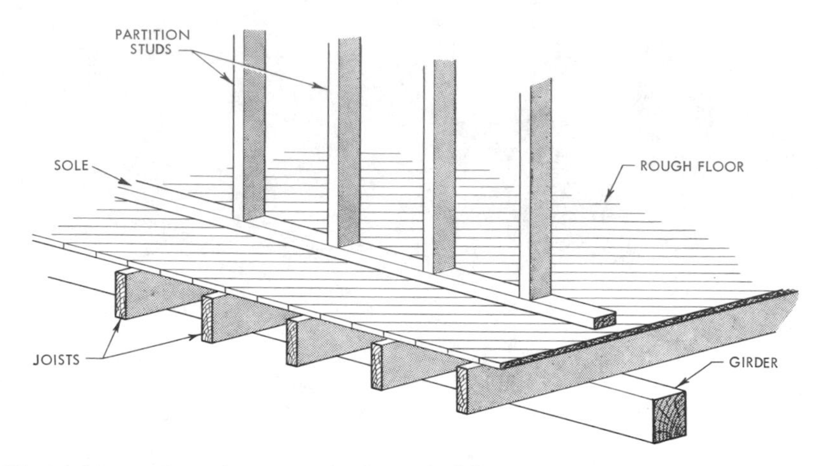

Fig. 3-27. Partition walls and support of a frame building.

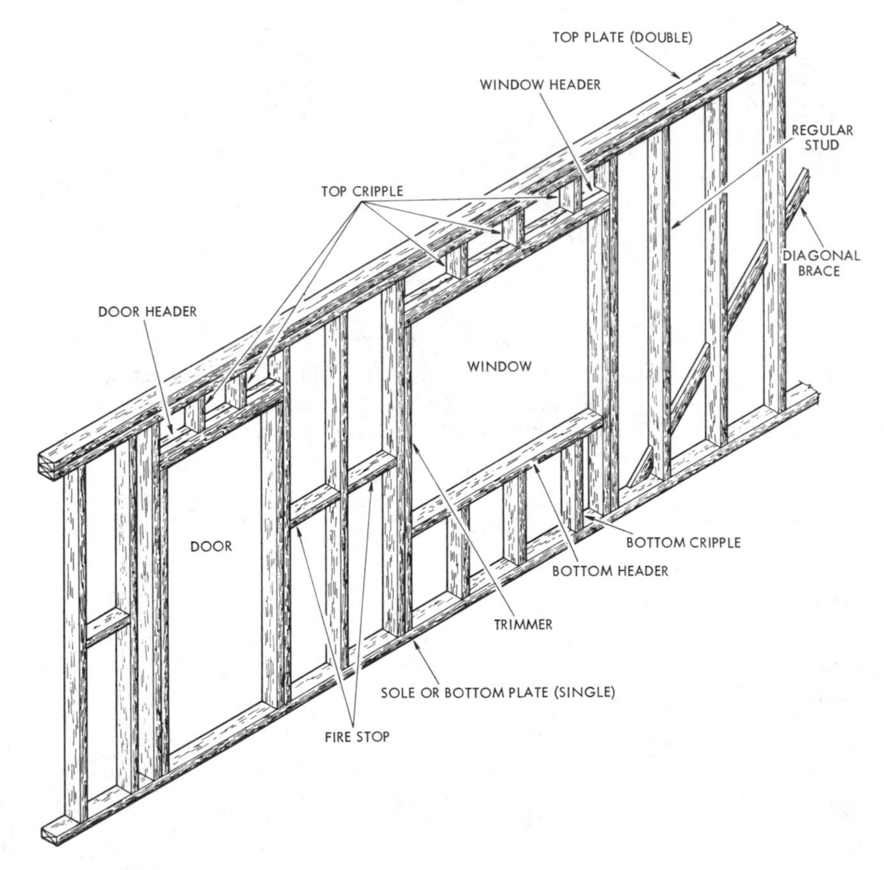

Fig. 3-28. An outside wall frame.

This is shown in Fig. 3-27. Note in Fig. 3-28 that a fire stop is located where a switch may be required. This may be notched for the switch box or the block moved enough to mount the box. If the block is moved slightly, it should be drilled for the cable or conduit before the box is mounted and becomes an obstruction to drilling. It has been found that a wiring job can be done more cheaply by going through the top plates and overhead to connect convenience outlets on opposite sides of the room. It requires more material but less labor. Studs and bot-

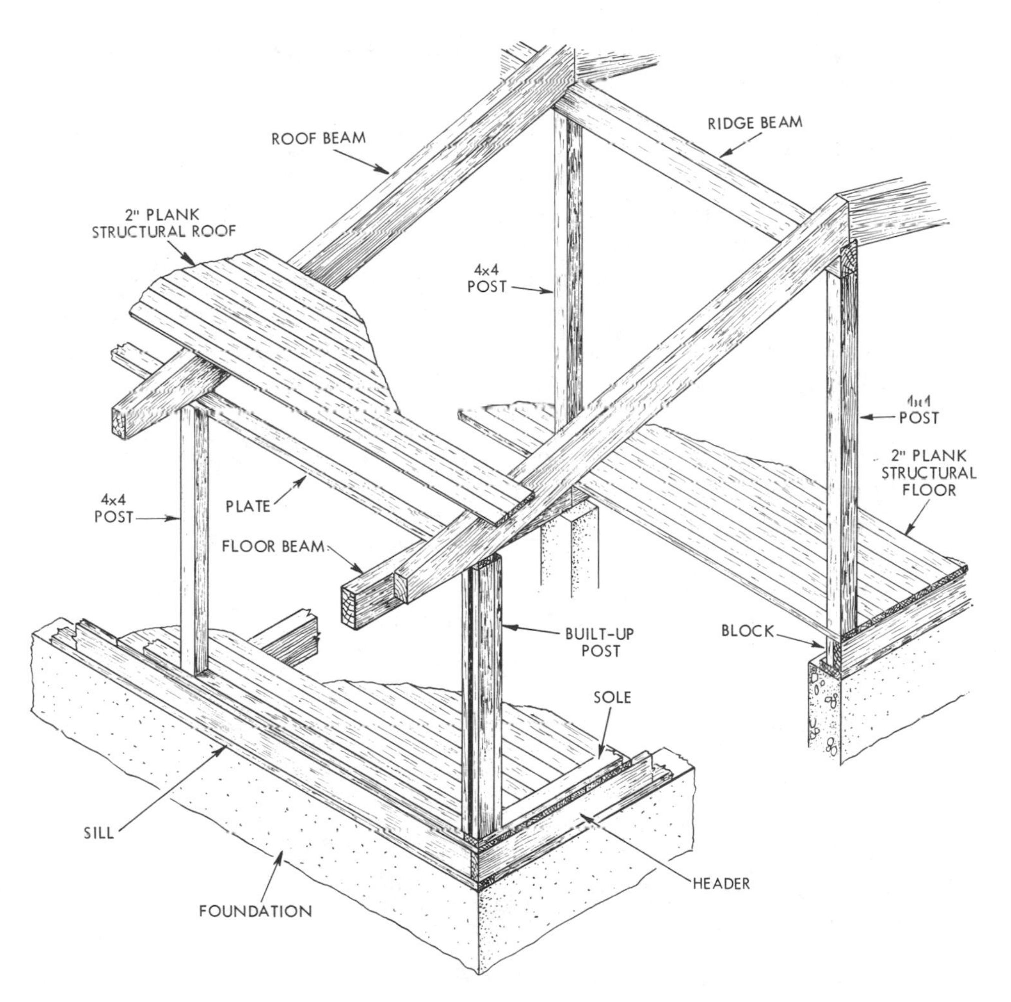

Fig. 3-29. Plank and beam framing.

tom cripples are drilled for connecting convenience outlets around a room.

Special problems may be encountered in plank and beam construction. In this construction, post and beam framing units are the basic load bearing members. Fewer framing members are needed, leaving more open space for functional use. See Fig. 3-29. Posts and beams may be of wood, structural steel or concrete. The roof deck can double as finished ceiling. Problems include the difficulty in concealing wiring and duct work, and the necessity for extra care in the choice of materials and in planning.

Structural Details

Closely related to architectural symbols and equally important to the electrician are structural details and conventions. Conventions are symbols representing things like doors, windows and vents. Some of the commonly used conventions are shown in Fig. 3-30. Construction details of the type shown provide information on the structural details of a building and are used in plans to illustrate how the various materials are to be assembled. Familiarly with the terminology of construction details and the items they represent is necessary for the electrician if he is to discuss intelligently the details of a job with the carpenters and other craftsmen.

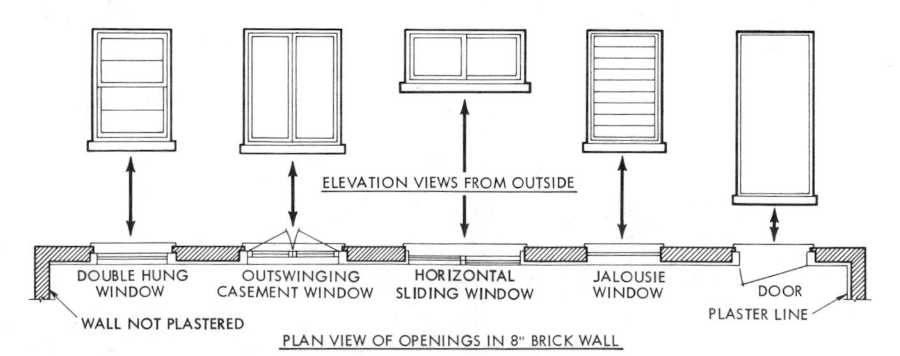

Fig. 3-30. Plan and elevation views in a brick wall.

Architectural Symbols

General structural information is also conveyed by symbols as shown in Fig. 3-31. Symbols and details used in architectural plans are ordinarily based on standards established by such an interested group as the American National Standards Institute. Many architects prefer, however to use their own symbols and notations to some extent. A knowledge of the commonly used symbols will assist in understanding some deviations on plans. The material symbols in Fig. 3-31 are used on plans and elevations, and for sections. The symbols shown are those commonly used to represent construction materials; references are usually made in the specifications for details.

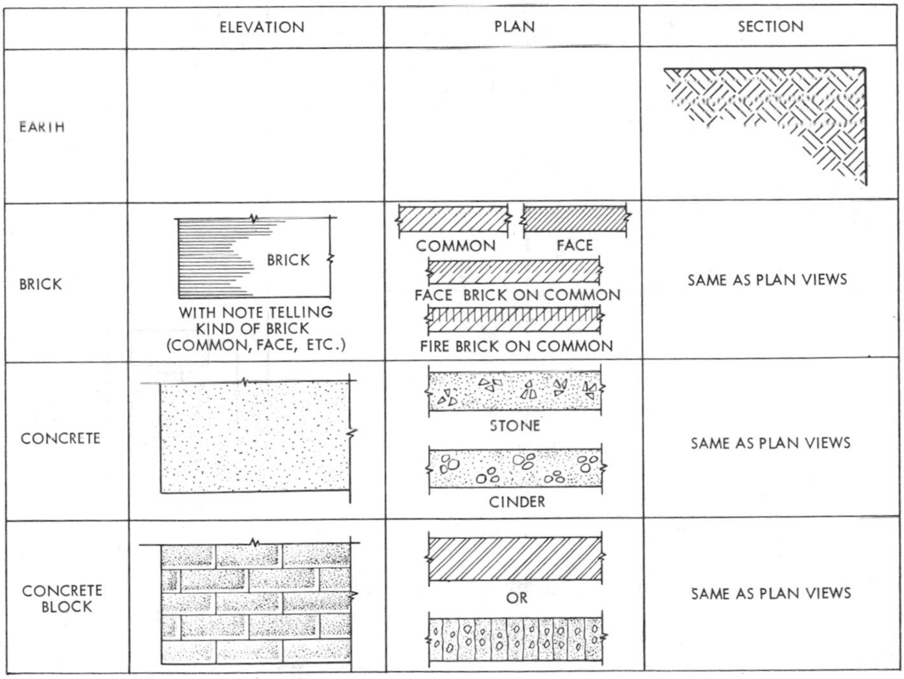

	ELEVATION	PLAN	SECTION
EARTH			
BRICK	BRICK WITH NOTE TELLING KIND OF BRICK (COMMON, FACE, ETC.)	COMMON FACE FACE BRICK ON COMMON FIRE BRICK ON COMMON	SAME AS PLAN VIEWS
CONCRETE		STONE CINDER	SAME AS PLAN VIEWS
CONCRETE BLOCK		OR	SAME AS PLAN VIEWS

Fig. 3-31. Architectural symbols.

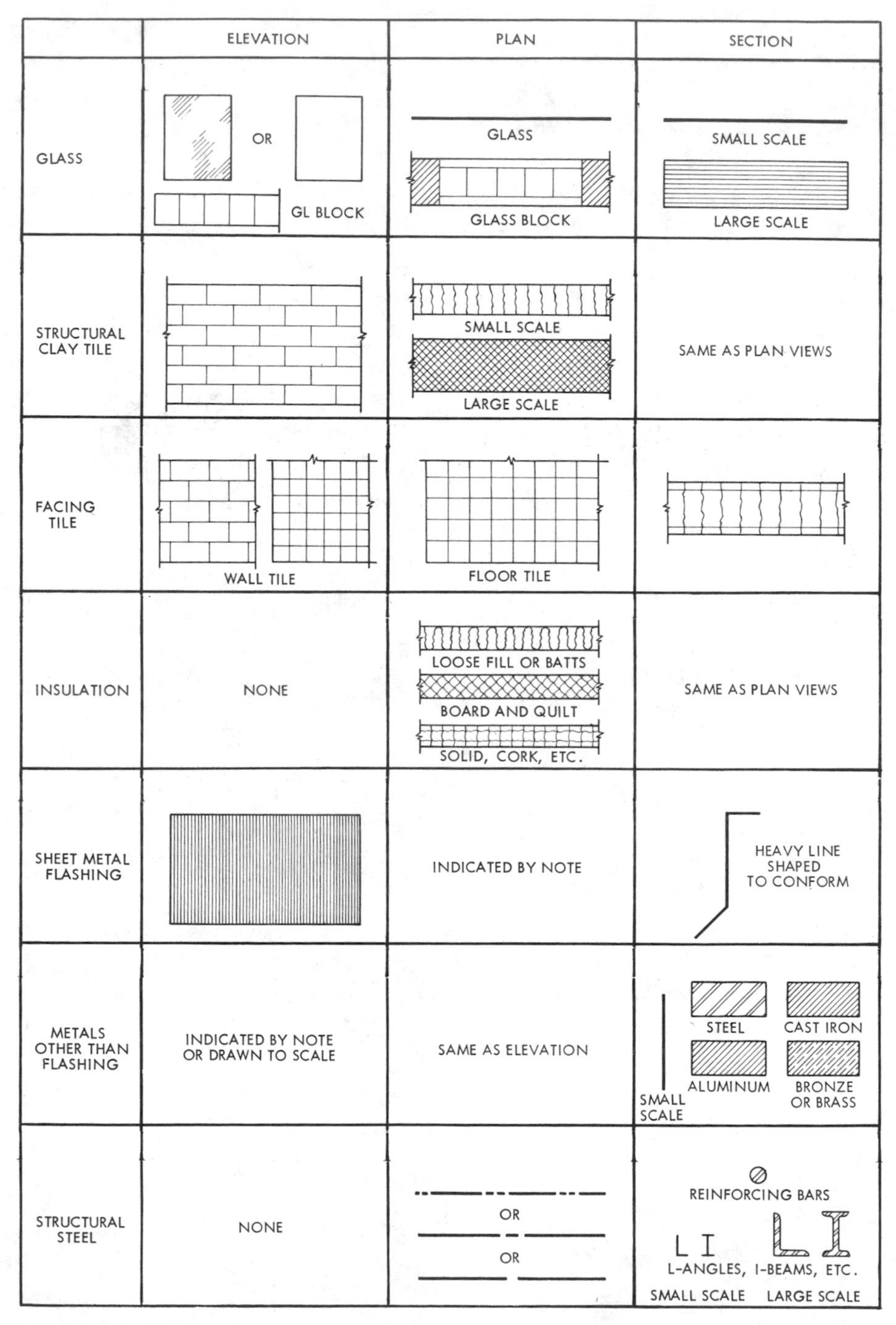

Fig. 3-31. Cont'd.

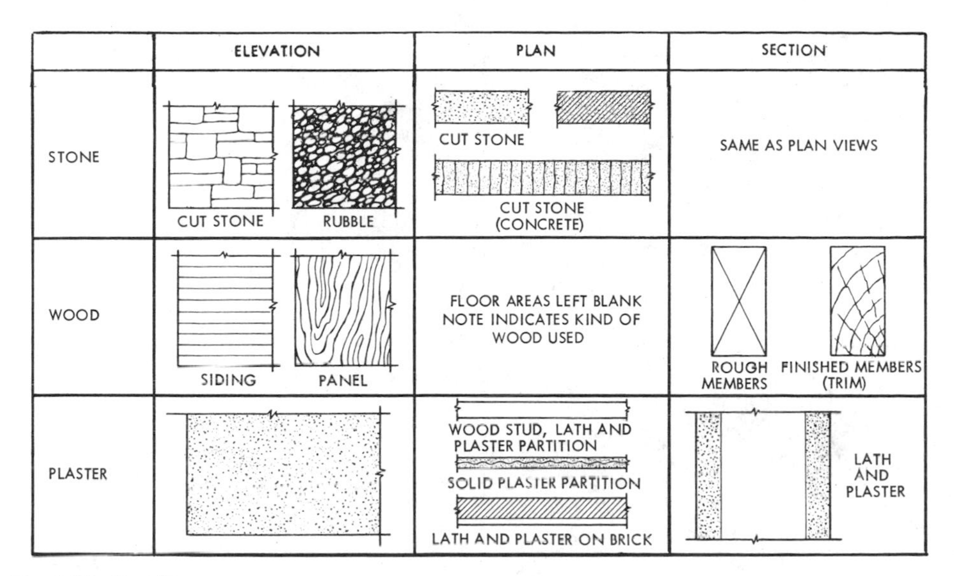

Fig. 3-31. Cont'd.

Symbols of Other Trades

Some of the symbols used in other trades closely resemble those used in electrical wiring. This similarity is particularly true of symbols used in plumbing, heating, and refrigeration. If confusion is to be avoided, electricians must be able to identify such symbols correctly. Some trade symbols that are often confused with electrical symbols are shown in Fig. 3-32.

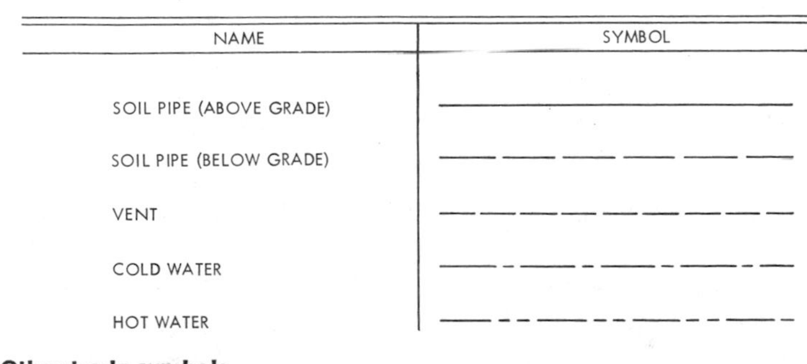

NAME	SYMBOL
SOIL PIPE (ABOVE GRADE)	————————————
SOIL PIPE (BELOW GRADE)	—— —— —— —— —— ——
VENT	— — — — — — — — —
COLD WATER	—— - —— - —— - —— - ——
HOT WATER	—— - - —— - - —— - - —— -

Fig. 3-32. Other trade symbols.

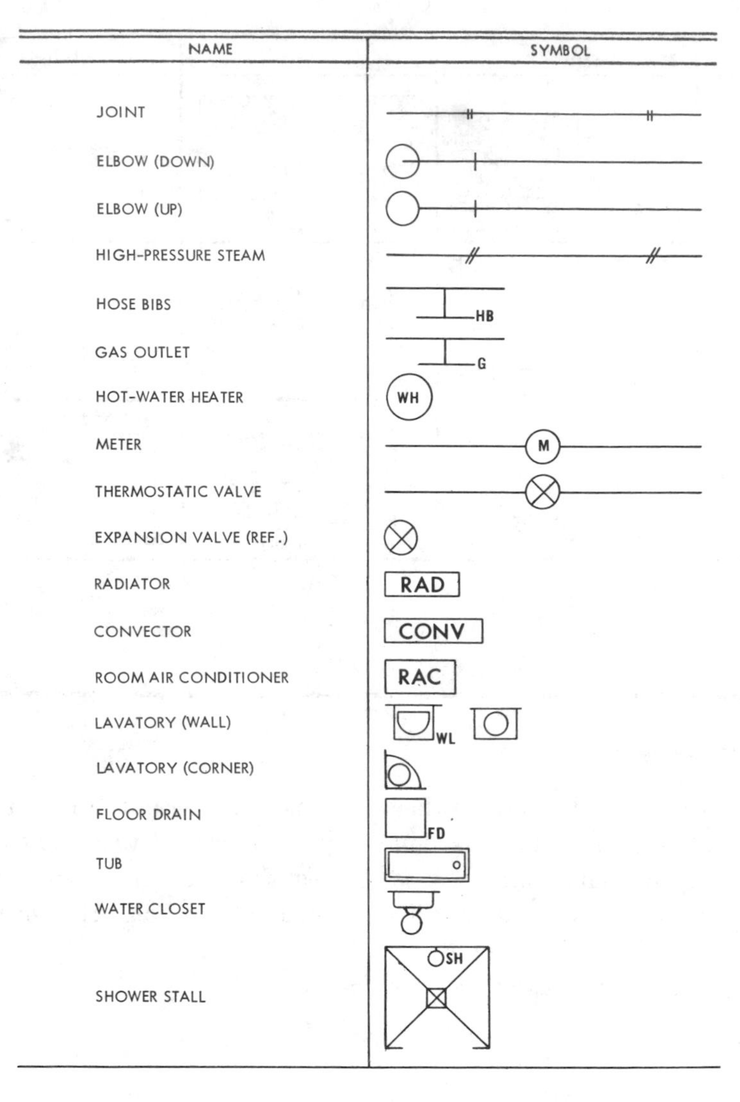

Fig. 3-32. Cont'd.

Abbreviations and Definitions of Construction Terms

Many trade terms and the names of material are abbreviated when used on plans. A list of those abbreviations most commonly used are shown in *Table I*.

Code books refer to nomenclature

TABLE I. ABBREVIATIONS USED ON PLAN VIEWS

ACCESS PANEL	AP
ACOUSTICAL TILE	AT
AIR CONDITIONING	AIR COND
ANCHOR BOLT	AB
ANGLE	∠
APARTMENT	APT
APPROXIMATE	APPROX
ARCHITECTURAL	ARCH
ASPHALT TILE	AT
BASEMENT	BSMT
BATHROOM	B
BATH TUB	BT
BEAM	BM
BEARING PLATE	BRG PL
BEDROOM	BR
BLOCKING	BLKG
BLUEPRINT	BP
BOILER	BLR
BRICK	BRK
BROOM CLOSET	BC
BUILDING	BLDG
BUILDING LINE	BL
CABINET	CAB.
CATCH BASIN	CB
CELLAR	CEL
CENTER	CTR
CENTER TO CENTER	C TO C
CENTER LINE	℄ TO CL
CINDER BLOCK	CIN BL
CIRCUIT BREAKER	CIR BKR
CLOSET	C, CL OR CLO
COLD WATER	CW
CONCRETE BLOCK	CONC B
CONCRETE FLOOR	CONC FL
CONDUIT	CND
CONSTRUCTION	CONST
CONTRACT	CONT
COUNTER	CTR
CUBIC FEET	CU FT
DETAIL	DET
DIAGRAM	DIAG
DIMENSION	DIM.
DINING ROOM	DR
DISHWASHER	DW
DOWN	DN
DRAWING	DWG
DRYER	D
ELECTRIC PANEL	EP
END TO END	E TO E
EXTERIOR	EXT
FINISH	FIN.
FINISHED FLOOR	FIN. FL
FIXTURE	FIX.
FLOOR	FL
FLUORESCENT	FLUOR
FLUSH	FL
FOOTING	FTG
FOUNDATION	FND
FRAME	FR
FULL SIZE	FS
FURRING	FUR.
GARAGE	GAR
GLASS BLOCK	GL BL
HOSE BIB	HB
HOT WATER HEATER	HWH
INSIDE DIAMETER	ID
INTERIOR	INT
KITCHEN	K
LAUNDRY	LAU
LAVATORY	LAV
LENGTH	L, LG OR LNG
LIGHT	LT
LINEN CLOSET	L CL
LINOLEUM	LINO
LIVING ROOM	LR
MAIN	MN
MATERIAL	MATL
MAXIMUM	MAX
MEDICINE CABINET	MC
MINIMUM	MIN
MISCELLANEOUS	MISC
ON CENTER	OC
OPENING	OPNG
OUTLET	OUT.
OVERALL	OA
PANTRY	PAN.
PARTITION	PTN
PLASTER	PL OR PLAS
PLUMBING	PLBG
PORCH	P
PRECAST	PRCST
PREFABRICATED	PREFAB
PULL SWITCH	PS
RANGE	R
RECESSED	REC
REFRIGERATOR	REF
REINFORCE OR REINFORCING	REINF
ROUGH	RGH
RUBBER TILE	R TILE
SCALE	SC
SCUTTLE	S
SECTION	SECT
SERVICE	SERV
SHOWER	SH
SINK	SK OR S
SOIL PIPE	SP
SPECIFICATION	SPEC
SQUARE FEET	SQ FT
STAIRWAY	STWY
STANDARD	STD

TABLE I. ABBREVIATIONS USED ON PLAN VIEWS (CON'T)

Term	Abbreviation	Term	Abbreviation
STORAGE	STG	UNFINISHED	UNF
SWITCH	SW OR S	UTILITY ROOM	URM
TELEPHONE	TEL	VENT	V
THERMOSTAT	THERMO	VINYL TILE	V TILE
TOILET	T	WASHING MACHINE	WM
TREAD	TR OR T	WATER CLOSET	WC
TYPICAL	TYP		

of frame construction and other details, such as cutting, notching and boring for the passage of electrical cables and conduits. The wireman must understand this nomenclature in order to do his job properly. Also, in his trade an electrician frequently comes in contact with workers in other trades. To communicate properly with these workers the wireman should be able to name the structural members of a building. Definitions of construction terms are given in *Table II*. (Chapter 4 contains additional structural information.)

TABLE II DEFINITIONS OF CONSTRUCTION TERMS

AREAWAY: AN OPEN SUBSURFACE SPACE AROUND A BASEMENT WINDOW OR DOORWAY, ADJACENT TO THE FOUNDATION WALLS.

BASEBOARD: THE FINISHING BOARD NEXT TO THE FLOOR.

BASE SHOE: A MOLDING FINISHING THE ANGLE BETWEEN THE BASEBOARD AND THE FLOOR.

BATT INSULATION: A TYPE OF SMALL-SIZED BLANKET INSULATING MATERIAL, USUALLY COMPOSED OF MINERAL FIBERS AND MADE IN RELATIVELY SMALL UNITS FOR CONVENIENCE IN HANDLING AND APPLYING. SOMETIMES SPELLED BAT.

BEAM: ANY LARGE PIECE OF TIMBER, STONE, IRON, OR OTHER MATERIAL, USED TO SUPPORT A LOAD OVER AN OPENING, OR FROM POST TO POST; ONE OF THE PRINCIPAL HORIZONTAL TIMBERS, RELATIVELY LONG, USED FOR SUPPORTING THE FLOORS OF A BUILDING.

BEARING: THAT PORTION OF A BEAM OR TRUSS WHICH RESTS UPON A SUPPORT; THAT PART OF ANY MEMBER OF A BUILDING THAT RESTS UPON ITS SUPPORTS.

BEARING PLATE: A PLATE PLACED UNDER A HEAVILY LOADED TRUSS BEAM, GIRDER, OR COLUMN, TO DISTRIBUTE THE LOAD SO THE PRESSURE OF ITS WEIGHT WILL NOT EXCEED THE BEARING STRENGTH OF THE SUPPORTING MEMBER.

TABLE II DEFINITIONS OF CONSTRUCTION TERMS (CON'T)

BEARING WALL OR PARTITION: A WALL WHICH SUPPORTS THE FLOORS AND ROOF IN A BUILDING; A PARTITION THAT CARRIES THE FLOOR JOISTS AND OTHER PARTITIONS ABOVE IT.

BENCH MARKS: A BASIS FOR COMPUTING ELEVATIONS BY MEANS OF IDENTIFICATION MARKS OR SYMBOLS ON STONE, METAL, OR OTHER DURABLE MATTER, PERMANENTLY FIXED IN THE GROUND, AND FROM WHICH DIFFERENCES OF ELEVATIONS ARE MEASURED.

BRACE: A PIECE OF MATERIAL USED TO STIFFEN THE FRAMEWORK.

BRACKET: A SUPPORT FOR A CORNICE; A PIECE OF WOOD NAILED AT THE INTERSECTION OF STUDS AND CEILING JOISTS TO FORM A COVE IN THE PLASTER.

BRIDGING: SMALL BRACES PLACED BETWEEN FLOOR JOISTS AND STUDDING TO STIFFEN THEM.

BUTTON BOARD: PERFORATED PANELS APPLIED TO INTERIOR WALLS TO SUPPORT PLASTER.

BUILDING LINE: THE LINE, OR LIMIT, ON A CITY LOT BEYOND WHICH THE LAW FORBIDS THE ERECTION OF A BUILDING; ALSO, A SECOND LINE ON A BUILDING SITE WITHIN WHICH THE WALLS OF THE BUILDING MUST BE CONFINED; THAT IS, THE OUTSIDE FACE OF THE WALL OF THE BUILDING MUST COINCIDE WITH THIS LINE.

BUILDING PAPER: A FORM OF A HEAVY PAPER PREPARED ESPECIALLY FOR CONSTRUCTION WORK. IT IS USED BETWEEN ROUGH AND FINISH FLOORS, AND BETWEEN SHEATHING AND SIDING, AS AN INSULATION AND TO KEEP OUT VERMIN. IT IS USED, ALSO, AS AN UNDERCOVERING ON ROOFS AS A PROTECTION AGAINST WEATHER.

CASING: A FINISHING BOARD USED TO COVER THE EDGE OF THE PLASTER AROUND DOOR AND WINDOW OPENINGS.

COLUMN: A SORT OF POST OR PILLAR.

CORNICE: THE OVERHANG OF A ROOF; THAT PART OF A ROOF PROJECTING BEYOND THE BUILDING.

CRAWL SPACE: IN CASES WHERE HOUSES HAVE NO BASEMENTS, THE SPACE BETWEEN THE FIRST FLOOR AND THE GROUND IS MADE LARGE ENOUGH FOR A MAN TO CRAWL THROUGH FOR REPAIRS AND INSTALLATION OF UTILITIES.

DIAGONAL BRACE: A PIECE OF LUMBER, LET INTO A FRAME AT AN ANGLE, EXTENDING FROM TOP TO BOTTOM PLATE.

FASCIA: A BOARD USED IN CORNICE CONSTRUCTION, USUALLY NAILED TO THE ENDS OF THE RAFTERS.

FINISH FLOOR: A FLOOR, USUALLY OF HIGH-GRADE MATERIAL, LAID OVER THE SUBFLOOR. THE FINISH FLOOR IS NOT LAID UNTIL ALL PLASTERING AND OTHER FINISHING WORK IS COMPLETED. ALSO CALLED FINISHED FLOOR.

FIRESTOP: A PIECE OF LUMBER OF THE SAME THICKNESS AS THE STUDDING, CUT IN HALFWAY BETWEEN FLOOR AND CEILING, BETWEEN THE STUDS.

FLASHING: PIECES OF METAL, USUALLY TIN, USED TO PREVENT LEAKAGE AROUND DORMERS, CHIMNEYS, GUTTERS, PLUMBING VENTS, AND CONDUITS.

FLOORING: THE MATERIAL USED FOR LAYING A FLOOR, AS FLOORING BOARDS, A FLOOR.

FOOTING: A FOUNDATION AS FOR A COLUMN; SPREADING COURSES UNDER A FOUNDATION WALL; AN ENLARGEMENT AT THE BOTTOM OF A WALL TO DISTRIBUTE THE WEIGHT OF THE SUPERSTRUCTURE OVER A GREATER AREA AND THUS PREVENT SETTLING. FOOTINGS ARE USUALLY MADE OF CEMENT AND ARE USED UNDER CHIMNEYS AND COLUMNS AS WELL AS UNDER FOUNDATION WALLS.

FORM: AN ENCLOSURE FOR HOLDING CONCRETE MIX IN PLACE UNTIL IT HAS SET.

FRAME: THE SKELETON OF A STRUCTURE; A FRAME THAT RECEIVES OR HOLDS SASHES OR CASEMENTS.

FURRING: STRIPS NAILED ONTO TIMBERS TO BRING THEM TO A DESIRED LINE.

GIRDER: A STRAIGHT HORIZONTAL BEAM USED TO SPAN AN OPENING OR TO CARRY WEIGHT.

GRADE LINE: THE LEVEL OF THE GROUND AT THE BUILDING LINE.

HAND RAIL: A HORIZONTAL BAR ON THE SIDE OF A STAIRWAY FOR HELP IN ASCENDING OR DESCENDING STAIRS.

HARDWARE: BUILDING WARE MADE OF METAL (SUCH AS LOCKS, BUTTS, WEIGHTS, NAILS, AND WIRE).

HEADER: A BEAM FITTED BETWEEN TRIMMERS AND ACROSS THE ENDS OF TAIL BEAMS IN A BUILDING FRAME. A JOIST OR JOISTS PLACED AT THE ENDS OF AN OPENING IN A FLOOR WHICH SUPPORTS THE SIDE MEMBERS.

INTERIOR FINISH: A TERM APPLIED TO THE TOTAL EFFECT PRODUCED BY THE INSIDE FINISHING OF A BUILDING, INCLUDING NOT ONLY THE MATERIALS USED BUT ALSO THE MANNER IN WHICH THE TRIM AND DECORATIVE FEATURES HAVE BEEN HANDLED.

JAMB: THE SIDE-POST OR LINING OF A DOORWAY OR WINDOW OPENING

JOISTS: THE TIMBERS THAT SUPPORT THE FLOOR OR CEILING OF A BUILDING

TABLE II DEFINITIONS OF CONSTRUCTION TERMS (CON'T)

LATHING: IN ARCHITECTURE, THE NAILING OF LATH IN POSITION; ALSO A TERM USED FOR THE MATERIAL ITSELF.

MOISTURE BARRIER: A WATERPROOFED MATERIAL USED TO RETARD PASSAGE OF VAPOR OR MOISTURE THROUGH AN INSULATOR PLACED ON THE WARM SIDE.

MOLDING: AN ORNAMENTAL BAR OF WOOD USED TO FINISH AN ANGEL, THE EDGE OF A CORNICE, OR THE LIKE.

MUDSILL: A SILL THAT RESTS ON THE FOUNDATION OF A BUILDING, ONTO WHICH THE FRAME IS FASTENED

ON CENTER: A TERM USED IN TAKING MEASUREMENTS, MEANING THE DISTANCE FROM THE CENTER OF ONE STRUCTURAL MEMBER TO THE CENTER OF A CORRESPONDING MEMBER, AS IN THE SPACING OF STUDDING, GIRDERS, JOISTS, OR OTHER STRUCTURAL MEMBERS. SAME AS CENTER TO CENTER.

PARTITION: AN INTERIOR WALL SEPARATING ONE PORTION OF A HOUSE FROM ANOTHER; USUALLY A PERMANENT INSIDE WALL WHICH DIVIDES A HOUSE INTO VARIOUS ROOMS. IN RESIDENCES, PARTITIONS OFTEN ARE CONSTRUCTED OF STUDDING COVERED WITH LATH AND PLASTER; IN FACTORIES, THE PARTITIONS ARE MADE OF MORE DURABLE MATERIALS, SUCH AS CONCRETE BLOCKS, HOLLOW TILE, BRICK, OR HEAVY GLASS.

PIER: ONE OF THE PILLARS SUPPORTING AN ARCH; ALSO, A SUPPORTING SECTION OF WALL BETWEEN TWO OPENINGS; A MASONRY STRUCTURE USED AS AN AUXILIARY TO STIFFEN A WALL.

PIER BLOCK: REDWOOD LAID ON TOP OF CONCRETE PIERS AS FOOTING FOR A POST

PILASTER: A RECTANGULAR COLUMN ATTACHED TO A WALL OR PIER; STRUCTURALLY A PIER, BUT TREATED ARCHITECTURALLY AS A COLUMN WITH A CAPITAL, SHAFT, AND BASE.

PLATE: A TERM USUALLY APPLIED TO A 2 X 4 PLACED ON TOP OF STUDS IN FRAME WALLS. IT SERVES AS THE TOP HORIZONTAL TIMBER UPON WHICH THE ATTIC JOISTS AND ROOF RAFTERS REST, AND TO WHICH THESE MEMBERS ARE FASTENED. ALSO, A FLAT PIECE OF STEEL USED IN CONJUNCTION WITH ANGLE IRONS, CHANNELS, OR I BEAMS IN THE CONSTRUCTION OF LINTELS.

PLASTER BOARD: PANELS OF GYPSUMPAPER COMBINATION NAILED TO STUDS AND CEILING JOISTS TO HOLD PLASTER

POST: A TIMBER FIXED IN AN UPRIGHT POSITION AS FENCE POST, PORCH POST, NEWEL POST, OR UNDERPINNING POST

RAFTER: ONE OF THE TIMBERS SUPPORTING ROOF SHEATHING AND SHINGLES

RIBBON: A NARROW STRIP OF BOARD LET INTO STUDDING.

RIDGE: THE HIGHEST PART OF A ROOF; THE COMB; RIDGE BOARD; THE BOARD ONTO WHICH THE RAFTERS ARE NAILED

RISER: THE VERTICAL PART OF A STEP THAT SUPPORTS THE TREAD

ROUGH FLOOR: A SUBFLOOR SERVING AS A BASE FOR THE LAYING OF THE FINISHED FLOOR WHICH MAY BE OF WOOD, LINOLEUM, TILE, OR OTHER SUITABLE MATERIAL.

SASH: THE FRAMING IN WHICH PANES OF GLASS ARE SET IN A GLAZED WINDOW OR DOOR, INCLUDING THE NARROW BARS BETWEEN THE PANES

SHEATHING: BOARDS OR OTHER MATERIAL NAILED TO RAFTERS OF STUDS TO PROVIDE A BASE FOR ROOF OR WALL FINISH MATERIALS

SHINGLES: PIECES OF MATERIAL LAPPING OVER EACH OTHER AND CARRYING AWAY RAIN OR SNOW WATER; A ROOFING MATERIAL.

SIDING: BOARD USED FOR COVERING THE OUTSIDE OF A HOUSE OR OTHER BUILDINGS.

SILL: THE LOWEST MEMBER BENEATH AN OPENING, SUCH AS A WINDOW OR DOOR; ALSO, THE HORIZONTAL TIMBERS WHICH FORM THE LOWEST MEMBERS OF A FRAME SUPPORTING THE SUPERSTRUCTURE OF A HOUSE, BRIDGE, OR OTHER STRUCTURE.

STRINGER: A LONG, HORIZONTAL TIMBER USED TO CONTACT UPRIGHTS IN A FRAME OR TO SUPPORT A FLOOR; ALSO, THE INCLINED MEMBER WHICH SUPPORTS THE TREADS AND RISERS OF A STAIR.

STOP: MOLDING FOR DOORS TO STRIKE AGAINST OR FOR SASH.

STUD: IN BUILDING, AN UPRIGHT MEMBER, USUALLY A PIECE OF DIMENSION LUNMBER, 2 X 4 OR 2 X 6, USED IN THE FRAMEWORK OF A WALL. ON AN INSIDE WALL THE LATH ARE NAILED TO THE STUDS. ON THE OUTSIDE OF A FRAME WALL, THE SHEATHING BOARDS ARE NAILED TO THE STUDS. THE HEIGHT OF A CEILING IS DETERMINED BY THE LENGTH OR HEIGHT OF THE STUDS.

TRIM: IN CARPENTRY, A TERM APPLIED TO THE VISIBLE FINISHING WORK OF THE INTERIOR OF A BUILDING, INCLUDING ANY ORNAMENTAL PARTS OF EITHER WOOD OR METAL USED FOR COVERING JOINTS BETWEEN JAMBS AND PLASTER AROUND WINDOWS AND DOORS. THE TERM MAY INCLUDE ALSO THE LOCKS, KNOBS, AND HINGES ON DOORS.

TRIMMER: THE BEAMS OR FLOOR JOISTS INTO WHICH A HEADER IS FRAMED.

VAPOR BARRIER: MATERIAL USED TO RETARD THE PASSAGE OF VAPOR OR MOISTURE INTO WALLS THUS PREVENTING CONDENSATION WITHIN THEM. THERE ARE DIFFERENT TYPES OF VAPOR BARRIERS, SUCH AS MEMBRANE WHICH COMES IN ROLLS AND IS APPLIED AS A UNIT IN THE WALL OR CEILING, AND THE PAINT TYPE WHICH IS APPLIED WITH A BRUSH.

Specifications

Specifications are a vital part of building plans. They give additional information to the information shown in the drawings or blueprints. They contain written instructions with necessary information pertaining to types and kinds of materials to be used, fabrication and installation, dimensions, colors, quality, finishes, and other details. Specifications which thoroughly cover a construction job leave little room for conflict of opinion or doubts since all of the details are made clear. There is little room for misinterpretation or misunderstanding of the plans. They also enable the contractor to make cost bids based on quality of material and labor required. Written specifications are invaluable in this respect because they lessen the chance of costly omissions. They are as binding in a contract as the drawing themselves.

"Specs" for a construction job are written in sections each pertaining to a different craft. The following excerpts from typical electrical specifications will give some idea of their scope.

4. *Scope of Work:* The electrical subcontractor shall furnish all labor, materials, equipment, appliances, and services necessary for completion of the electrical work as indicated on the drawing and specifications herein.

5. *Wiring System:* The complete wiring system shall be installed in galvanized, flexible conduit with the exceptions of low voltages control. Low voltage wiring shall be plastic insulated, 300 volts minimum. Where exposed to the weather, conduit shall be made of rigid steel which has been galvanized. The service shall be 115/230 volts 3 wire and all outlets shall be connected for 115 volt operation, except where otherwise noted. Service and branch-circuit wire shall conform to the local electrical code.

6. *Substitution of Material:* Whenever an article is mentioned by trade name, either in the specifications or on the drawings, such designation is intended to establish a standard of merit and design. When an item is specified by manufacturer and catalog number and is then followed by the words "Or Equal By" and the names of the manufacturers, it is presumed that the manufacturers' items are equal. Such items may be used if, in the opinion of the owner, these items are equal. The judgement of the owner as to equality will be final.

7. *Cleanup:* The electrical contractor shall leave clean all parts of the electrical system, including panels, fixtures, switches, and plates, fixture lamps, and the like.

"Specs" give a running description of the installation of electrical equipment on a job. This is very beneficial to the learner and enables him to see how a job should go together. A full understanding of all requirements of a job cannot be gained until all drawings and the specifications are thoroughly understood.

QUESTIONS FOR DISCUSSION OR SELF STUDY.

1. What other names are used in reference to blueprints?

2. What are blueprints?

3. On a separate piece of paper draw electrical symbols so that they correctly match their written identification:

Wall Outlet
Ceiling Outlet
Wall Fan Outlet
Range Outlet
Ceiling Blanked Outlet
Clock Outlet
Ceiling Junction Box
Garbage Disposal

4. How do you designate outlets which are to be permanently attached and which do not require receptacles?

5. What does the following Symbol represent?

6. On a separate piece of paper draw electrical symbols so that they correctly match their written identification:

Single-Pole Switch
Three-way Switch
Double-Pole Switch
Four-way Switch

7. With a scale 1/4″ = 1′. How long are the following lines in feet?

(A) 

(B) ______________________________

(C) __________________________

(D) _______________________

8. With 1/8″ scale, how long are the following lines in feet?

(A) ____________________________

(B) _____________________

(C) ______________________________

9. In Fig. 3-11, how many chimes and push buttons are shown?

10. How many 3-way switches are shown in the kitchen in Fig. 3-11?

11. How many convenience outlets are shown in the living room in Fig. 3-11 that *are* switched? How many are *not?*

12. In Fig. 3-11, what does the broken line marked C.O. leading from a switch in front represent?

13. Why shouldn't meters normally be installed on bedroom walls?

14. What does the following symbol represent?

15. Which plan assists the wireman in locating a point of entry or service head for the electrical service?

16. What does "Section A-A" of Fig. 3-10 show?

17. Why is it important to seek out wall or ceiling thickness or finishes?

18. Wiring and crawling clearance under a structure may be found on which plan?

19. How may mounting heights of convenience outlets over kitchen counter tops be determined?

20. Describe the following Abbreviations:

21. Lay out circuits, in Fig. 3-11, to conform with local or the National Electric Code. Use example in the back of the

National Electric Code as a guide. Figure service wire size and circuit breakers or fuses for branch circuit wiring, add necessary outlets should the plan not conform to code.

22. When plans, specs or the code calls for convenience outlets to be spaced so that no place along a wall is 6 feet from an outlet, what is the maximum spacing of outlets?

23. What are electrical specs?

24. What are special purpose outlets shown on the left side of the kitchen in Fig. 3-20?

25. How are the number of conductors indicated in a wiring run?

Chapter 4

Methods of Wiring Nonmetallic Sheathed Cable

Nonmetallic sheathed cable is used where low cost is a factor and its use is permitted by the code for the area where it is to be installed. The ease with which it may be installed and the relatively lower material costs are factors which make this system less expensive than others.

The absence of protection against mechanical injury, and the fact that it is only moisture-resistant, tend to restrict the conditions under which this system of wiring may be used. The local department of building and safety should be consulted regarding restrictions. Information is also available in the National Electrical Code.

Local electrical codes and regulations should always be consulted for exact requirements of usage.

Fig. 4-1. This nonmetallic sheathed cable is defined as 12-2 AWG. No. 12 is the size and the second number, 2, refers to the 2 conductors. (12-3 cable would have 3 conductors). Conductor material is available in copper and aluminum. (Crescent Insulated Wire & Cable Co.)

Fig. 4-2. Another type, NM nonmetallic sheathed cable, comes with a ground conductor. Conductors are insulated with a color coded polyvinyl chloride (PVC) compound. The cable is covered with a tough, moisture and abrasion resistant PVC jacket.

Nonmetallic sheathed cable consists of two or three insulated conductors, either with or without an additional bare conductor for grounding purposes. The trend is to eliminate all cables without grounding wires. Fig. 4-1 shows the construction of a nonmetallic sheathed cable.

The NEC recognizes two types of nonmetallic sheathed cable in No. 14 AWG to No. 2 inclusive: Type NM and Type NMC. The outer covering on Type NM is flame-retardant and moisture-resistant. The insulation on Type NMC, besides being similar to that on Type NM, offers additional protection against fungus and corrosion. Both of these cables are available with an uninsulated conductor which is used for grounding purposes only. See Fig. 4-2.

Type NM cable may be used for both exposed and concealed work in normally dry locations. It may be installed in the voids of concrete block or tile walls, if not subject to excessive moisture or dampness. It cannot be used where corrosive vapors or fumes are present, and cannot be imbedded in concrete, cement, or plaster.

The NMC cable may be installed in the same locations as NM cable. In addition, NMC cable may be installed in moist, damp or corrosive locations; and in outside or inside walls of masonry block or tile. It may be run in the shallow chase of a masonry wall that is covered with plaster; if within 2″ of the finished surface, however, it must be protected by a ¾″ wide steel plate which is not less than 1/16″ thick. This may be a local code problem and local codes should be checked.

Neither the NM or NMC types of non-metallic sheathed cable may be used for service entrance, nor should they be installed in commercial garages or other hazardous areas. All applications and essential requirements are governed by the electrical code. The latest prevailing code book should be consulted for specific details.

General Cable Installation

The electrician must notch or bore through existing structural members for the installation of cable. Since electrical code books refer to the structural members by name, it is essential that the electrician be familiar with those structural members and have an understanding of the building he is wiring. Fig. 4-3 gives the names of structural members in a typical frame house. A study of this illustration should help in understanding the code book references. This knowledge will also help in communicating between the crafts and will assist in the reading of the building plans. (Consult Chapter 3, Figs. 3-27, 3-28 and 3-29 for additional structural information.)

Outlet locations are marked, boxes and fuse panel fastened in place, and holes bored where necessary for the running of circuit wires. Boxes are fastened in place, cable is pulled through holes or stapled to wooden members, as the case may be, and secured in the boxes, where this is required. Splices are made, joints taped, circuits run into the fuse panel. Feeder and service conductors are installed. The NEC should be consulted to determine what one can and cannot do with nonmetallic sheathed cable.

Generally speaking, before any holes are bored in the frame for cable installation, the outlet and switch locations and elevations are marked by the electrician on the studs with chalk or crayon. Locations are determined from the electrical floor plan. Ceiling lights are "laid out" on the sub-floor. After the service entrance panel has been properly located, holes are bored through studs, plates, fire blocks and whatever is necessary to facilitate a route of cable installation from fuse or circuit breaker panel to switches and outlets. (In boring and notching be sure to conform to code regulation and minimum standards.)

The outlet boxes should then be nailed to their proper locations on studs or ceiling joists. The outlet boxes may obstruct the boring process if installed first; however, some wiremen prefer to nail boxes before drilling to better line up holes with boxes. All cables are then generally pulled into place and stapled. Joints or splices are then made. Wires are folded back into the entrance panel and outlet boxes in preparation for wall lathing and plastering or finish.

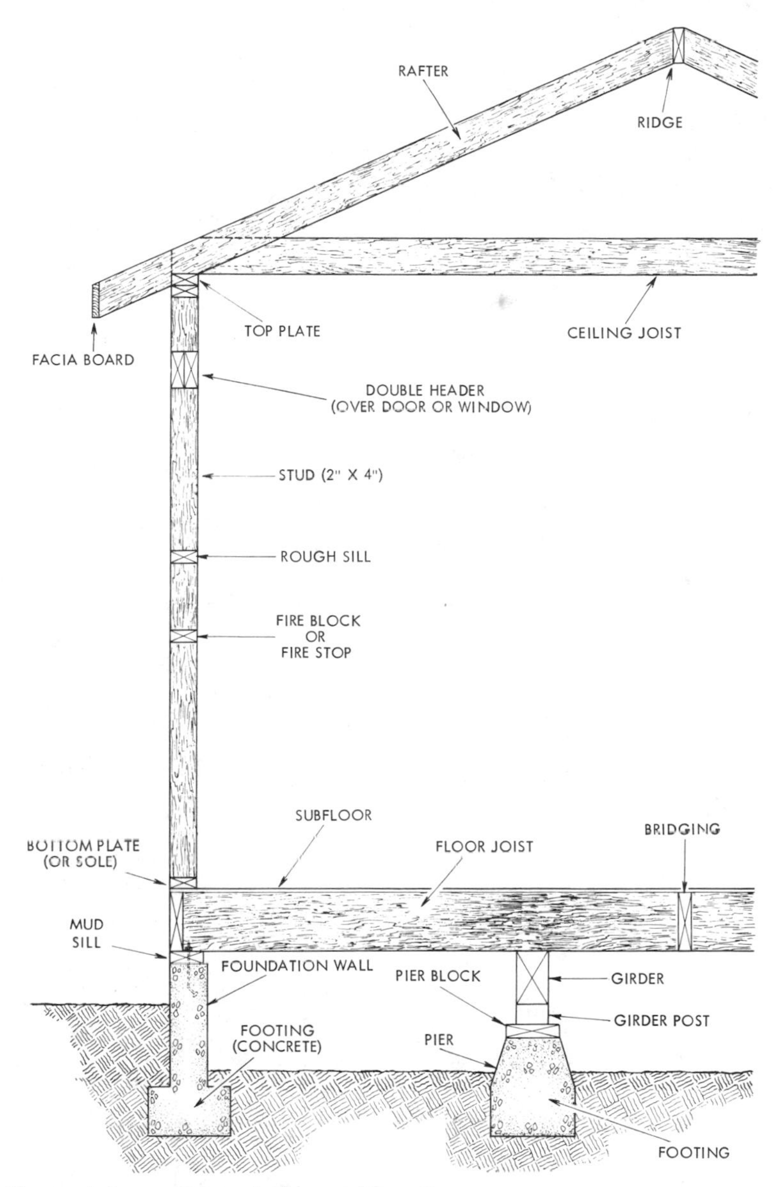

Fig. 4-3. Nomenclature and layout of a wood frame.

After the final coat of paint has dried, lighting fixtures are hung, and switches, receptacles, circuit breakers and plates are installed. This is called putting on the "finish" or "trimming out a house."

Boring Holes

Holes must be near the centers of

joists and studs, or at least 2″ from the nearest edge. Cable may also be laid in shallow notches cut in wooden members if it is protected at these points against the driving of nails. A 1/16″ steel plate is sufficient for this purpose. Holes are usually bored in a line so the pulling of the cable is somewhat easier than when the holes are askew. See Fig. 4-4. The most common method is to use an electric drill (sometimes with a bit exten-

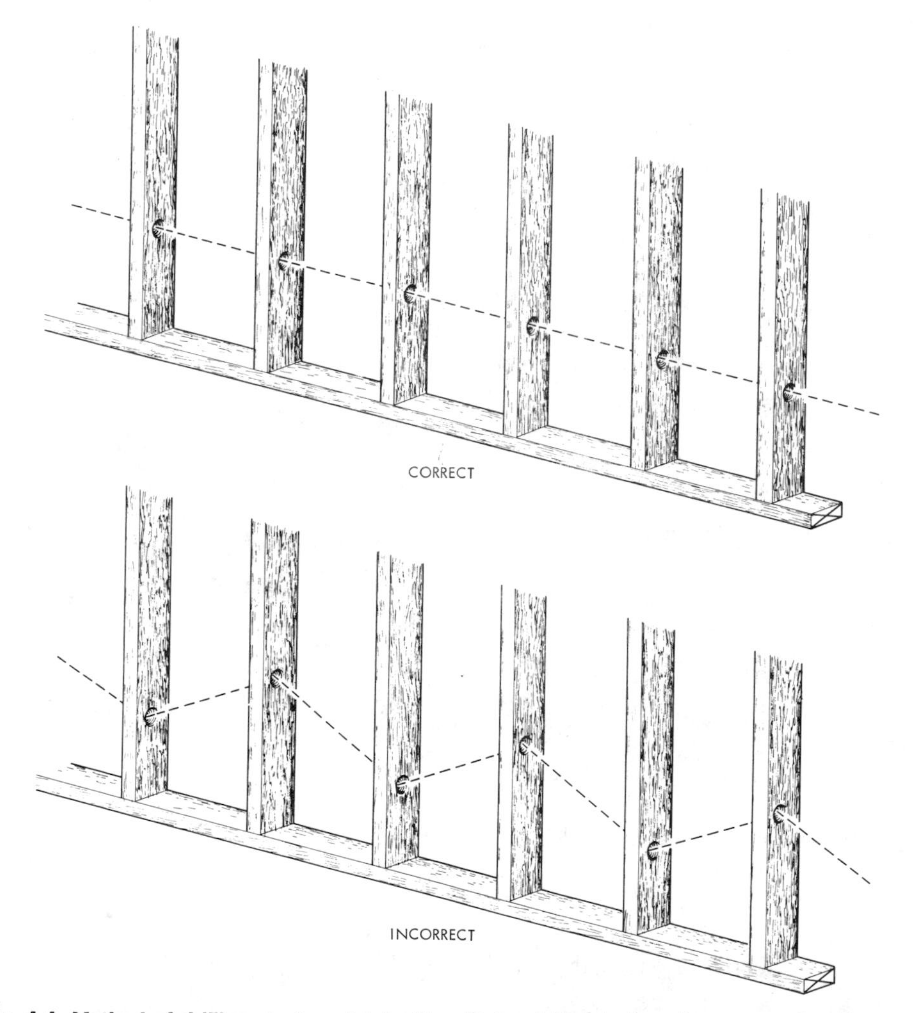

Fig. 4-4. Method of drilling studs or joists. Top: Holes drilled in line gives ease of pulling and saves material. Bottom: Holes *not* drilled in line, or on a gradual slope when changing elevation, makes cable pulling difficult.

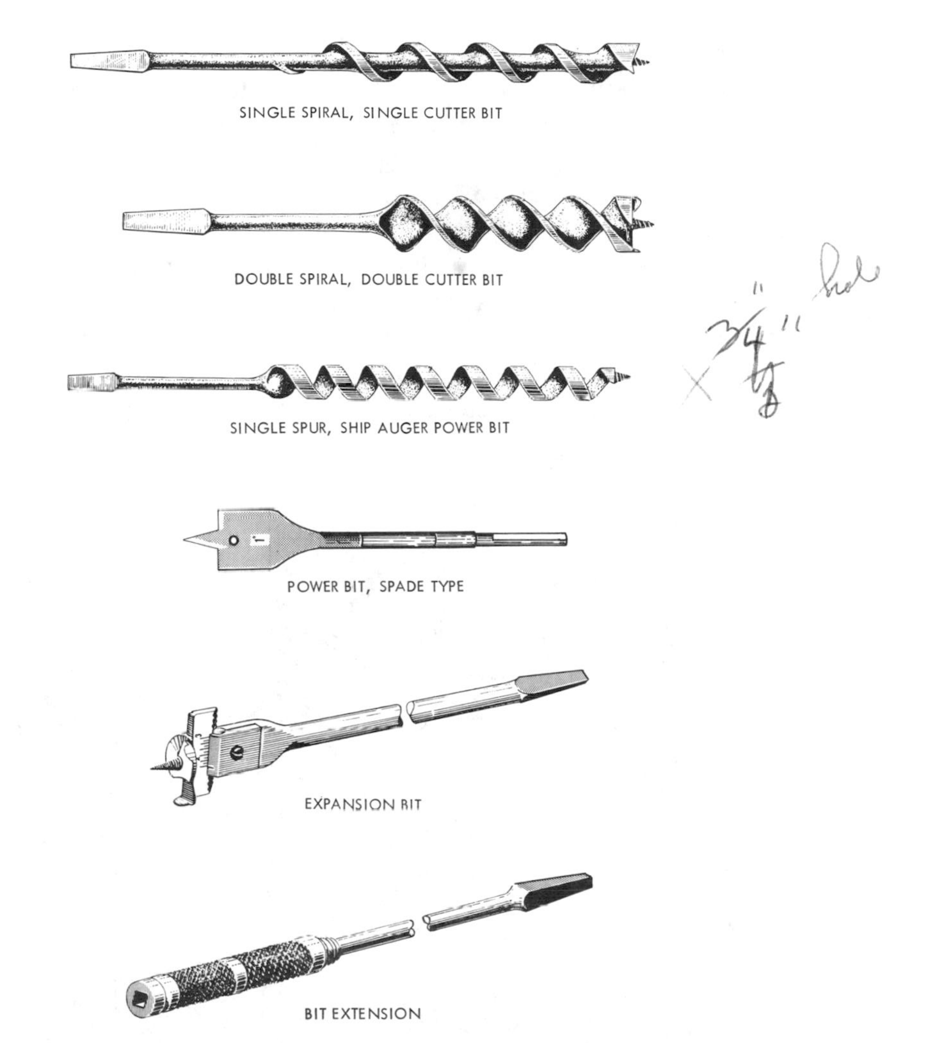

Fig. 4-5. Bits used for drilling for electrical wiring. The ship auger power bit is the most popular for all rough boring; it has a hexagon shank for positive drive without slippage. The expansion bit is used for drilling a few large holes with a manual brace. The bit extension is used for difficult to reach places.

sion) for drilling joists and studs. A bit extension with a slight angle is very popular on the job. Wiremen doing housing work frequently refer to it as "the wand." Holes drilled in this manner are, for a given size of cable, bored one or two sizes larger than the cable. Brace and bits are not commonly used to drill wooden frames for rough wiring as they are too time consuming. Electric drills are commonly used.

The standard ship-auger with no spur (Fig. 4-5, 3rd from top) is best for boring holes in old house wiring. Nails are often struck in this kind of work because there is little choice in the matter of hole locations. This style of bit is not damaged so easily as other types, and it is somewhat easier to put in shape again after striking a nail.

Wiremen on new construction work seldom take the time to look on the other side of timbers to be bored. Therefore, the standard ship auger bit with no outlining spur has become popular on new construction work also.

The spade type bit (Fig. 4-5, 3rd

Fig. 4-6. Drill extension fits onto regular ½″ drill.

from bottom left) is also used in power drills. The expansion bit (Fig. 4-5, 2nd from bottom) is used for boring large sized holes. It has an adjustable cutter. When adjusted to size it can be checked for accuracy on a piece of scrap lumber. Both the expansion bit and the bit extension (Fig. 4-5, bottom) are used in hand braces. The bit extension is used to extend the length of the bit and is used mostly in remodeling wiring to drill through fireblocks and bracing. Fig. 4-6 shows a bit extension used on a ½″ electric drill.

An auger bit of the proper size should be selected. Usually an inch and one eighth (1⅛) or an inch and one quarter (1¼) for ½″ flex is used with an electric drill, ¾″ to 1″ is used for cable wiring. In drilling the holes through the studs, care must be taken, in order to prevent accidents, to get a secure footing as well

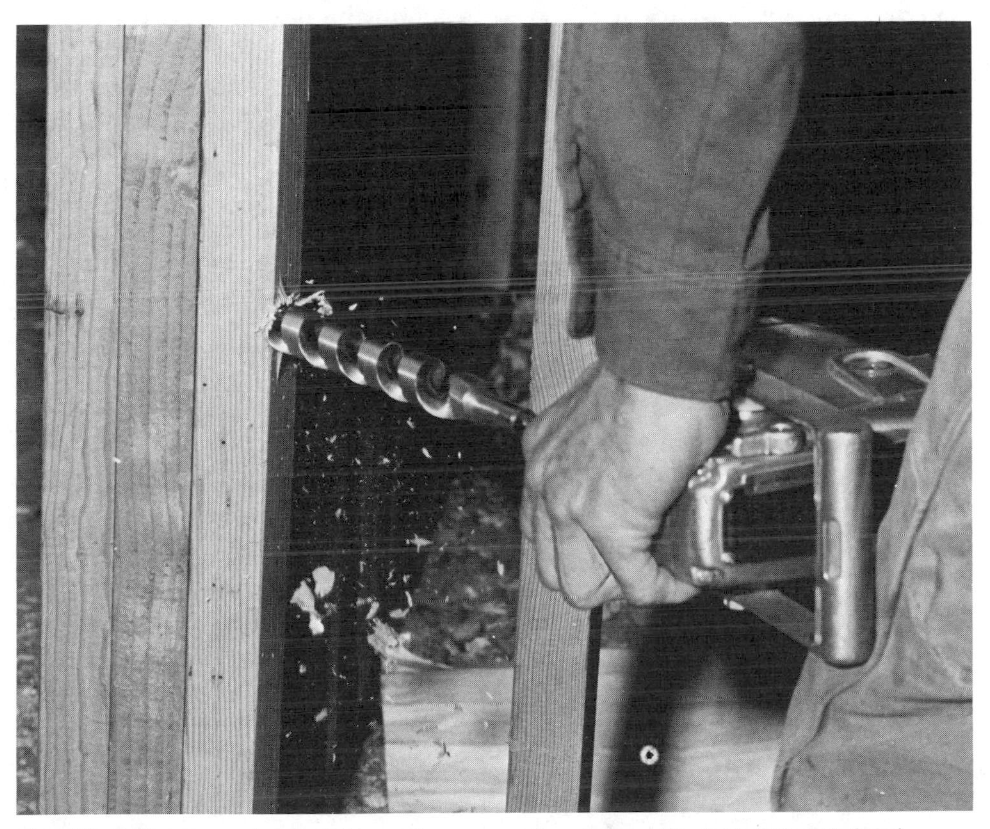

Fig. 4-7. Drilling around corners in a safe manner. (Los Angeles Trade-Technical College)

as a safe angle. The drill operator should be braced securely at all times. In the event that the bit hits a nail or other insurmountable obstacles, there will be a tendency for the drill to throw the operator. Fig. 4-7 shows one way of bracing the drill. When drilling around corners, the same smooth in-line holes should be maintained in order to make it easier to install the conduit. This is the same for cable wiring or flexible conduit systems.

When the electrician is drilling and hits an obstruction, a spike or a nail, he must stop and disengage his bit. Ordinarily, he would use an auger which is furnished by the contractor, and it must be kept sharp. The bit is sharpened with a square or ½ round file or an auger bit file. Fig. 4-8, top, shows the two cutting edges that are sharpened on an auger bit. The chip lifter cutting edge is sharpened in the throat or the bottom part. See Fig. 4-8, middle. The side lip is sharpened on the inside cutting edge. See Fig. 4-8, bottom. Use a No. 0 cut, 6 inch file as shown. If an auger bit is used which has an outlining spur, sharpen on the front inside of the spur from the base to the top.

A bit with an outlining spur bores a smooth hole as compared to the hole of a ship auger, such as the one illustrated in Fig. 4-8. When working in a building frame or rough construction an electrician normally does not need smooth holes because they are covered up and the holes are used for conduit or cable. However, whenever a smooth hole is desired, an outlining spur should be used.

The best electric drill for drilling a house on rough wiring is one with a reversing switch. Should the drill stop in heavy drilling, it is easily removed by reversing the drill motor. However, drills are furnished by the electrical contractor and not much choice is left in the matter for wiremen.

A good driller can make it very easy on the next man who will pull cable or flexible conduit. He will drill them as much in line as possible and when he changes angles he will drill a sloping angle with large radius, instead of drilling abrupt turns. This also makes it easier on the wiremen pulling wire or cable.

After the holes are bored, cable or flexible conduit is pulled from box to box through the properly routed holes that the electrician should follow. If the wireman sees several holes and one cable or conduit is following on the outside continuously when pulled in, he should continue to do this and not criss-cross them. Criss-crossing would no doubt change the circuitry when the wires or cables are pulled and connected (putting different outlets on different circuits, which is not the intent). Holes which are expressly drilled for one in-line run are the holes that should be used

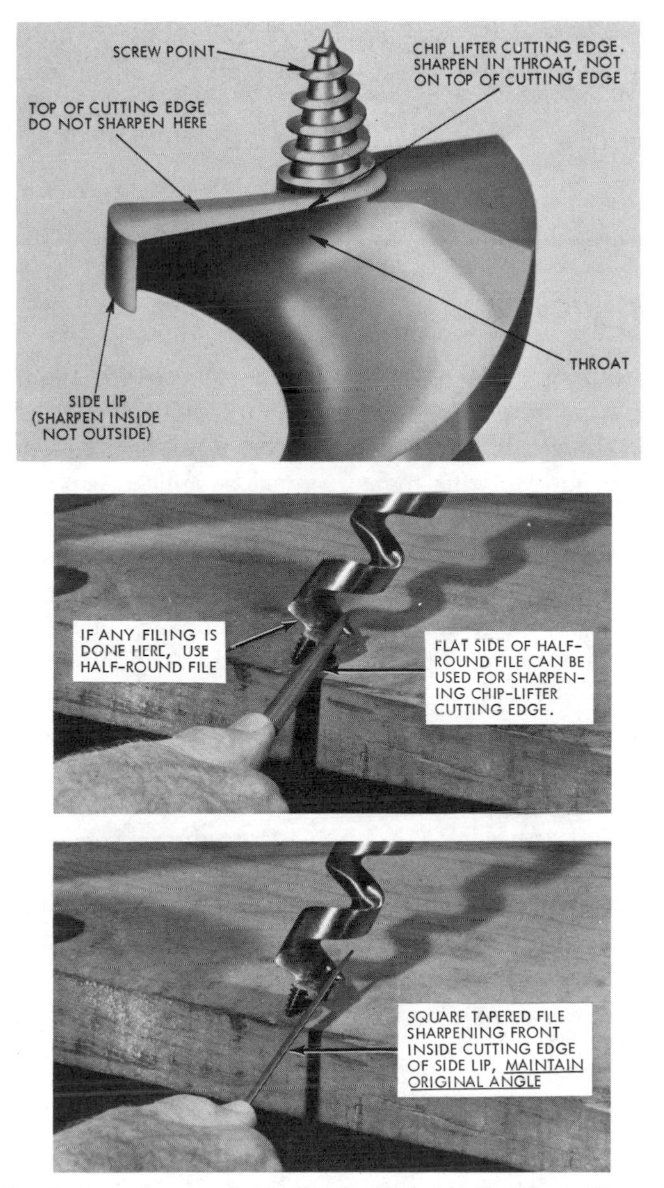

Fig. 4-8. Top: Terminology of an auger bit. Middle: Sharpen chip lifter cutting edge with No. 0 cut 6″ square file or the flat side of half-round file. Maintain same angle in the throat. *CAUTION: Do NOT file on top of cutting edge.* Avoid filing groove at base of screw point. If necessary, use half-round or round file to maintain radius in throat under the screw point. Bottom: File front edge of side lip on same angle as a new bit, using a No. 0 cut 6″ square file. Maintain a square corner between side lip and chip lifter cutting edge. *CAUTION: Cutting edge of side lip must be even with chip lifter cutting edge. Do not file on the outside of the side lip as this reduces the diameter of the bit.* (Greenlee Tool Co.)

for that run only. A next row of holes, if there are parallel runs, is for a different run and is not to be criss-crossed for this reason.

Local codes should be consulted for restrictions on boring or cutting studs, joists or other wooden structural members.

Working With Cable

Although wire may be taken from a coil for long runs by grasping the inner turn and drawing it out, cable may not be unwound in this manner without some difficulty. The reason lies in the fact that cables would twist and become hard to manage. Unrolling is necessary and this may be done by turning the end of the cable like a crank as it is withdrawn from the box. The end may also be whirled about the head to unwind the coil. Be careful of other workers if the latter method is used.

Because of the nature of the material, runs of nonmetallic sheathed cable (also armored cable) must be cut at each outlet point, and a new run started from there. As a result, more splices are required than in knob-and-tube or in conduit work, since straight-through wires too, must be spliced.

The outer sheath of nonmetallic cable can be stripped off with a knife, or a cable ripper. (See Fig. 4-9 for a cable ripper.) Then, the wires are skinned in the same manner as ordinary single conductor, by removing the insulation like sharpening a pencil with a knife—if a knife is used. (See Fig. 4-2.)

Procedures For Locating Ceiling Outlets. Fig. 4-10 represents a plan view of a bedroom which requires a center light. There are two basic methods: One is to mark the center point on the floor, and then transfer

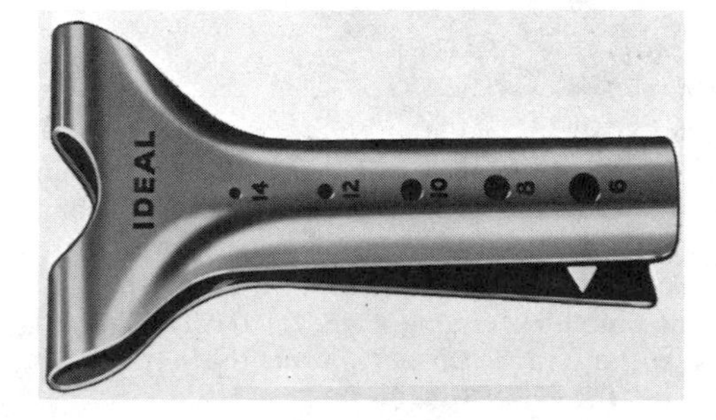

Fig. 4-9. Cable rippers with wire gage. Cuts ⅝″ O.D. (or smaller) nonmetallic sheathed duplex or lead covered cable. Just squeeze onto cable and pull. Overall length 3¾″. (Ideal Industries, Inc.)

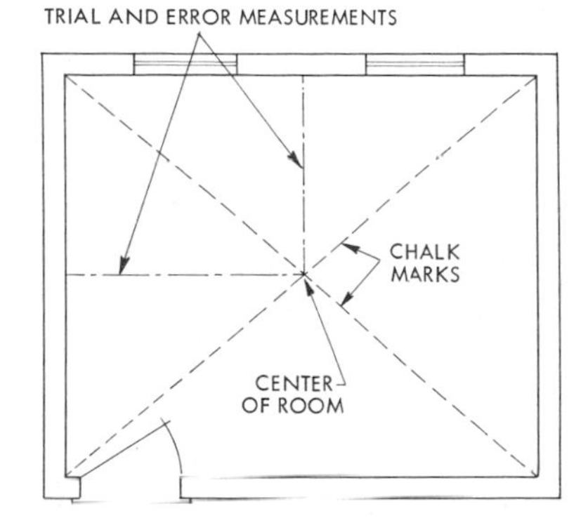

Fig. 4-10. Finding center of the room.

this point to the ceiling by means of a plumb line. The other is to do this work at the ceiling line. Fig. 4-11 illustrates a plumb bob being used. (A plumb bob is a weight with a center point. The heavier the plumb bob the greater the wind resistance against swaying back and forth when establishing a vertical line.) When ceiling outlet locations have been established by marking the outlet on the sub-floor, this location is transferred to ceiling joists for the nailing of ceiling outlet box as shown.

The center of the floor area can be found in two ways. Chalk lines may be drawn between diagonal corners of the room, the crossing point being the desired location. The second choice is to measure between opposite walls, finding the spot by trial and error. The latter method would seem more time consuming, but if there are three or more rooms of identical dimensions, distances found for one room may be marked on a light stick and transferred quite rapidly to the others.

So that a straight run of holes may be drilled in ceiling joists, or studs, care should be taken to sight properly from one end of the run to the other before and while drilling.

Supports. Conductors should be supported at intervals not exceeding 4½ feet and within 12 inches of outlet boxes, cabinets, or fittings. This rule does not apply, of course, to fished work in existing buildings. Cable runs must be continuous from outlet to outlet, as no splice is permitted in the cable itself. It is necessary, therefore, to make all circuit splices in outlet boxes, junction boxes, switch boxes, or other outlet points.

Nonmetallic sheathed cable is fastened to wooden members by means of staples, such as in Fig. 4-12. To avoid damage to cable care should

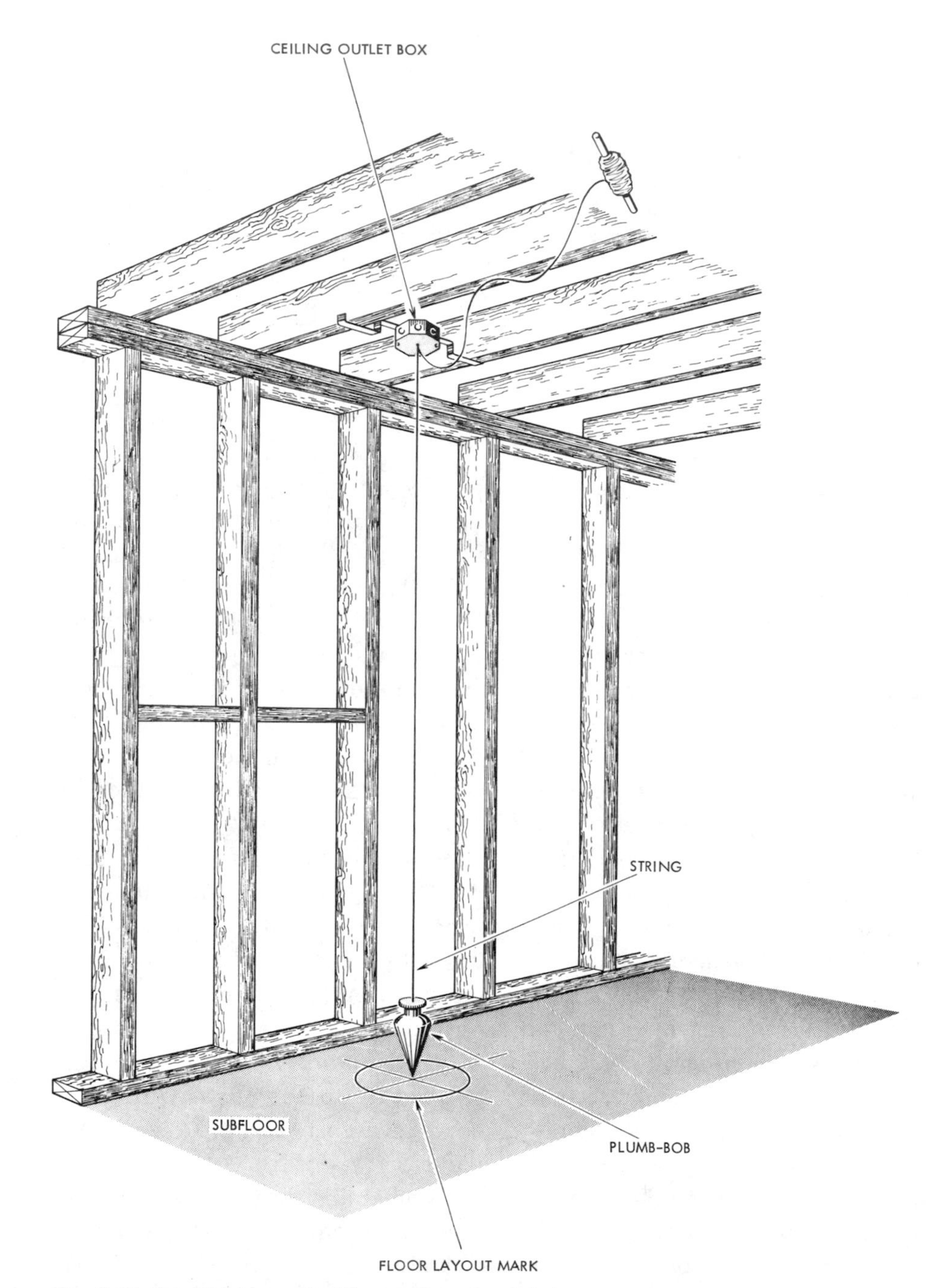

Fig. 4-11. Establishing vertical lines with a plumb bob.

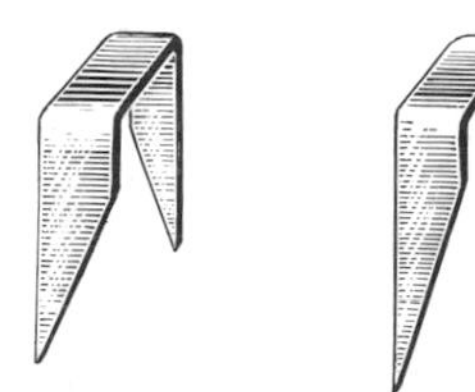

Fig. 4-12. Staples for securing cable.

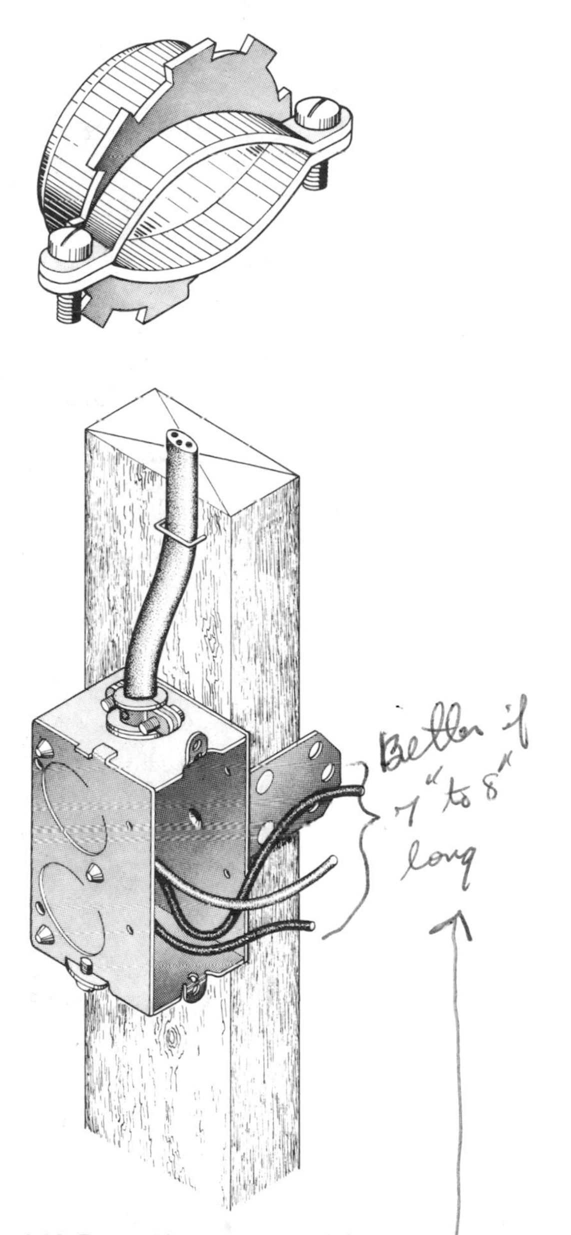

be exercised not to drive staples too tightly. Cable is attached to metallic boxes by cable clamps which are included in the box or by means of separate outlet box connector, as shown in Fig. 4-13.

Nonmetallic boxes may be employed with this cable. Fig. 4-14 illustrates some of the important requirements connected with the use of nonmetallic sheathed cable. Fig. 4-15 shows the method of laying cable at right angles to the attic joists. Two guard strips should be used. In general, where permitted, exposed cable should closely follow the surface of the building finish or run on running boards. See Fig. 4-16. In unfinished basements the smaller sized cable may be passed through bored holes or may be fastened to running boards for protection. See Fig. 4-17. In most cases larger sized cables (see Fig. 4-17 for specific sizes) may be fastened to the underside of floor joists in an unfinished basement.

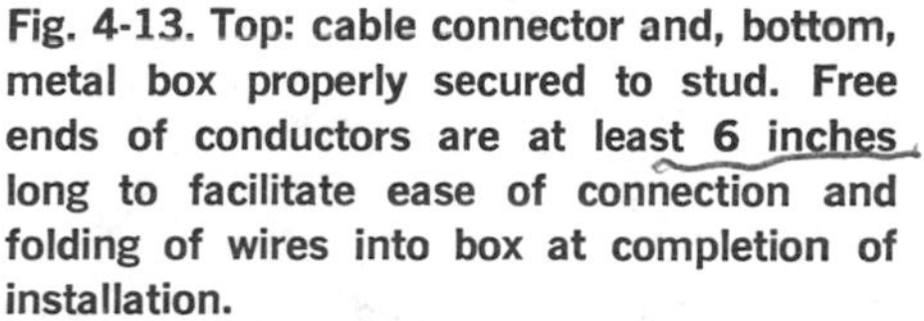

Fig. 4-13. Top: cable connector and, bottom, metal box properly secured to stud. Free ends of conductors are at least 6 inches long to facilitate ease of connection and folding of wires into box at completion of installation.

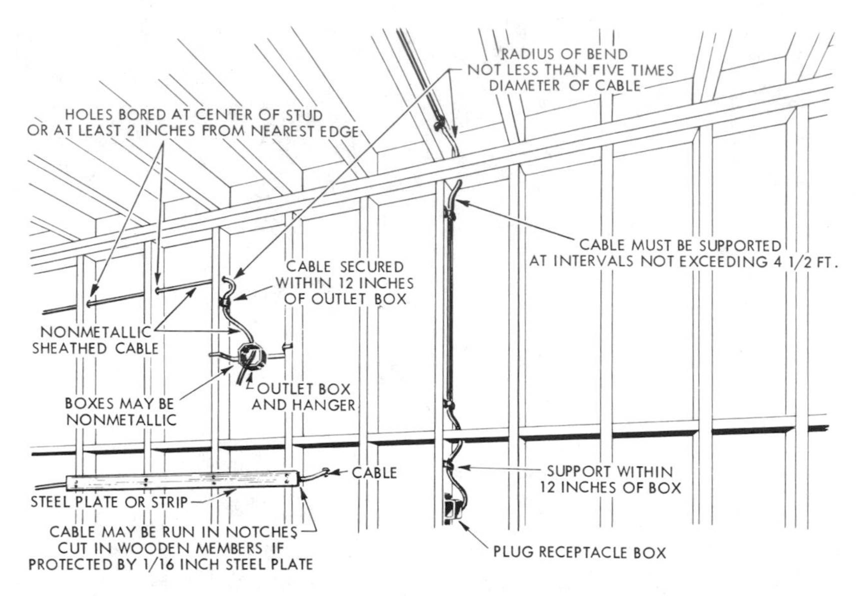

Fig. 4-14. Typical nonmetallic sheathed cable installation.

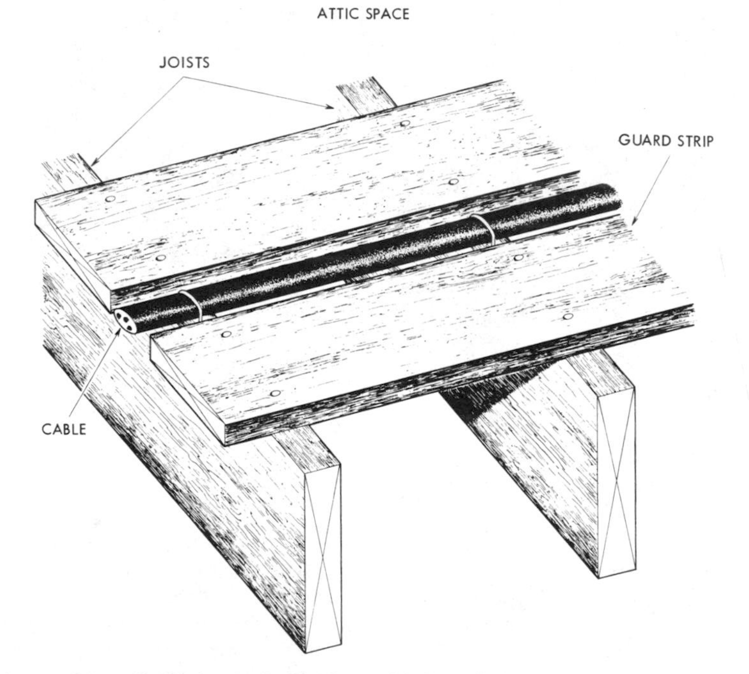

Fig. 4-15. Approved method of guard stripping cable in attic space.

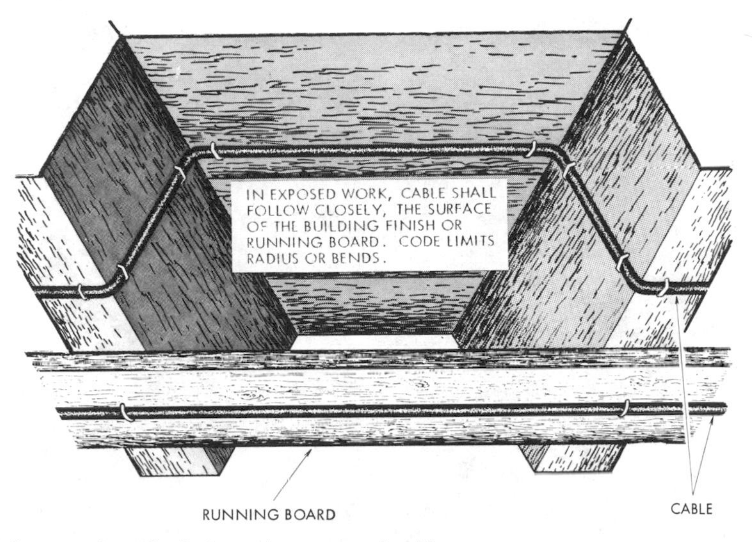

Fig. 4-16. Approved method of running exposed cable.

When passing through a floor, nonmetallic cable should be enclosed in rigid conduit or pipe extending at least 6 inches above the floor to prevent damage to the cable.

Box Mounting Heights

Heights for switches and plug receptacles, given in the plans, are from finished floor to the center of the outlet, unless otherwise noted. Proper allowance should be made, when roughing in, for the kind and thickness of finished floor. Where the height given for plug receptacle outlets is low, failure to provide for floor thickness may result in the baseboard overlapping the outlet where this type of construction may still be used. The same thing can happen from neglecting to check width of the baseboard which is to be installed.

Commonly specified heights for wall outlets are 12″ for plug receptacles (convenience outlets) and 48″ for switch boxes. The specifications for a job may indicate desired positions. Outlets for laundry, utility rooms and garages are generally placed 4 feet above the floor. These dimensions are not extremely critical. Many wiremen use their hammer handles as measuring gauges for locating heights of wall outlets. The handles are so cut as to locate the center of the boxes at 12″ elevation. Greater care should be taken when locating switches and outlets above counter tops in kitchens and bath-

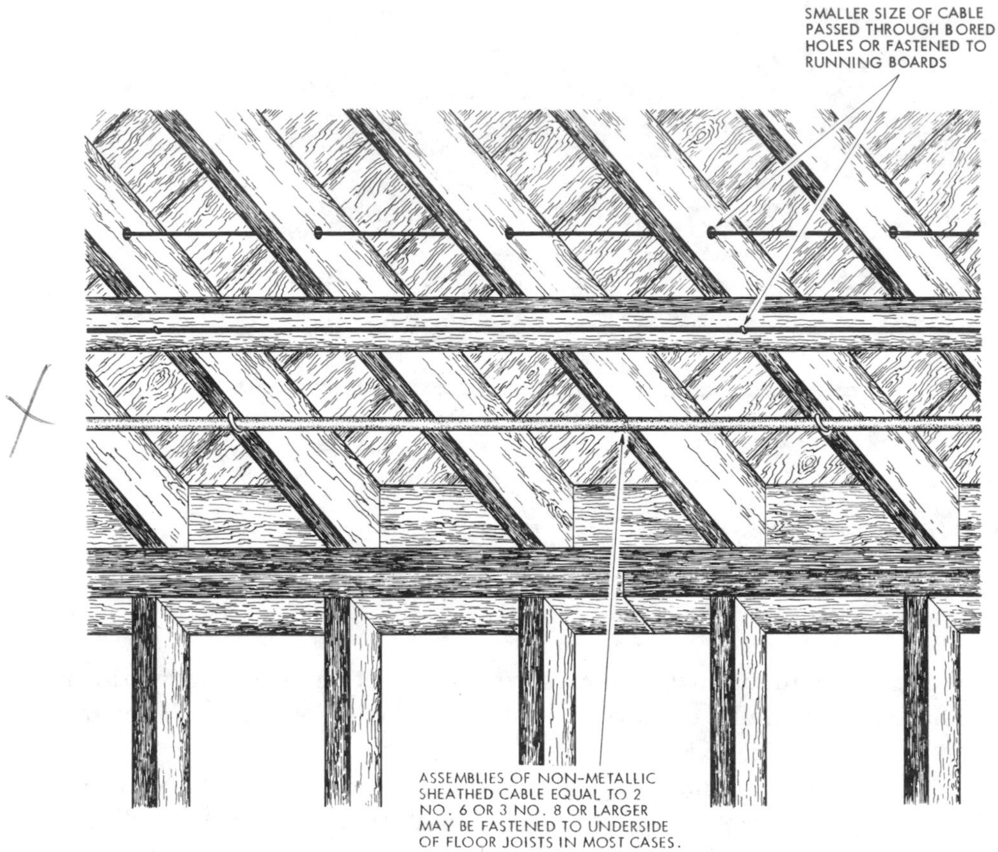

SMALLER SIZE OF CABLE PASSED THROUGH BORED HOLES OR FASTENED TO RUNNING BOARDS

ASSEMBLIES OF NON-METALLIC SHEATHED CABLE EQUAL TO 2 NO. 6 OR 3 NO. 8 OR LARGER MAY BE FASTENED TO UNDERSIDE OF FLOOR JOISTS IN MOST CASES.

Fig. 4-17. Nonmetallic sheathed cable run at angles to joists in unfinished basement.

rooms. The height of ceramic tile splash panels and counter finished tops dictate proper locations of boxes. Fig. 4-18 illustrates a problem that may be encountered. The box elevation could be mistakenly mounted so that half would finish in the tiled area, and half out. Elevation drawings are useful here. Deeper than usual plaster rings are also used in these areas, due to thicker wall finishes.

Since the use of aluminum cable may involve a wire size larger, for the same current carrying capacity, than copper, this should be considered when selecting the type and size of box outlets and devices. Nonmetallic boxes, or metallic boxes with outside clamps or cable connec-

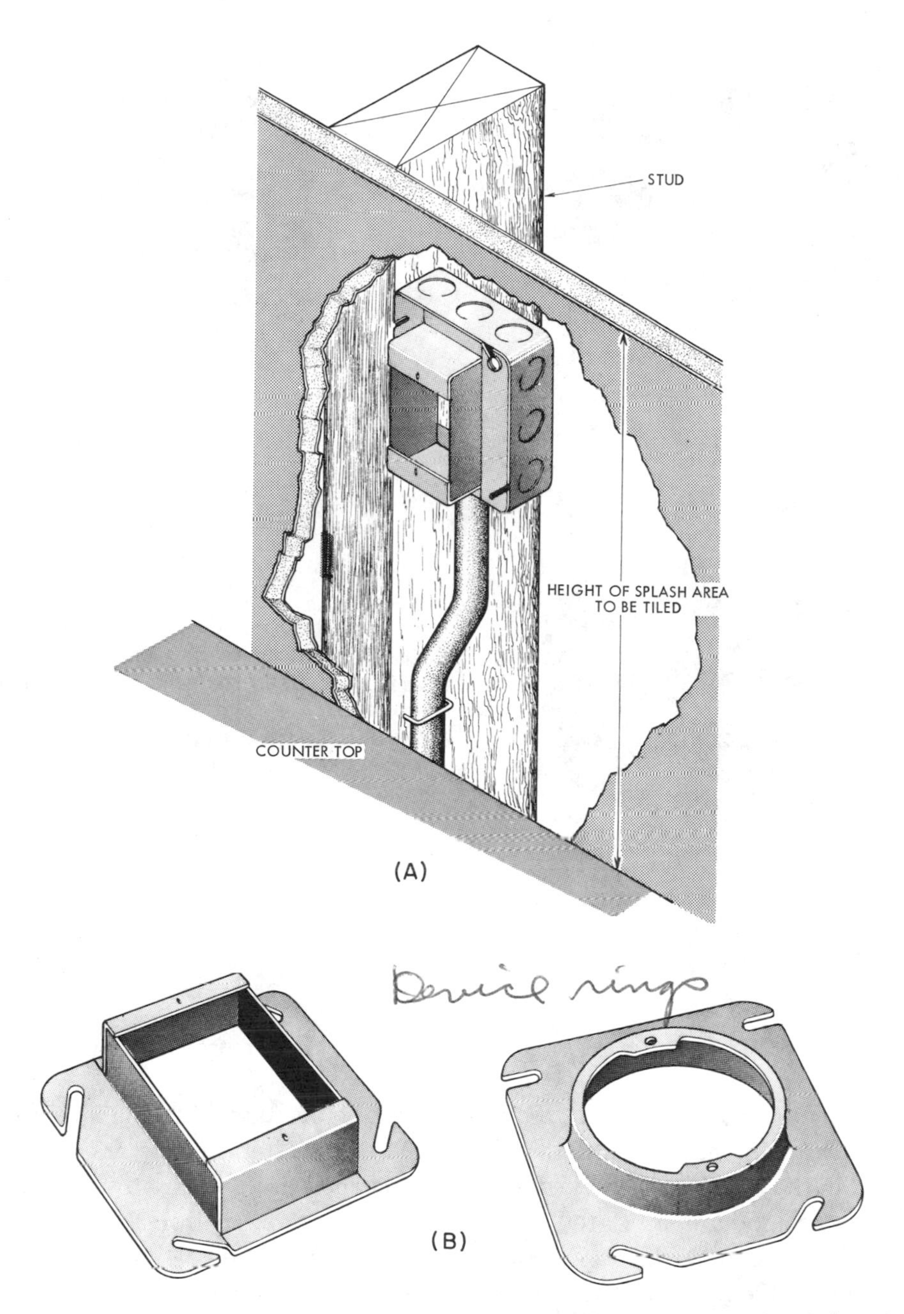

Fig. 4-18. (A) Details of construction are found on the building plans for locating outlets properly. Note tile wall box cover, or ring, bringing finished receptacle plate out to surface of finished tile. (B) Plaster rings and box covers are used to bring the wiring enclosure to finished plaster and tile lines and to adapt boxes to wiring device mountings.

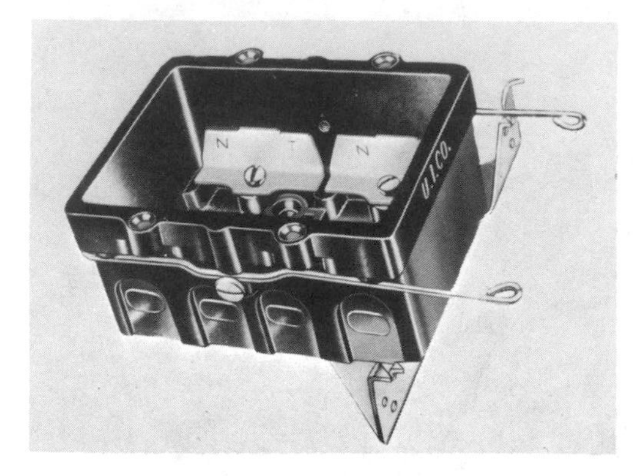

(A)

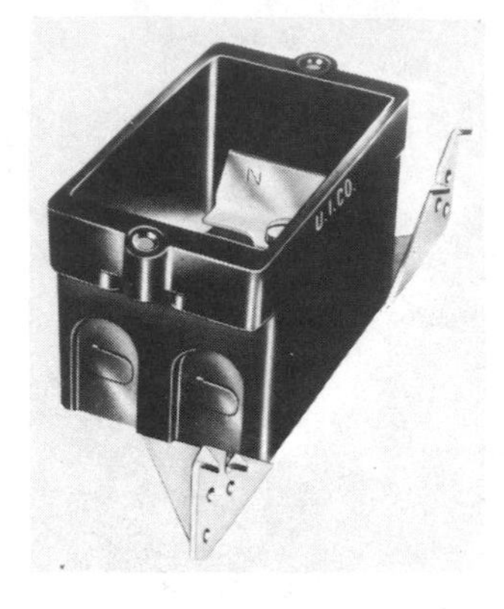

(B)

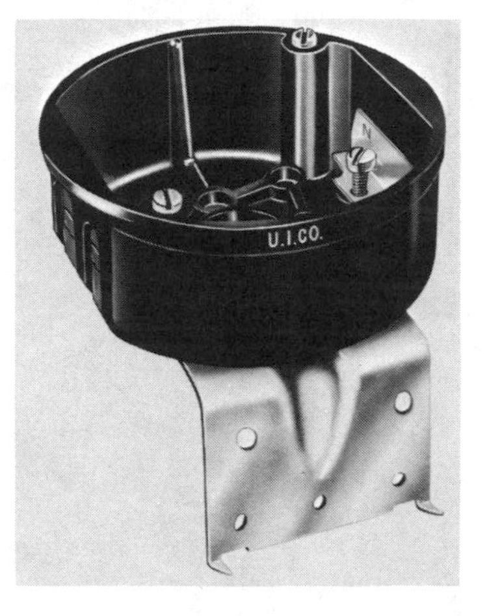

(C)

Fig. 4-19. Insulated outlet boxes. Note cable clamping devices inside boxes. (A) Two gang switch box. (B) single gang switch or duplex receptacle box, and (C) light outlet box. Metal boxes are similar; some have clamps inside and some use cable connectors outside.

tors, are preferred, as they do not affect the number of conductors allowed in the box. Fig. 4-19 illustrates plastic—compound insulated boxes sometimes used with non-metallic sheathed cable wiring systems. These boxes are tough and not as easily destroyed with hammer blows as were the original bakelite boxes. Note cable fastening clamps inside the boxes. Metal boxes are similarly constructed. Some are used with cable connectors installed by the wireman outside the box as in Fig. 4-13, left. Grounding connection requirements may vary with the locality and should be checked with local codes.

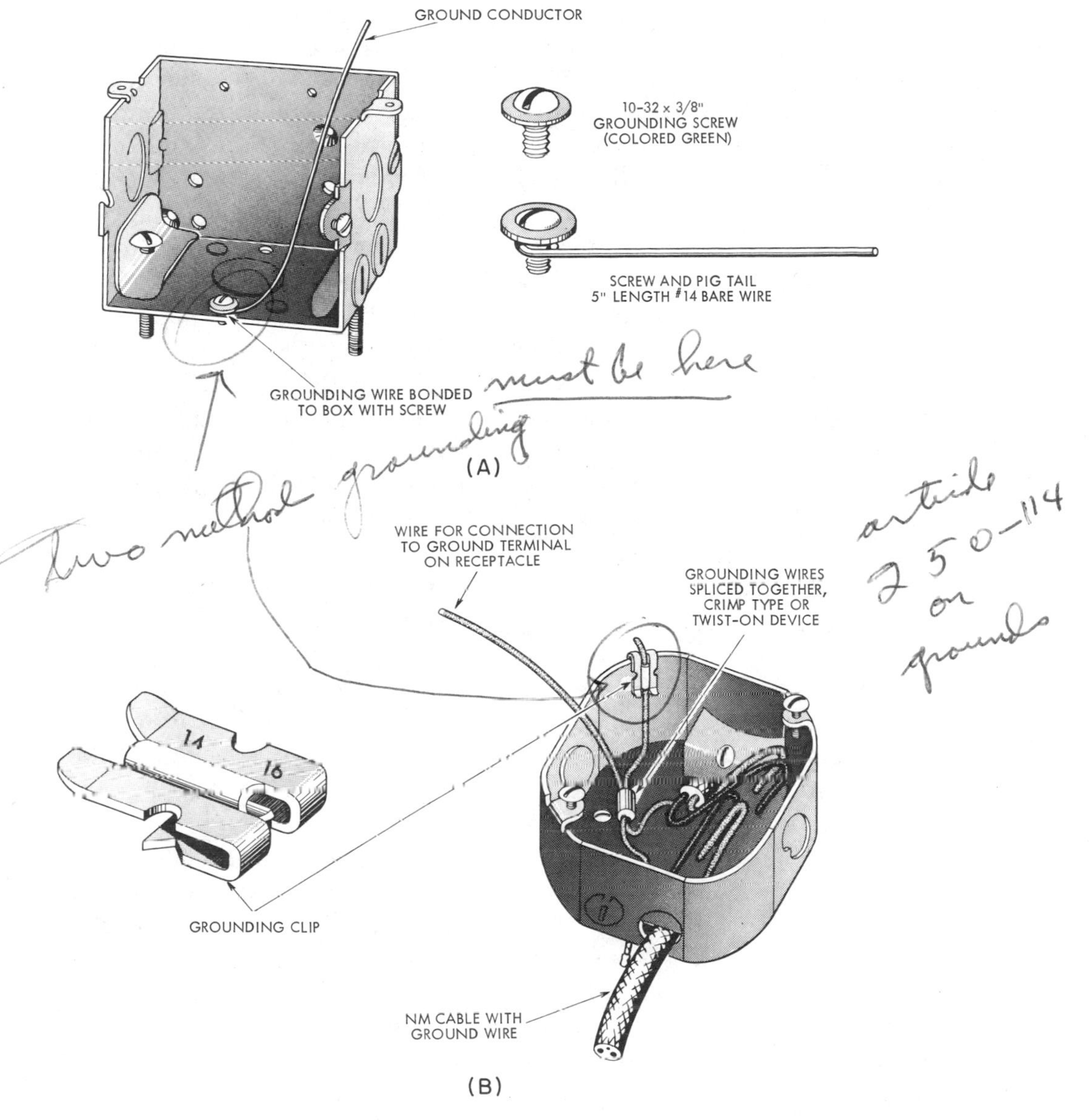

Fig. 4-20. Grounding methods for metal outlet and switch boxes.

GENERAL ELECTRIC 75C 3 COND.

Fig. 4-21. Unarmored service entrance cable. (General Electric Co.)

Grounding Devices

Effective grounding of electrical equipment is provided by means of a bare conductor in each cable of a non-metallic sheathed cable system. The grounding wires are usually twisted together and crimp-connected — one of which is extended and attached to the metal box by a grounding clip or grounding screw. See Fig. 4-20. Grounding screws on receptacles and in boxes are green in color. Some boxes are grounded externally to a cold water pipe. This method saves one less wire inside the box so that larger size boxes are unnecessary due to too many wires inside the box. The code restricts this to remodeling wiring only.

Grounding receptacle devices are grounded to the metal box in a similar fashion for portable plugged-in appliances.

Chapter 12, "Grounding for Safety," gives more detailed information on grounding.

Service Entrance Cable

Service entrance cable is employed in lieu of wire and conduit for connecting the power company service drops to the service disconnecting means for the building. Fig. 4-21 shows one kind of service entrance cable. Type SE is used for overhead services, while Type USE is for underground.

Service entrance cable may be employed for interior wiring systems only when all circuit conductors, including the neutral, are of the rubber-covered or the thermoplastic type. Service entrance cable with uninsulated grounding conductor may be used to supply electric ranges, built-in cooking devices and clothes dryers when it originates at the service entrance, or it may serve as a feeder from service cabinet to another building, providing cable has a final nonmetallic outer covering and alternating current supply circuit does not exceed 150 volts to ground. Chapter 11 gives more details on service equipment.

Underground Feeder and Branch-Circuit Cable

Electrical Codes list Type UF cable as acceptable for underground feeders and branch circuit uses. The outer braid of this cable is suitable for direct burial in the ground, being moisture, fungus, and corrosion-resistant. Where single-conductor cables are used, all wire of the particular circuit, including the neutral wire, shall be run together in the same trench or raceway.

Fig. 4-22 illustrates an example of the use of direct burial cable feeding an outdoor landscape or walkway light. No conduits are necessary to encase the cable. The cable should, however, be buried at a suitable protective depth not less than 18″ generally.

This cable may be employed for

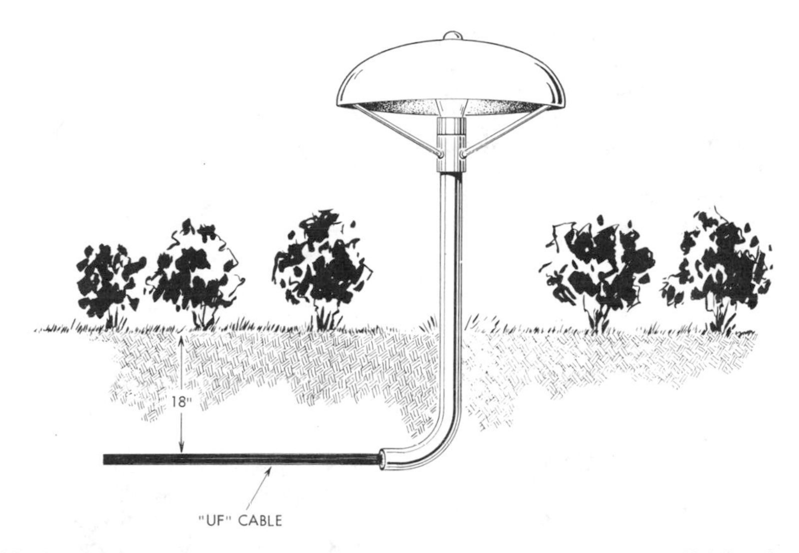

Fig. 4-22. A common application for underground cable feeding outdoor lighting for lighting landscaping, illuminating walkways, trees and shrubs. Cable is buried directly in earth protected by its plastic compound covering.

interior wiring systems in dry, wet or corrosive locations. It may not be used for service entrance, nor in commercial garages or other potentially hazardous locations. (It is similar in appearance to Fig. 4-2.)

QUESTIONS FOR DISCUSSION OR SELF STUDY

1. What is the skeleton of a wood building called?
2. When drilling horizontally in a wall to pull in cable, what common structural member must be drilled?
3. What member must be drilled in going from wall to attic?
4. What members support the sub-floor?
5. What structural member is over a window or door?
6. What members support the lath and plaster above the rooms?
7. How are the outlets and switch locations laid out to begin wiring?

8. Outlet boxes are installed generally before or after drilling? Why?
9. Trimming out houses means what?
10. A 14/2 cable means what?
11. Of what materials are type NM conductors made?
12. How are cables supported?
13. Why is guard stripping cables necessary?
14. Why is it necessary to ground metal boxes and outlet receptacles?
15. What is a plumb bob used for?
16. Describe two methods of grounding metal boxes.
17. Which wood bit is best for drilling in rough work? Why?
18. How should holes be properly drilled in studs and joists?

QUESTIONS ON THE NATIONAL ELECTRICAL CODE

1. Can Type NM cable be used for exposed work?
2. Can Type NMC cable be imbedded in concrete?
3. Is it permissible to fish Type NM in an interior concrete block wall?
4. What type of nonmetallic cable would be installed in a dairy barn?
5. Where installed in holes bored in timber, how far should NM cable be from the nearest edge of the joist or stud?
6. What should be the maximum distance between staples where NM cable is fastened to the side of a joist?
7. What should be the maximum distance from an outlet box that a staple may be driven?
8. How is NM cable protected from driven nails when notched into studs?
9. What is the minimum radius of bends for NM cable?
10. In exposed work, generally, what precautions must be taken?
11. What is the requirement for free length of conductors at outlet boxes?
12. What section or table of the code makes provisions for the number of conductors permitted in a box?
13. What is the code definition of "grounded?"
14. To what extent is the NEC modified by your local code in regard to the use of nonmetallic sheathed cable?
15. How may direct burial cable be used?
16. What are the approved AWG (American Wire Gage) maximum and minimum conductor sizes for copper? for aluminum?
17. What is the requirement for plaster ring depths?
18. In what instance may cable clamps or connectors be omitted?

Methods of Wiring Metal Clad Armored Flexible Cable

Chapter

5

The installation of metal clad cable, also called armored cable, is relatively simple and is similar to that of non-metallic sheathed cable. Much of the equipment and material is also the same.

Metal clad armored cable is a fabricated assembly of insulated conductors in a flexible metallic enclosure. (Flexible conduit is another method of wiring and is explained in the Chapter 13, "Methods of Conduit Wiring.") The NEC recognizes four types of flexible armored cable: AC, ACT, ACL (are common for residential use) and MC. (MC is mainly used in commercial and industrial.)

This type of indoor wiring consists of two or more insulated wires protected by a wound, galvanized steel strip cover. This metal winding forms a flexible tube. Frequently, armored cable is referred to as "B-X". Because of its flexible covering, this type of cable can be used as an extension to a previously installed conduit system. Armored cable is considered acceptable by most building codes. It offers protection similar to rigid conduit.

Armored cable was introduced under the trade name of *BX* by one manufacturer, and is still referred to as such, regardless of the brand. Fig. 5-1 shows the construction of Type AC cable. It can be obtained with either one, two, three, or four conductors. Single-conductor cable is used mostly as a grounding wire.

In order to provide flexibility, the steel covering on BX was not wound so tightly as in older types of cable. This "loose" construction introduced

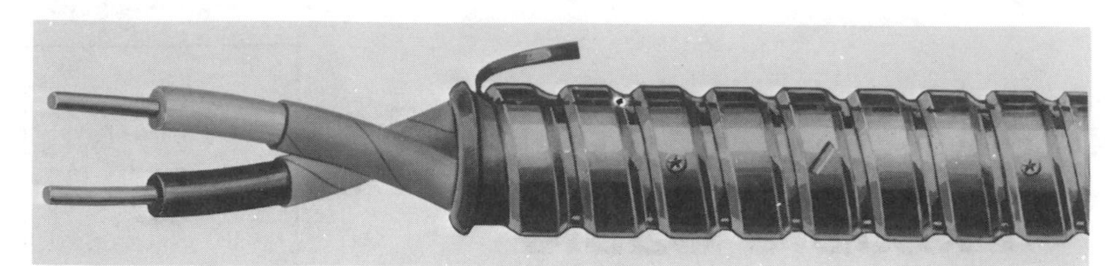

Fig. 5-1. Type AC armored cable. (Crescent Insulated Wire & Cable Co.)

a comparatively high resistance into the steel covering so that it did not provide acceptable grounding continuity. Manufacturers, therefore, placed a copper or aluminum bonding strip underneath the armor and in intimate contact with it. The NEC now provides that Types AC and ACT shall have a copper or aluminum bonding strip throughout their entire length.

It is worth noting the correct way of designating wire size and number of conductors in cable. Cables having two No. 14 wires are called "fourteen-two," written as 14/2; those with three No. 12 wires are called "twelve-three," written as 12/3; those with three No. 10 wires are called "ten-three," written as 10/3, and so on throughout the various combinations.

Armored cable is delivered in coils which are ordered by wire size and the number of conductors desired. Lengths are standard, running from 250 feet for 14/2, 12/2 and 10/2, to 200 feet for larger sizes, down to a
200 feet for larger sizes, down to a
100 feet for the largest size. The coils are held together by iron wire bindings; once the bindings are removed the coil is ready for use. The cable is removed off the coil from the center and unwound to avoid kinks while installing. This procedure is similar to that used with nonmetallic sheathed cable.

Applications

A general analysis of the conditions of approval for a wiring system will show that one of the conditions of approval is based on the ability of a system to resist possible damage from the inherent conditions of the place where the system is installed.

In dry locations, Types AC and ACT cables may be used for either exposed work or for concealed work. They may be fished in the air voids of masonry walls where the walls are not exposed to excessive moisture. They may be used also for underplaster extensions.

Type ACL cable, Fig. 5-2, may be imbedded in concrete or masonry,

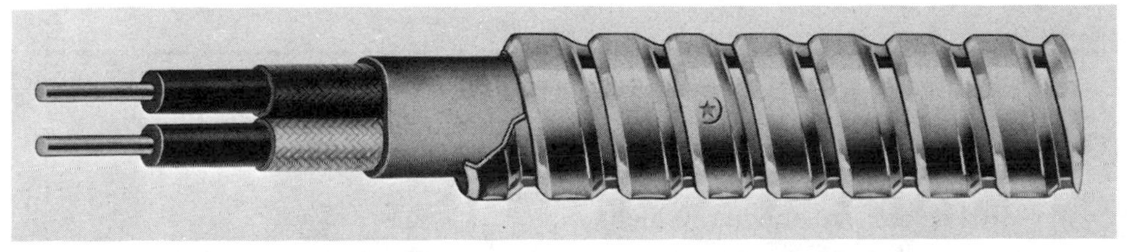

Fig. 5-2. Lead covered armored cable type ACL. (Crescent Insulated Wire & Cable Co.)

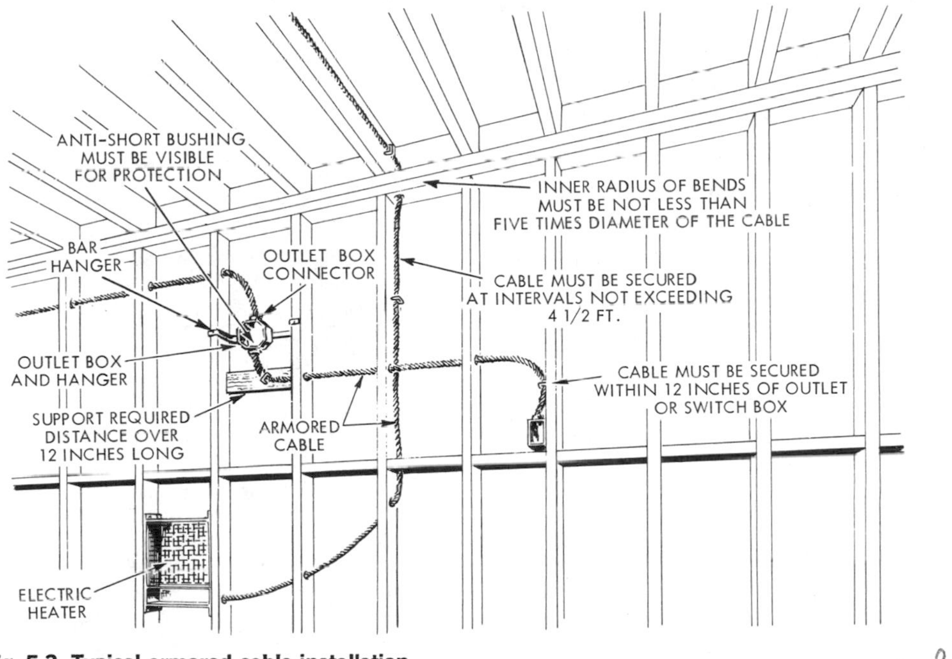

Fig. 5-3. Typical armored cable installation.

run underground, or used where gasoline or oil is present.

Essential Requirements. Fig. 5-3 shows that the same general rules apply to installation of armored cable as to nonmetallic sheathed cable. Distance between supports should not exceed 4½′, and the cable should be supported within 12″ of an outlet box. Armored cable is supported in a manner similar to nonmetallic sheathed cable, using staples. In accessible attics this type of cable should be protected in the same manner as nonmetallic sheathed cable.

Outlet Boxes

An outlet box or the equivalent must be inserted at every point in the system where access to enclosed wires is necessary. They fall into three general classes: square, octagon, and rectangular. See Fig. 5-4. Each of these comes in various, widths, depths, and knockout arrangements. There are knockouts in the sides and bottoms for readily making cable or conduit entries. A knockout is a round indentation, punched into the metal of the box, but left attached by a thin edge or by narrow strips. It can be removed easily with a hammer or pliers. Bushings and locknuts are used to secure this type of box to a run of conduit. Connectors are used to secure cables. Outlet boxes are obtained in galvanized finish and can be used for either concealed or open work where there are no volatile, explosive or flammable vapors, dust or similar hazards present, and if not exposed to the weather.

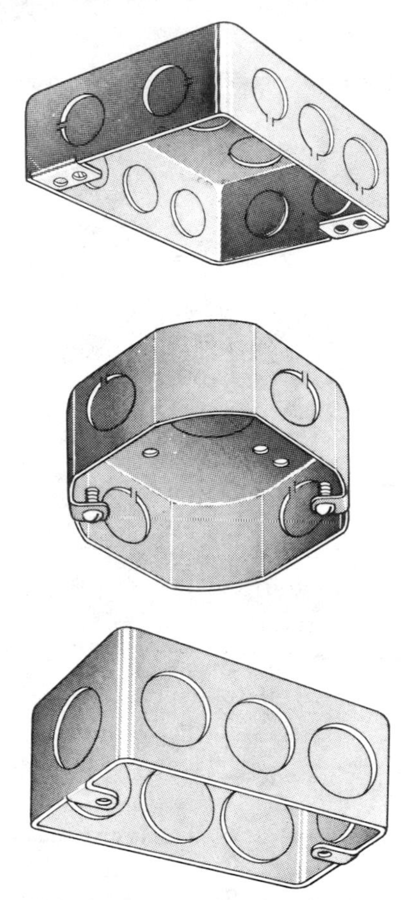

Fig. 5-4. Square, octagonal and rectangular switch or outlet boxes.

Although there is no hard and fast rule as to the kind of box needed for a specific purpose, the general practice is to use the octagon type for lighting outlets, the other two for switches or receptacles. Outlet boxes are also used for junction and pull boxes, or in place of an ell, when making a 90° turn when using conduit.

The size and number of conductors regulates the cubic inch capacity of a box. In other words, the greater

TABLE I. DEEP BOXES

Box Dimension, Inches Trade Size or Type	Min. Cu. In. Cap.	Maximum Number of Conductors				
		#14	#12	#10	#8	#6
4 x 1¼ Round or Octagonal	12.5	6	5	5	4	0
4 x 1½ Round or Octagonal	15.5	7	6	6	5	0
4 x 2⅛ Round or Octagonal	21.5	10	9	8	7	0
4 x 1¼ Square	18.0	9	8	7	6	0
4 x 1½ Square	21.0	10	9	8	7	0
4 x 2⅛ Square	30.3	15	13	12	10	6*
4 11/16 x 1¼ Square	25.5	12	11	10	8	0
4 11/16 x 1½ Square	29.5	14	13	11	9	0
4 11/16 x 2⅛ Square	42.0	21	18	16	14	6
3 x 2 x 1½ Device	7.5	3	3	3	2	0
3 x 2 x 2 Device	10.0	5	4	4	3	0
3 x 2 x 2¼ Device	10.5	5	4	4	3	0
3 x 2 x 2½ Device	12.5	6	5	5	4	0
3 x 2 x 2¾ Device	14.0	7	6	5	4	0
3 x 2 x 3½ Device	18.0	9	8	7	6	0
4 x 2⅛ x 1½ Device	10.3	5	4	4	3	0
4 x 2⅛ x 1⅞ Device	13.0	6	5	5	4	0
4 x 2⅛ x 2⅛ Device	14.5	7	6	5	4	0
3¾ x 2 x 2½ Masonry Box/gang	14.0	7	6	5	4	0
3¾ x 2 x 3½ Masonry Box/gang	21.0	10	9	8	7	0
FS — Minimum Internal Depth 1¾ Single Cover/Gang	13.5	6	6	5	4	0
FD — Minimum Internal Depth 2⅜ Single Cover/Gang	18.0	9	8	7	6	3
FS — Minimum Internal Depth 1¾ Multiple Cover/Gang	18.0	9	8	7	6	0
FD — Minimum Internal Depth 2⅜ Multiple Cover/Gang	24.0	12	10	9	8	4

* Not to be used as a pull box. For termination only. SOURCE: NATIONAL ELECTRIC CODE

the number of wires entering the box the larger and/or deeper the box requirements. See Table I as a sample illustration of sizes and conductor numbers. The NEC should be consulted for specific information of this nature. Usually one conductor allowance is subtracted from the table for a fixture stud within the box or a cable clamp or flush device mounted within.

Fixture Studs and Hangers. In addition to provision for conduit or cable entries the octagon or square outlet box has, at the back, a central knockout for a fixture stud. Fixture studs must be provided where the outlet is intended for a heavy lighting fixture. One type of stud in the boltless type is held in place by a locknut. See Figs. 5-5 and 5-6. Two kinds of fixture studs are shown in Figs. 5-5 and 5-6—one in combination with a straight-bar hanger, the other with an offset hanger. Some city codes require a fixture stud in each ceiling outlet. Another type is fastened with bolts. See Fig. 5-7. This type is sometimes called a "crowfoot."

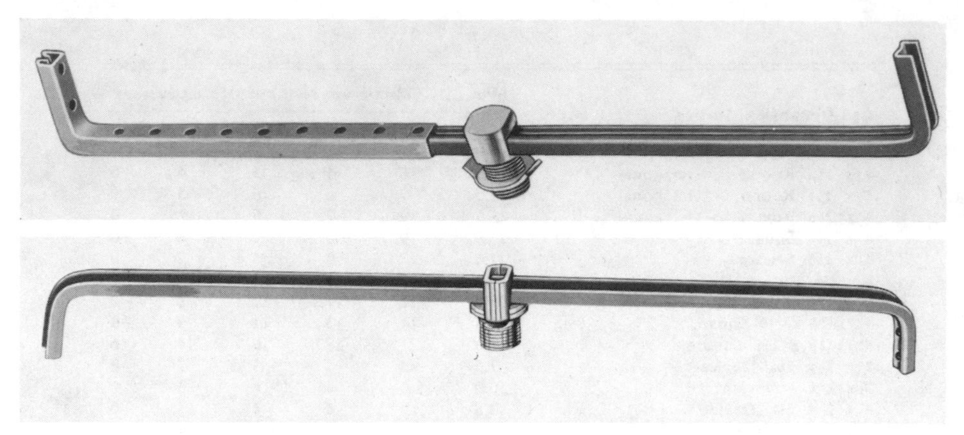

Fig. 5-5. Two straight bar hangers.

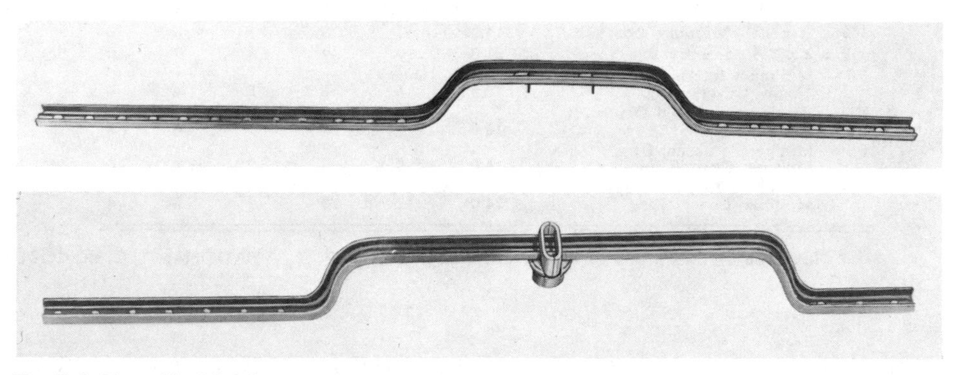

Fig. 5-6. Two offset bar hangers.

Outlet-Box Covers and Plaster Rings. Covers for boxes are of numerous forms, each having a particular use. The *blank* type closes the outlet entirely. The *plaster rings* shown in Fig. 5-8 are used on wall and ceiling boxes. Octagon boxes do not always require plaster rings; their need is governed by structural conditions on the particular job and size of fixture canopies which are to be used. Plaster ring covers should be used only where necessary because they hinder the work of installing wires and making splices. Square boxes used for lighting outlets must always have plaster rings to adapt them to the outlet. Special rings can

Fig. 5-7. Fixture stud. (Raco All Steel Equipment Inc.)

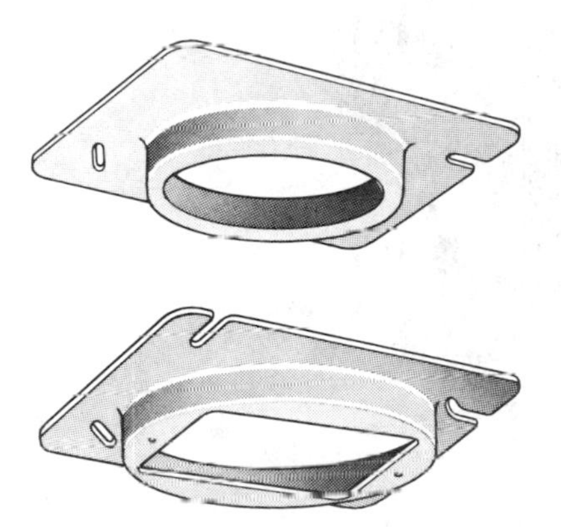

Fig. 5-8. Plaster rings. Top: Raised, 4″ square ceiling outlet. Bottom: Raised, 4-S, single device.

be obtained with internal ears drilled and tapped so that switches or other devices can be mounted on them.

The "switch ring," shown in Fig. 5-8 bottom, is used on square boxes for switch and receptacle outlets. The raised opening in the cover permits the use of devices having rather deep bodies, and serves also as plaster ground, where needed. These covers can be obtained with various amounts of projection, from 1/4″ to 1 1/2″, to accommodate all different thicknesses of wall finish.

Plaster rings are not made for rectangular boxes because these are intended only for switches and receptacles and are set according to wall thicknesses initially. Tapped holes allow switches or receptacles to be attached directly to the box. Box covers normally refer to closing of boxes on surface mounted construction.

Extension Rings. Octagon and square outlet box extension rings, shown in Fig. 5-9, come in a variety of depths from 1 1/2″ up. They are similar to the outlet boxes in every way, except that instead of being closed in the back, they are open, having only a narrow flange to act as a seat when installed on a box. These rings are used to bring the edge of the outlet box out to the face of the plaster in case the box has been mounted too far back, or to provide additional space when an extra deep box is required. They are also used in some remodeling work. When it is screwed to an existing outlet box, surface wiring can be continued.

Connectors and Boxes. Cables are secured to boxes by clamps which are an integral part of the box, Fig. 5-10, left, or by means of separate connectors, Fig. 5-10, right, and Fig.

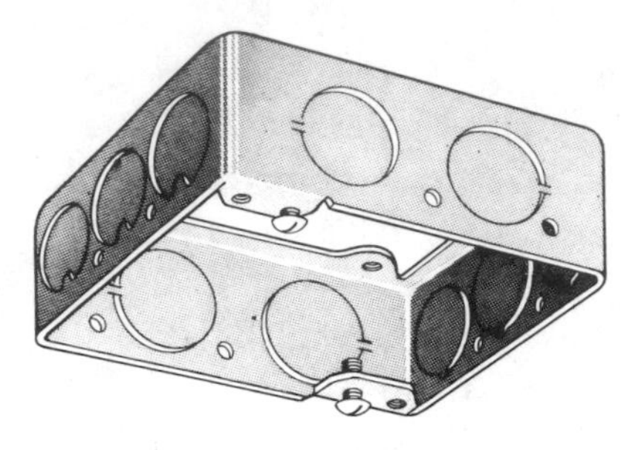

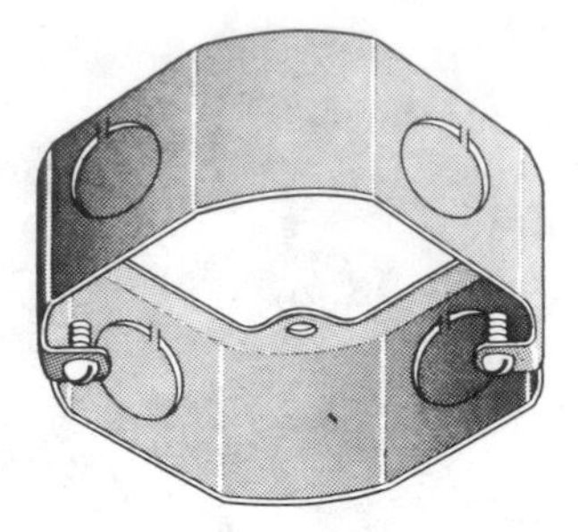

Fig. 5-9. Square and octagon outlet box extension rings.

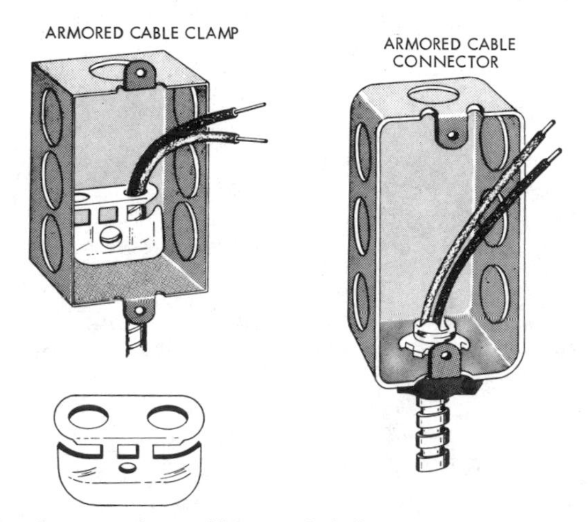

Fig. 5-10. Methods of connecting cable to switch boxes.

5-11. The connectors have a standard conduit thread and a locknut at one end, so that they may be attached to standard boxes. The clamp at the opposite end of the connector body is made in various forms, some of them applicable to both nonmetallic sheathed cable and armored cable. The connectors shown in Fig. 5-11 are very similar to the connectors used for nonmetallic sheathed cable, but sturdier. However the connectors for armored flexible cable have a peephole so the red color of the anti-short bushing can be seen. (Fig. 5-14A illustrates an anti-short bushing.)

Round outlet boxes are usually called "Ceiling Boxes" since they are primarily installed in ceilings (or overhead) and contain the wiring for lighting purposes and also provide a

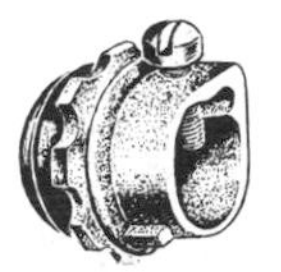
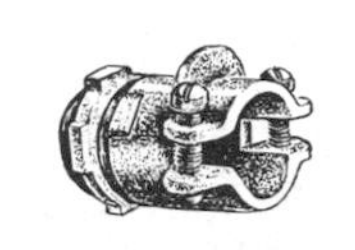

Fig. 5-11. Outlet box connectors.

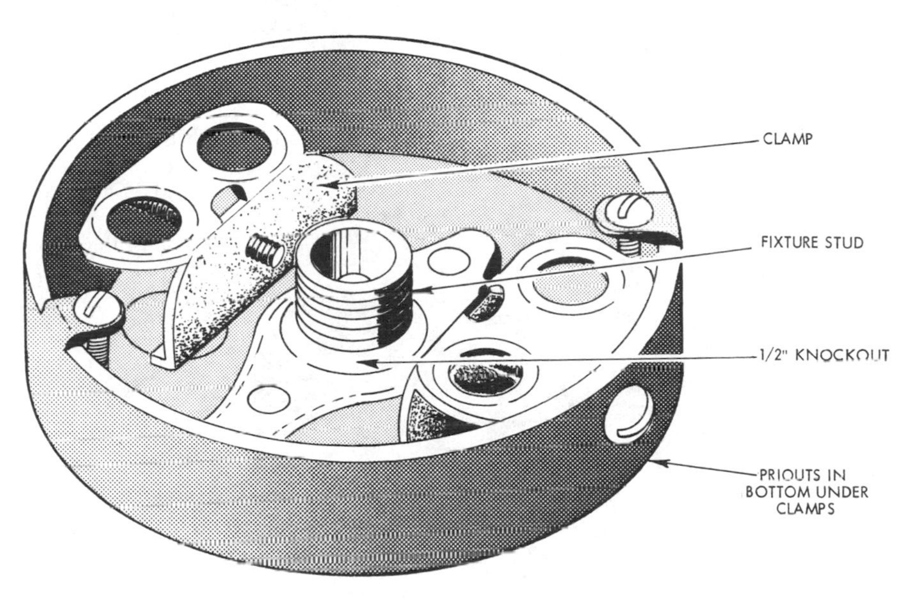

Fig. 5-12. Ceiling fixture box generally used for remodeling work. (Steel City Division, Midland-Ross Corp.)

support for fixtures. Also used for old work where fixture canopy must cover box. See Fig. 5-12.

Due to narrow depth of 3/4" all wiring entrances (priouts) are located in the bottom of the box. The center knockout locates the integral fixture stud and, as shown in Fig. 5-12, has two clamps designed to accept and hold armored cable systems.

Round boxes for armored cable are 3 1/4" in diameter.

Fig. 5-13, top, shows location of the clamps for armored cable when used with a square outlet box. The center 1/2" knockout may be used to field-mount a fixture stud or bar hanger. (See Fig. 5-13, bottom.) The box can be nailed directly to a stud or joist. Boxes with mounting brackets are also available.

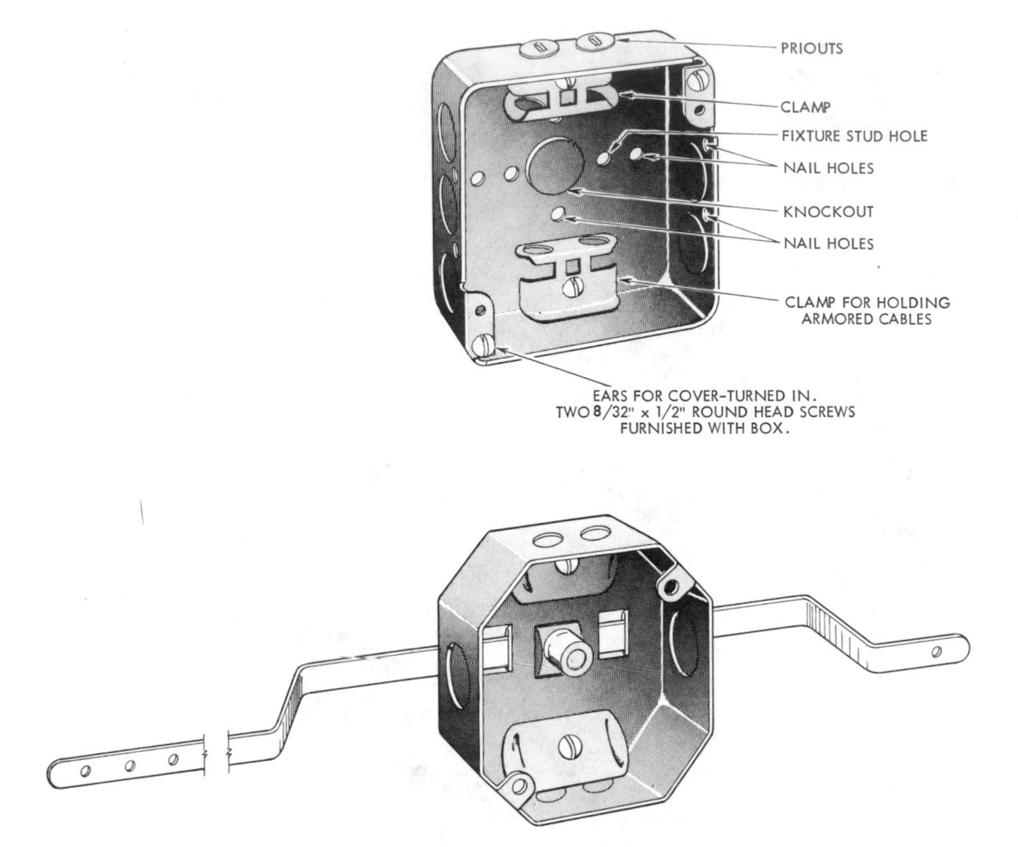

Fig. 5-13. Top: Square outlet box, 1½" deep, with internal cable clips. Bottom: Offset bar hanger used for mounting fixture boxes (4" octagon box). (Steel City Division, Midland-Ross Corp.)

Grounding Armored Cable

On armored cablework, equipment grounding is done in the same manner as for conduit (see Chapter 13), except that box connectors are used to secure the cable to the enclosure instead of bushings and locknuts. The code requires that armored cable, except type ACL, shall have an internal bonding strip, either copper or aluminum, to lower continuity resistance of the outer covering. This wire or strip, Fig. 5-14A, runs the entire length of the cable just inside the armor; it must be connected to

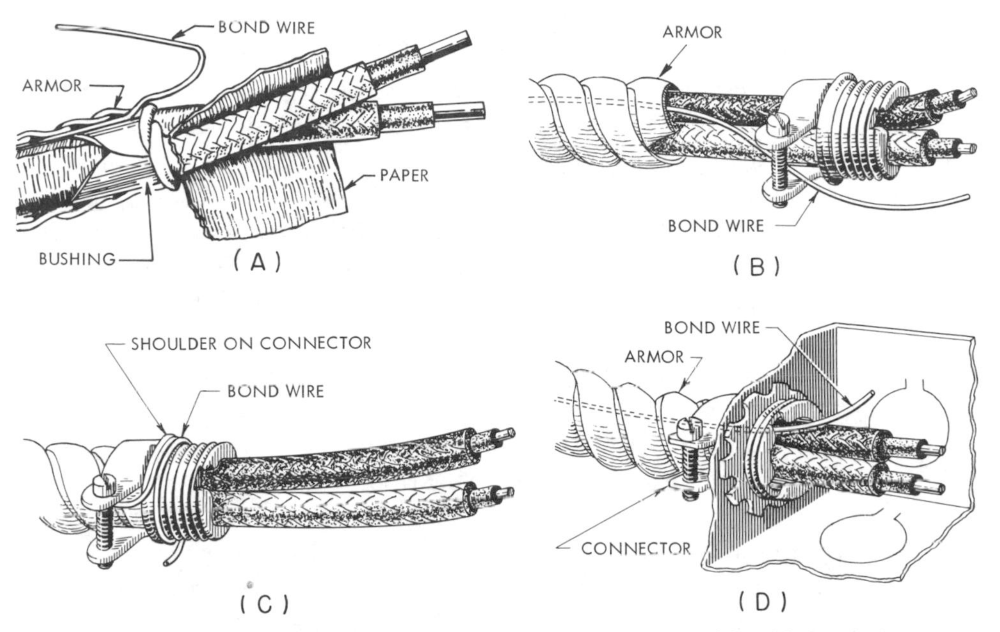

Fig. 5-14. Methods of maintaining continuous grounded enclosures with cable bond wires.

the enclosure or other receptacle into which the cable is entered. There are three ways of making connections.

1) By drawing the bond wire out through the clamping ears of the box connector, and twisting it around the threaded end of the connector, Figs. 5-14B and 5-14C. In this position it is held fast between the enclosure and the shoulder of the connector as the locknut is set up. (2) By running the bond wire into the enclosure along with the wires, Fig. 5-14D, wrapping it around the threaded end of the connector and clamping it between the enclosure and the locknut. This method requires a washer back of the locknut to make sure that the wire is not damaged by the locknut. (3) In some jurisdictions it is permissible to fold the wire back over the outside of the armor, letting it rest under the clamp, then setting up the latter and cutting off the surplus length.

Any of these methods can be used with setscrew types of box connectors. With methods (1) and (3) the anti-short bushing is slipped into place under the bond wire, Fig. 5-14A, but in method (2) the anti-short bushing is slipped over the bond wire.

In Fig. 5-14D the bond wire may ground the box and receptacle outlet device in a similar manner as shown for nonmetallic sheathed cable. More grounding information is given in Chapter 12, *Grounding for Safety*.

Polarity Grouping. All wires of an alternating-current circuit when encased in conduit or in armor, must be within the same enclosure. If not so installed, inductive heating may result. This is explained in detail in Chapter 6, *Switching Circuits*. The point here is to use a 3 conductor cable when three wires are needed in a circuit. Do not use one 2 conductor cable and one conductor of another cable.

Preparing the Cable. Armored cable may be cut readily with a hacksaw. The best method is to cut through one of the convolutions about 6″ to 8″ from the end, and then break it off by twisting back and forth or as shown in Fig. 5-15, right.

Care should be taken not to cut too deeply to avoid damage to conductor insulation. The use of protective gloves are also useful.

The outer wrapping may then be removed and the anti-short bushing installed as in Fig. 5-16. Tear the paper off back underneath the metal sheath to make room for the bushing.

The connector or clamp can now be attached, and secured to the box. Cable may be fastened to wooden

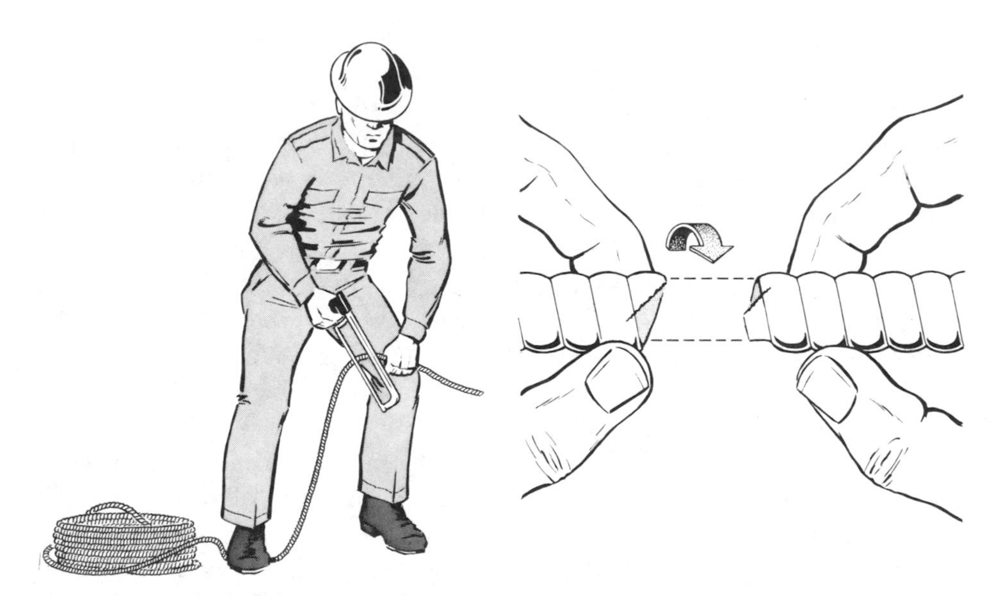

Fig. 5-15. Left: Cutting the armor which is held taut by stepping on it and holding it over the knee. Right: Note that the armor is cut at an angle to the lay.

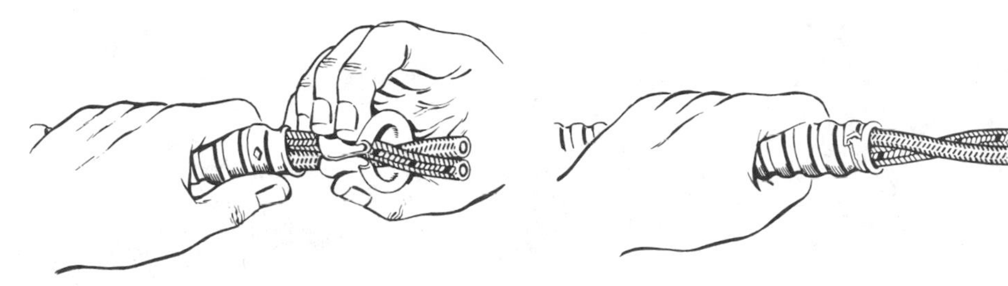

Fig. 5-16. Installing antishort bushing.

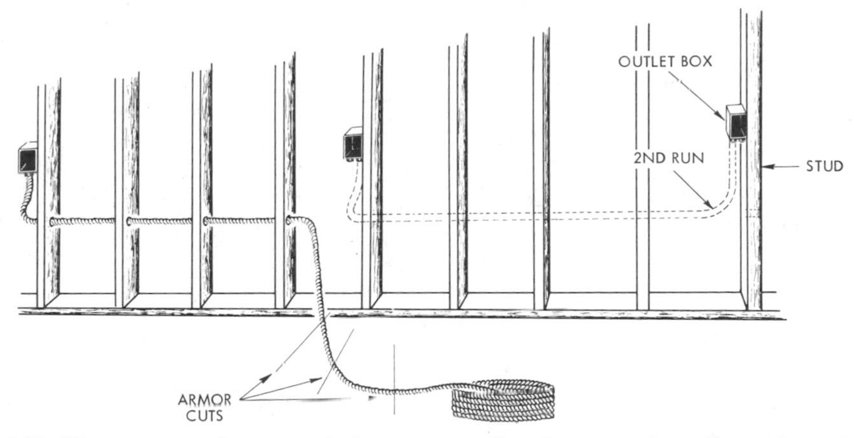

Fig. 5-17. Three armor cuts are made in sequence; the wires are cut on the center cut. It is now ready for 2 way makeup.

members by staples, similar to those for nonmetallic sheathed cable, or by armored cable straps. To avoid twisting and kinking, armored cable must be taken from the coil by unrolling or by unwinding it. A cable reel is useful.

When a cable end is made up to an outlet box after being pulled through bored holes, and after a measurement for a cut on the coil end is taken, three different armor cuts may be made 6 to 8 inches apart. See Fig. 5-17. The end length of the run is the center cut and includes free length of conductors. Here conductors are cut also with a hacksaw or high leverage diagonal cutting pliers after the armor has been cut with a hack saw. Fig. 5-18

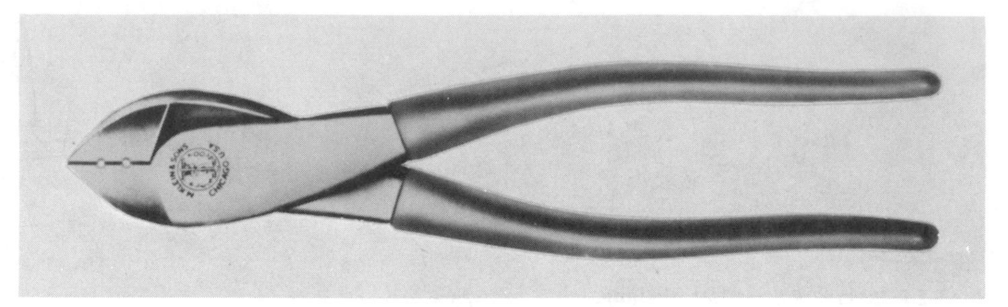

Fig. 5-18. High leverage oblique cutting plier. Note the two skinning holes for 12 and 14 gage wire. (Mathias Klein and Sons)

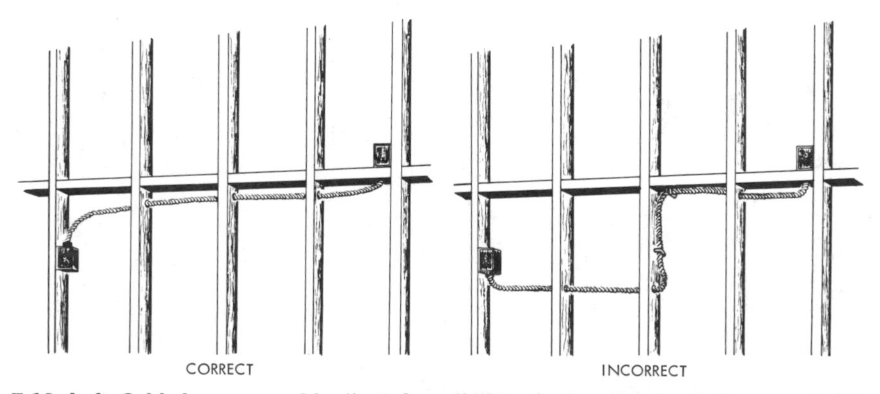

Fig. 5-19. Left: Cable is conserved by "running wild" to shorten distance between outlet boxes. Right: Cable length is wasted for this concealed job.

illustrates the cutting pliers that are used. The cable, after skinning of the armor on both ends, is now ready for make up to box connections.

All cable must be a continuous length without splices between outlets and fittings. It must also be properly secured in place at intervals not greater than those specified by the code.

It need not be run parallel or at right angles to structural members when the cable will be covered up or concealed. Remember the shortest distance between two outlet boxes is a straight line. This method is more economical in saving cable length. See Fig. 5-19. However, for better appearance and workmanlike manner, cables, like conduit, should be run parallel and at right angles to structural members for exposed work.

Electricians' Tools

An electrician who does work for an electrical contractor is expected to furnish his own hand tools as may be seen in Fig. 5-20. In the following some of the uses are described. *Slip-joint (or pump) pliers* (Fig. 5-20A) are commonly used in making cable installations to tighten box connectors and locknuts. Enough leverage is possible to tighten together small conduits. *Side-cutting pliers* (B) are used for cutting wire and for stripping wire by crushing insulation directly behind the hinge. The handles can be used to tap in the outlet box knockouts and remove them with the plier gripping jaws also. High leverage *diagonal pliers* (C) are useful in cutting wires where it is difficult to use side cutting pliers. For example, as for cutting armored cable conductors after the metal armor has been cut. They are also useful in skillfully looping wires for screw terminals, in the absence of long nose pliers.

Long nose pliers (Fig. 5-20D) are generally used for making eyelets in wires for screw terminations and positioning objects in tight places.

The 10″ and 14″ *pipe wrenches* (Fig. 5-20E) are used to tighten conduits together. They have also served in emergencies as a pipe vise, for threading conduit, by clamping the conduit in opposite directions and forming a tripod with wrench handles and conduit. The *adjustable wrench* (F) is used on such positive sized items as hex-head cap screws, bolts of knockout punches and others.

Long-nosed *electrician's hammers* (Fig. 5-20G) are useful should nails be required inside the back of outlet boxes. *A hand brace* (H) is commonly used with a wood bit to bore holes. It is also used to turn a conduit reamer in deburring conduit.

The slight curvature of the *skinning knife* blade (Fig. 5-20I) is useful in reaching around wires and cables for removing insulation. A couple of different *chisel* sizes (J) for notching work makes the job easier.

The electrician is required to have a *hacksaw* frame (Fig. 5-20K). The contractor furnishes blades for the hacksaw frame. Those best for cutting rigid conduit have 18 to 24 teeth per inch; those for cutting thin-wall conduit and armored cable have 24 to 32 teeth per inch. *Key-hole saws* (L), with blade assortments, are used to enlarge holes and to notch structural frame members for cables and conduits.

Miscellaneous size *screwdrivers*

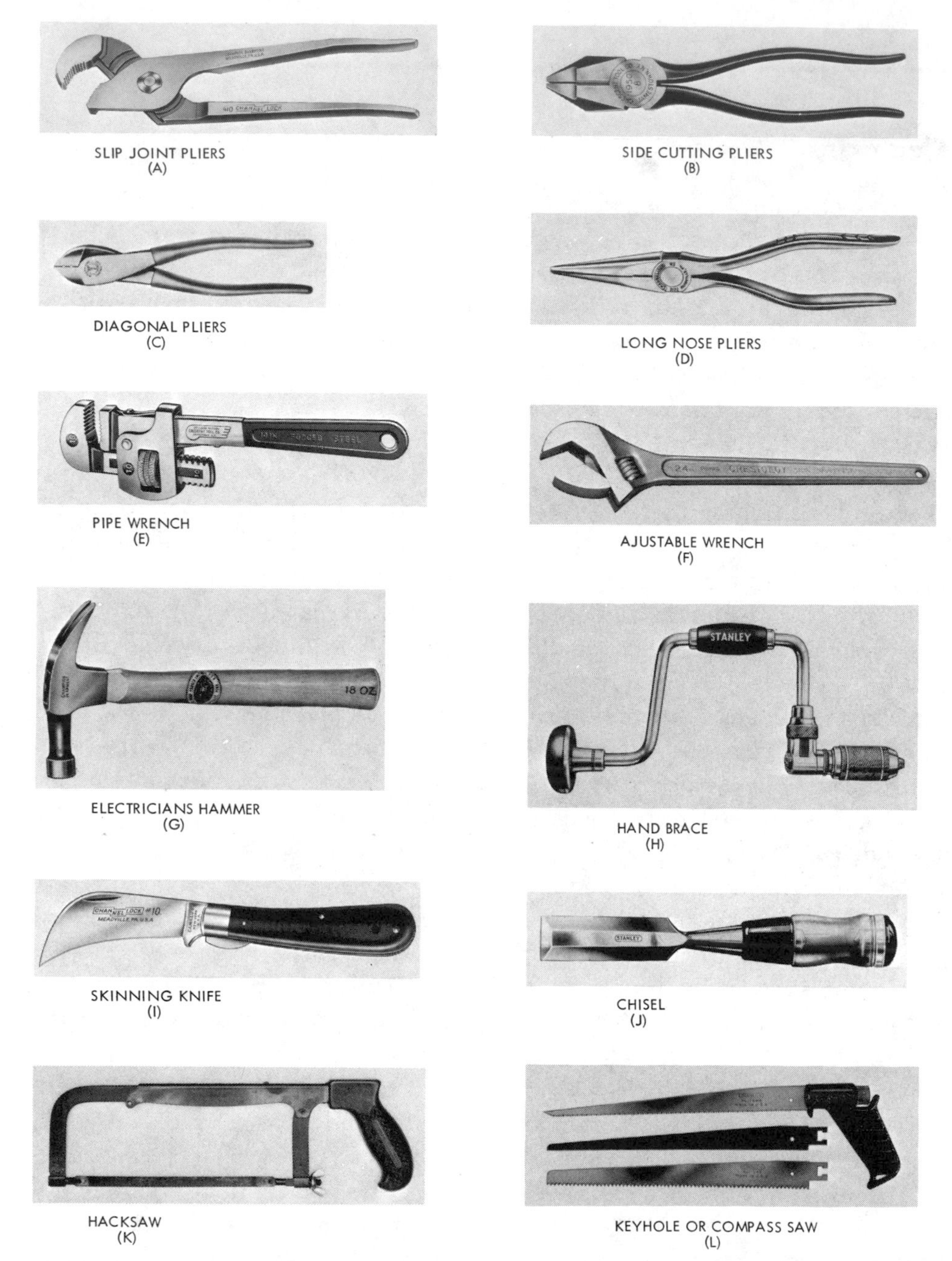

Fig. 5-20. Electrician's hand tools. (A, D, G and I, Champion DeArment Tool Co.; B, E, F and K, Cresecent Tool Co.; C, Mathias Klein and Sons; H, J, K and M, The Stanley Works; O, Lufkin Rule Co.

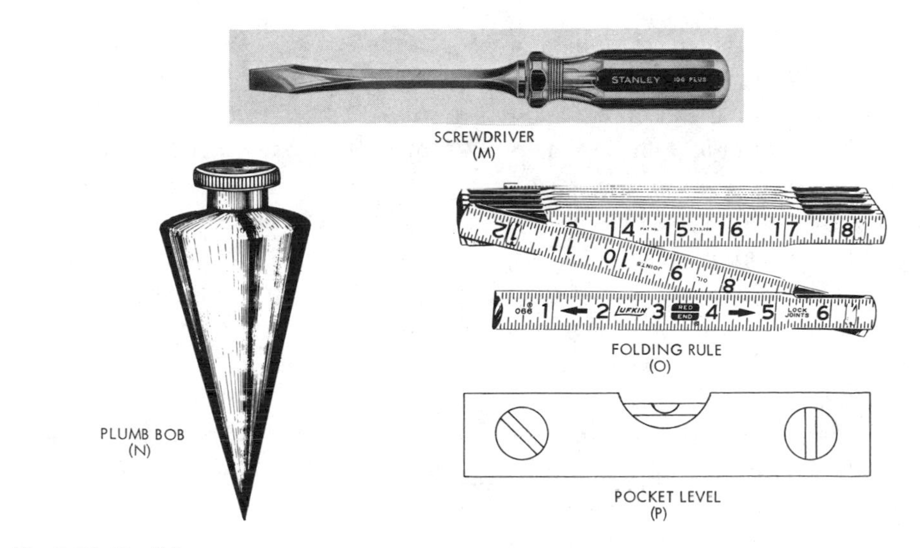

Fig. 5-20. Cont'd.

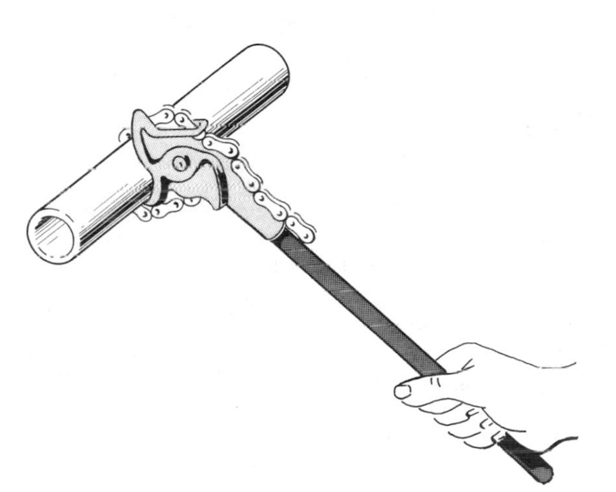

Fig. 5-21. Chain tongs.

(Fig. 5-20M) are convenient for tightening different size screws. They are sometimes misused for digging plaster and concrete from outlet boxes.

A *plumb-bob* (Fig. 5-20N) is used to establish vertical lines by dropping a pointed weight on a cord. A measured mark on the floor can be transferred to ceiling joists. It is a very valuable tool in establishing vertical lines for precision surface conduit and box installations.

A folding rule (Fig. 5-20O) is recommended over a steel measuring tape. Folding rules are made of wood or plastic for electricians. They should be inside reading so that they can be laid flat when measuring a flat surface. Besides a measuring tool they can be used for protractors in matching angles of bends for conduit work. The standard length is 6 feet. Swivel joints should be kept oiled. Should a break occur at a joint, it can be repaired with a good glue. Folded outside numbered surfaces should be protected from pocket or tool pouch wear with clear varnish or lacquer.

The *pocket level*, or *torpedo level*, (Fig. 5-20P) is used to level, or plumb, vertically or horizontally, objects as boxes on the surface, conduits, plates and others.

Chain Tongs. These wrenches are of the chain and lever variety. They are made in sizes from 12″ to 48″. See Fig. 5-21. They are used to tighten conduits.

QUESTIONS FOR DISCUSSION OR SELF STUDY

1. Generally speaking how does wiring with armored cable compare with the non metallic sheathed cable wiring system?
2. How are the cut ends of armored cable protected when installed in an outlet box?
3. What type of methods are used to connect armored cable to outlet boxes?
4. How should cable be run between outlets for concealed work?
5. How should cable be run from outlet to outlet in exposed work?
6. Why should three armor cuts on the coil end be made when running from the outlet to outlet?
7. How are armored cable conductors cut from the coil when dead ending to an outlet box?
8. How is the cable held when cutting?
9. How are continuous grounds maintained?
10. What are fixture studs?
11. Name the three general classes of outlet boxes.
12. In case an outlet box is mounted 1″ too far back into a partition, what simple corrective measure is possible?
13. What is a "crowfoot"?
14. Fig. 5-8 shows a plaster ring or box cover?
15. When nailing boxes without brackets, how many nails are recommended?
16. In grounding armored cable, how is it secured to the enclosure and how is the bonding strip attached?

QUESTIONS ON THE NATIONAL ELECTRICAL CODE

1. Type "AC" armored cable is used for what circuits?
2. May metal clad cable be installed in exposed or concealed work, or both?

3. May type "AC" cable be used in the following locations: dry; embedded in plaster; damp or wet locations; all of the preceding?
4. In what location is type "ACL" generally used?
5. By what method must armored cable be supported?
6. What is the required distance between supports and from outlet boxes?
7. What are the exceptions to Question No. 6?
8. What is the minimum radius requirement for cable bends?
9. What is the maximum radius requirement for cable bends?
10. What are the maximum number of bends permitted for armored cable installations?
11. How are wires protected from abrasion at cable terminal points?
12. How is electrical continuity maintained in metal enclosures?
13. How many wires are subtracted from the table of maximum number of conductors in a box if an outlet box contains a fixture stud?
14. What is the maximum number of #14 conductors allowed in a 4″ x 1½″ box with a fixture stud contained?
15. What is the general minimum depth outlet box allowed?
16. What are the requirements for unused openings in outlet boxes?
17. What are three uses for the tool shown in Fig. 5-20(B)?
18. What are the advantages of the inside reading (folding) rule?
19. What is the required free length of conductors at outlets and switch points?
20. How should boxes be set in walls and ceilings with regard to the finished surface?
21. What is the maximum number of #12 conductors allowed in a 1½″ x 4″ square outlet box with inside cable clamps?
22. What is the maximum ampacity of 14-2 copper conductor, TW insulated cable?

Switching Circuits

After cables are pulled and made up to outlet boxes, proper splices must be made to insure correct control of circuitry and operation of lights, receptacles and other wiring devices. This is also true for conduit wiring systems. After the conduit system is installed, wiring is pulled into place. These methods are also common to commercial and industrial systems of like capacities. Switching, or proper control of electrical energy, is vital to the satisfactory operation of all electrical systems.

Polarity Wiring

The figures in this chapter show *polarity of wiring*. This is where one of the wires entering an outlet box is pictured as a dashed line and all others are shown as heavy lines. The dotted line indicates the *neutral* or identified conductor which is grounded to earth; it is always white or natural gray for residential circuits. The neutral wire is connected to the screw shell of the lamp socket, thus making it impossible for anyone touching the shell or the lamp base to receive a shock. An important rule is that the neutral wire should never be run to a switch; only one of the *hot* wires should be used for this purpose. The neutral wire should never be fused. It should not be switched or disconnected unless all the supply wires are disconnected simultaneously.

The hot wires, indicated by heavy lines in the illustrations, may be red or black. Other colors, except white or green, are sometimes allowed. Where two or more wire ends are brought together within an outlet box, a pigtail splice is made.

Reasons for Polarity Wiring

Neutral conductors should be run in the same conduit or cable as "hot" conductors of the circuits to which they are connected. There are two reasons for this practice: (1) to prevent overloading of neutral wires through wrong connections, and (2) to avoid harmful effects of induction. The first of these principles may be understood with the help of Fig. 6-1. Conductors for the two two-wire circuits are No. 14 Type TW copper. The hot wires of circuits *1* and *2* each carry a load of 15 amps, which represents their maximum allowable current-carrying capacity (ampacity).

To Prevent Overloading of Neutral Wires. If the lamps are properly connected, Fig. 6-1A, each neutral wire carries 15 amps. But, if the lamps of circuit *2* are connected accidentally to the neutral conductor of circuit *1*, as indicated in Fig. 6-1B,

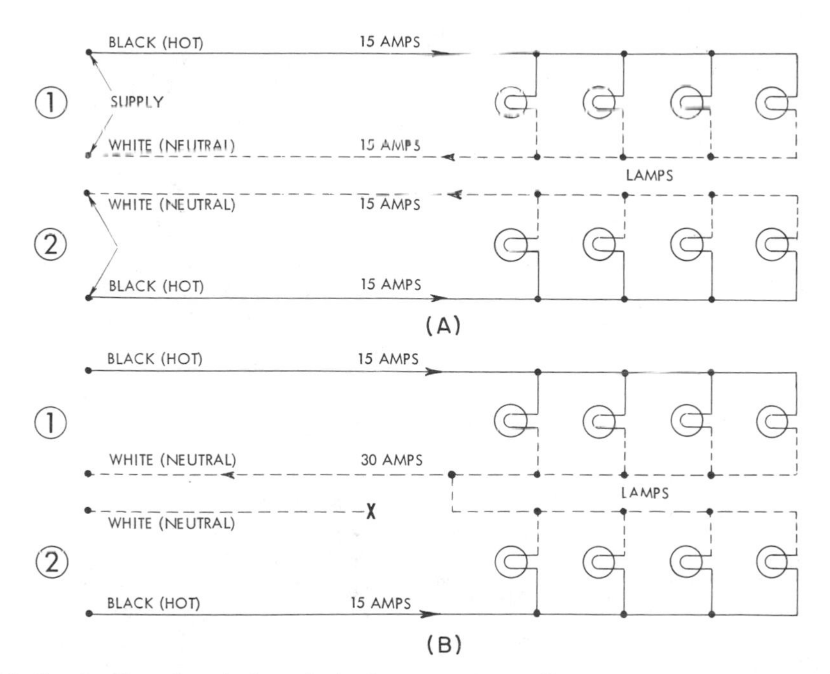

Fig. 6-1. Overloading of neutral conductor by wrong connection.

this wire is forced to handle 30 amps. Such an overload damages insulation on this particular wire, and the high temperature of the overheated copper may damage insulating coverings on adjacent conductors.

When the neutral conductor is run with the hot conductor of the circuit, this sort of accident is not so likely to happen. Overload may still occur, as a result of carelessness, where two or more circuits are placed in the same conduit. Nevertheless, such a dangerous situation is not so likely to arise as when neutral conductors are installed without regard to pairing them with the hot ones.

To Avoid Harmful Effects of Induction. In order to grasp the second principle, the term *induction* must be understood. Fig. 6-2A shows two conductors which make a loop around a piece of iron, an A-C current flowing in them. The iron core, through magnetic action, creates a voltage which opposes supply, thus increasing voltage drop in the circuit, and second, severely heating the iron itself.

Fig. 6-2B illustrates a pair of circuit conductors inside an iron conduit. Note that the iron is outside the wires instead of being enclosed by them. Also, the magnetic flux set up by current flowing in one conductor is offset by current which flows in the opposite direction through the second conductor. Magnetic effects are thus cancelled out, and induction cannot take place.

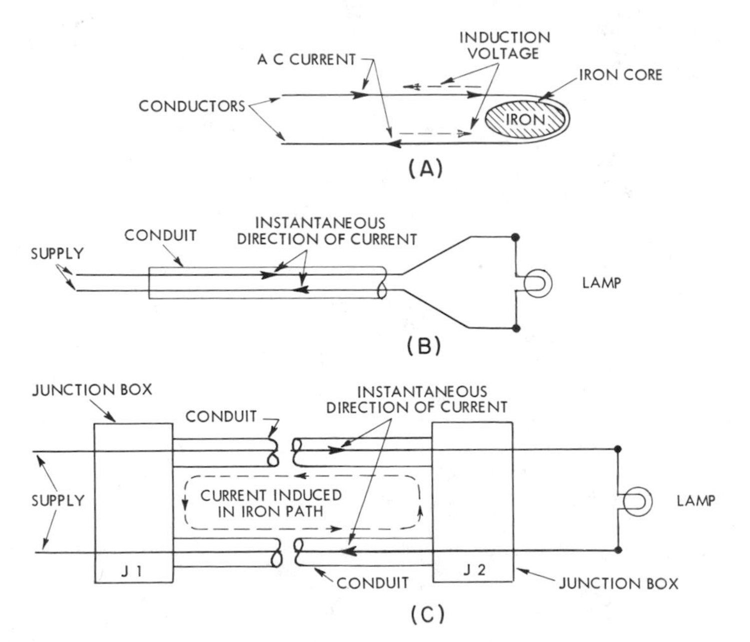

Fig. 6-2. Effect of induction.

Where the hot conductor is in one conduit and the neutral conductor in another, between junction boxes *J-1* and *J-2*, Fig. 6-2C, the wires enclose an iron core which is formed by conduits and junction boxes. As a result, induction takes place. Voltage drop in the circuit is increased, and the iron members are heated. The greater the amount of current in the circuit, the greater the voltage drop and the heating. With currents of a few amperes, heating effects are not serious. But with larger currents, say upward from 30 amps, dangerously high temperatures may be reached.

Single and Double Pole Switches

Single-Pole Switches

A simple, single-pole *switch-loop* is shown in Fig. 6-3. The outlet to be controlled is at the end of the supply circuit. There are three 2-wire pigtail splices in the outlet box, one connecting the neutral supply wire with the fixture white wire coming from the screw shell of the socket, one connecting the hot supply wire to the *hot leg*, and one connecting the *return leg* from the switch with the colored fixture wire coming from the center contact of the socket. On a conduit job, one splice may be saved by pulling through enough wire of the hot leg to reach the switch.

The current flow will be traced from the source, through the box to the hot leg to the switch, through the switch in the "on" position, up the switch return through the outlet box to the lighting fixture, through the lamp filament and out through the neutral wire, completing the circuit satisfactorily. The circuit would not operate with accidental "grounds," "opens" or "shorts." ("Grounds" here means the hot or unidentified conductor in contact with a grounded member of equipment; "opens" means a non-complete circuit, such as a poor splice. "Shorts" means here that the conductors of opposite polarity are mistakenly touching each other at uninsulated surfaces thereby creating a current path, or short cut, back to the source of energy without going through the load or lamp.)

A similar hot leg is shown in Fig. 6-4, except the supply circuit to the outlet does not stop there but goes on to additional outlets which are independent of the switch. The two wires from the outlet box to the switch are called the *switch loop*. The wire feeding the switch is the *hot leg*. The wire from the switch to the light is the *return leg* or *switch leg*. There are two 3-wire pigtail

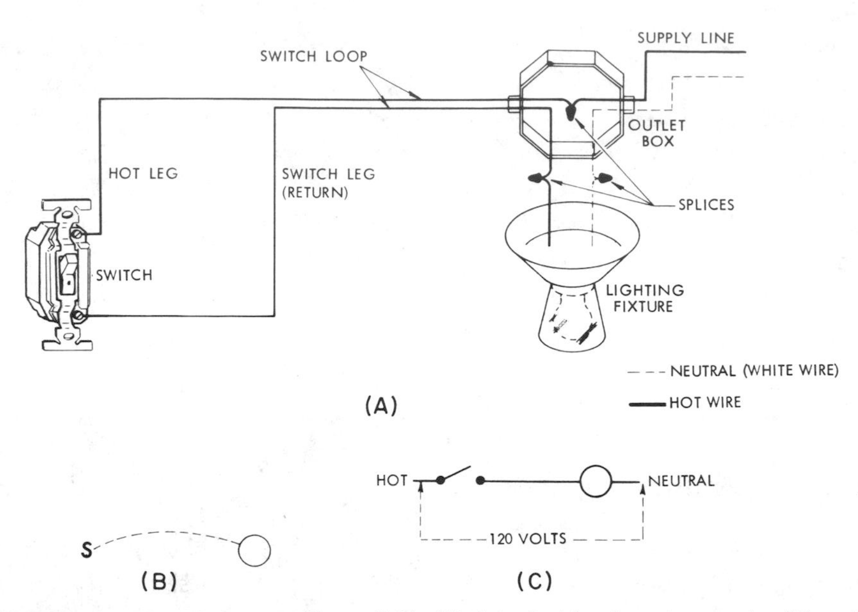

Fig. 6-3. Single pole switch controlling a light. (A) Actual wiring layout example, (B) architectural plan symbol, (C) elementary diagram.

splices and one 2-wire splice in the outlet box. The reason for two of the splices being 3-wire is that two wires of the circuit must continue on to supply additional outlets. The incoming neutral, the outgoing neutral, and the fixture wire coming from the socket screw shell are joined. The incoming hot wire, the outgoing hot wire, and the switch hot leg are connected together. The return leg from the switch and the fixture wire coming from the socket center contact are joined.

Special Note: As a general rule, a neutral conductor (white wire) should not be switched or used as one leg in a switch loop. This rule is readily applied to conduit work.

However, armored and nonmetallic sheathed cable, whether 2-, 3-, or 4-conductor, have a white conductor. The Code now states that cable having an identified conductor may be used for single-pole, three-way, or four-way switch loops where connections are made so that the *unidentified* wire is the return conductor from switch to outlet. This rule nullifies the old requirement that the identified conductor must be painted at the ends to destroy identification.

Two-Gang Switch. A two-gang switch, that is, two switches in one box under one faceplate, each controlling a separate outlet, is shown in Fig. 6-5. In this case the supply circuit enters outlet box No. 2 and

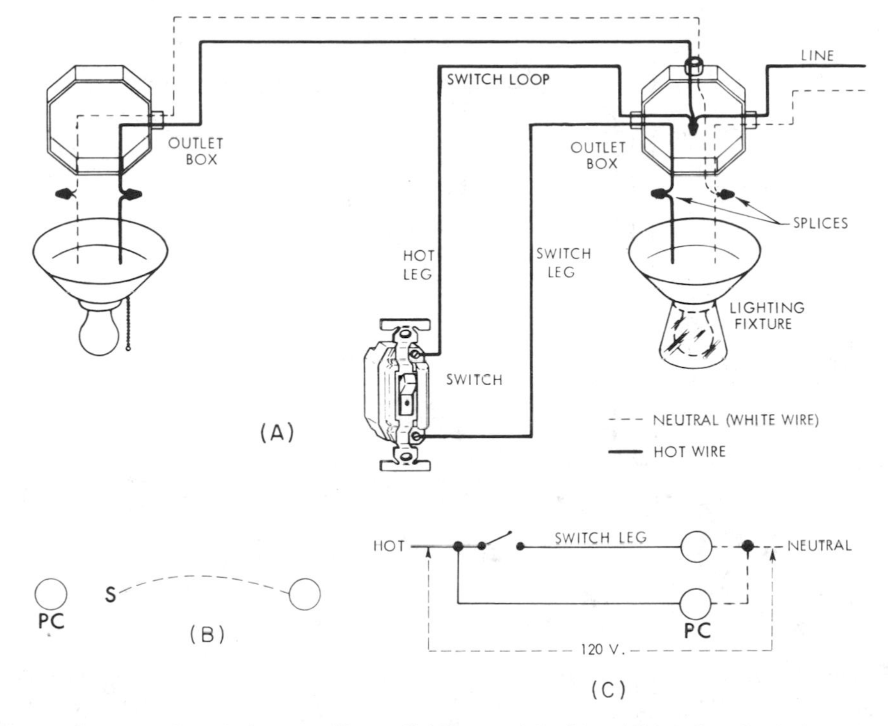

Fig. 6-4. Single pole switch controlling a lighting outlet with additional outlet independent of switch. (A) Actual wiring layout example, (B) architectural plan symbol showing switched lighting and pull chain, (C) elementary diagram.

from there goes into outlet box No. 1. In box No. 2 there is a 3-wire splice for the neutrals, a 2-wire splice for the incoming and outgoing hot leg of the circuit, and a 2-wire splice for the return leg from the switch. The return leg must pass through outlet box No. 1 to reach the switch; a 2-wire splice is shown in this box. This splice can be saved by pulling through enough wire to reach either the switch or outlet box No. 2, depending on the direction of pull. Outlet box No. 1 has, in addition to a splice for the return leg from switch 2 to outlet box No. 2, two 2-wire splices for the wires of the switch loop. There is also a 2-wire splice for the neutral and load wires.

A somewhat different arrangement is shown in Fig. 6-6, where the switch loop is taken from outlet box No. 2, instead of No. 1, as in the preceding case. There is one additional 2-wire splice in outlet box No. 2 but only two 2-wire splices in box No. 1, a saving of one 2-wire splice.

Three-Gang Switch. A 3-gang switch for the control of three separate outlets fed by a circuit should be arranged in the same manner, except that there would be three return

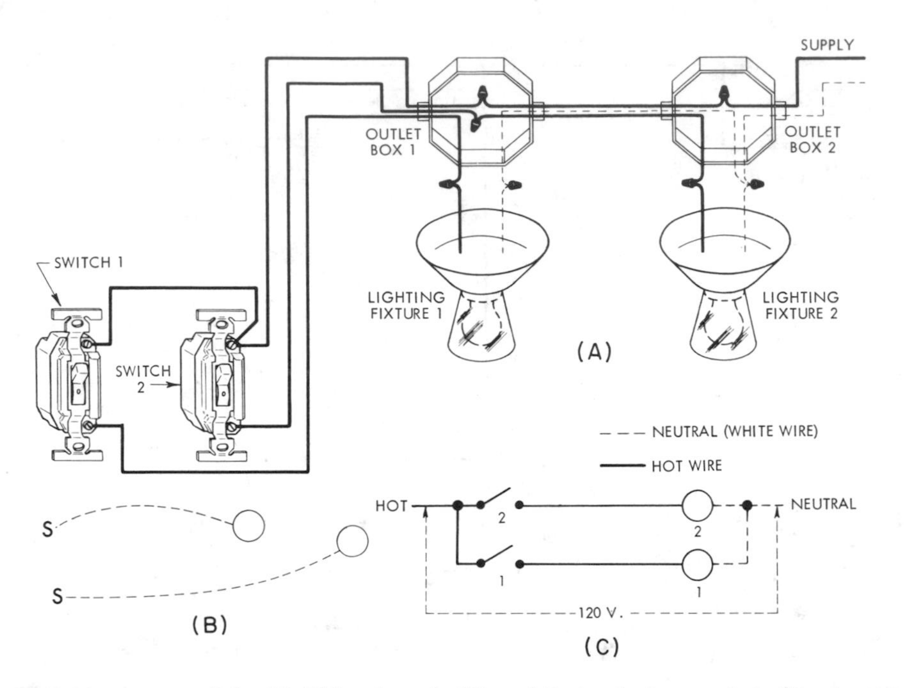

Fig. 6-5. Two gang switch. (A) Wiring layout, (B) architectural plan symbol, (C) elementary diagram.

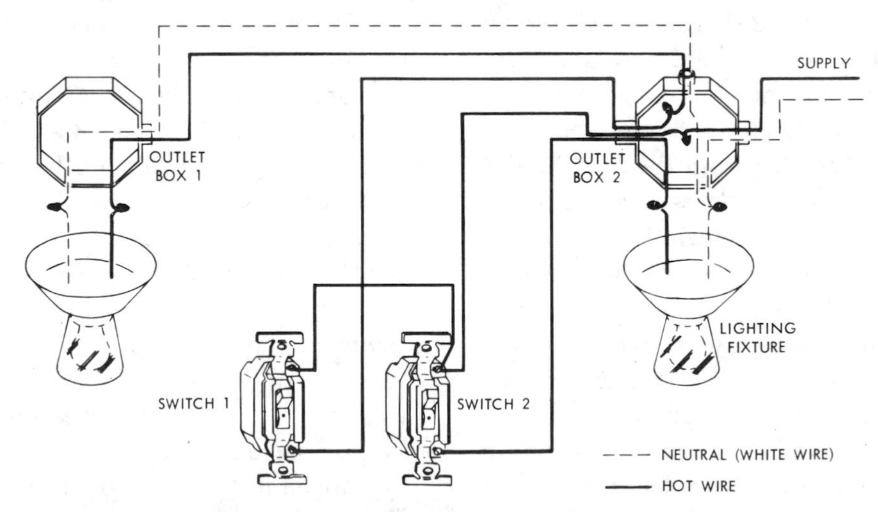

Fig. 6-6. Two gang switch with switches placed between outlets.

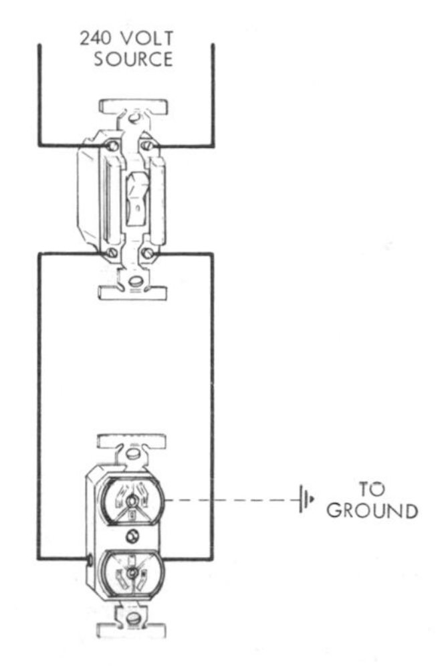

Fig. 6-7. Double pole switch used to control both line wires supplying a convenience outlet.

legs from the switches. The third return leg runs to the third outlet box through each of the other two boxes.

Double-Pole Switches

For many switching jobs, the single-pole unit is adequate because it is connected so as to interrupt current flow in the hot wire and leave the fixture *electrically dead* when it is in the "off" position. Sometimes it may be desirable to break both sides of the line, in which case a double-pole switch must be used. As an example, Fig. 6-7 shows a 3-wire convenience outlet and a double-pole switch that controls both hot line wires of a 240 volt circuit.

Many plug-in types of appliances use double-pole switches as a safety measure because the manner of inserting the plug determines which shall be the hot wires inside the device. This is a "polarized" plug and can only be connected to the receptacle in one manner.

Three-Way and Four-Way Switches

Three-Way Switches

There are times when it is desirable to control one or more lamps from two different points; for example, in a long corridor, a stairway, or a room with two separate entrances. This is done by means of two 3-way switches. Instead of hav-

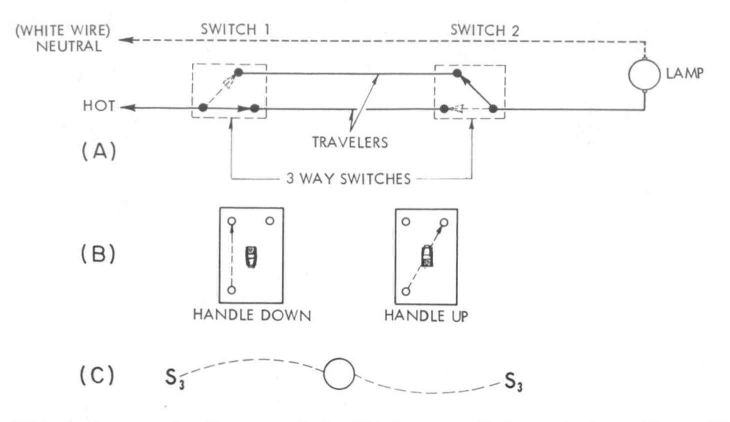

Fig. 6-8. (A) Wiring diagram for 3 way switch; (B) 3 way switch contact positions. The lamp is off, throwing either switch will cause it to go on. (C) Architectural plan symbols for 3 way switches, used for example for a hallway or a room with a switch at two doors.

ing a single terminal at both top and bottom, there is an additional terminal at one end. In order to identify these terminals throughout this discussion, the single terminal at one end of the switch is called the *hinge point* or *common*. Some manufacturers use a black oxide screw here for purposes of identification. Some use a copper screw here with two brass screws for traveler terminals to assist in distinguishing terminal identification. The hot leg is connected to the black oxide or copper screw. The terminals at the other end of the base, Fig. 6-8, are connected to similar terminals of the other control switch by two wires which are called *travelers*.

Blades of the switch mechanism are so arranged that the hot terminal is connected to either one of the traveler contacts as the handle is moved from one position to the other. That is, the switch itself has no "off" position. The hinge point of one switch is connected to the hot side of the supply circuit; the hinge point of the other switch is connected to the center contact of the lamp to be controlled. (*Hinge point* means the terminal that is common to the other two terminals in different positions. One position is shown internally in the switch by a broken line in the *on* position.) The screw shell of the lamp is connected to the neutral supply wire.

The illustration 6-9 is a diagram of the connections for a single outlet controlled by two 3-way switches.

No matter in which position the switch is turned, one or the other of the traveler terminals is made live.

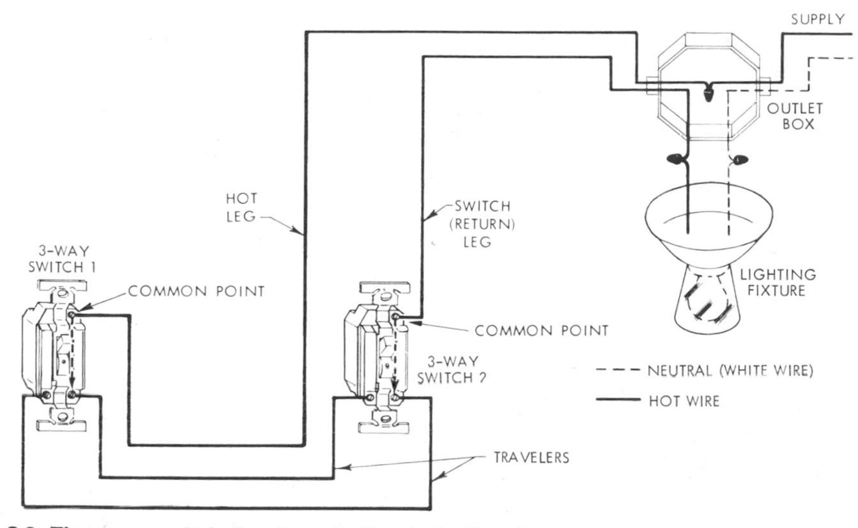

Fig. 6-9. Three way switch circuit controlling a single outlet.

In the switch unit the electrical connections are changed in a similar manner by moving the handle of the switch. Fig. 6-8A shows solid lines for one switch position, broken lines for the other. Fig. 6-8B shows this in another manner.

The path for the current, Fig. 6-9, may be traced from the *hot* supply wire in the outlet box to the common point of switch *1*, then to the right-hand traveler terminal, and through the upper traveler to the left-hand traveler terminal of switch 2, where the circuit is broken. Because both switch levers are in the same position and there is no connection from the left-hand traveler terminal to the common post, the lamp does not burn. But, if the lever of switch *2* is moved to the other position, current will flow from its left-hand traveler to the common point, through the return leg to the center contact of the socket, and through the lamp to the neutral. The lamp will now burn.

If the switch levers are returned again to the position shown in Fig. 6-9, the lamp goes out. Suppose the lever of switch *1* is moved to the other position. From the common point the flow is to the left-hand terminal, through the bottom traveler to the right-hand terminal of switch *2* and to the common point, through the return leg to the center contact of the socket, and through the lamp to the neutral, causing the lamp to burn. Thus, the movement of either switch lever controls the lamp.

The arrangement of wires, Fig. 6-9, must be followed if the wiring is done in conduit or armored conductors. However, if nonmetallic

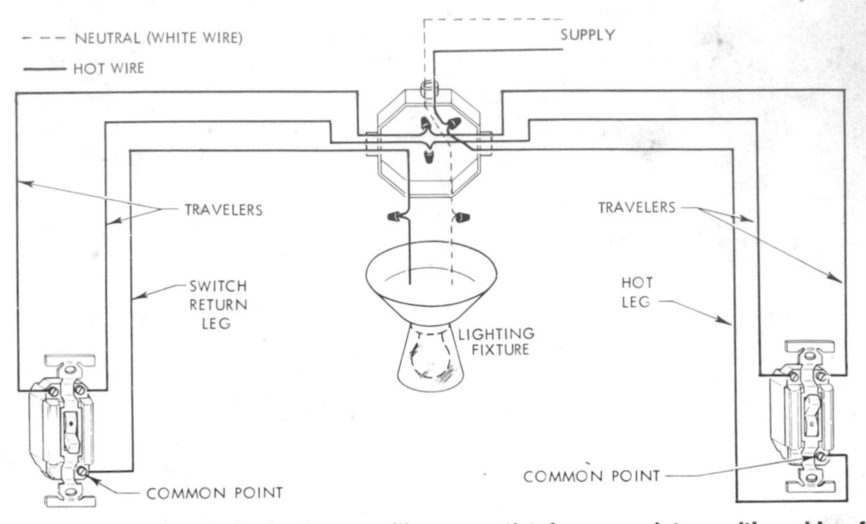

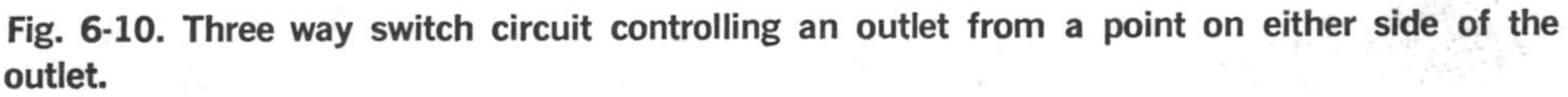

Fig. 6-10. Three way switch circuit controlling an outlet from a point on either side of the outlet.

sheathed cable is used, the hot switch leg can be run direct from switch *1* to the outlet box instead of by way of switch *2* if this is an advantage in saving material.

The scheme for controlling a lamp from two points at opposite directions from the outlet is shown in Fig. 6-10. Three wires are run to each of the two switches, the travelers being connected in the outlet box.

Controlling Several Outlets. A method for controlling several outlets from two points is shown in Fig. 6-11, the supply circuit entering outlet box 3. One of the switch loops is taken out of box 1, the other out of box No. 4 because these two are close to the switches.

The circuit path may be traced from the point of entry in box No. 3 through all the outlets and switches, and back to the starting point. The neutral goes to the screw shell of each of the sockets, and the *hot* leg through boxes No. 3 and No. 4 to the common point of switch *2*. Assume that in both switches, at the moment, common points are making contact with right-hand traveler terminals. "Traveler" wires are the wires connected between switches only, as shown in Fig. 6-8A. The path continues from the common point of switch *2* to the right-hand traveler which is the uppermost conductor, Fig. 6-11, through the four outlet boxes. Descending toward switch *1*, this wire connects to the left-hand

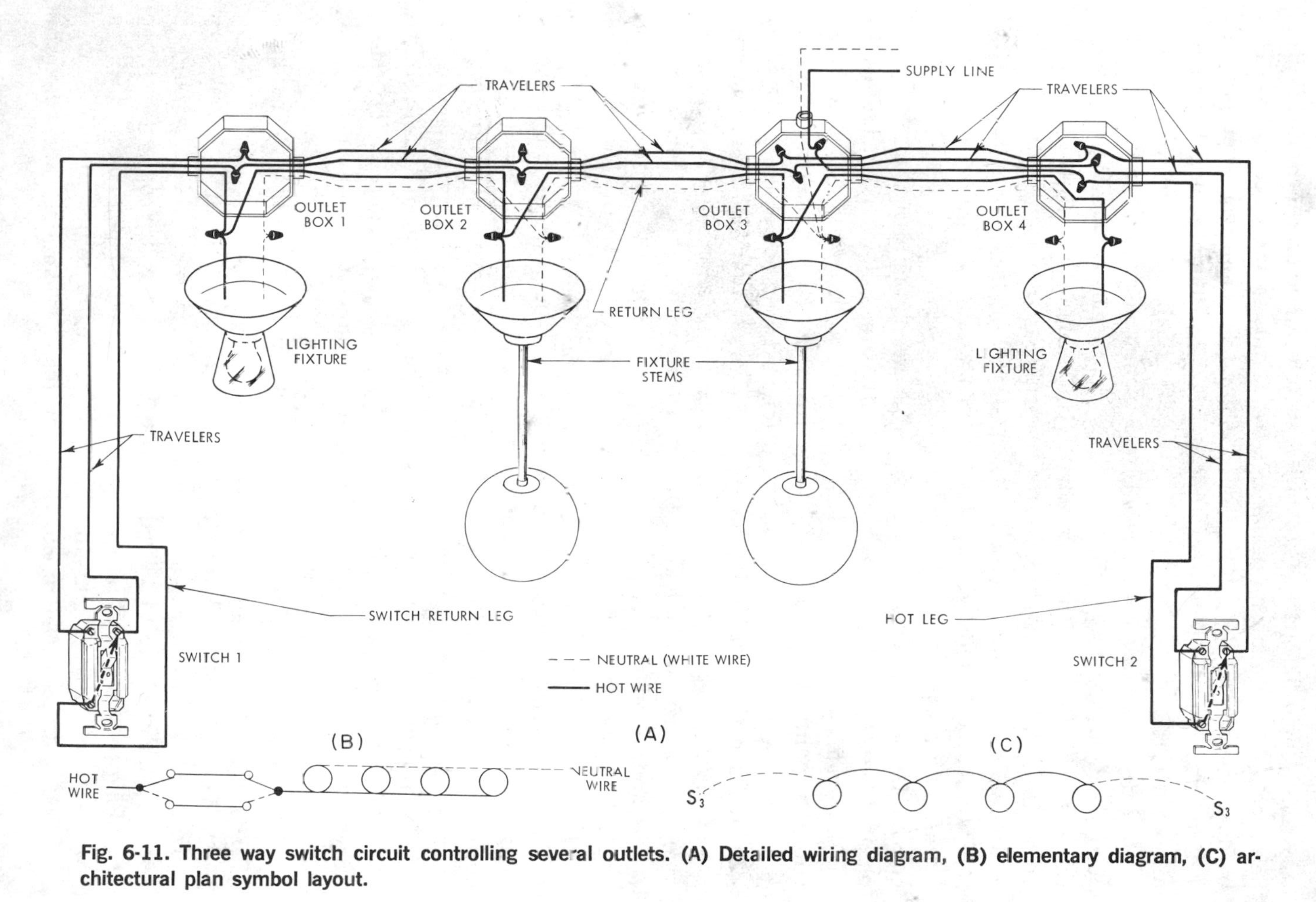

Fig. 6-11. Three way switch circuit controlling several outlets. (A) Detailed wiring diagram, (B) elementary diagram, (C) architectural plan symbol layout.

traveler terminal. The path is broken here, making the circuit incomplete, because the common point of the switch is making contact with the other traveler.

If switch *1* is now manipulated, current may flow from switch *2* over the upper traveler to the common point of switch *1*, through the return-switch leg to the center contacts of the four sockets, and through the lamps to the neutral leg of the supply circuit. Or, leaving switch *1* in its original position (lamps not burning), operate switch *2* to connect its common point with the left-hand traveler. Current then flows over this wire through the four outlet boxes to switch *1* and to the common point, over the return switch leg to the lamps, and through the latter to the neutral supply wire, causing them to burn.

In a metal-clad system, all wires must be run in the same conduit or armor from outlet to outlet. On nonmetallic cable installations, wires can be run in any desired manner. For example, it may be convenient to carry the travelers direct from switch to switch instead of through the four outlet boxes.

Where conduit is employed, two splices can be eliminated in each of the outlet boxes, Nos. 1, 2, and 3, by travelers in these boxes. This is done by pulling the travelers through the boxes without cutting them. In outlet box No. 4, the travelers and the switch can be drawn through in this way, thus saving three splices. This should be done whenever possible, that is, without paying too high a price in labor. For example, the outlets may be far apart or isolated from each other by various partition walls, so that more time would be lost returning from one outlet to the other for pulling back the unused slack than would be required for making splices.

Note in Fig. 6-11 that there are only three hot wires (including the travelers) from box No. 3 to boxes No. 1 and No. 2; whereas, there are four between box No. 3 and box No. 4. This additional leg means not only more wire and larger conduit, but also more labor than for the other two spans. This is the reason for selecting switch 2 as the *hot* line point, rather than switch *1*.

When rough wiring is installed in conduit systems or cable systems, the switches cannot be connected at this time. Switches, with the rest of the "finish" or "trim," are installed after walls are painted. Therefore some common method must be utilized on "rough wiring" to identify individual wires so that the wireman will know which wire to connect to which terminal of a three-way switch. This method is shown in Fig. 6-12. The travelers have a common distinction by being wrapped with a switch leg or return wire. Four way switch travelers are identified by

Fig. 6-12. Method of identifying wiring for 3 way switch connection on rough wiring installations. Wires are readily identified for proper switch terminal wiring after walls are plastered. Four way travelers are usually twisted or wrapped in pairs and connected to either terminal screw in pairs on the switch when ready to install.

pairing the wires going to a specific three way switch. This may be done by twisting the 4-way terminal end pairs of wires together and wrapping wires as shown in Fig. 6-12 for three-way switches.

Four-Way Switches

When it is desirable to control outlets from three locations, a 4-way switch is connected in the circuit between the two 3-way switches. A 4-way switch is similar in size and appearance to a 3-way of the same general type, except that it has four terminals. Many of the 4-way switches have the terminals arranged in the same relative positions as those of a double-pole switch, Fig. 6-7. The latter is always recognized by the words ON and OFF that mark positions of the handle. Three-way and 4-way switches have no handle markings. There is no standard for internal configurations of 4-way switches.

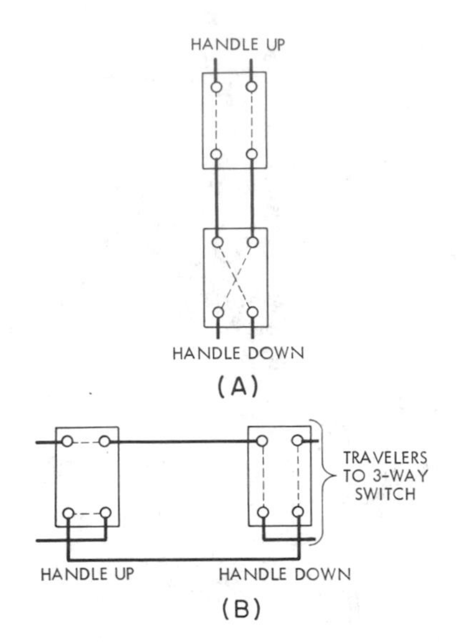

Fig. 6-13. Two types of four way switches. Top: Through wired type. Bottom: Crossed wired type (less common).

There are in general, two classes of 4-way tumbler switches. See Fig. 6-13. One is known as the "through wired type," the other, the "crossed wired type." In the through wired type, travelers from a 3-way, or another 4-way switch, would connect to the top and travelers to the other 3-way to the bottom. When the handle of the switch is actuated, electrical connections inside the switch change from straight across to diagonally across, or vice versa, as shown in the top illustration of Fig. 6-13.

In the crossed type, Fig. 6-13, bottom, one of the travelers coming in from the left goes to a switch terminal on that side. The other one crosses over to the terminal on the right. Also, on the outgoing travelers one connects on the right side, while the other crosses over to the left. Operation of the switch handle changes inside connections from straight across to up-and-down, or vice versa.

The wireman will not know which switch he is installing unless he checks the packing box for a diagram, shown in Fig. 6-14. Some boxes do not have diagrams. If there is no diagram the switch should be tested for continuity in both positions to make sure what has to be done. Most switches are of the through wired type.

Operation of Controls at Three Points. In order to understand operation of controls from three points, it is well to spend some time analyzing the circuit in Fig. 6-14, right, at

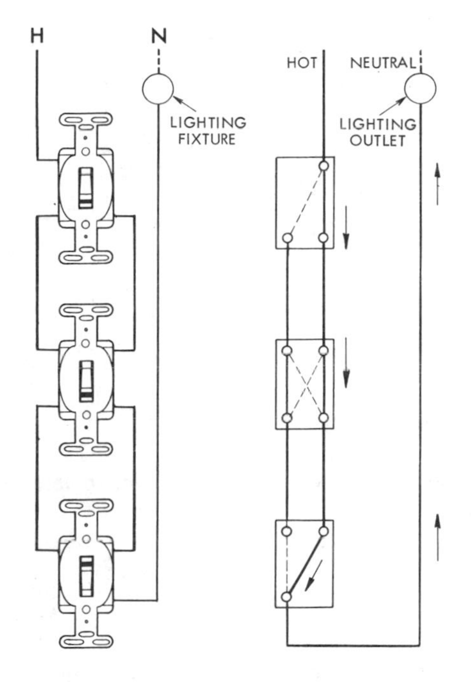

Fig. 6-14. Controlling outlets from three switch locations. The broken lines in the switches on the right indicate alternate switch positions.

the moment current flows through wires and switches as indicated by the arrows, causing the lamp to burn.

Suppose that the upper 3-way switch in Fig. 6-14 is operated. Mark this connection on a piece of paper and trace the circuit through the switches. Note that the lamps go out because there is no connection at the right-hand traveler wire to the lamps. Next, suppose that the 4-way switch is operated making connection in the criss-cross position as shown by the broken line. Tracing the circuit, after operating the two switches, current flows to the common, or hinge point, of the top 3-way switch, diagonally through the switch to its left terminal, and down through the traveler wire to the 4-way switch. The path continues diagonally through the 4-way to the lower right terminal shown by the broken line. Then the current continues through the 3-way making a complete circuit and lighting the lamp.

If the bottom 3-way is operated, current flow ceases. The circuit may be restored, however, by turning either of the remaining switches. A profitable exercise for the learner is to make rough sketches, similar to Fig. 6-14, of resulting current paths

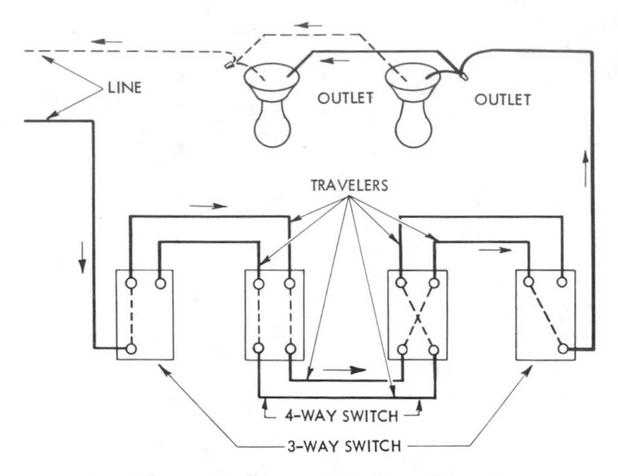

Fig. 6-15. Wiring arrangement for controlling outlets from four places.

when the three switches are operated in the various possible ways and combinations.

A wiring diagram involving four control points is shown in Fig. 6-15. Two 4-way switches and two 3-way switches are employed. Following the arrows, at the instant shown in the diagram, current flows through the first 3-way switch to the upper traveler, to the 4-way switch, then down to the traveler, and along it to the second 4-way switch. Here, it continues to the left contact of the second 3-way switch, then to lamps that are in parallel, and through them to the neutral wire, completing the circuit.

If any one of the four switches is turned, the path is broken. Continuity may be restored by operating any one of the other three. In this instance, too, the learner can profit by making rough sketches of current paths when these four switches are operated in numerous sequences.

To control a circuit or a group of lamps from any greater number of points than shown in the examples, all that is required is to add 4-way switches. Thus, three 4-way switches are required if there are five control points, four if there are six points, and five if there are seven points. There must be a 3-way switch at each end of the control circuit, with the required number of 4-way switches in between. This may also be seen in Fig. 6-13 where more 4-ways may be added in series with 3-ways at each end of the switching circuit.

Other Types of Switches

Dimming Switches

"Electrolier switches" (a type of switch used for multiple switching) were originally used with fixtures that had two or more lamps, or groups of lamps, in the dining rooms or similar locations where different intensities of illumination were wanted for various occasions. The switch mechanism was actuated by repeated manipulation of the button or handle.

These multiple circuit wiring devices gave way to the use of the popular semiconductor or solid state device. The device shown in Fig. 6-16 has three settings: high, low and off. The high setting allows full voltage, full current and full light. The low position internally inserts a semiconductor diode that reduces the voltage and current to one half of full. This position allows approximately 30 percent light and increases normal incandescent light bulb life many thousands of hours. However, special long and slender incandescent lamps with extended filament lengths may have a reduction in operating life due to vibra-

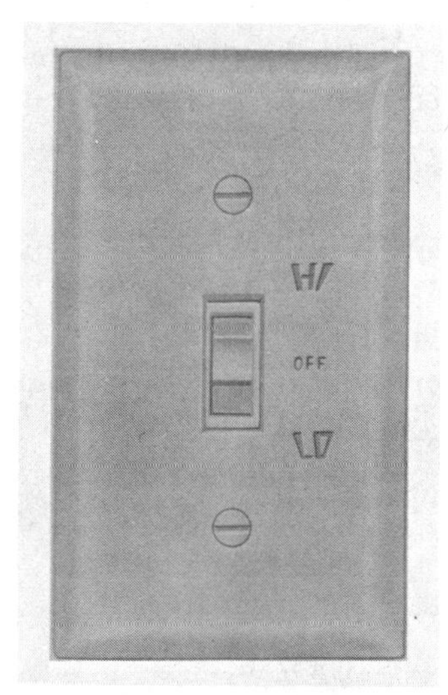

Fig. 6-16. Three position light dimming switch installed the same as a single pole switch. (Slater Electric Inc.)

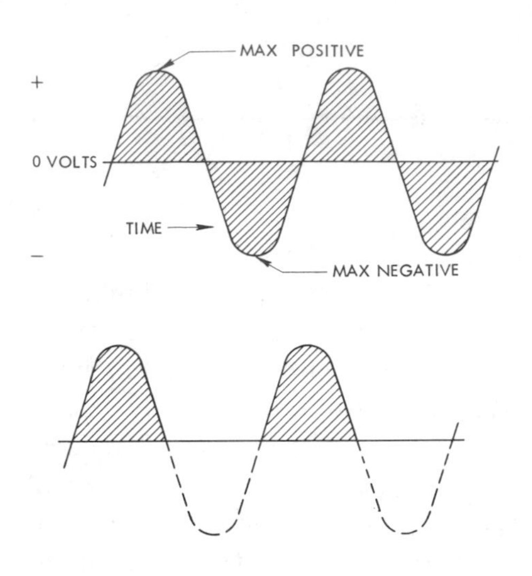

Fig. 6-17. Graphical view of lighting control. Top: Full voltage, full light; Bottom: one-half voltage, dimmed light.

tions of the filaments caused by effects of the current pulses shown in Fig. 6-17.

These dimming switches are installed the same as an ordinary single-pole switch with only two wires to connect.

Rotary Dimmer Switches

With the advent of SCR's (silicon controlled rectifiers) and similar semiconductors, dimmers have become very accessible to the general public. Rotary dimmer switches provide a tap on-off switch and any desired lighting level. Its full range control is continuous from 0 percent, absolute darkness, to 100 percent, full illumination. When rotated in a clockwise direction, light intensity increases. Lights dim when the control knob is rotated in a counterclockwise direction. This switch is mounted in any standard single gang switch box in the same manner as standard single pole switches. Switches can be ganged in boxes for multiple installation. Rotary dimmer switches connect in the circuit and mount as standard single pole switches do. A standard finish switch plate is used to complete the installation. Fig. 6-18 illustrates a rotary dimmer switch without the cover plate.

Rotary dimmer switches are available for three-way lighting control systems also. The dimmer merely replaces one standard 3-way switch with 3-way control where dimming is desired. The standard 3-way switch can turn lights on and off but cannot control the intensity of light.

A tap on and off switch allows presetting of light at any desired level. In 3-way installations, lights can be

Fig. 6-18. Rotary dimmer switch for single gang box and plate. (Ideal Industries, Inc.)

turned off at either of 2 alternate locations. Control must be set above 0 percent light when switched off. When the ordinary 3 way switch is switched on, lights will go on at intensity previously set.

A semiconductor or controlled dimmer works quite simply. First, let us look at the sine wave of the 60 hertz A.C. current which is supplied to 99 percent of today's electrified homes. This is shown in Fig. 6-17A. The current and voltage go from 0 to a maximum positive, then down to 0, and then to a negative maximum. This is one hertz. In one second this occurs 60 times and it is called 60 hertz frequency.

With SCR's or similar semiconductors used in dimmers the amplitude of the voltage or current sine wave is not changed. The circuit is a timing circuit. It prevents the current from flowing in the positive or negative half of the wave until some specific point is reached. This point is determined by the potentiometer setting decided upon by the user. He merely rotates the knob until the desired light level is obtained.

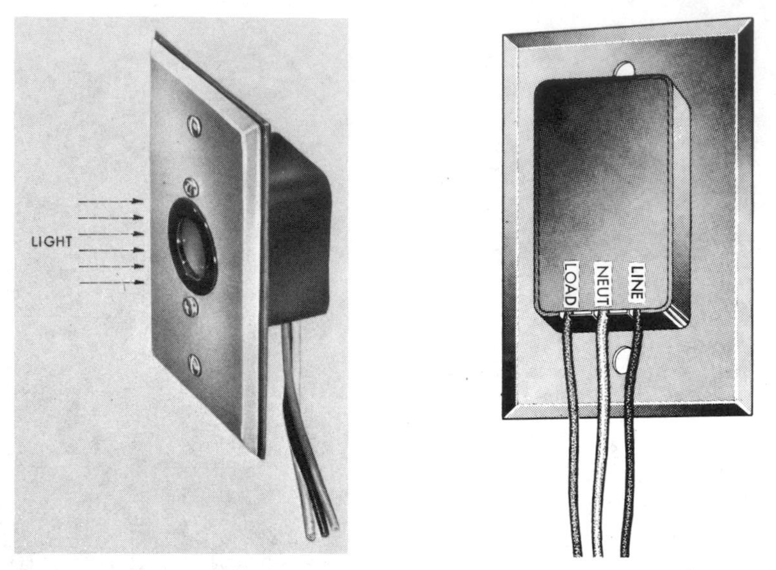

Fig. 6-19. Photoelectric control switch for mounting outdoors in weatherproof switch box with gasket for automatic light or sign switching. Left: Front shows round window containing semiconductor light sensitive material. Right: Back illustrates wiring connections in circuit: line, neutral and load. (Slater Electric Inc.)

Outside Photoelectric Control Switch

Photoelectric switches automatically turn lights on at dusk and off at dawn. See Fig. 6-19. They promote safety, convenience and security in and around all types of buildings. They discourage prowlers from unoccupied buildings or property.

Residential applications include garage or porch night lights, post lanterns, patio and garden lights, swimming pool lights and landscape lighting. They are also used commercially for entrances, sign and display lights, passageways and other automatic lighting switching purposes. Fig. 6-20 illustrates the wiring for a typical photoelectric switch.

The semiconductor internally conducts, or is "fired," to turn the light on when enough light ceases to strike the sensitive material that triggers the switch. The light will remain on as long as not enough sunlight shines into the device. The "electrical bias" of the triggering mechanism is changed when light is applied, thus turning the light off automatically.

Some photoelectric switches have adjustment screws for different light intensities. Some have adjustable light "hoods" to reduce or increase the light striking the switch. Others

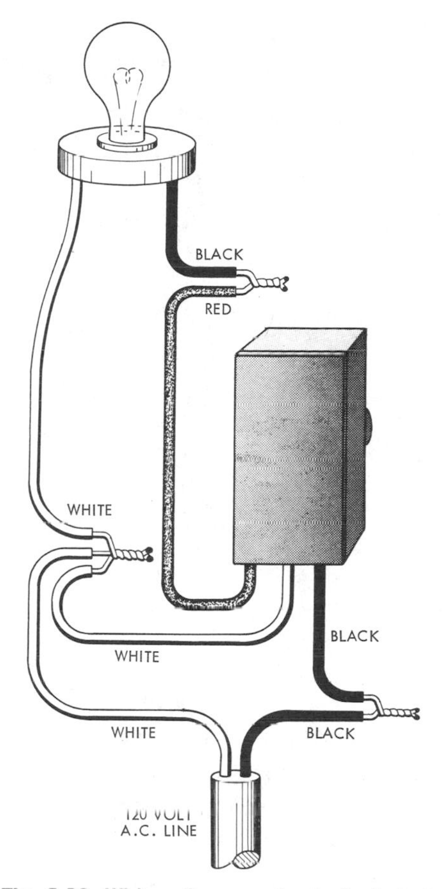

Fig. 6-20. Wiring diagrams for typical photoelectric control switch.

Fig. 6-21. Swivel photoelectric switch. Adjustment allows unit to be aimed upward for "late turn on" and downward for "early turn on." (Slater Electric Inc.)

are positioned to the light or away from it. See Fig. 6-21. The "eye" should be mounted facing a northernly direction if possible, with unobstructed natural light, for best operating results. The "electric eye" unit must be positioned so as to be unaffected by beams from the light controlled, auto headlights, window lights, signs or other light that may affect satisfactory operation of the control unit. Electric clocks are used also for switching lights automatically.

Time Delay Toggle Switch

Time delay switches are used for business, industry and residential purposes. They look like and are installed as ordinary single pole switches, and they turn lights *on* as ordinary switches. However, when the switch is turned to delay, which is the *off* position, a delayed action keeps the light on for a convenient

FLIP SWITCH TO "DELAY" (THIS IS OFF POSITION). DELAYED ACTION KEEPS LIGHT ON TEMPORARILY. LIGHT GOES OUT AUTOMATICALLY AFTER SAFE INTERVAL OF TIME.

Fig. 6-22. Operation of a time delay toggle switch.

amount of time. Normally the time is non-adjustable, however. More complicated controls are available with adjustable delay.

A typical time delay switch is controlled by a sealed aluminum chamber. A vacuum is created when the switch is flipped to "on" position. When toggle is moved to delay, the vacuum is dissipated at a safely timed interval and the light automatically goes off. Fig. 6-22 illustrates a typical use of the time delay switch.

Pilot Light Switches

These switches are used for controlling lamps in pantries, laundries, furnace rooms, or other infrequently entered locations where light or load is out of sight. They are sometimes employed on electric irons, hot plates, and such devices which may be damaged if connected accidentally for too long a period. The pilot light burns as long as the switch is closed (on). When the pilot light is on we know that the light or load is also on. We need the pilot light because the load or light is out of sight. Fig. 6-23 shows how one of these switches controls a lighting outlet. The connections to a convenience outlet are exactly the same. To help in making proper connections some pilot light switches have the neutral terminal identified by a white or silver-colored screw, and the hot one by a colored screw or brass. A pilot wiring circuit is shown in Fig. 6-23.

Some switches have small incandescent lamps in their handles for

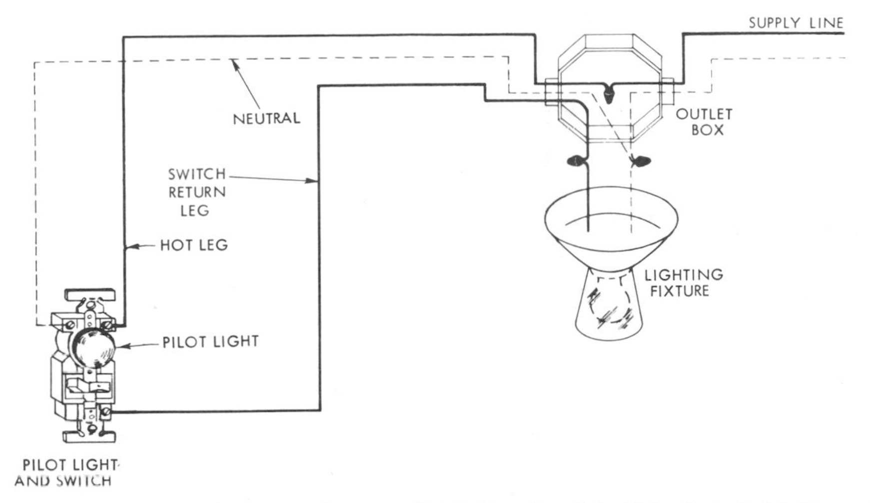

Fig. 6-23. Wiring circuit of an incandescent pilot light and switch. Note that pilot light requires neutral wire.

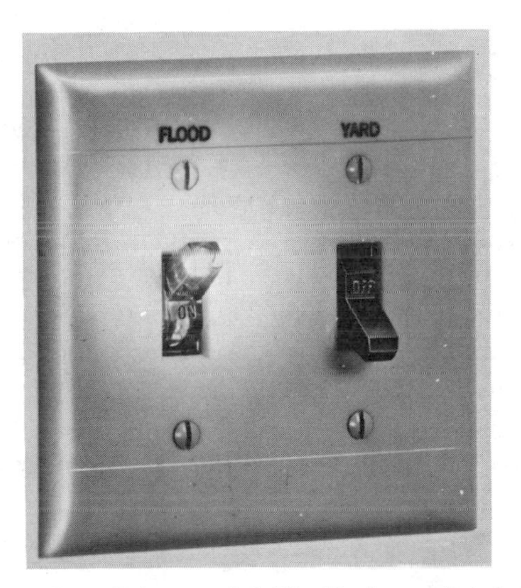

Fig. 6-24. Pilot lighted handle switches are brightly lit when switch is "on" indicating that a remote light or load is operating.

pilot lights. Other switches have neon lights in their handles. See Fig. 6-24.

Switches are also available with a long life neon lamp inside the translucent toggle handle which emits a

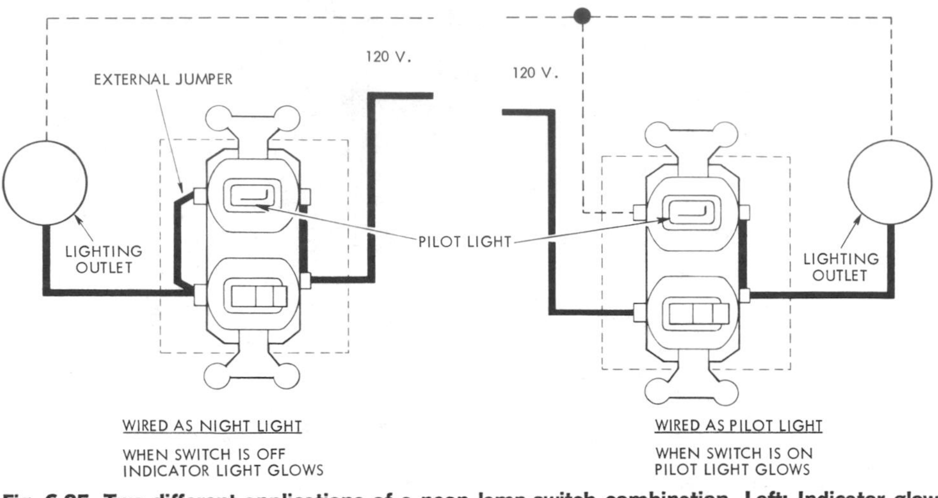

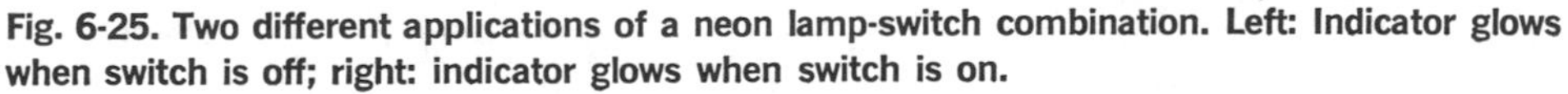
Fig. 6-25. Two different applications of a neon lamp-switch combination. Left: Indicator glows when switch is off; right: indicator glows when switch is on.

soft glow when the switch is in the *off* position. See Fig. 6-25. This reverses the normal procedure. When the switch is *on*, the toggle is not lighted. They generally give just enough light to locate switches easily in the dark, but not enough to be objectionable. Otherwise, they are constructed and connected the same as single pole toggle switches. Fig. 6-25 illustrates two different uses of a neon lamp switch.

Duplex Switches

These are contained in a single device and are similar to duplex receptacles. See Fig. 6-26. Some duplex switches require the same type finish plate as a duplex receptacle. They are also available with or without a terminal jumper break-off tab. Therefore, this device can be installed with a common feed for both switches or a separate feed for two independent circuits, with the break-off tab removed. Fig. 6-27 illustrates methods of connecting a duplex switch.

Combination Wiring Devices

A switch that includes a switch and a receptacle is also used. See Fig. 6-28. They can be used in any of three alternate methods: (1) the switch can be wired to control the outlet to be mounted in the same box, (2) the switch can control a light with the receptacle hot, or (3) the switch and outlet can be wired for independent circuits. Fig.

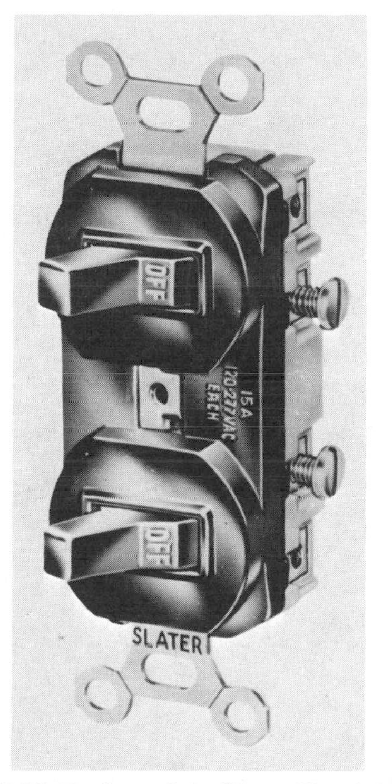

Fig. 6-26. Duplex switch. (Slater Electric Inc.)

6-29 shows ways for connecting single-pole switch combinations. Three-way combinations are also used.

Split-Wired Duplex Receptacle Switches

Local control or switching is not generally required for receptacle outlets, although such control is sometimes desirable and recommended. A commonly used technique for plug outlet control is to *split-wire* a duplex receptacle so the top outlet is controlled by a wall *switch* and the bottom outlet is always hot, as shown in Fig. 6-30, top.

Convenience receptacle outlets are available with an "easy break-off" feature of a tie bar fin between terminal screws permitting use of duplex receptacles for either single

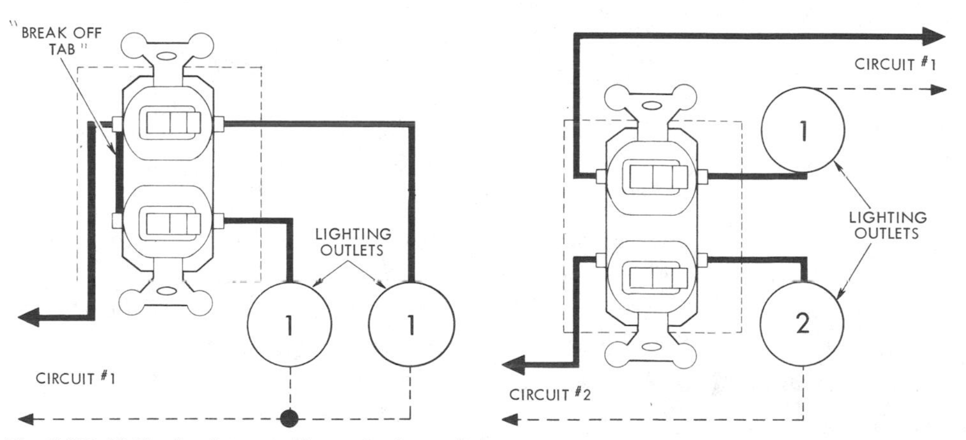

Fig. 6-27. Methods of connecting a duplex switch.

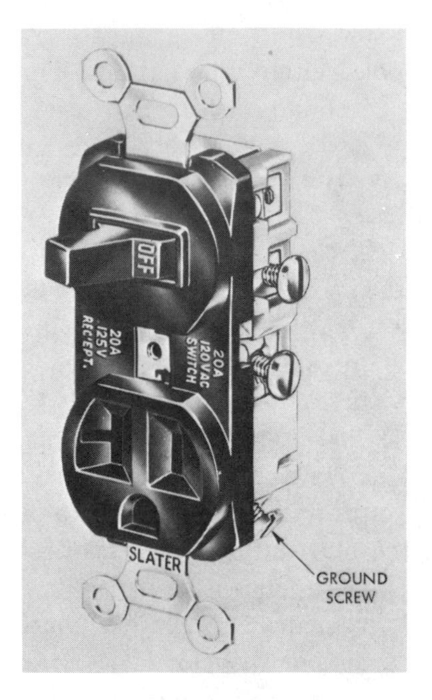

Fig. 6-28. Combination switch and receptacle. (Slater Electric Inc.)

or two circuit applications. The fin is simply raised or lowered a few times with screw driver or pliers to break it off. For separate circuits with one common neutral, one fin is broken off. For two separate circuits, both fins are broken off. This is shown in Fig. 6-30, bottom.

Closet or Pantry Light Switching

Fig. 6-31 illustrates a common method of mounting a switch outside the closet, or small room to be lit, with the lighting fixture inside over the door, or on the ceiling if permissible. Fig. 6-31 also shows the connecting method of a convenient automatic door switch. The closet light goes on when door is opened. When a door switch is used, the wall switch is omitted. For small closets, lighting fixtures can be mounted just outside the closet on the ceiling to provide additional general lighting as well as closet lighting.

Key Operated Switches

Many switches require rigid control of electrical circuits, especially when children are present. Key operated switches are generally used for this purpose. A switch and key is shown in Fig. 6-32. The only difference in this switch from others is the manner of operation.

A.C. Quiet Switch

With the turn of the century

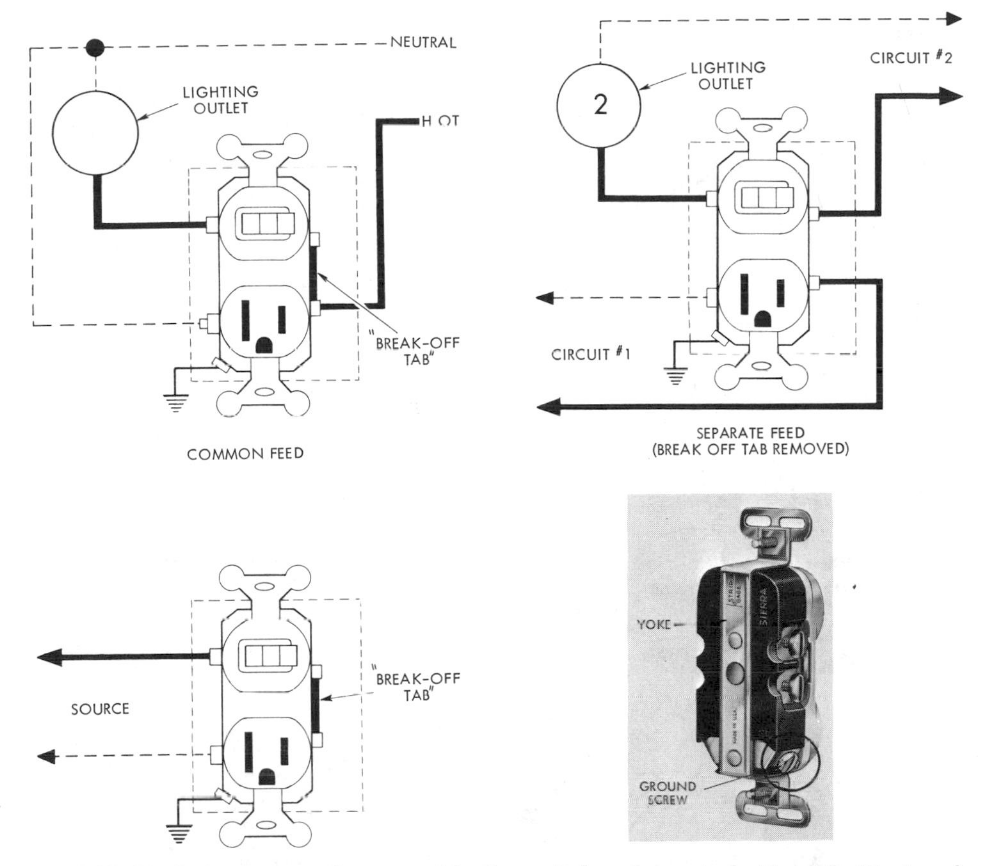

Fig. 6-29. Methods of connecting a combination switch and receptacle. Top left: Single pole switch controls light only. Power outlet for grounded appliances. Top right: Single pole switch and grounded power outlet on separate circuits. Bottom left: Grounded power outlet controlled by single pole switch. Bottom right: Duplex receptacle showing grounding screw (identified by hexagonal shape and green color), and yoke sometimes used for grounding receptacle on surface installations. (Sierra Electric Corp.)

nearly all lighting installations were 110 volts direct current. D.C. produces a "hot arc" when a circuit is interrupted. One of the first switches to accommodate this type of service had to "break" fast and also had to have a large "gap." This strong, fast action resulted in a "loud snap."

Over the years size and strength of the switch components used were reduced. This caused the device to come close to the lower limit of acceptability. Some of these switches barely passed Underwriters' Laboratory Tests.

A "T" rated switch, referred to in

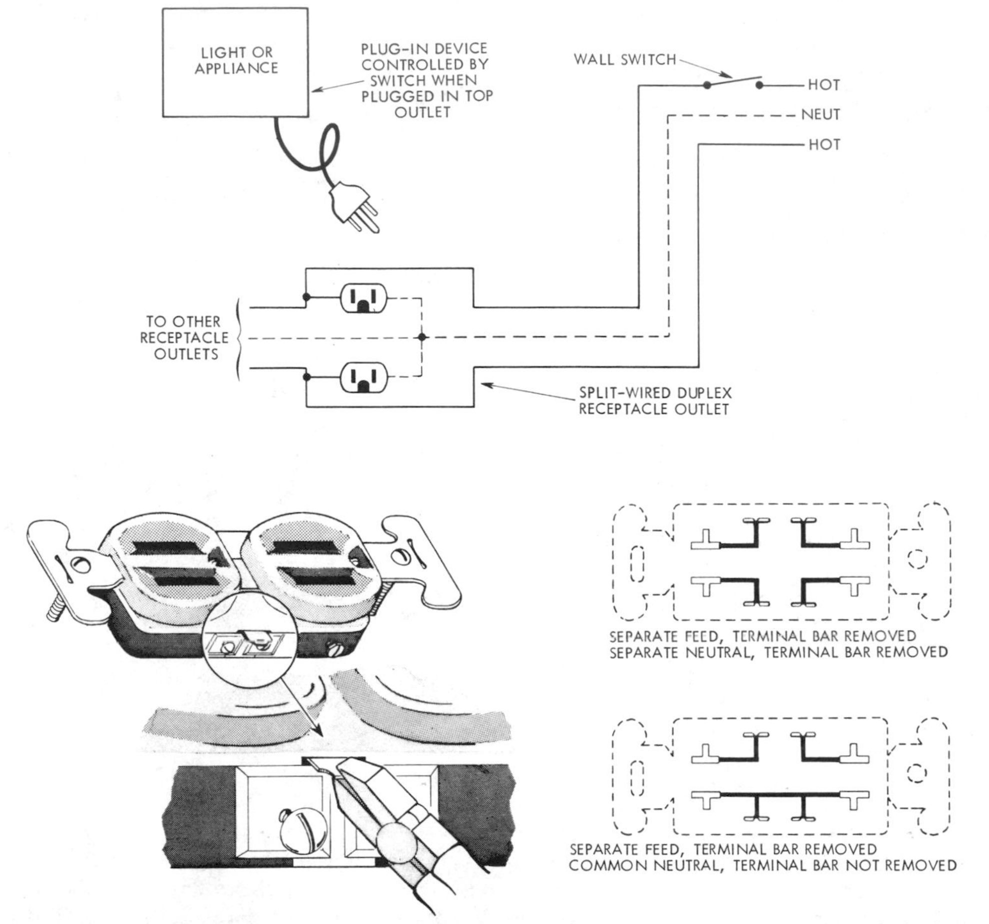

Fig. 6-30. Top: Split-wiring control of receptacles. Bottom left: Removing terminal tie bar either by prying out with a screwdriver or pulling out with pliers. Bottom right: Different circuits obtained by removing terminal tie bars.

the NEC and some job specifications, is essentially the same as a non "T" rated or ordinary A.C. switch, with the exception that all components have to have a greater degeree of precision and strength in order for them to pass a more severe test. The "T" stands for Tungsten, meaning the tungsten filament found in incandescent lamps for which use this switch is designed to control. When non "T" rated switches were introduced, lighting was accomplished with carbon lamps. The major difference between a carbon and Tungsten lamp, in reference to the effect on the switch, is that when the contact on a carbon lamp is closed, the

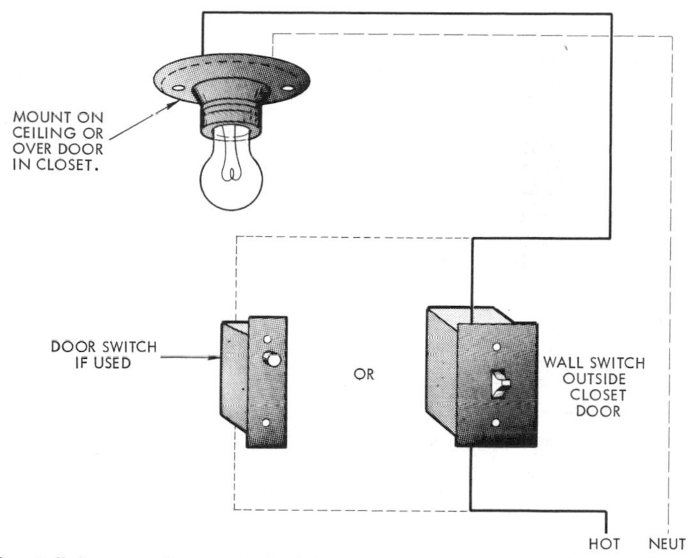

Fig. 6-31. Closet lights can be controlled by standard single pole wall switches or by low voltage automatic door switches.

Fig. 6-32. Key operated locking type switch with key; used where rigid control of the electrical circuit is required. Keys are normally supplied with each switch. (Sierra Electric Corp.)

Fig. 6-33. A.C. quiet touch (push button) switch. A.C. quiet switches are controlled with toggle handle action also. Both push button and toggle handle switches are finished with the same normal switch plate. (Slater Electric Inc.)

current does *not materially* change except to increase slightly *after* the lamp has reached full brilliance. (Carbon lamps are no longer normally used for illumination). With a tungsten incandescent lamp, a very severe inrush of current takes place when the contact is closed. In fact, the initial inrush of current is almost 10 times higher than when the lamp is at full brilliance.

With the use of A.C. current growing universally, a new type of switch, the A.C. quiet switch, was proposed and finally accepted by the Underwriters' Laboratories, specifically designed for A.C. use only. This switch did not require a fast "make and break," therefore eliminating the loud snap action—resulting in the origination of the name A.C. Quiet switches, instead of just A.C. switches. These quiet switches are practically in full use today. Fig. 6-33 shows an A.C. quiet switch.

Wiring Enclosure Finish Plates

Switch or wall plates are installed by the wireman after switches are connected and installed. Wall plates are made of plastic and metal materials with a wide range of colors and finishes. They are packed with

Fig. 6-34. Large plates have an extra 3/4" in height and width. This king-size plate provides ample area for covering most damaged wall surfaces, such as chipped tile or plaster, and over-sized holes in wood paneling or concrete block. (Sierra Electric Corp.)

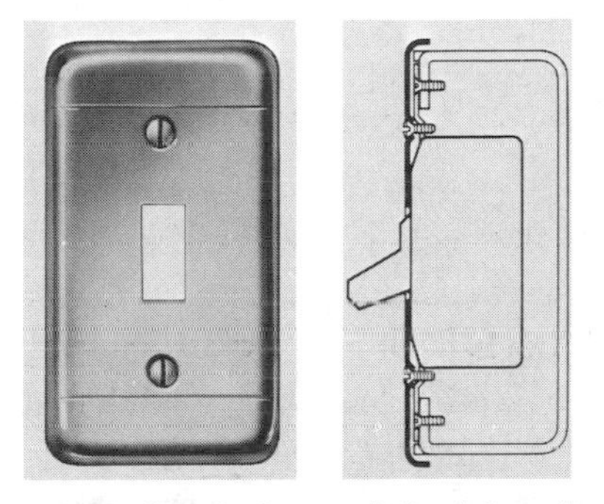

Fig. 6-35. Handy box switch plates give a finished appearance and style to surface mounted boxes. These plates are made for standard devices. Plate conceals edge of box. They are made in a complete range of most-used wall plate opening sizes. (Sierra Electric Corp.)

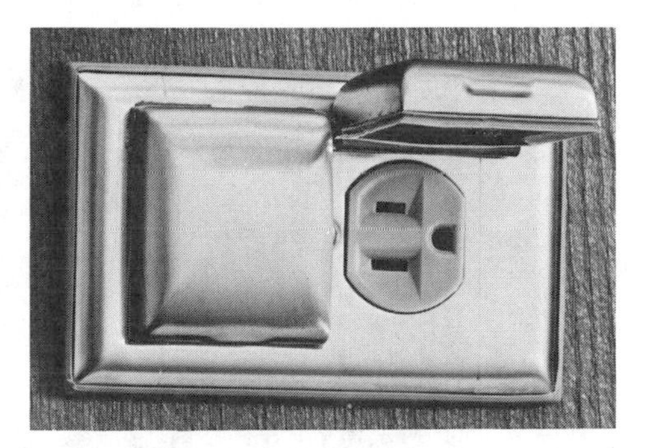

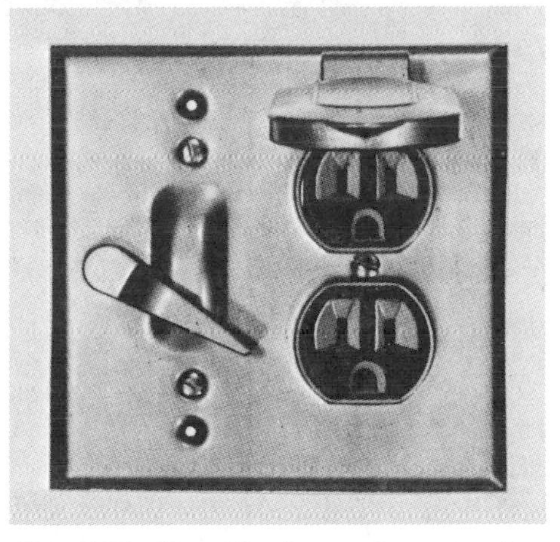

Fig. 6-36. Top: Weatherproof cover plate mounted on duplex flush receptacle. Bottom: Combination switch and receptacle weatherproof plate for mounting with gasket on 4", two gang box. (Slater Electric Inc.)

mounting screws to match. Larger than normal plates may be obtained. Fig. 6-34 shows the advantage in using the larger size plate. These plates are commonly called "jumbo" plates in the trade. They are convenient for covering damaged wall areas around switch boxes and for protecting more of the wall surfaces around switches from soiling.

Surface mounted switch boxes (boxes mounted on the wall surface) require smaller plates as shown in Fig. 6-35. They cover the switch box only as shown in side views.

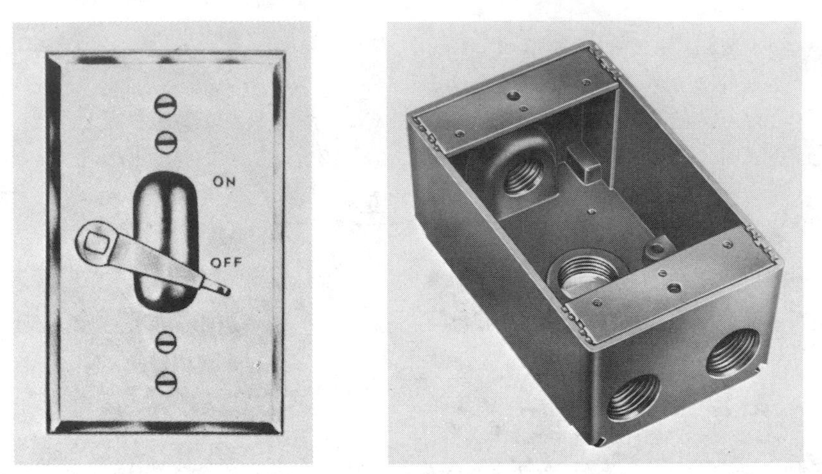

Fig. 6-37. Left: Weatherproof single gang switch plate for mounting on weatherproof box (right) with gasket. Note threaded hole openings on box for use with rigid conduit or electrical metallic tubing, weatherproof fittings. (Slater Electric Inc.)

Weatherproof flush plates are shown mounted in Fig. 6-36. A gasket is mounted under the plates and the hinged covers contain gaskets for protecting receptacles also.

Surface mounted switch boxes require the same switchplate as the flush mounted. However, a special box is required as shown in Fig. 6-37, right. These boxes contain threaded openings for conduit or electrical metallic tubing, weatherproof fittings. Plate gaskets are also required for this type of box. Boxes and plates are available for multi-switch ganging. The switch operating levers on the plates must be properly engaged with the switch handle when mounting plates.

Miscellaneous Switching Circuits

Electrical control or switching of other lights, convenience outlets, appliances or household equipment is the responsibility of the wireman. The following paragraphs give a few final instructions for the installation and promotion of electrical conveniences. (Chapter 8, "Remote

Control Wiring," is another major common system of controlling outlets, and additional information may be found there.)

Bell or Buzzer Wiring

Fig. 6-38 shows a simple bell circuit which consists of a power source, generally a low voltage transformer, a pushbutton and a bell wired together.

The push button is in series with the bell and the power source represented with the unbroken line.

When the normally open contacts of the button are pressed together, current flows from the power source through the closed contacts of the button, through the connecting wire to the bell and returns to the power source completing the circuit and ringing the bell.

Any number of push buttons may be connected in parallel as shown by the dashed lines for multiple control of the bell.

In the return-call system, Fig. 6-39, there is a bell and a push button at point *A*, and a like combination at point *B*. This method of connection is used for signalling between points which are remote from one another. If the button at *A* is pressed, the bell at *B* rings but not the bell at *A*. Likewise, when the button at *B* is pressed, only the bell at *A* rings.

The paths may be traced readily. Starting at *B*, when the button is pressed, current from the right hand terminal of the power source flows through the wire to the button, through the closed contacts and the connecting wire to bell *A*, then through the bell and wire to the left hand terminal of the power source. The bell at *B* does not ring because its circuit is open at the push button contacts of *A*. In like manner, when the button at *A* is pressed, current flows out of the right hand terminal of the power source through bell *B* and connecting wire to push button *A*, the contacts of which are closed. The current then returns through the wire to the left hand terminal of the power source.

The circuit of Fig. 6-40 makes use of double-contact push buttons in a return-call arrangement. These push buttons are constructed so that the lever is normally held against an upper contact by means of a flat spring. When the button is pressed, the lever opens the upper contact,

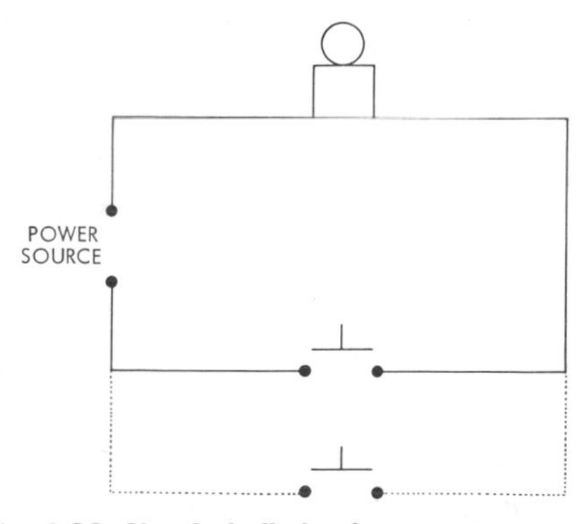

Fig. 6-38. Simple bell circuit.

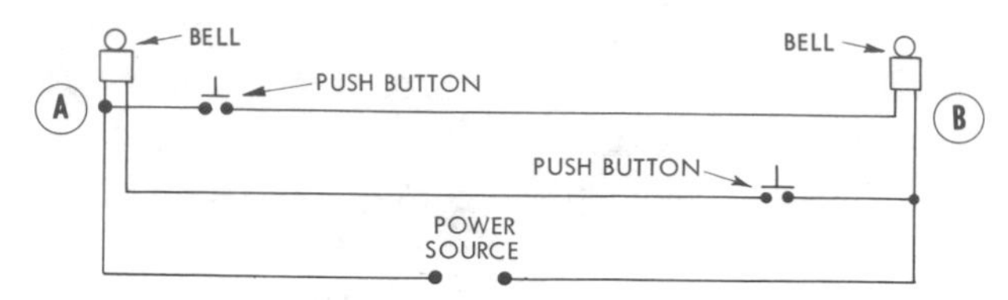

Fig. 6-39. Return-call bell system.

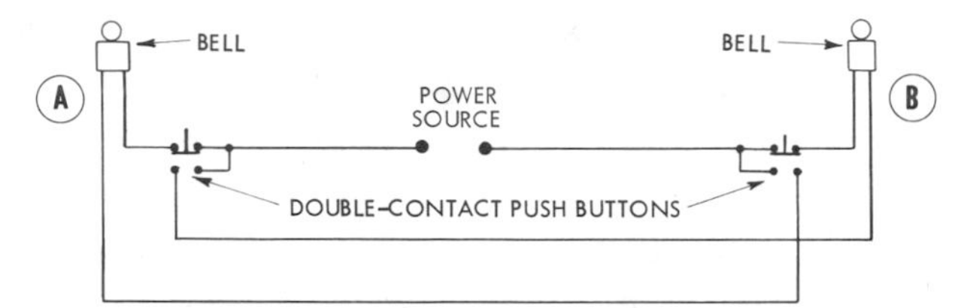

Fig. 6-40. Return-call bell system using double contact push buttons.

and makes the lower one. When button *B* is pressed, current flows out of the right hand terminal of the power source through the connecting wire and lever to the lower contact. After passing through the wire to bell *A*, it returns through the upper contact of button *A*, the lever which is still pressed against the contact, and the circuit wire, to the left hand terminal of the power supply. Bell *B* cannot ring because its circuit is broken when push button *B* is pressed.

When push button *A* is actuated, current flows out of the right hand terminal of the power source through the contact lever of button *B*, upper contact of bell *B*, connecting wire, and the lower contact of button *A*, to the lever which is being held in contact with it, and from there to the left hand terminal of the power supply.

Note only three wires are required between call stations. Bell return call systems have largely become displaced in the home due to introduction of music—intercommunication combination systems.

Chime Circuits

Old style bells have been displaced in the modern home by door chimes. In the more common type, two distinct tones are provided so that the householder may know whether the caller is at the front door or at the back. The front button usually brings forth a succession of tones, or a chord, while the back door button gives rise to a single tone.

The diagram, Fig. 6-41, shows two

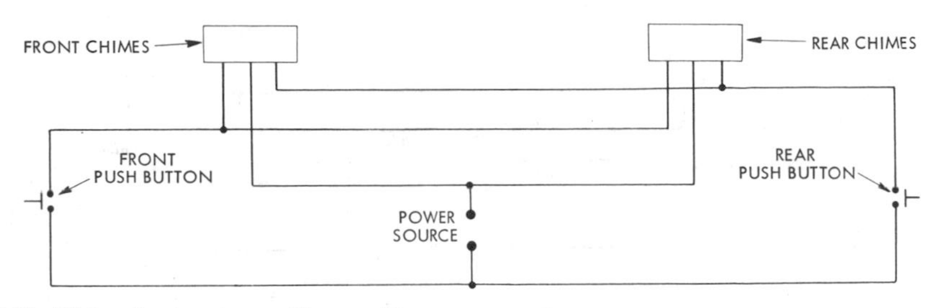

Fig. 6-41. Wiring for two-tone chimes with power supply.

sets of chimes, one at the front and one at the back. Although this is a common arrangement, a single chime, centrally located so that it may be heard all through the house, is sometimes employed. The arrangement is the same in either case, the two units being connected in parallel as shown. In fact, three or four chimes may be so connected, when the power supply is large enough to handle them.

The chime unit has three lead wires, the center one of which connects to both of its magnetic coils or solenoids. When the front button is pressed, current flows out of the lower terminal of the battery through the button, the front door solenoid of the chime, and the connecting wire, to the other terminal of the power supply. Since two units are connected in parallel, the front door notes will be sounded by each of them. If the rear button is pressed, current from the lower terminal of the power source will flow through the back button, the rear-door solenoid in both chimes, and the common wire, back to the other terminal of the power source.

Fig. 6-42 illustrates a circuit which employs a transformer power supply. Connection of the chimes is exactly the same as in Fig. 6-41, but there is an additional item, the house number light, which is frequently used.

The house number light is connected permanently across the terminals of the transformer. There is no switch in this circuit, so that the light burns continuously. The globe consumes very little power, and its output is so feeble that it is unnoticed in daylight hours.

Combination Music / Intercom Systems

Combination units such as that in Fig. 6-43 have become popular. The system consists of a master station and a number of remote stations in various rooms. Stations are often provided in vestibules or entries so that the householder may talk to a caller without opening the door. The

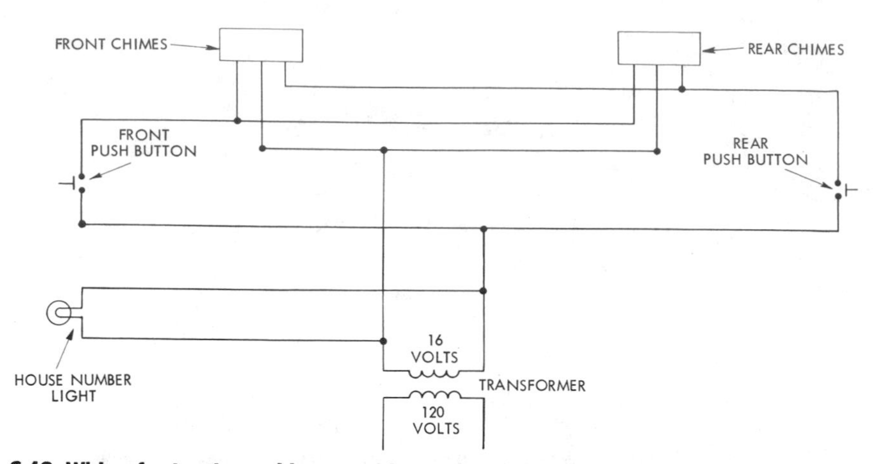

Fig. 6-42. Wiring for two-tone chimes and house number light with transformer power supply.

Fig. 6-43. Modern combination music and intercom unit. (NuTone, Inc.)

system illustrated here combines choice of radio or photograph music with a selective-talking intercom system. Connection of these devices by the wireman is relatively simple because the manufacturer furnishes color-coded wire, numbered terminal blocks, and detailed instructions. The principles involved are nearly the same for other manufactured combination music-intercom systems.

Home Protection Devices

There are several varieties of electronic devices on the market for home protection. They are primarily designed to protect the home when the occupants are absent. Most of these are sensing devices—they sense through sound and motion and announce the presence of an intruder. Some electronic systems signal an intruder's entry by alarms, by switching on lights, or by phoning the police. Microphones can be installed in the house which will be activated by the detection of sound. Electric eyes can be installed which will detect movement. Windows can also be rigged with magnetic contact which will detect movement and sound an alarm.

In addition to detection, devices are available which switch lights or TVs on and off at odd intervals.

Wiring information on these devices and on devices even more sophisticated, may be obtained from equipment manufacturers. While these devices may be complicated in themselves, the wiring installment of most of this equipment into the residence is relatively simple.

QUESTIONS FOR DISCUSSION OR SELF STUDY

1. What is "polarity of wiring" and how is it shown?
2. What is meant by the term "switch leg"?
3. What is a switch loop?
4. How is the neutral connected?
5. How many wires pass from an outlet box to a two gang switch box? Can you identify each?
6. How can three-way switches be used to control plug receptacles?
7. What is the normal color of a neutral wire and neutral device terminals?
8. What are "travelers"?
9. How can one tell the difference between a double pole switch and a four-way switch?
10. How could a double pole switch be used to control two circuits?
11. How could a four-way switch be used as a three way switch?
12. What switches are used to control a light from two points?
13. How must the ground wire be connected to a plug receptacle?
14. How many wires pass in a conduit from a three-way switch to a four-way switch? Name them.
15. How many three-ways and how many four-ways are required to control a light from four places?
16. Where are pilot-light switches used?
17. How can ordinary push-buttons be used in a return-call bell system?

18. How many wires are required between stations of a return call system?
19. How are "travelers" identified in switch boxes in rough wiring so they may be readily connected on finish?
20. What is the effect of induction on armored cable when only one conductor is used to carry current?
21. What does two-gang mean?
22. What is a polarized receptacle outlet?
23. How are modern residential light dimming circuits controlled?
24. Besides a pilot light, how else are neon lit toggle handles used?
25. What is the difference between a "flush" and "surface" switch plate?
26. What is a "T" rated switch?

How to Make Electrical Connections

Chapter **7**

The importance of making good splices and taps (wire connections) in the wiring systems cannot be overemphasized. No connection will be as good mechanically as an unspliced wire, but approved methods of splicing, taping, soldering, or installing solderless connectors will produce a satisfactory joint. On the other hand, connections made in outlet boxes and at switches or receptacles by an inexperienced workman are often the weakest parts of an interior wiring system.

All splices are required to be mechanically and electrically secure. The tape for a splice should be sufficient to provide insulation equivalent to that which was removed.

Preparing the Wire

After the wiring is pulled into place on a construction job, all wire to wire connections or splices are made. The first step in making any kind of wire connection is to remove the insulation, a process commonly known as *skinning*. For this purpose, use a fairly sharp knife. Do not circle the wire with the blade at right angles, because in most cases this produces a groove in the conductor

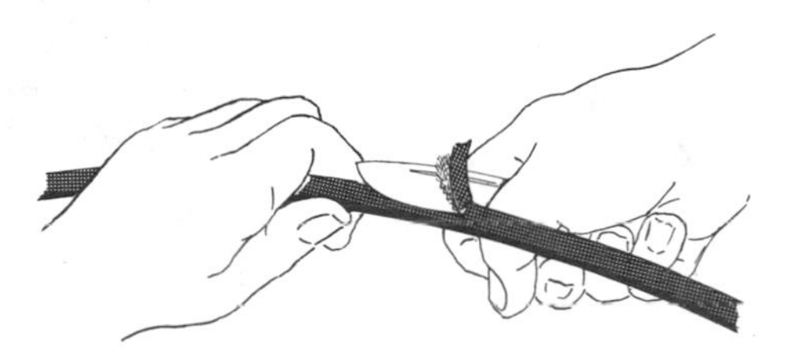

Fig. 7-1. Removing insulation. Cut the insulation at an angle of about 30 degrees. (Do *not* circle the wire with the blade at right angles to it. Usually, this produces a groove or nick in the wire which will cause it to break easily when bent at this point. Also a nick in the wire may reduce its capacity to carry current.)

which may cause the latter to break when bent at this point. It may also reduce its capacity to carry current. Instead, whittle the insulation away in a manner similar to that of sharpening a pencil, at an angle of about 30 degrees as in Fig. 7-1.

High-grade rubber and plastic insulations are tough and do not cut easily. Draw the knife across the material instead of straight along it. The higher the quality of insulation, the sharper the knife must be.

Insulation may also be removed from the ends of small wires by means of a tool called a *wire stripper*. Fig. 7-2 illustrates a wire stripper and shows how it is used.

A side cutting pliers such as the one shown in Fig. 7-3 is generally used for twisting the wires, cutting, and for making other connections, such as crimping small bullet shaped solderless connectors (see enlargement of pliers, Fig. 7-3). The side cutting pliers is one of the most ver-

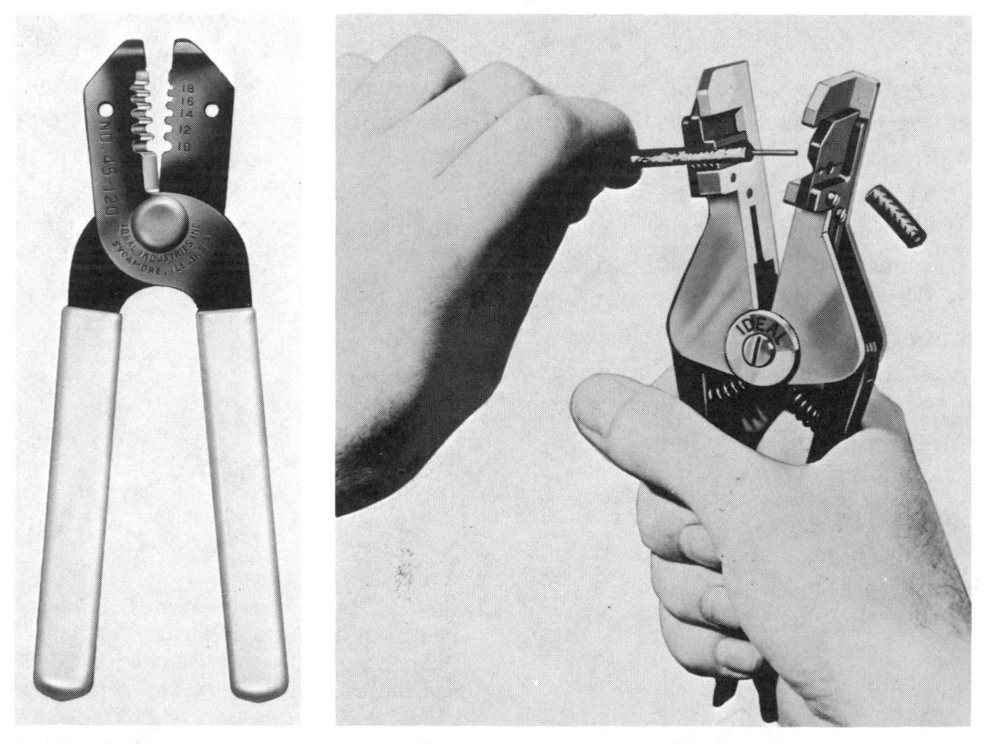

Fig. 7-2. Left: A handy, flat wire stripper carried by many wiremen. It strips sizes No. 18 to No. 10 easily. Right: Wire stripper in use. It is used more for bench work or for mass production jobs. (Ideal Industries, Inc.)

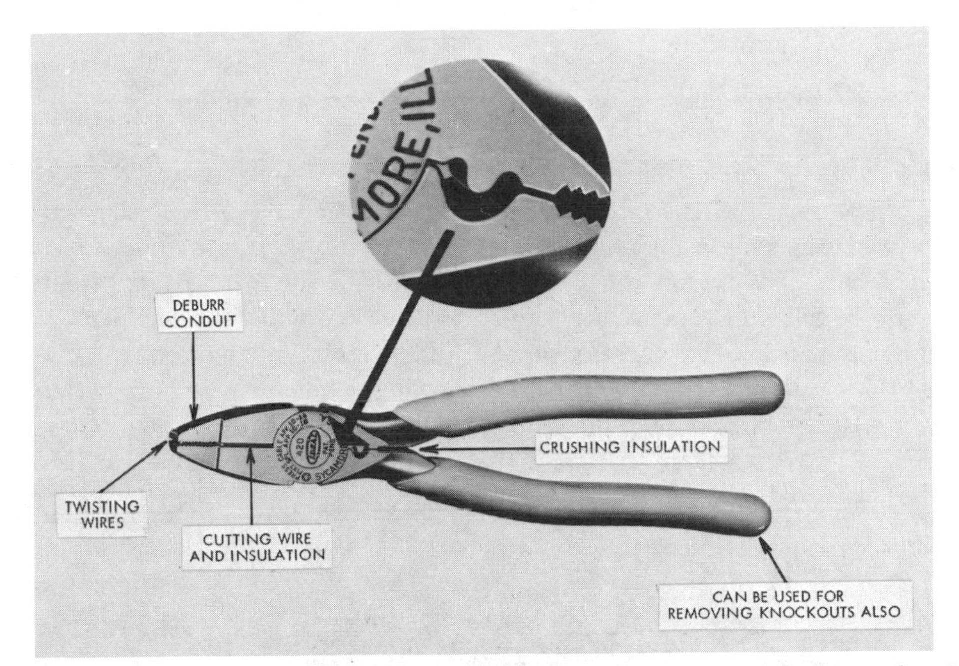

Fig. 7-3. Side cutting pliers: one of the most versatile and useful tools of a wireman. Insert shows enlargement of crimping die built in for use with patented wire connectors. This tool can also be used to crush wire insulation, for stripping, for twisting wires, cutting wire and insulation, removing knock-outs, deburring conduit, and other uses. (Ideal Industries, Inc.)

satile and useful tools of a wireman. It can be used to crush wire insulation for stripping, to twist and cut wire, to remove knock-outs from boxes, to deburr conduits, etc.

Removing Weatherproof Insulation. Insulation on weatherproof wire, which is used for outdoor open work, is tough and hard to skin, especially after becoming thoroughly weathered. Electricians lay the end of the wire on the pavement or hard surface and pound it with a hammer; this cracks the braid and loosens it from the conductor. There is nothing wrong with this method when care is exercised so that the face of the hammer strikes the wire squarely and yet not hard enough to deform the conductor. The asphalt, with which the braid is impregnated, does not readily come away from the copper even after the braid has been removed; however, a rag and some gasoline (or non-combustible, non-toxic cleaning fluid) will help. White gasoline, which does not contain lead, is preferable over premium fuels.

Soldered Splices, Copper Wires

Soldering can be done with the aid of a soldering iron, a soldering gun, or a torch. Fig. 7-4. (Electric soldering irons and guns are slower and more cumbersome on construction jobs.) A most common method is to use a methane, prestolite or propane torch. The joint is first thoroughly cleaned, covered with a thin layer of soldering paste, heated in the blue flame of a torch, and then treated with a piece of wire solder. The solder melts upon contact, metal flowing readily along and between the conductors.

The copper joint must be clean

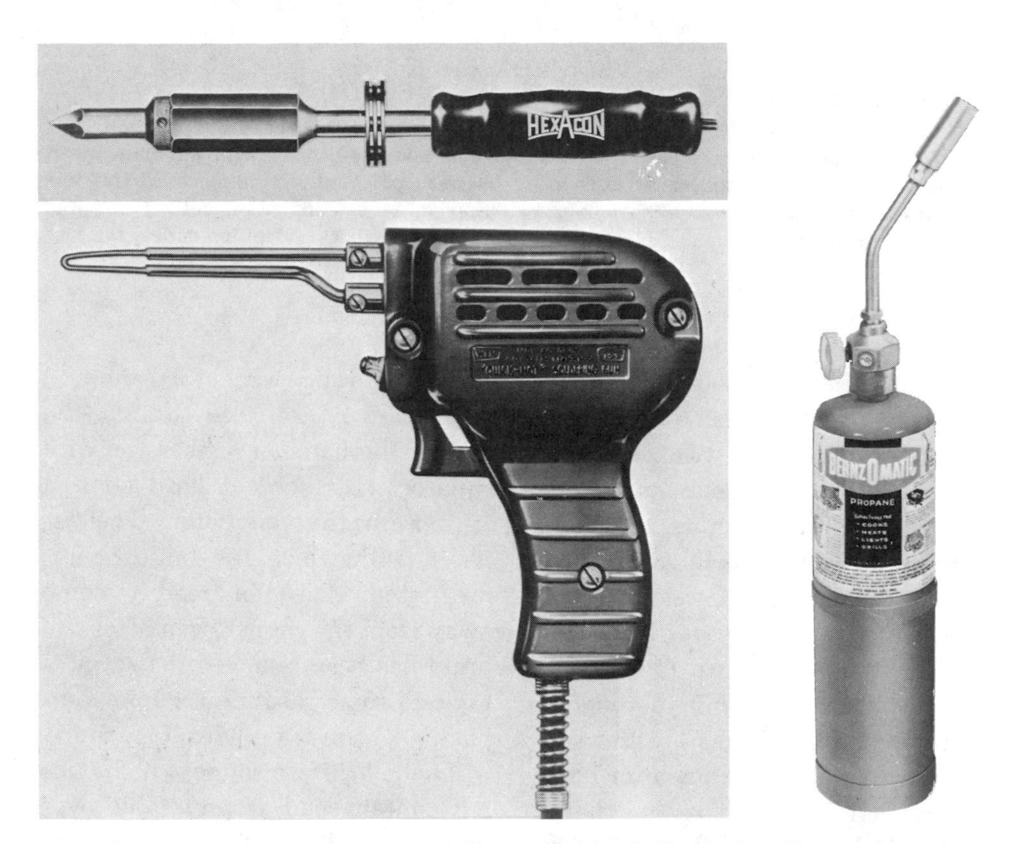

Fig. 7-4. Soldering iron, soldering gun and torch. (Hexacon Electric Co.; Wen Products, Inc.; Otto Bernz Co., Inc.)

before solder will flow evenly and adhere. The soldering paste is an aid to flowing, molten solder. If there is a secret to a good solder joint it is a clean joint. Also: be careful to heat the *wire* and not the solder. The heat of the copper wire itself should melt the solder. Be sure the iron is hot and clean so the wire will heat quickly. (If the iron is too cold the copper wire will heat slowly and conduct the heat under the insulation creating the possibility of insulation damage.)

Paste and Solder for Copper Wire. Soldering paste should be non-corrosive. Ordinary commercial pastes, Fig. 7-5, top, satisfy this requirement. Solder commonly used by electricians is 50/50 grade—50 percent tin and 50 percent lead. Various combinations of these metals are obtainable, such as a 40/60 mixture, but the 50/50 alloy, Fig. 7-5, bottom, flows more readily, and does not call for so much heat as do some other grades.

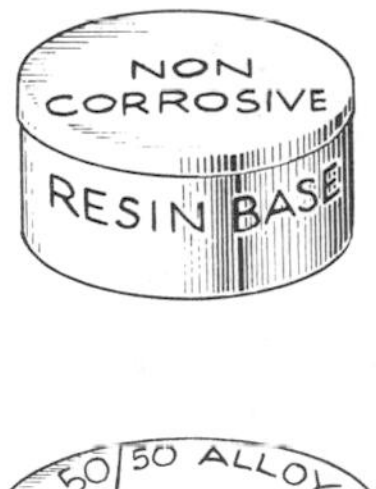

Fig. 7-5. Soldering materials. Top: Paste to aid even solder flow; bottom: wire solder.

Common Soldered Splices

There are three major kinds of soldered splices: (1) the *pigtail*, used for connecting the wire ends in outlet boxes; (2) the so-called *Western Union splice*, used principally on solid conductors which must carry their own weight for a considerable span or of a size too large for a pigtail; (3) the *tee tap*, used to connect one wire to a continuous run of another wire.

There are many kinds of ready-made wire connectors used universally today; however, the skill of being able to splice solid or stranded wire and cable when necessary, without the aid of manufactured connectors, is still necessary.

Making a Pigtail Splice. Skin the wires an inch or so, as in Fig. 7-6A. Cross the ends as in Fig. 7-6B. Hold them in position with thumb and fingers of one hand while twisting them together with pliers for six or eight turns, as in Fig. 7-6C. Use pliers, such as the one shown in Fig. 7-3, to twist the wires. Solder the splice as in Fig. 7-6D. Two, three or four wires can be joined in this manner. After the wire has cooled, wipe it with a damp cloth and insulate with tape.

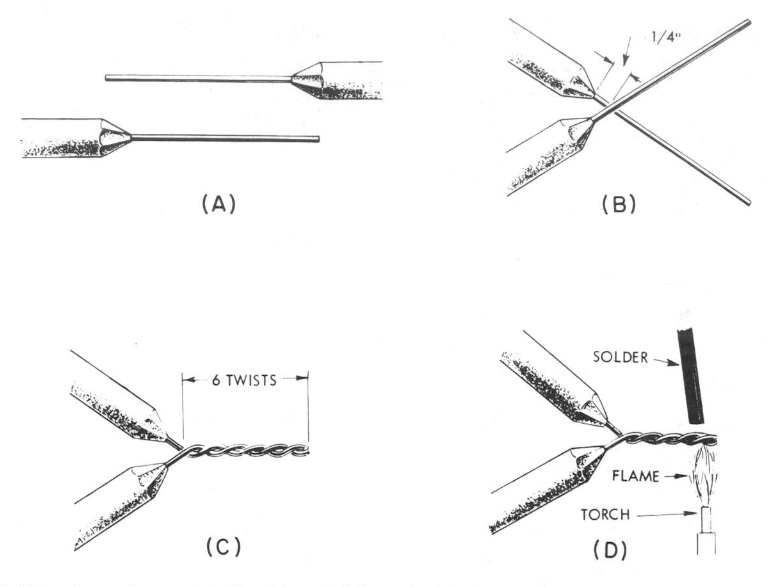

Fig. 7-6. How to make a pigtail splice. Folding the ends back is commonly omitted.

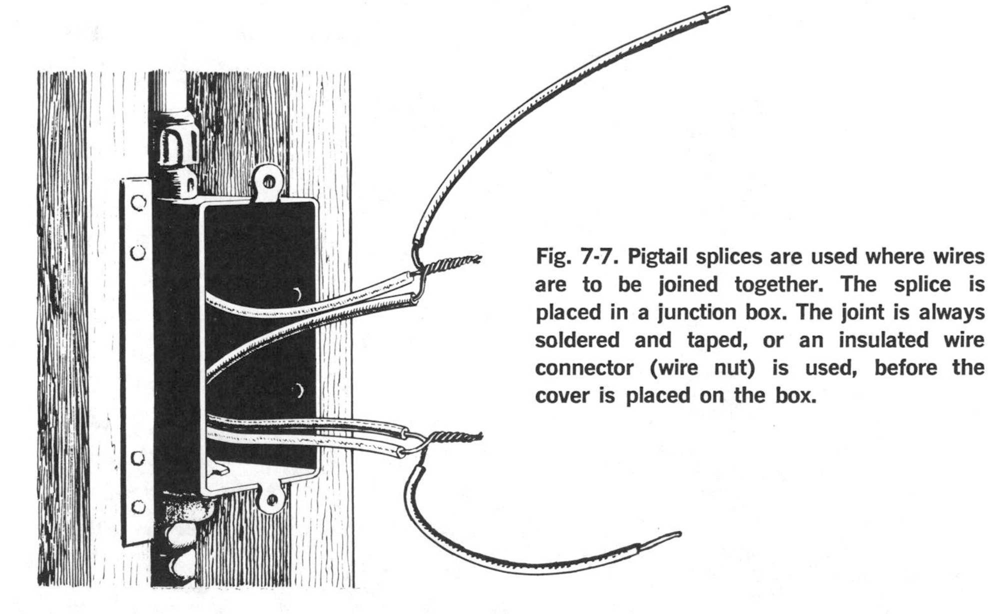

Fig. 7-7. Pigtail splices are used where wires are to be joined together. The splice is placed in a junction box. The joint is always soldered and taped, or an insulated wire connector (wire nut) is used, before the cover is placed on the box.

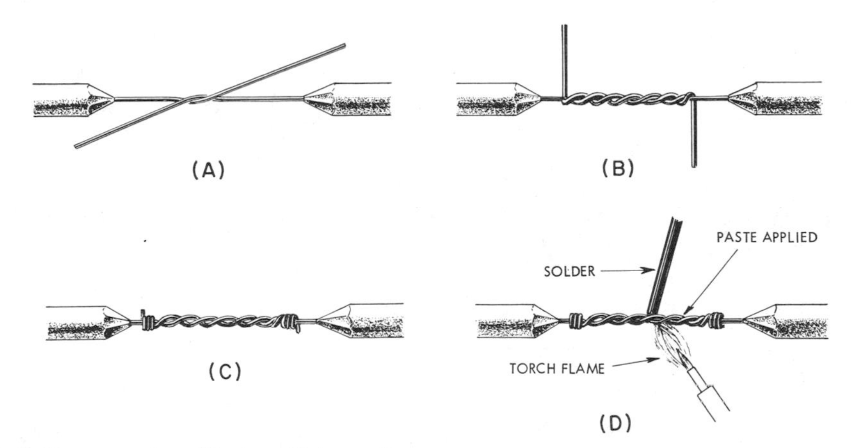

Fig. 7-8. How to make a Western Union splice.

Fig. 7-7 illustrates one of the most common uses of pigtail splices in a junction box.

Making a Western Union Splice. Skin insulation from the wires at a distance; the distance depends on their size, varying from about 3″ with No. 14 AWG to perhaps 6″ for No. 10. Cross the wires at their middle as in Fig. 7-8A. Twist the ends in opposite directions four to six turns, Fig. 7-8B. Now twist each end sharply, at right angles to the run of the splice, and wind three full turns, Fig. 7-8C. Cut off the excess ends, and solder, Fig. 7-8D. After the wire has cooled, wipe it with a damp cloth and insulate the splice with tape. Care should be exercised in not wiping a hot joint too soon. Hot solder may spatter and cause serious burns. The live steam generated may do the same, especially on a larger wire and cable splices.

Making a Tee Splice with No. 14 Wires. Remove the insulation on the main wire for a distance of 1¼″, and that on the tap wire 2″, as in Fig. 7-9A. Cross the wires as in Fig. 7-9B. Twist the tap wire around the main wire, Fig. 7-9C, and draw the turns tight by using pliers, Fig. 7-9D. Cut off excess wire and solder, Fig. 7-9E. After the wire has cooled, insulate the splice with tape.

Fig. 7-10 shows how to splice on to an existing circuit. The splice merely taps onto a feeder or circuit without the need of cutting the wire. Fig. 7-10, top, shows its use in open wiring. It is useful for low voltage, remote controlled wiring. Fig. 7-10, bottom, shows a common use of this splice in gutters. This situation oc-

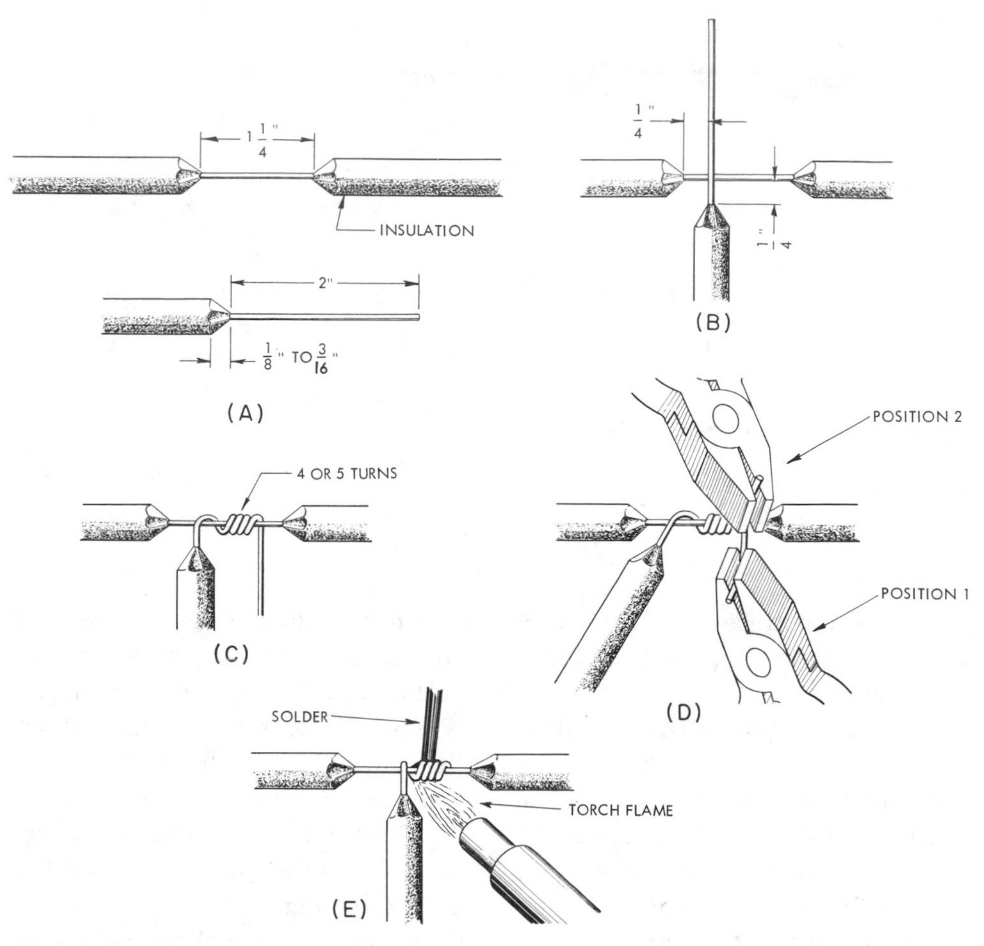

Fig. 7-9. How to make a tee splice in open and gutter wiring. Note Position 1 in D is used to start twisting wire and Position 2 is used to gain leverage with pliers across horizontal wire to tighten final splice and turns.

curs in many multi-family dwellings and in commercial and industrial applications.

Both splices must be insulated, of course, with tape.

The *tee* splice is also referred to in the trade as a *tap* splice. Either use, *tee* or *tap*, should be understood. Both are accepted.

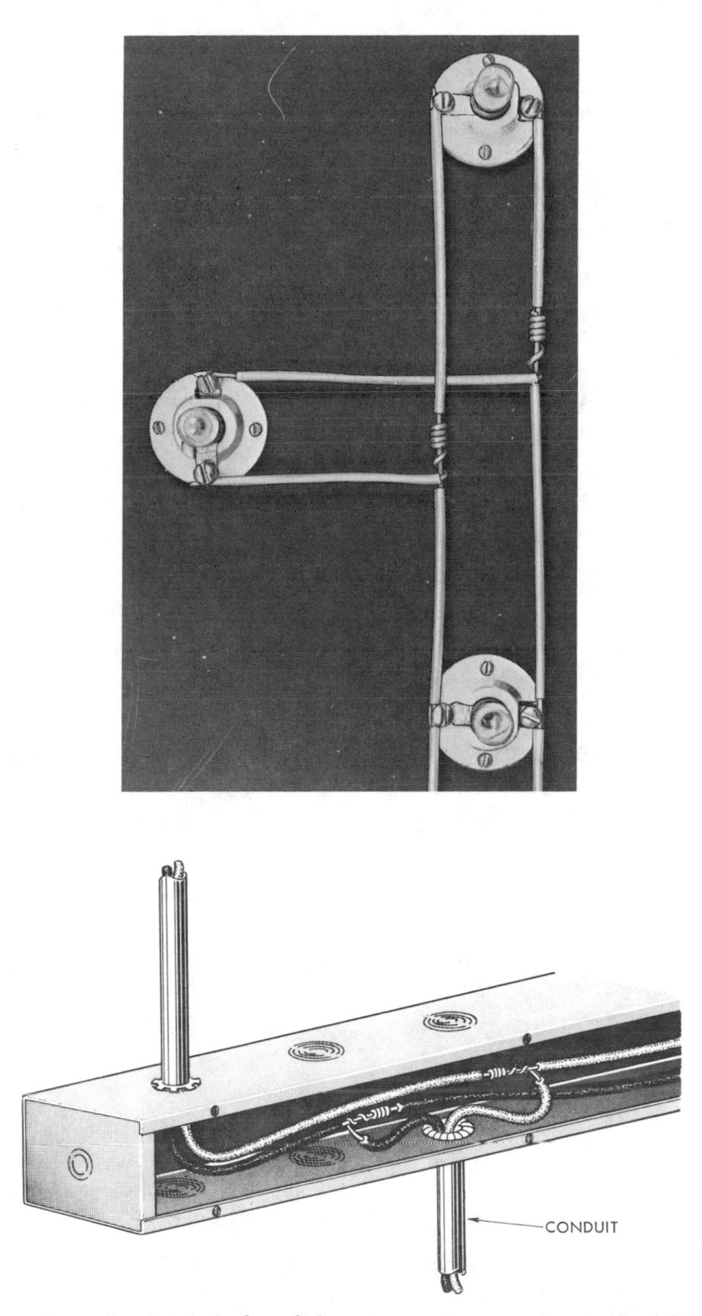

Fig. 7-10. A tee tap splice is used where it is necessary to connect onto an existing circuit in an open wiring system. The tee tap makes it possible to splice onto the wire without cutting it. Then it is insulated. Top: Open wiring with a tee splice. Bottom: A tee tap splice in a gutter wire enclosure. Splices must, of course, be insulated.

Other Soldered Splices: Stranded Fixture Wire and Portable Cords

Lighting fixtures generally have stranded wire that must be spliced to solid building wire. All wire joints to be soldered must be mechanically tight and clean before soldering. Fig. 7-11 shows a method of joining stranded fixture wire to solid building wire. The solid wire is bent over the stranded wrap before soldering the splice.

For safety purposes, portable extension cords with deteriorated insulation should be destroyed or discarded. Cords accidentally cut or damaged should have additional cord caps and bodies installed so that they may be coupled together safely. Fig. 7-12 shows a temporary splice used for extreme emergencies only. Splices are staggered, soldered and adequately taped. Splices in portable cords are not as good as approved cord end coupling devices. They should be changed as soon as possible. Electrical codes do not generally approve portable cord splices.

Splicing Stranded Building Wire: For splicing stranded copper building wire, fan out individual strands evenly as shown in Fig. 7-13, top. Intersect strands of each cable as evenly as possible, as in Fig. 7-13,

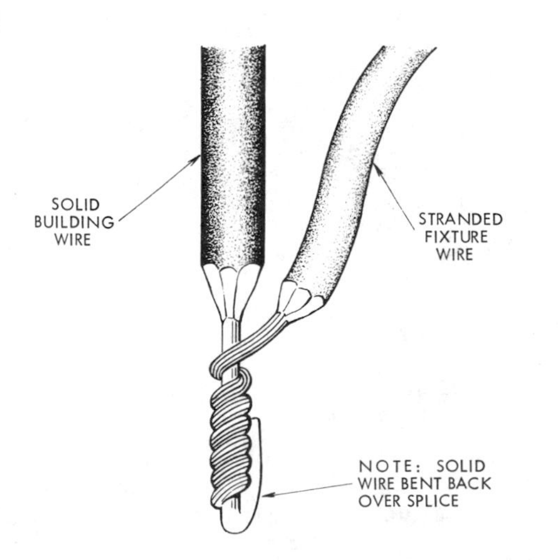

Fig. 7-11. Method of splicing stranded fixture wire to solid wire for a solder splice. Solid wire bent over splice keeps it mechanically tight.

Fig. 7-12. Portable cord splice for emergency purposes only. Cords should not be spliced usually.

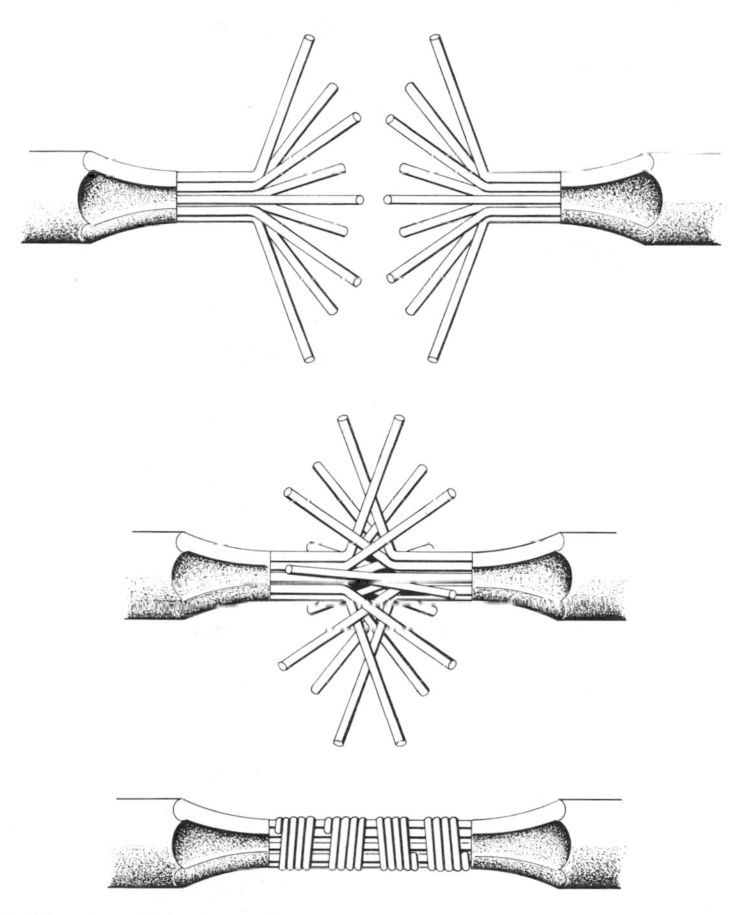

Fig. 7-13. Splicing stranded copper wire.

center. Then wrap one or more strands, depending upon cable size, around the assembly and repeat with strands of the opposite wire, wrapping in opposite direction so that they may be tightened. Follow through with alternate strands until each strand has been wrapped as

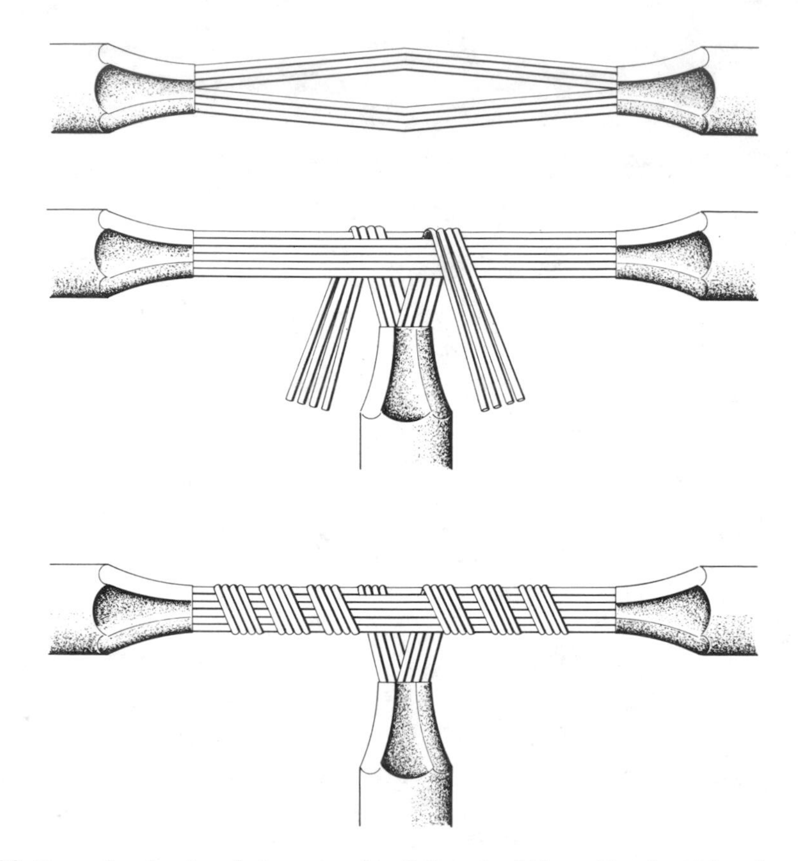

Fig. 7-14. Tap splice for stranded copper wire. Splice should be soldered and taped.

shown in Fig. 7-13, bottom. Splice is now ready for soldering.

A tap splice for stranded wire is shown in Fig. 7-14. To make good electrical and mechanical contact the horizontal strands are separated in the center of the conductor, Fig. 7-14, top. The tapping, stranded conductor is paired and inserted through the horizontal separation as seen in Fig. 7-14, center. The loose ends are then wound tightly onto the horizontal conductor in opposite directions. Fig. 7-14, bottom. The splice is then soldered and adequately insulated with tape. Factory made compression splicing devices are available which do not require the skill of a craftsman. Fig. 7-15 shows a method of splicing larger stranded wire or cable. The cable strands are fanned out as shown in Fig. 7-13, top, and then interwoven and connected making good electri-

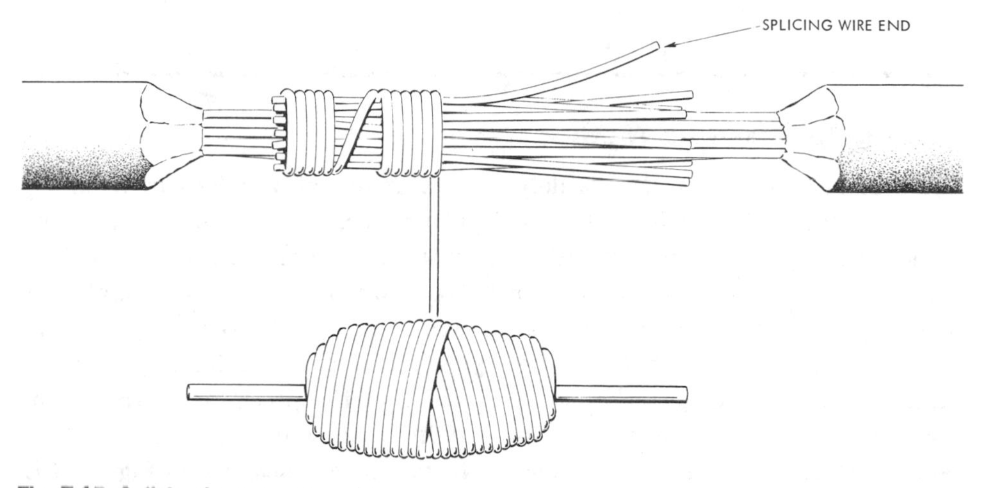

Fig. 7-15. Splicing large copper cable.

cal and mechanical contact to each other.

The cable strands are not turned upon themselves as before but are held securely with wraps of an independent wire. (See Fig. 7-15.) When the cable ends are laid together they are first tightened by tapping with a 2 x 4 size piece of wood and squeezed with a slip joint type of pliers.

If a hammer is used, care should be taken not to distort the strands. The cable ends are then held securely and tightly with turns of a smaller gauge of copper wire. Pull tightly with each wrap. A small pigtail splice is made with both ends of the splicing wire upon completing the last wrap. This pigtail splice is then tapped down into a smooth contour. Soldering paste is then applied and the joint is heated with a sufficient size torch for soldering. Molten solder on the joint can be "wiped" around and into strands and splicing wires for a smooth joint. A dry rag or cotton pad should be used for wiping.

When the joint is cooled it may then be taped.

This splicing method in Fig. 7-15 is called "mousing" a joint. The copper wire used on the cable is called "mousing wire," 16 or 14 gauge depending upon the size cable to be spliced. These splices are generally acceptable to cable with 600 volt ratings.

Taping Splices—Two Methods

There are, in general, two widely accepted methods for insulating a soldered splice, the purpose in either case being to cover the joint with protective material equivalent in electrical strength to that on the original conductor. One method calls for the use of rubber and friction tape; the other, for plastic tape. The older method, which is applicable to a Western Union splice, will be explained first.

Use of Rubber and Friction Tape. Cut a 4″ piece of rubber from the roll. There is a strip of glazed cotton tape on one side of the rubber to prevent adjacent layers from sticking together. Remove this material, and stretch the rubber to a longer length.

Starting at the left-hand end of the splice, Fig. 7-16A, make a turn around the wire to secure the tape. Now, wrap the joint diagonally toward the right, keeping the tape stretched and overlapping half the width of the preceding turn. At the right-hand end, the direction of travel is reversed, as in Fig. 7-16B, the wrapping continuing until two layers have been applied or more depending upon size or thickness of insulation removed. Continue wrapping procedure with any excess tape. The end should be pinched against the joint with thumb and forefinger, Fig. 7-16C, the heat of the fingers causing the rubber to stick to itself. A layer of friction tape is now applied in the same manner,

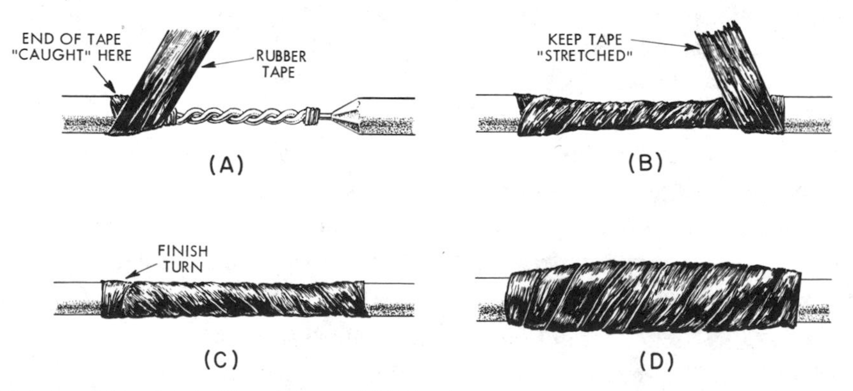

Fig. 7-16. Insulating joint with rubber and friction tape.

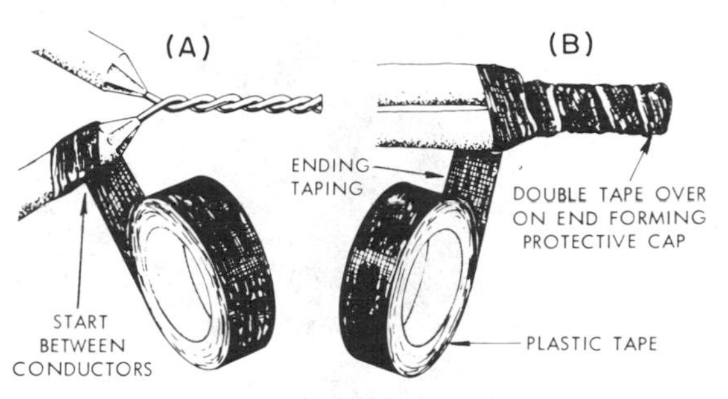

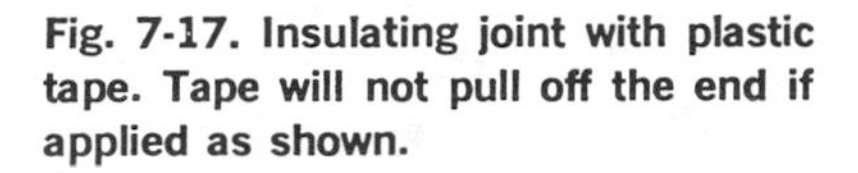
Fig. 7-17. Insulating joint with plastic tape. Tape will not pull off the end if applied as shown.

as shown in Fig. 7-16D. Since some friction tape tends to lose its adhesiveness and unwind as it ages, a quick-drying insulating varnish is sometimes added to prevent such occurrence.

Friction tape by itself is not considered insulating. It is used for mechanical protection of insulating rubber tape.

Use of Plastic Tape. Plastic tape has come into general use because of the saving in labor and in space, over the rubber-friction method. Fig. 7-17 shows the taping of a pigtail splice with this material. As may be seen in Fig. 7-17A and Fig. 7-17B. Secure tape on leg as shown in Fig. 7-17A. Then close legs together, tape down wire splice past the end and tape back, as shown in Fig. 7-17B. Cover all bare copper adequately. Although the manufacturer states that a single layer is sufficient for voltages up to 600 volts, it is well to apply two layers or more of half-lap tape for added mechanical protection.

Solderless Splices

Wire nuts were first used for splicing fixture wires to circuit conductors. A wire nut, illustrated in Fig. 7-18, consists of a cone-shaped spiral spring inside a molded insulating body. When screwed onto the wires, the spring presses them firmly together, and the insulating body protects the wires from external contact, so that no tape is required. Fig. 7-19 shows a cut away illustrating how the wire nut works.

Some construction jobs require that unsoldered pigtail splices be made before applying a wire connector to prevent any splices from "losing" a wire. See Fig. 7-20. When this method is employed it is necessary to twist the wire in a clockwise direction also.

Wire nuts or connectors can be used to splice wires up to size No. 6 AWG.

Fig. 7-21 illustrates a wireman making solderless splices using insulated wire nuts (connectors).

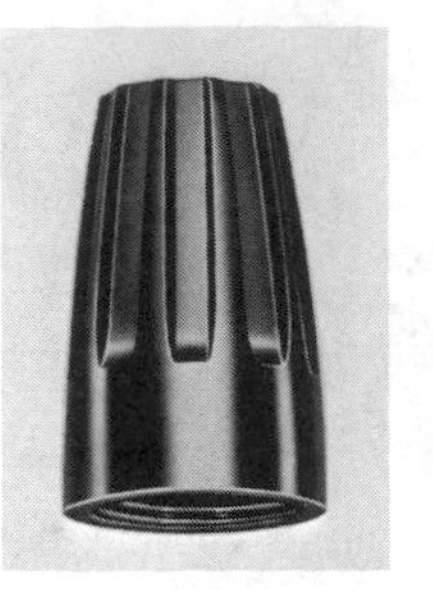

Fig. 7-18. Wire nut. (Ideal Industries, Inc.)

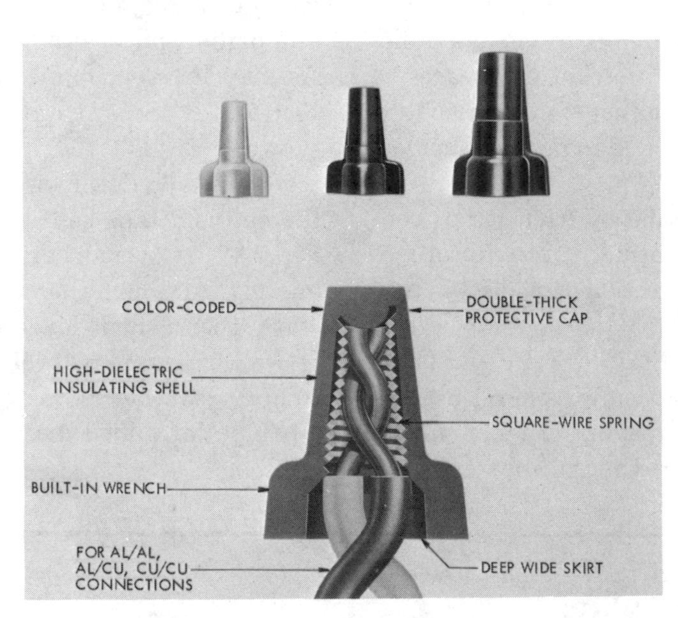

Fig. 7-19. Cross section of solderless wire connector twisted clockwise onto pigtail splice. Connectors are used on aluminum to aluminum wire splices, aluminum to copper, and copper to copper wire splices. This connector is sometimes called a "wing nut" because of built-in wrench. "Wire nuts" refer to the same connector without the "wings."

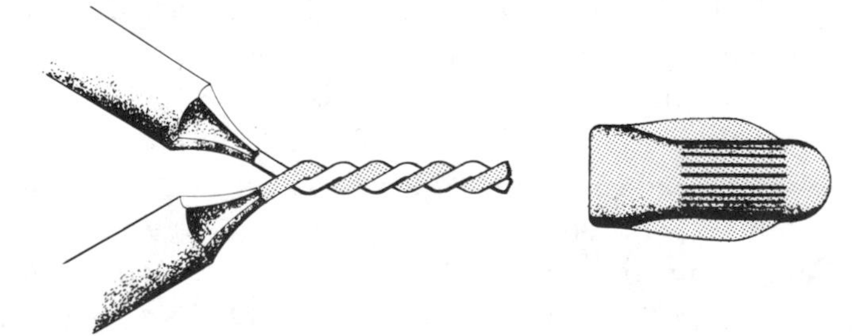

Fig. 7-20. Unsoldered pigtail splices are sometimes made before applying wire connectors.

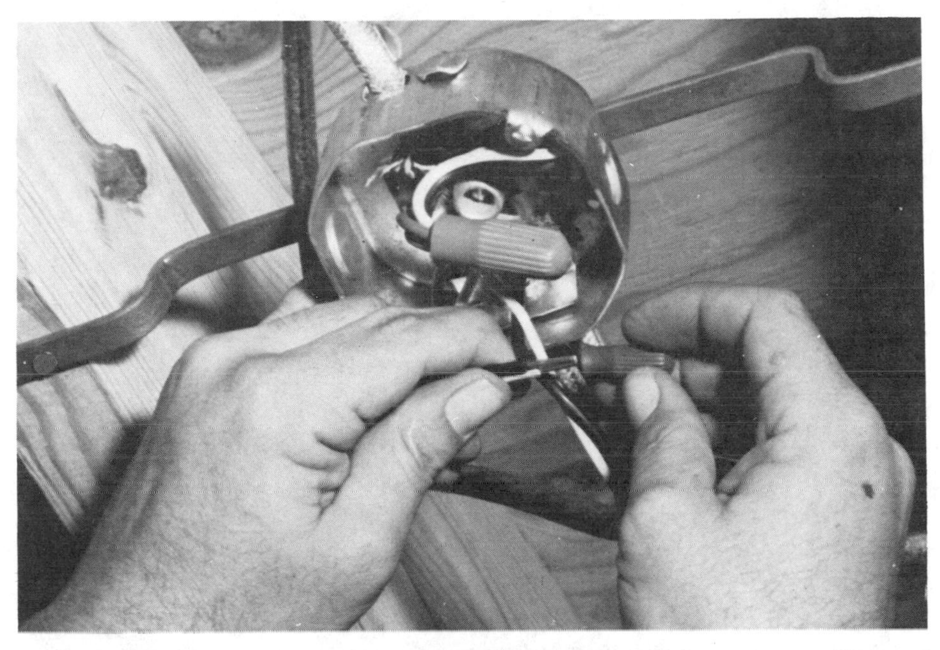

Fig. 7-21. Making a splice in a ceiling outlet box with insulated wire connector. (Minnesota Mining and Manufacturing Co.)

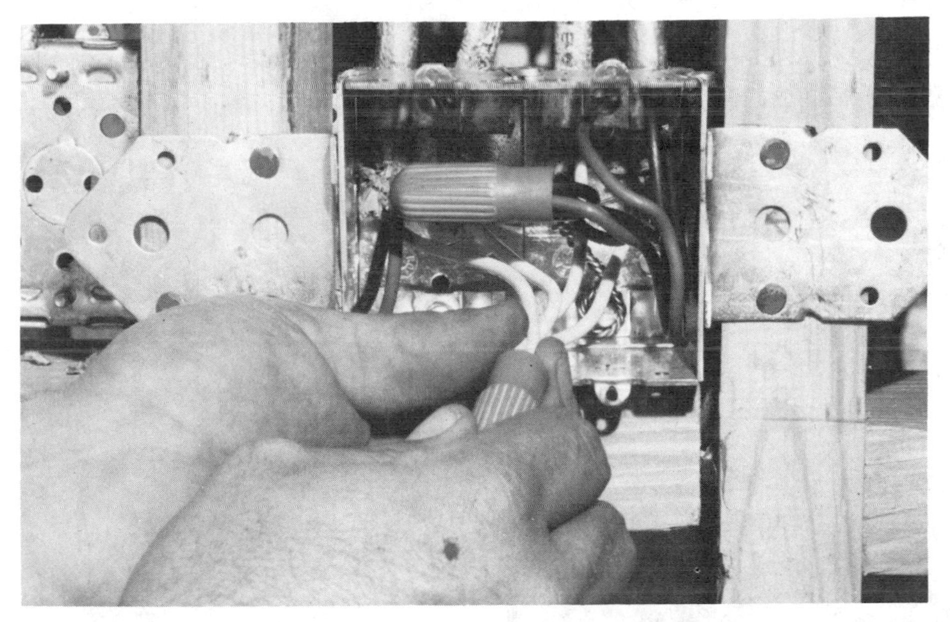

Fig. 7-22. Wire splices are also made in switch boxes and at convenience outlets. (Minnesota Mining and Manufacturing Co.)

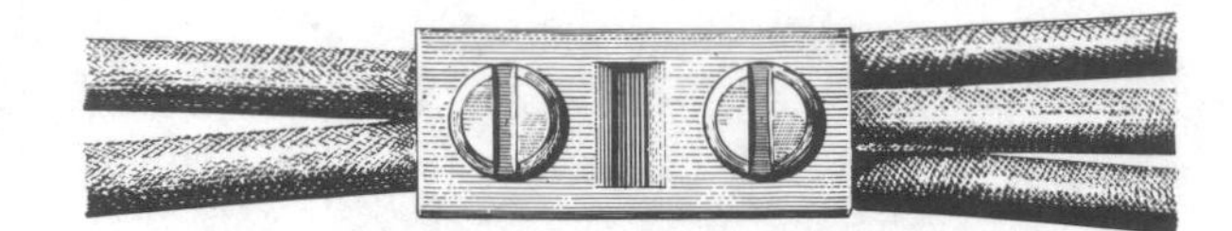

Fig. 7-23. Fixture connector. (H. B. Sherman Manufacturing Co.)

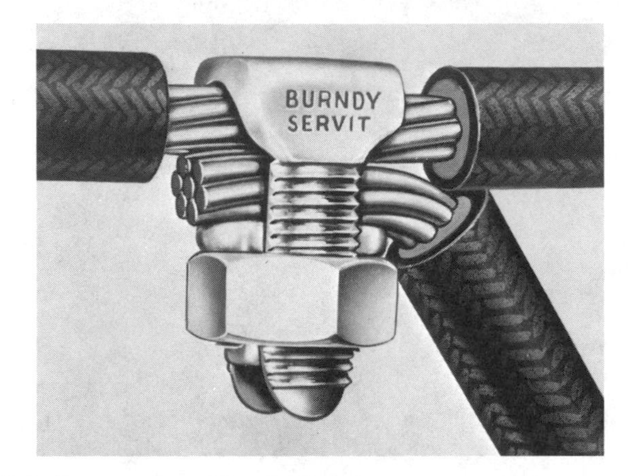

Fig. 7-24. Split bolt connector. (Burndy Corp.)

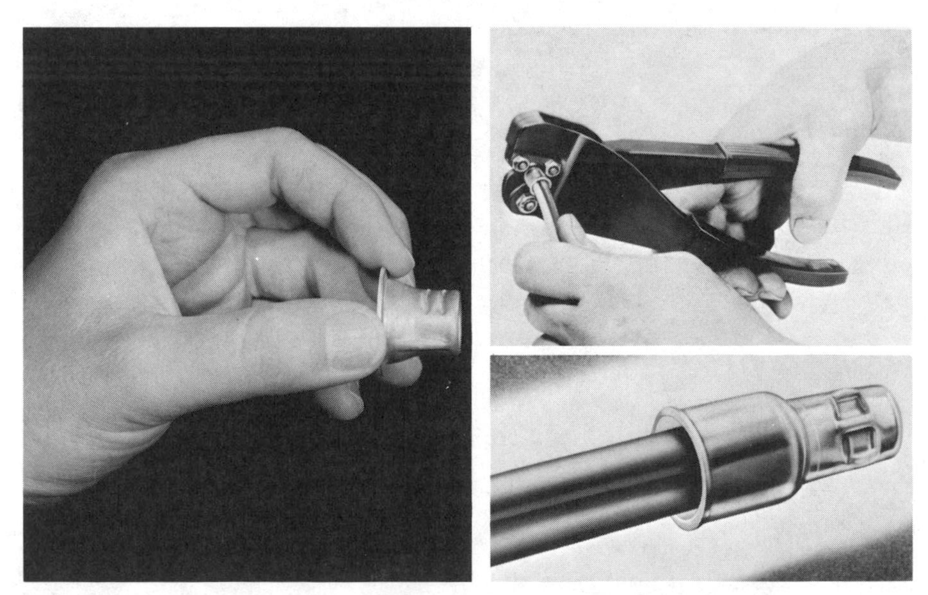

Fig. 7-25. Installing crimp type solderless connector. (Buchanan Electric Products Corp.)

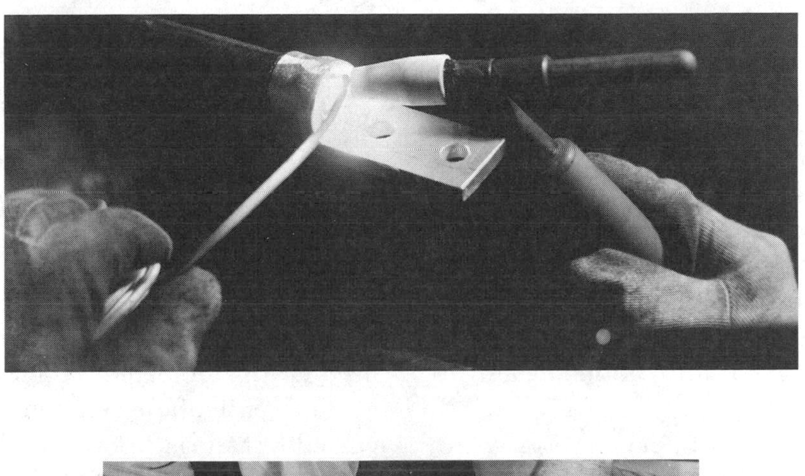

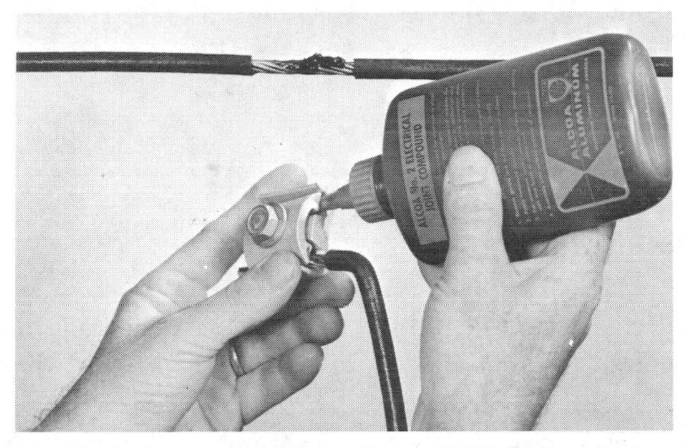

Fig. 7-26. Terminating aluminum conductors. Top: Welding, center: compression, bottom: pressure.

Common applications are splicing conductors in ceiling outlets, as shown, and in any box where splices are to be made. Fig. 7-22 shows splices being made in a two-gang switch box. Note "jumper" cable looping from the top of one box in another room to the one being connected. This cable contains the hot and neutral wires continuing a lighting circuit.

Fixture connectors, Fig. 7-23, make use of set-screws to hold the wires in place. With this device taping is, of course, necessary. Fig. 7-24 shows a split bolt connector suitable for heavier wires such as with service equipment.

The crimp type of solderless connector, Fig. 7-25, is also quite popular. The wires are first twisted together and a copper thimble slipped onto the joint. A special, plier-like tool is used to crimp thimble and wires together. The joint may then be taped or fitted with a vinyl insulating cap, as shown in the illustration.

Splicing Aluminum Conductors. Aluminum conductors can be soldered with special solder and flux and high heat, but they are usually joined by other methods involving pressure, compression, or welding, as in Fig. 7-26. Experience has shown that the best results are obtained when aluminum fittings are used with aluminum conductors. More recently, however, plated copper alloy fittings have proved suitable for use with aluminum. In any case, a special compound should be applied to surfaces of conductors and lugs at points of contact in order to prevent corrosion and to ensure good electrical contact. When copper and aluminum conductors are connected together, the splicing device should have a separate slot or compartment for each type of conductor. Attempts to hold them together under the same screw or compression nut may lead to ultimate trouble on account of corrosive galvanic action of two dissimilar metals acting like a small battery.

Terminal Connections

Many faulty connections can be avoided if a simple method is followed when fastening wires to switches, receptacles, and sockets. Since wiring to electrical equipment is always attached with right-hand screws, the wire should be bent around the screw in a clockwise direction so that tightening of the screw will draw the wire in and not

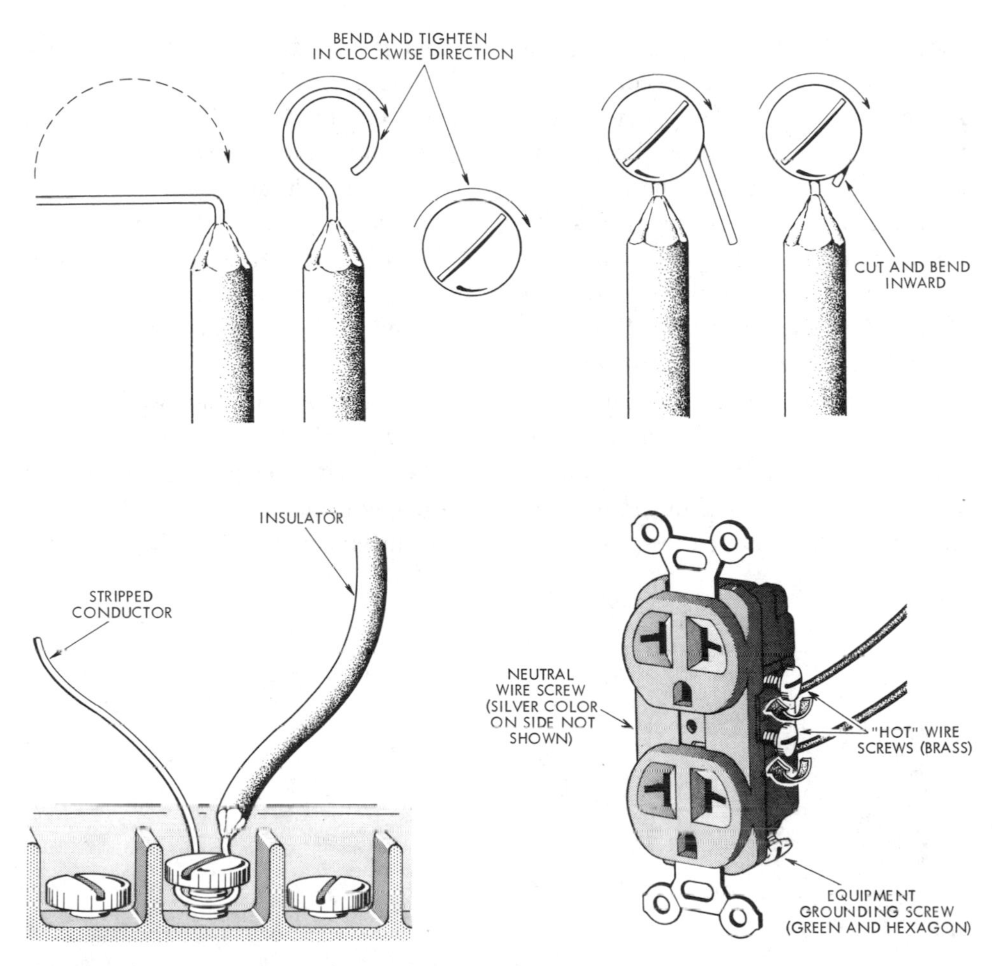

Fig. 7-27. How to terminate wires at screw terminals.

push it away. Fig. 7-27 shows the correct method of placing wires under screws prior to tightening.

Fig. 7-27, top left, shows the two steps in bending an eyelet in solid wire: first the wire is bent at a right angle, and second the end of the wire is then grasped by nearly any kind of pliers and bent into a loop of proper screw size. Fig. 7-27, top right, shows an alternate method of making the loop by bending around the screw and allowing the turning of the screw to form a loop.

When connecting to less accessible wiring terminals (see Fig. 7-27, bottom left), 6 or 8 inches of insulation may be stripped from the conductor. A loop may be partially made near the end of the insulation. Using the two conductor ends as handles the partial loop may be placed under the screw head and the screw may be closed. If insulating barriers around terminal screws prevent close cutting of excess wire, the wire may be broken off by bending it back and forth or in a rapid circular motion. Wire metal fatigue should occur after several motions and the wire should break off near the screw.

Screws are tightened in clockwise direction, Fig. 7-27, bottom right, to prevent opening of the eyelet. Fig. 7-28 illustrates how the screwdriver blade should fit into the screw slot. Some wiring devices have access holes with internal pressure plates for quick rear wiring. Wires are stripped and inserted straight into holes. One should check the code when using this method with aluminum wire. Screw terminals are acceptable with both aluminum and copper wiring.

Manufacturers of wiring devices

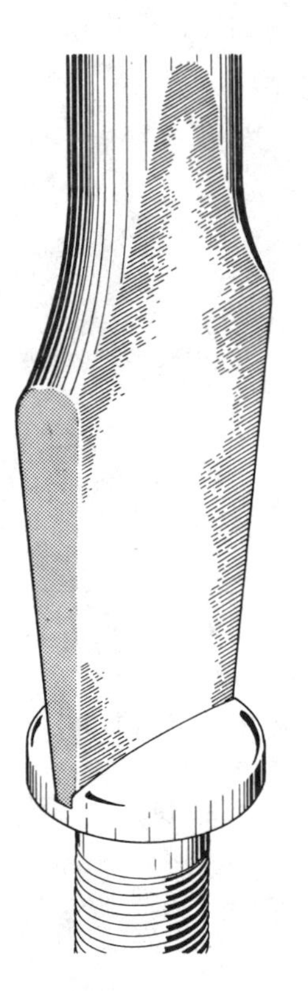

Fig. 7-28. Screwdriver blade should fit fully and evenly into screw slot.

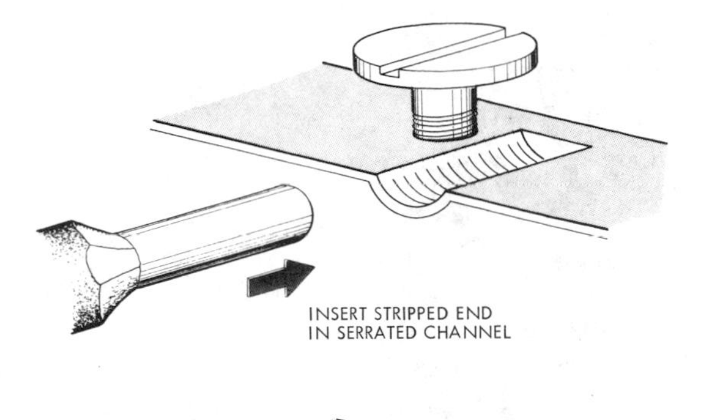

Fig. 7-29. Serrated channel pressure terminal wiring.

TIGHTENING SCREW
LOCKS WIRE IN PLACE

have built-in features for terminal wiring which they call speed wiring. Underwriters' Laboratory and code approval should be sought before these methods are used if there is any doubt about their passing inspection. For example, aluminum wire may not be approved for use with all terminal connections. Most wiring devices include screw terminals also for conventional wire looping.

Fig. 7-29 illustrates a wire terminating method of the pressure type. A stripped wire is held in place by tightening a screw over a serrated wire channel. No wire looping is necessary here. This method may be found on some circuit breakers and other devices.

Fig. 7-30, top, illustrates insulation strip length and method of back wiring a switch by inserting wire into wire clamping hole in the back of the switch. Fig. 7-30, bottom, shows how wires may be released from device by inserting screwdriver to relieve pressure of a clamping device. Some wiring devices contain provisions for inserting two wires in a hole thereby saving a splice.

Fig. 7-31, top left, illustrates a feature which provides wiring with partial looping of wire. The wire is stripped, inserted in a hole in the terminal and looped around the screw. The screw is then tightened for a good pressure connection. Fig. 7-31, top right, shows a side-wired

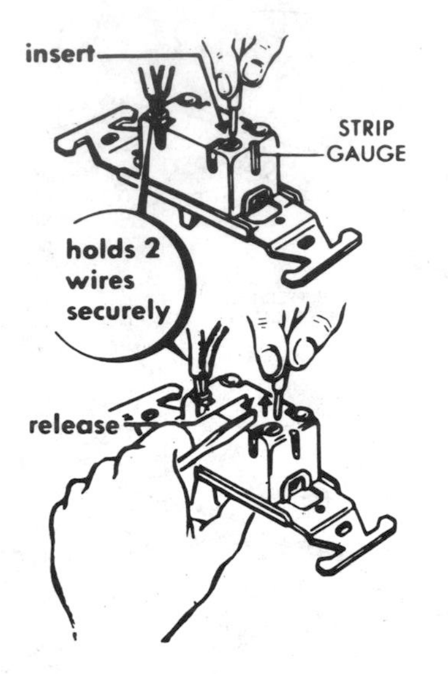

Fig. 7-30. Top: Illustrates insulation strip length and method or back wiring. Bottom: Wires may be released from device by inserting screwdriver into release slot. (Slater Electric Inc.)

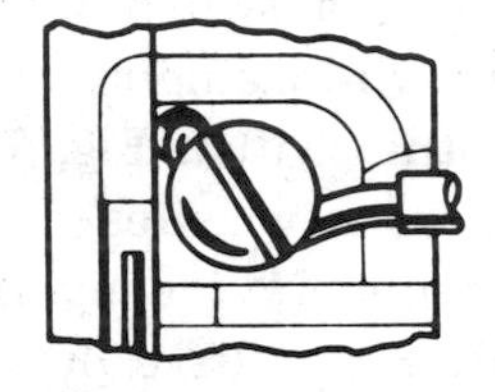

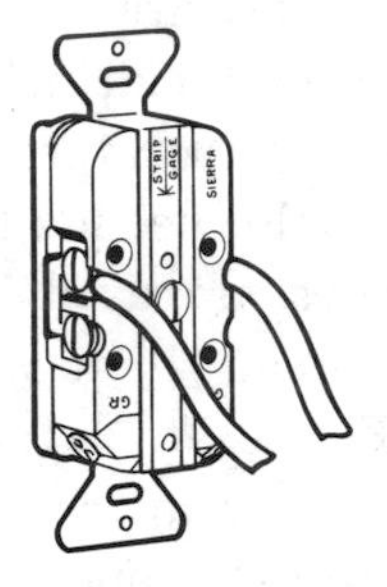

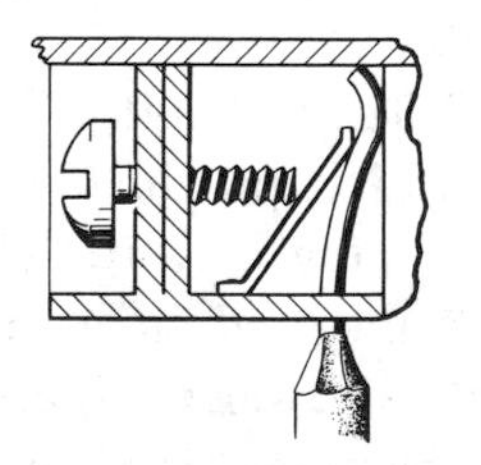

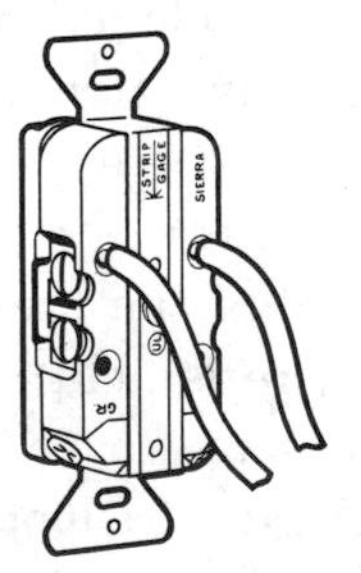

Fig. 7-31. Top: A partial wire loop running from hole and around screw. Bottom: A back wired receptacle; tightening the screw wedges the wire securely in place. (Top and bottom right: Sierra Electric Corp.)

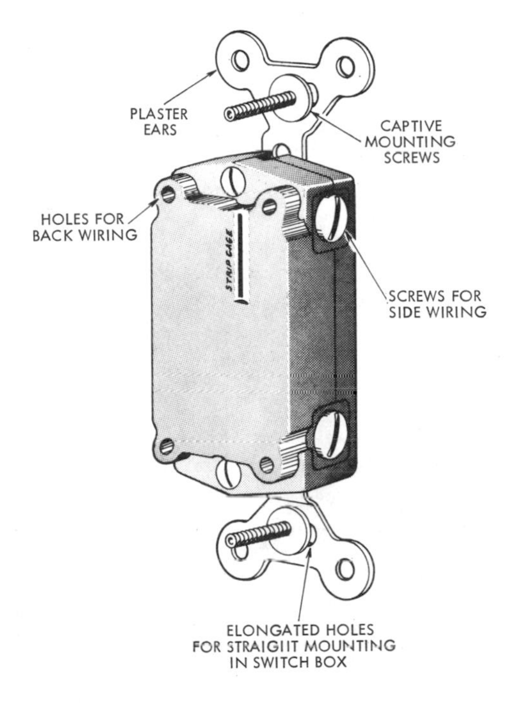

Fig. 7-32. Switch features. Note the "Strip Gage." This shows the length to strip insulation to insure proper depth for insertion into holes for back wiring.

switch or receptacle using either complete wire looping method or partial wire looping under a tightened screw. Fig. 7-31, bottom, shows the back, optional-wiring method of inserting stripped wires into holes.

Fig. 7-32 shows a switch ready for mounting which can be either back wired or side wired. In this case mounting screws are provided and are held in place with fiber washers. Plaster ears as shown are to hold the switch in its proper depth position in the event the plaster is thicker than the box setback. Elongated mounting holes are provided so that the switch may be mounted straight should the box be mounted tilted.

Fig. 7-33 shows a screw pressure type of wire splicing device. Some single pole switches have an extra binding screw that is not connected to the switching mechanism. Stripped wires are slipped into the slots under the loosened binding screw, which is then tightened. This saves making a bulky splice.

Solderless Terminals. The solderless terminals of Fig. 7-34 have largely replaced small solder lugs. One type, as seen in the illustration, is fastened by crimping, or indenting by means of a special tool.

Solderless Lugs. There are three general types of solderless lugs: *set-screw*, *compression*, and *indented*,

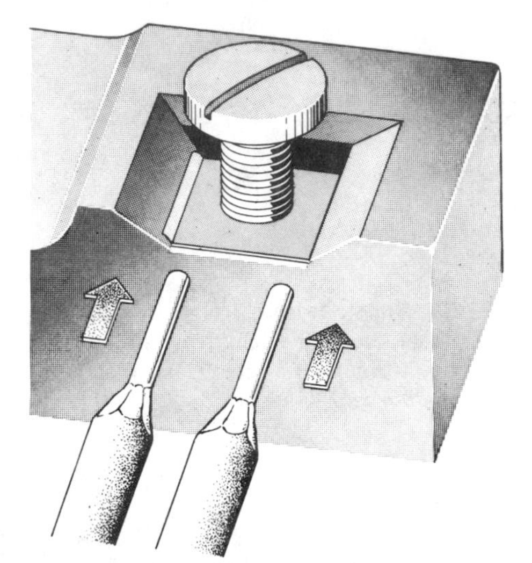

Fig. 7-33. Wires are inserted into slots and the screw is tightened making a splice.

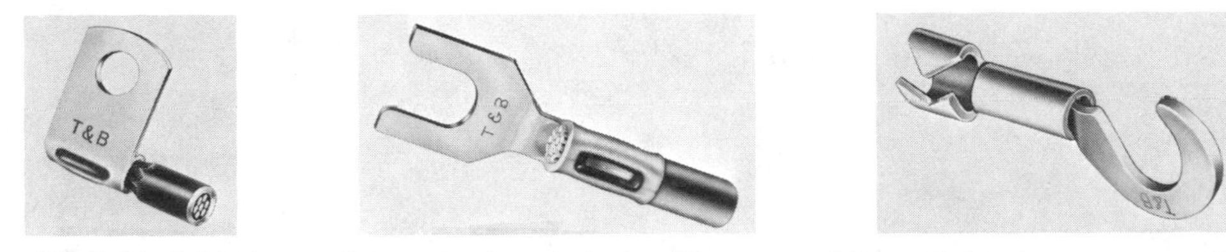

Fig. 7-34. Solderless, crimp type wire terminals. (Thomas and Betts Co.)

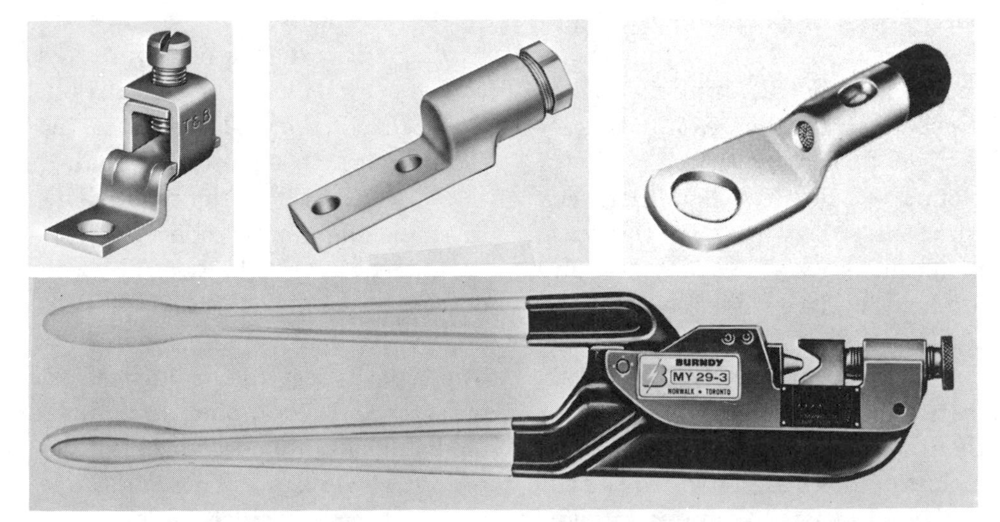

Fig. 7-35. Lugs and indenter for large conductors. (Thomas and Betts Co., Penn-Union Electric Co., Burndy Corp.)

Fig. 7-35. The latter are installed by means of hydraulic pressure devices, or by tools which employ compound leverage. Welding is also used to some extent in making solderless connections.

QUESTIONS FOR DISCUSSION OR SELF STUDY

1. State the most likely weak point of an interior wiring installation.
2. How would you remove weatherproof insulation from copper wire?
3. What does the expression 50/50 mean, with respect to solder?
4. How can you protect the sharp end on a pigtail splice?
5. How is the Western Union splice used?
6. How are wire nuts installed?
7. How is rubber tape used in conjunction with friction tape?
8. How are solder joints heated?
9. What is the most common method of making splices?
10. What is a compression connector?
11. What is a crimp connector?
12. Why should an electrician know that solder splices with factory connectors are available?
13. How are aluminum wires joined?
14. How are copper and aluminum wires joined?
15. Why is a joint compound used on aluminum terminals?
16. What kind of terminals are used with aluminum conductors?
17. What should be done with rubber tape as it is being applied?
18. What term describes the manner in which tape should be applied to a joint?
19. Where are wire nuts used?
20. What device holds the wires together in the small crimp connector?
21. How can one insure against the possibility of losing a wire when using wire nut connectors?
22. In which direction should wire nuts be turned onto wires?
23. How much insulation should be applied to a newly soldered splice?
24. Why is it necessary to bend the solid wire back over the stranded when splicing to fixture wire?
25. Why is it necessary to fan out strands before bringing together cable joints?
26. Why should a solid wire tap be turned onto a through conductor by using pliers with leverage as in Fig. 7-9D, Position 2?
27. Why use soldering paste when soldering?
28. List common uses of the line or side cutting plier.
29. What does "wiping" a joint mean?
30. Plastic tape has higher insulating qualities than rubber—how much is generally required in comparison?

Chapter

8 Remote Control Wiring

To cope with demands for greater convenience and greater flexibility of control, a low-voltage, remote switching system has been perfected. The purpose of remote control in an electrical circuit is to govern the operation of a current consuming device at some point removed from the device. A wall switch, in one sense, is a form of remote control for a ceiling outlet, because the light is turned on and off at a point that is some distance away.

In the remote control wiring system, relays located in each outlet, or elsewhere, perform the actual *switching* of the current. The relays are controlled by small switches which operate on 24-volt current, which is stepped down by a transformer from the 115-volt normal house current. Since the switches operate at a low, safe voltage, lower voltage similar to that used for door chimes may be used. In addition, because these relays only require a momentary impulse to change from on to off or off to on, the control switches are momentary contact switches, that is, the current only flows for the length of time the switch is pressed.

The great advantage of a remote control system is that, within reason, as many switches as desired may be easily and inexpensively installed, and the outlet controlled by these switches may be turned on and off by any one of these switches, wherever located.

Principles of Remote Control Wiring

The remote control devices are different from ordinary switches in that low voltage is used. To show how this is accomplished, examine Fig. 8-1. Suppose this is an ordinary toggle switch. The switch knob, however, is made of soft iron. If a coil is placed above the switch knob, the knob will be drawn upward when the coil is energized due to electromagnetic attraction. Here, a battery is used for simplicity as the source of power supply for the remote control unit. The coil is energized by closing the push button, which can be placed at a distance from the switch. When the coil is energized, the switch moves to the ON position and remains there, so that it is no longer necessary to energize the coil.

Another circuit can be used to turn the switch off. These two circuits are shown in Fig. 8-2A. When the upper push button is pressed, the switch is turned on; when the lower push button is pressed, the switch is turned off. This experimental circuit can be greatly simplified by eliminating one battery and combining the two push buttons into one unit, as in Fig. 8-2B. Inspection of this circuit will show that, when the ON button is pressed, current flows through the upper coil, turning the switch on. The switch is turned off when the circuit through the lower coil is energized.

Use of Remote Control Units in Wiring Systems

The first step in applying remote control to a system is to convert

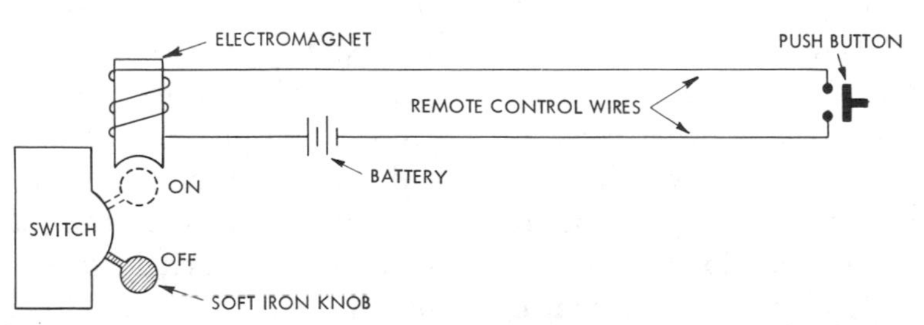

Fig. 8-1. Wiring diagram shows principle of low-voltage remote control for "on" position.

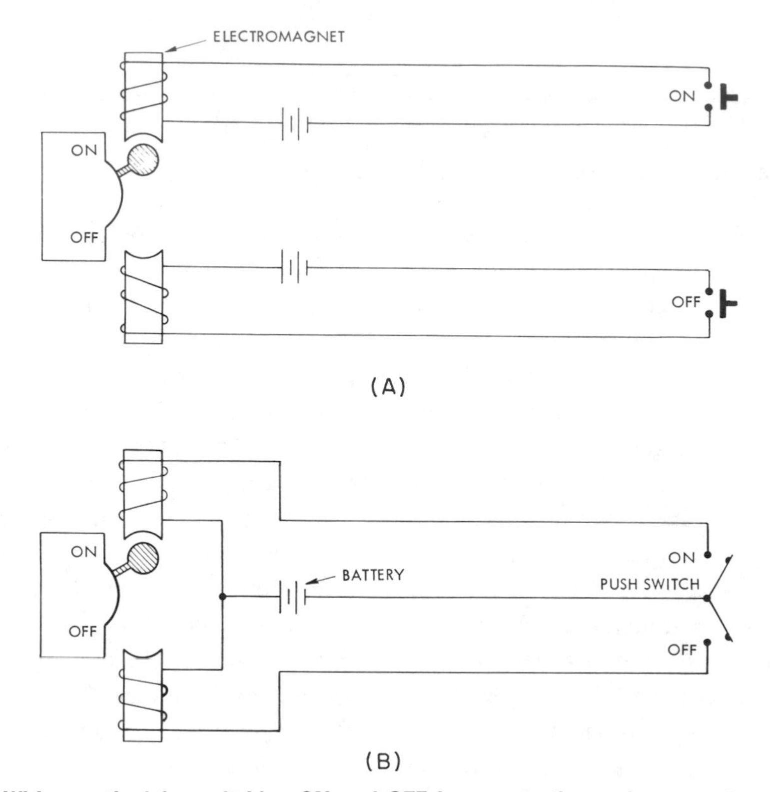

Fig. 8-2. Wiring method for switching ON and OFF by remote three wire control.

it to alternating current operation; otherwise, necessity for battery replacement would be undesirable. A step-down transformer is used to reduce the voltage from 120 volts to 24 volts. There is an additional refinement in the alternating current unit, namely, the exciting coils are combined into one unit with a center tap. Simplifying the switching results in a reduction in size.

The basic circuit for alternating current operation is illustrated in Fig. 8-3. By connecting one side of the transformer to the center tap of the coil, either half of the coil can be excited at will. For example, pressing the ON push button energizes the right-hand half of the coil, moving the plunger to the right. This closes the contacts in the lamp circuit, causing the lamp to light.

This simple switching arrangement can be extended so that an outlet can be controlled from a number of different points, by placing the

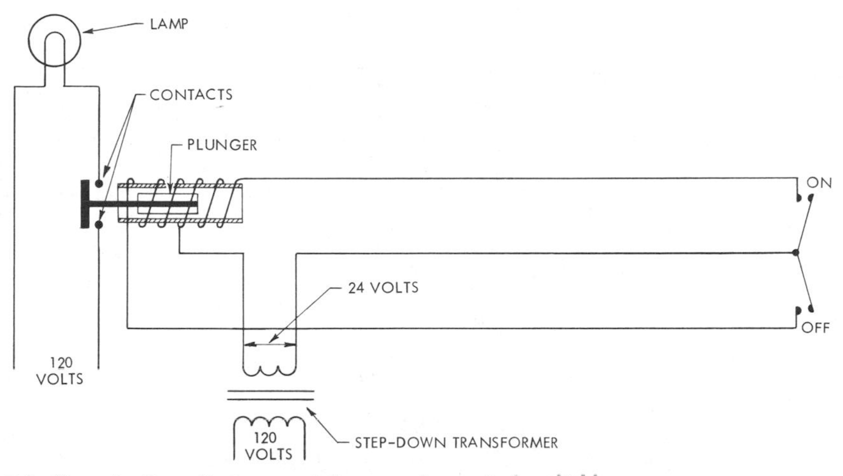

Fig. 8-3. Use of alternating current for remote control switching.

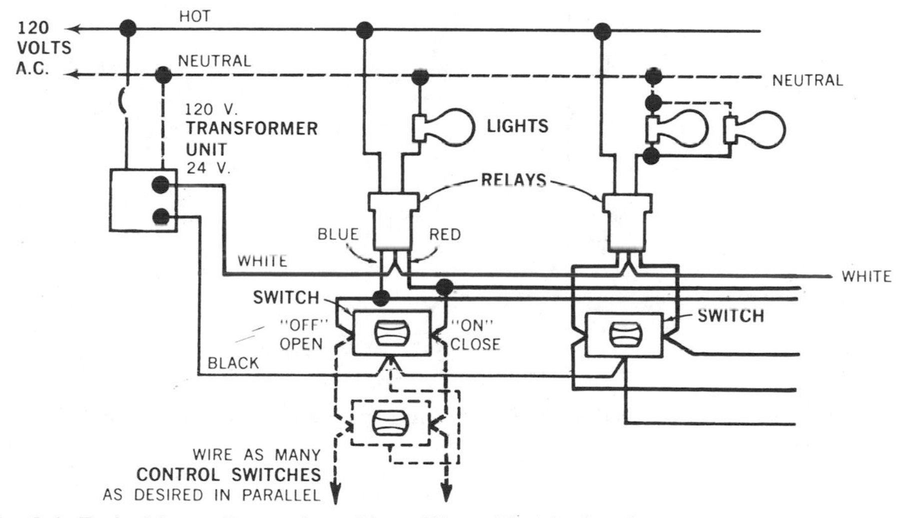

Fig. 8-4. Typical low voltage relay wiring. (Sierra Electric Corp.)

desired number of push buttons in parallel. This may be seen in Fig. 8-4, a typical arrangement of low voltage relay wiring.

Different circuit arrangements for two sets of push buttons controlling different circuits are shown in Fig. 8-5. The first outlet is controlled by three push buttons in parallel. This outlet may be a single light or a

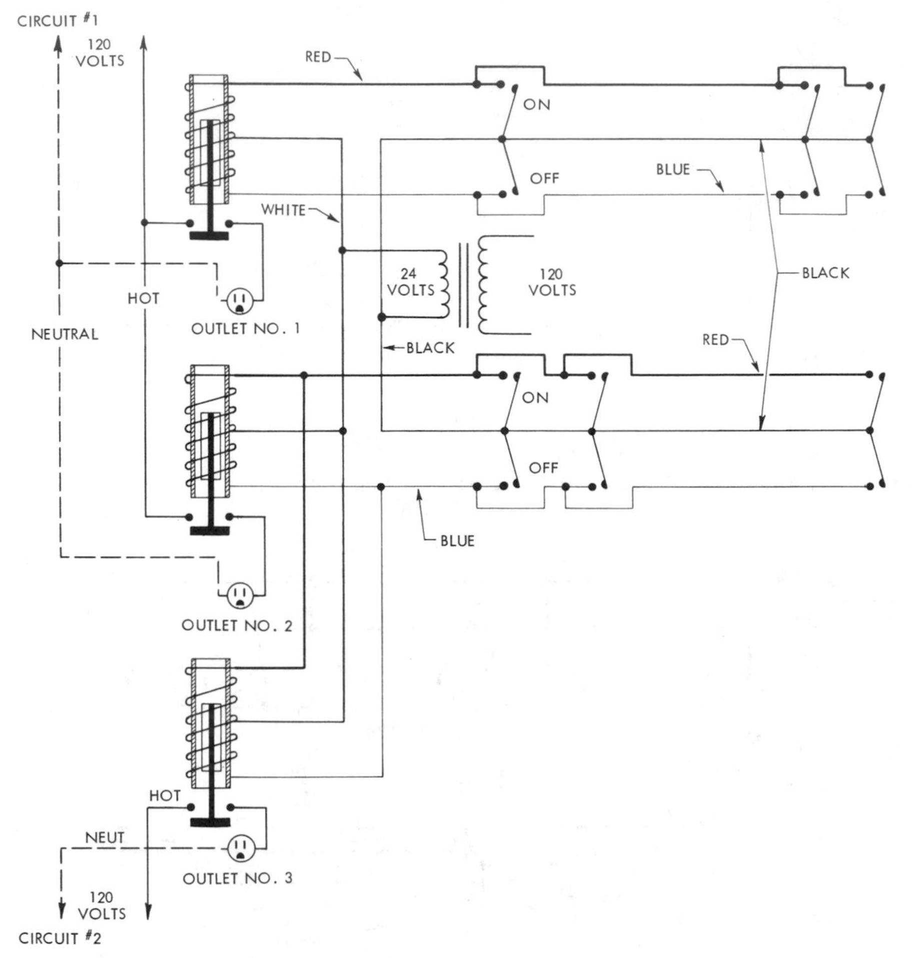

Fig. 8-5. Diagram of remote control circuit for control from various points and control of two different circuits from the same switches.

whole circuit. The two outlets indicated in the lower portion of the illustration can also be controlled by several push buttons. The fact that they need not be on the same branch circuit greatly increases the flexibility of this wiring system.

Advantages of Remote Control Wiring

Some of the advantages of this system are already evident, but there are still others worth noting. A major consideration in any wiring system is safety. The low voltage system results in safer operation by removing high voltages from control points. A further advantage is the ease of switch relocation. There is also a savings in labor and material when this system is used for many switching stations.

In addition, a selector switch can be arranged to turn off all lights from a single location. In this way there is no likelihood of having lights on all night in some remote part of the home because they were overlooked.

Equipment for Remote Control Wiring

Transformer

The transformer used in remote control wiring, Fig. 8-6, left, has a 120-volt primary winding and a 24-volt secondary winding. The mounting plate is constructed so that it can be attached to a standard four inch square outlet box, Fig. 8-6, right. The primary, or 120-volt, leads come out through the mounting cover, which permits them to be connected directly to the supply within the box. The secondary terminals can be seen on top of the transformer case. These transformers have an additional advantage that is worth noting. This is the current limiting feature, which pro-

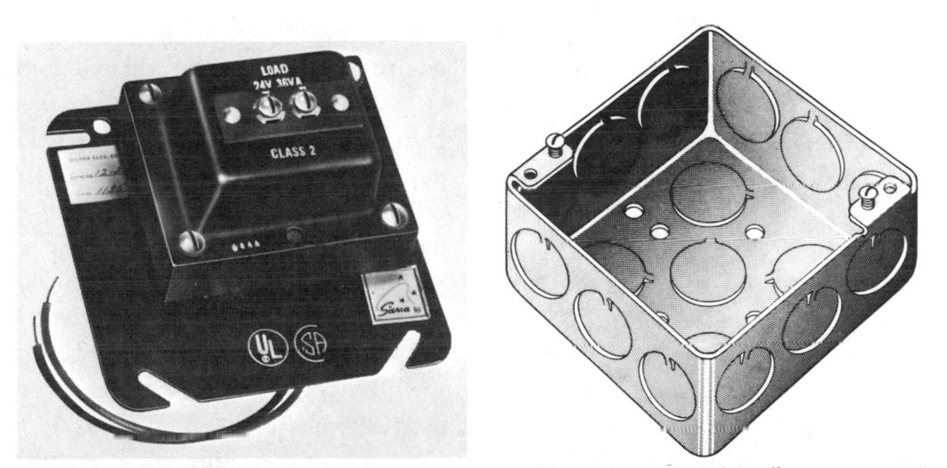

Fig. 8-6. Left: Low voltage transformer for remote control wiring. Right: A 4" square box for transformer mounting. (Left: Sierra Electric Corp.)

tects the windings from overload. Therefore, no overcurrent protection is needed in the secondary circuit; even a direct circuit will not cause the transformer to overheat.

These transformers are built to operate several relays at one time, the number depending upon the size of the particular unit.

The run from the push button to the relays must be considered. The capacity of the transformer and the current drain of the relays must be taken into account when arranging a wiring plan. Manufacturers of remote control equipment offer this information. As an example, three or four relays may be operated from a distance of about 100 feet. Accordingly, a lesser number of relays can be operated from a greater distance until the limiting distance for operating a single relay is reached. Generally, this is in the vicinity of 1000 to 2000 feet, depending upon the individual transformer, relay, and conductors. This distance, however, is usually great enough to handle any switching problem in the home or on the farm.

Rectifiers

Semi-conductor rectifiers are connected at the transformer in the control circuit of low voltage, remote control systems to improve performance of residential installations by eliminating any A.C. *noise* effects on the relay. They are used in other installations where *noise* may be an objectionable factor. For more extensive *noise control,* power supplies are available that produce full wave rectification for even better efficiency. Acoustics in a room may be so well controlled that rectifiers for direct current to operate relays may not be needed. The sound may only be momentary but sometimes objectionable.

Rectifiers convert alternating current to direct current and operate by offering a low resistance path or conduction in one direction, the forward direction, and a high resistance path in the reverse direction. They conduct when a positive instantaneous voltage is connected to the anode terminal and a negative voltage to the cathode terminal.

Relay

The relay used in remote control wiring is a single-pole switch that is set in the ON or OFF position by exciting the proper coil element. Once the relay is set in either position, it remains there until the other coil element is energized. For this reason it is unnecessary to maintain excitation after switching is accomplished. Therefore, the switches that control the exciting current need be only momentary-contact units, or push buttons. The 24-volt leads and the 120-volt leads come from opposite ends of the unit. The barrel of the relay is inserted through a ½″

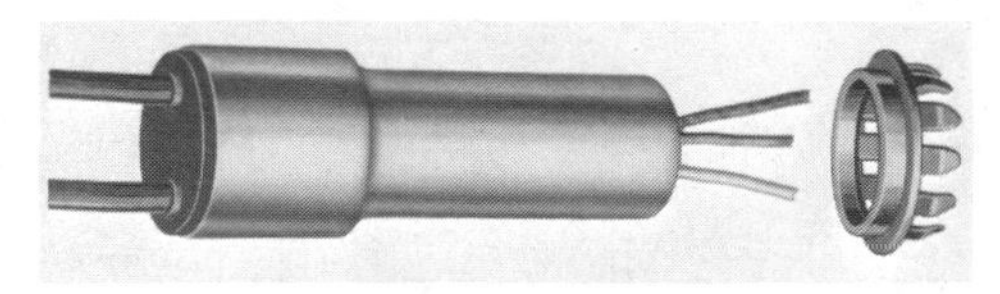

Fig. 8-7. Low voltage remote control relay with mounting ring. (Sierra Electric Corp.)

knockout from the inside of the outlet box.

This arrangement isolates the low-voltage wiring from the high-voltage wiring, as required by the National Electrical Code. Switching in the high-voltage circuit is handled within the box, while low-voltage control equipment is on the opposite side of the separator formed by the outlet box. This construction allows the relay to be placed at the desired switching point without additional mounting and isolating devices and running heavier load wires to switches.

Fig. 8-7 shows the construction of one particular relay. The high-voltage leads and their associated contacts are built into the portion of the relay remaining within the outlet box. The solenoid and low-voltage leads are set in the opposite end. Note larger size wire on 120 volt leads.

Control Switches

The control switches, which are constructed for flush or surface mounting, are actually two push buttons built into one unit, as seen in the diagrams. These momentary contact switches provide several important advantages.

1. An unlimited number can be wired in parallel, since none are actually "in" the circuit except during the short period that they are being pressed.

2. One type of switch performs the functions of a single-pole switch, a 3-way switch, and a 4-way switch with no complicated wiring.

3. Switches can be added to existing circuits at any time in the future *without* changing the wiring.

A plaster ring box cover is used to support flush mounted switches, but the surface mounted units can be installed almost anywhere. This is especially advantageous when switches are to be placed on thin partitions, brick filled walls, or other mounting locations having limited depth.

Fig. 8-8 illustrates common single and three gang voltage switches. Note Fig. 8-8, bottom left, method of mounting switches, which shows a mounting strap for a single switch. switch, Fig. 8-8, top left, is mounted in hole in strap shown in Fig. 8-8, bottom left. With a screwdriver the tab is indented into the slot seen on the switches, Fig. 8-8, top left. This is done with twisting motion of the screwdriver. Fig. 8-8, right, shows three switches mounted in a finish plate.

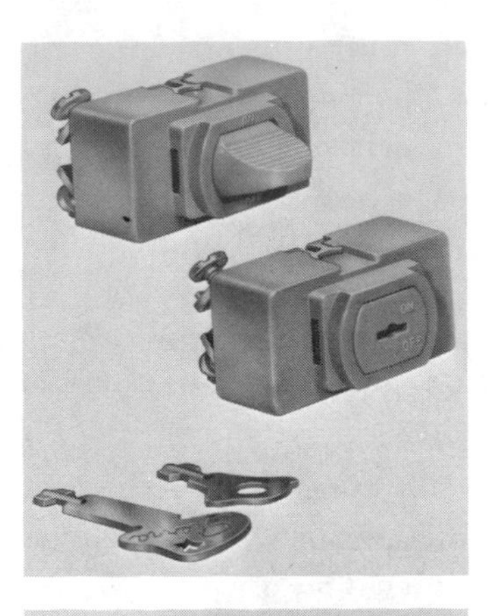

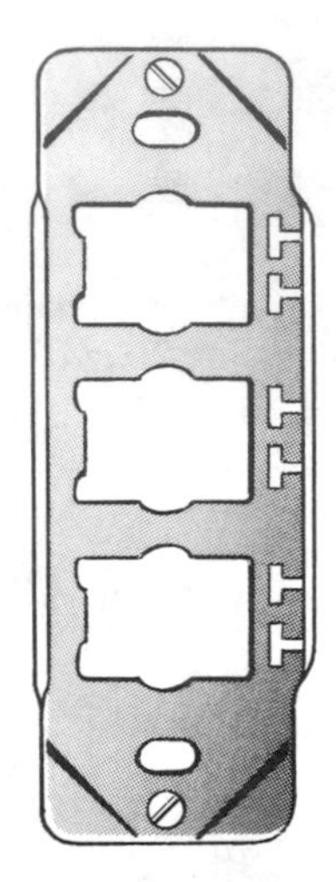

Fig. 8-8. Control switches. Top left: Low voltage switches: toggle handle and key type. Bottom left: Three low voltage switches mounted with finish plate. Right: Standard, interchangeable, single-device strap used in mounting switches. (Left: Sierra Electric Corp.)

Master Selector Switch

The selector switch provides one of the outstanding features of low voltage, remote control wiring. It permits control of any or all relays from a centrally located position; for example, the bedroom. This selector switch, Fig. 8-9, top left, has the usual push button, and, in addition, it has a selecting device. The selecting device inserts the push button in circuit with any one of a number of different relays or groups of relays. Fig. 8-9, top right, shows details of the selecting device. The bottom part of Fig. 8-9 shows a pictorial of the wiring for the selector switch circuit.

The selector switch allows all circuits to be turned on or off from this point. Depressing the push button in the OFF position and rotating the selector through the several positions accomplishes this result. With proper wiring arrangements, it is possible to turn particular circuits off and others on with a single motion. This is done by reversing leads of units that are to be turned on when the OFF button is pressed. These switches are also available with 12 circuits and the manual master selector switch operations may also be

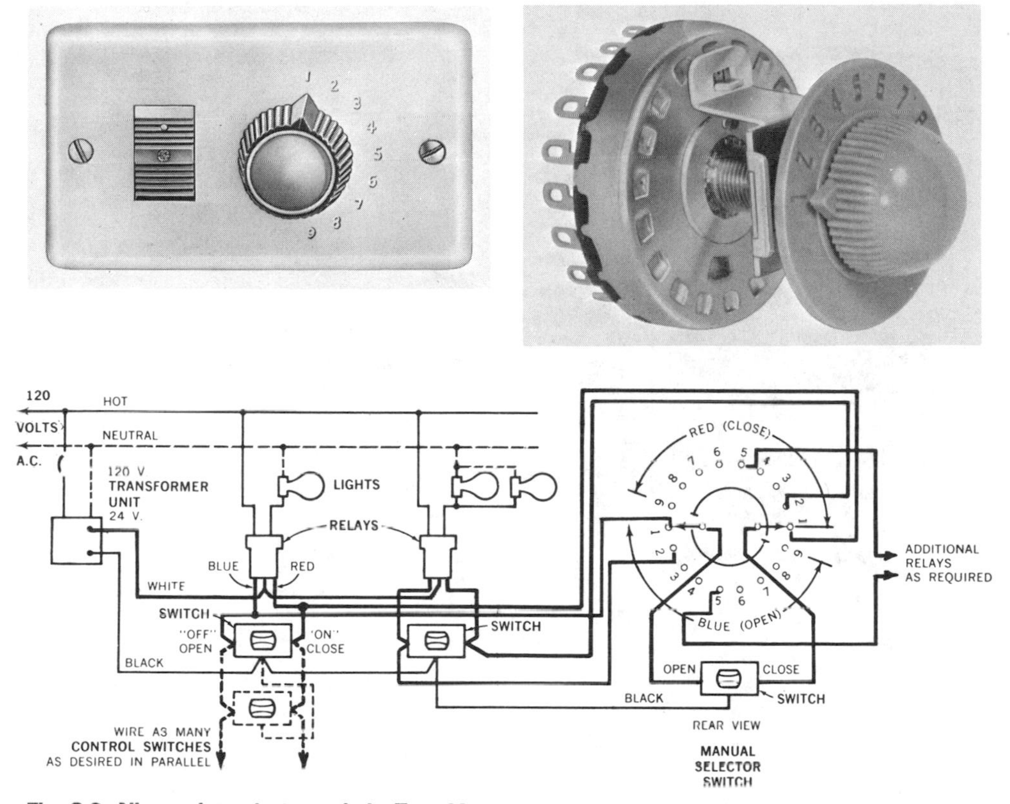

Fig. 8-9. Nine point selector switch. Top: Master controls. Bottom: Wiring for manual master position (9 position). (Top left: General Electric Co. Top right and bottom: Sierra Electric Corp.)

accomplished by a motor driven master control. Motor driven master controls are available with 23 positions. Fig. 8-10 shows a typical 23 position, motor driven master controller. Fig. 8-11 illustrates the wiring diagram for a motor driven controller. Note that manual switches are wired parallel with motor controlled wiring terminals. By placing the switches at a remote point and using a motor to do the sweeping, users can get 23-circuit control simply by touching one of the control switches that can be located at any desirable location.

Wire

Remote control wiring is generally done with No. 16 or 18 TF (thermoplastic covered) wire or equivalent. Better check code for requirements. Generally, no additional protection is necessary. These conductors are

Fig. 8-10. Motor master controller with 23 positions. (Sierra Electric Corp.)

run in the open through partitions, floors, and ceilings.

The push buttons have a marked ON-and-OFF position so that the proper section of the relay solenoid is excited to open or close the high voltage contacts. For this reason, there must be consistency in the wiring procedure. One recommendation is to employ color coding. Examples of this color code are noted in Figs. 8-4 and 8-5. A second method codes the wire itself. A ribbed section on the outside of one of the wires of the group. Fig. 8-12, serves to identify all three conductors. Thus, the ribbed wire may be taken as the common one, the middle wire the closing circuit, and the lower wire for the opening circuit. Multiple conductor cable to 23 conductors is available for such applications as connecting centrally ganged and located box to manual or motor driven master selector switches. Fig. 8-13 illustrates a multi-conductor cable. (Fig. 8-20 shows how multi-conductor cable is run through ceiling joist and partitions.)

Heavy current-carrying conduc-

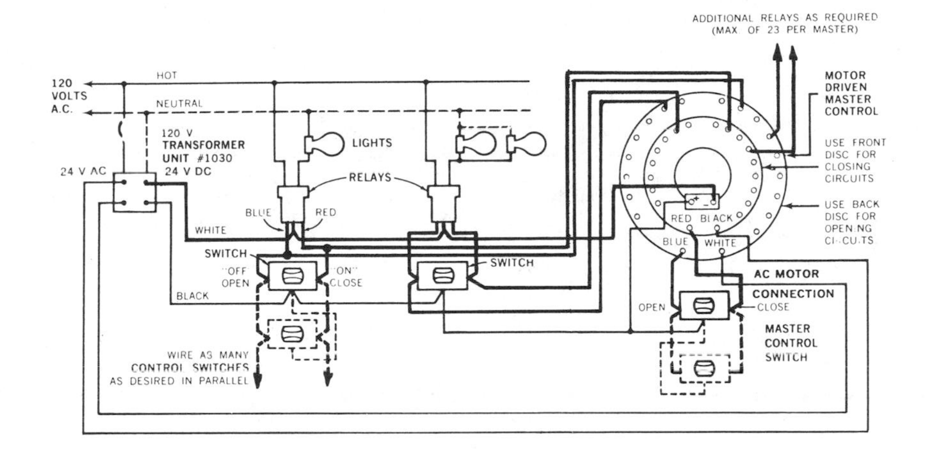

FOR AC OPERATION OF RELAYS
install jumpers included as shown. No direct connection is required to operate motor of Master Control.

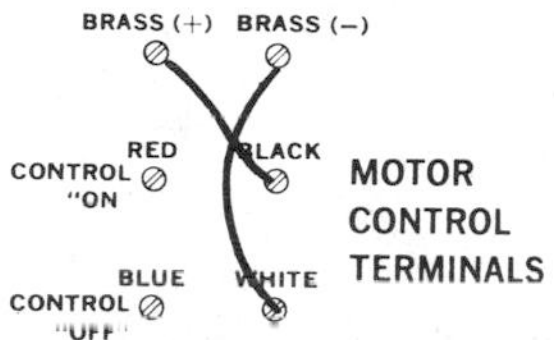

Fig. 8-11. Wiring for motor driven master control (23 position). Motor on master control is A.C. only. For D.C. relay operation connect black and white terminals of master control to 24 v A.C. terminals on transformer. Connect relay system to 24 v D.C. terminals of transformer. For A.C. relay operation use jumpers included, connecting brass (+) to black and brass (—) to white terminals. See wiring diagram, bottom. (Sierra Electric Corp.)

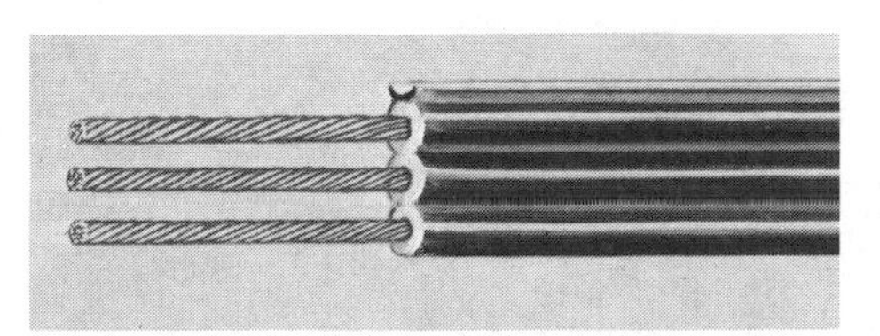

Fig. 8-12. Three conductor cable, showing identifying rib on upper conductor. (General Electric Co.)

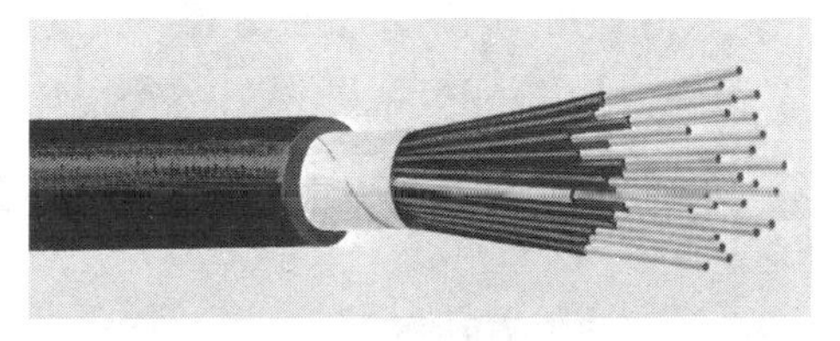

Fig. 8-13. No. 18 multi-conductor cable, color coded. (Sierra Electric Corp.)

tors do not have to loop down to the switches or make several runs across a long distance to permit several points of control. Instead, the conductors go straight to the box where the relay can be located, while the control switches use light, inexpensive wire that make the many loops to all the switches. This eliminates the need for more expensive current-carrying conductors and minimizes line-voltage drop.

When wiring, leave 5″ or 6″ of extra wire length at the relay. This allows enough slack to replace faulty relays without difficulty.

Planning Remote Control Wiring

As in any wiring project, the desired operation of electrical facilities should be determined in advance. The flexibility of remote control wiring offers a wide choice of controls in switching arrangements.

Switches placed at the entrance of the home can control selected lights inside, so that entry is made into a well-lighted building. The front entrance can be lighted from additional points in the building; for example, the kitchen. Radio or television can be turned off from the telephone location. Reassurance is immediately gained in moments of necessity by illuminating home and yard from a switch in the bedroom. Exhaust fans as well as heating units can also be controlled from one or more remote locations.

Locating the Relay

There are three general methods of arranging remote control relays. Each has distinct advantages in the type of service to be rendered.

Outlet Mounted Relays. This is probably the simplest method from the standpoint of installation. A saving in high voltage wiring is also effected by placing the relay at the point of control. With this method, however, the individual relays are not readily accessible if servicing should be necessary. Fig. 8-14 illustrates this type of mounting.

Gang Mounted Relays. Here, all of the relays are mounted in one conveniently located panel. This scheme provides the simplest arrangement for servicing, which is an important consideration when a large number of units are to be operated. There is an additional feature to gang mounting, namely, it provides completely silent operation. With the panel box in an isolated location, it is impossible to hear the tripping of the relays. This mounting is shown in Fig.

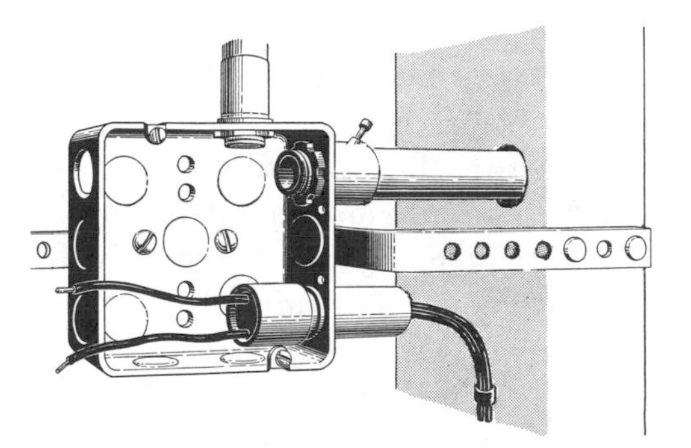

Fig. 8-14. Method of mounting outlet boxes with remote control relays between studs or ceiling joists.

Fig. 8-15. Gang-mounted relays installed in attic. (General Electric Co.)

8-15. Note that 115 voltage supply wiring runs from relay to outlet.

Zone Grouped Relays. Arranging the relays in several groups affords some advantages. The residence, or other installation, is divided into areas in which the control of outlets is limited to four relays. This reduces the high voltage runs as compared with the system employing a single panel box, yet retains most of the accessibility gained with the single panel. The usual procedure is to mount the relays in a centrally located housing, the size of which is determined by the dimensions of the particular relays. The box, in this case, provides the necessary shielding between high-and-low voltage sections.

Locating Transformer

Location of the transformer is not the same for the three relay arrangements. When gang mounting is employed, the transformer is situated at the panel, Fig. 8-15. However, when either outlet mounting or area grouping is employed, the transformer should be more centrally located. This is usually at some convenient and readily accessible point in the basement or attic.

Locating Selector Switch

Probably one of the most convenient locations for the selector switch is in the bedroom, since all lights may be properly set for the night before retiring. Variations in wiring the selector switch, as already noted, will turn on lights that should burn all night—gangway porch, hall, or bathroom—and turn off all other lights. A second desirable location for a selector switch is the kitchen, permitting control to be extended to other portions of the building without interrupting common household duties. The family room is a third location that may warrant a selector switch, because it is usually the center of activity.

Wiring Procedures

Installing Outlet Boxes

Outlet boxes are mounted in the conventional manner, since the use of low voltage remote control does not affect supply wiring procedures. It does, however, simplify the switching of circuits. Fig. 8-14 illustrates one acceptable method for mounting an outlet box. The box is held in place by a mounting strap attached to the joists. This illustration also shows two sections of conduit entering the box, and the remote control relay in place.

Controlled Load Wiring Installation

Fog. 8-16 illustrates a typical installation using non-metallic sheathed cable or armored cable. The construction here is with gang mounted relays centrally located. Uncontrolled receptacle outlets are not shown. Note how load wiring lighting circuits are brought to relay box for switching. This method may use a little more wire but will save in splices and box sizes and even plaster rings in some cases.

When metallic flexible conduit or electrical metallic tubing is used the outlet runs are very similar to the conventional wiring job, with the omission of the switch legs. All 120-volt lighting runs start at the relay gang box, providing switched return

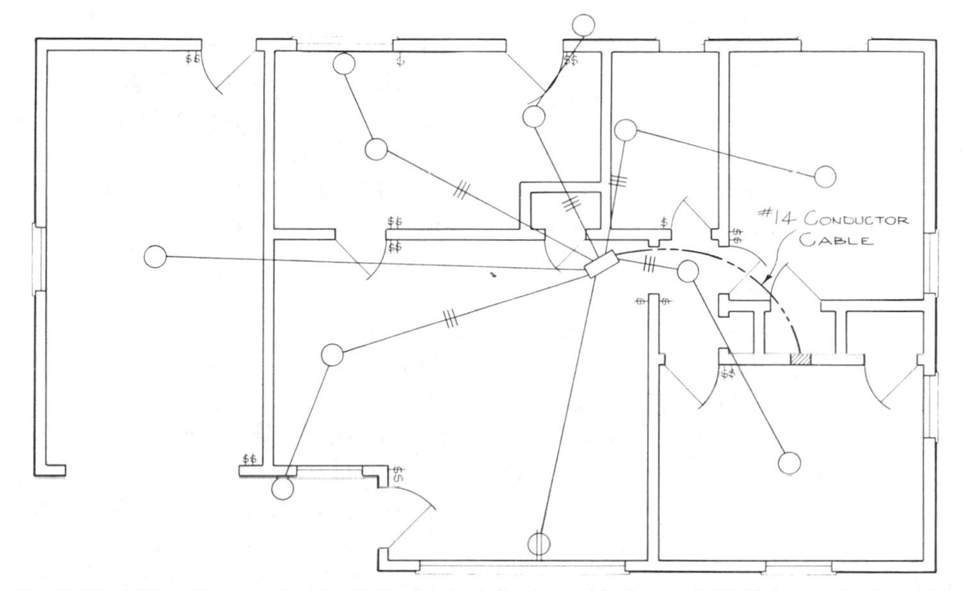

Fig. 8-16. 120 volt runs when both 2 wire and 3 wire cable is used. Multiple conductor cable is run from relay gang box to master control center.

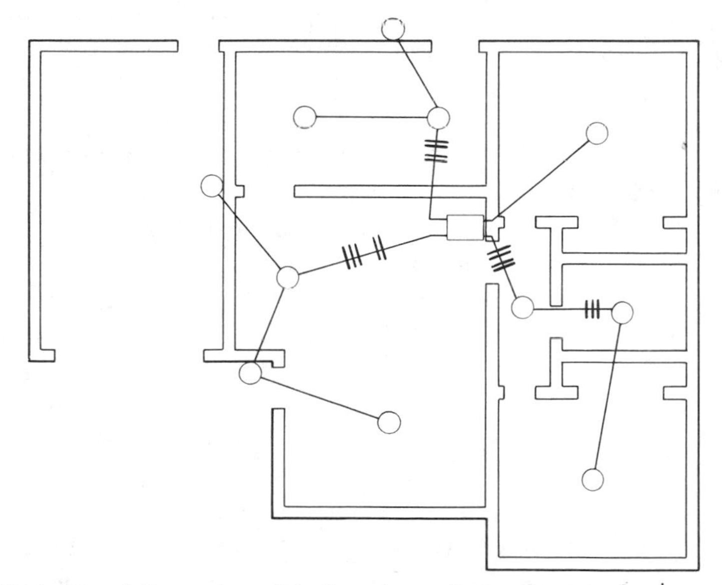

Fig. 8-17. Typical conduit runs to outlets from low voltage relay gang box.

legs from each relay and neutrals to all relay controlled outlets. The cable or conduit runs are shown more simply in Fig. 8-17.

One leg from a relay and a pigtail lead from the neutral for fixture connection is dropped off at each outlet box.

The relay gang box is fed from the service distribution panel. In large residential construction, relays are zone grouped. Relays are strategically located in zones of more than one gang box to save long conduit runs.

When the relays and gang box are first mounted the relays should be identified by the circuit they will feed. A suggested identification is as follows:

1. Kitchen Light
2. Front Porch Light
3. Living Room Light
4. Family Room Light
5. Etc.

Space for identification can be found inside the box cover.

As the high voltage wires are pulled into the relay unit, they should be made up (connected) at the time they are pulled. The circuit being pulled is known, the identification chart can be checked for the proper relay and wire connected to it. This avoids having to work with a lot of loose wires, saves confusion and mistakes, and also saves labeling the wires with labels that are apt to be lost or become illegible before the wires are connected. Relay-circuit identification is vital to the wireman who may be called to troubleshoot electrical failures within the home years after construction completion.

Installing Switches

Switch installations in the low-voltage remote control system is greatly simplified as compared to the conventional wiring system. The switch can be mounted on a plaster ring box cover attached to the studding, Fig. 8-18. This ring will accommodate up to three low-voltage switches. The stud may have to be notched slightly for 3 switches or vertical ring mounting. A nail is partially driven directly behind the ring opening to wrap or tie low-voltage wire until ready for connecting switches or to hold the wire while pulling. The wire or cable may also be stapled into position here.

Installing Relays

Relays can be gang-mounted on a metal strip, in a box, Fig. 8-15, which serves also to isolate high- and low-voltage circuits from one another. The method of installing outlet-mounted relays is shown in Fig. 8-14. Up to four relays may be installed in this manner for "zone location" of relays also. However, a check of local requirements should be made to

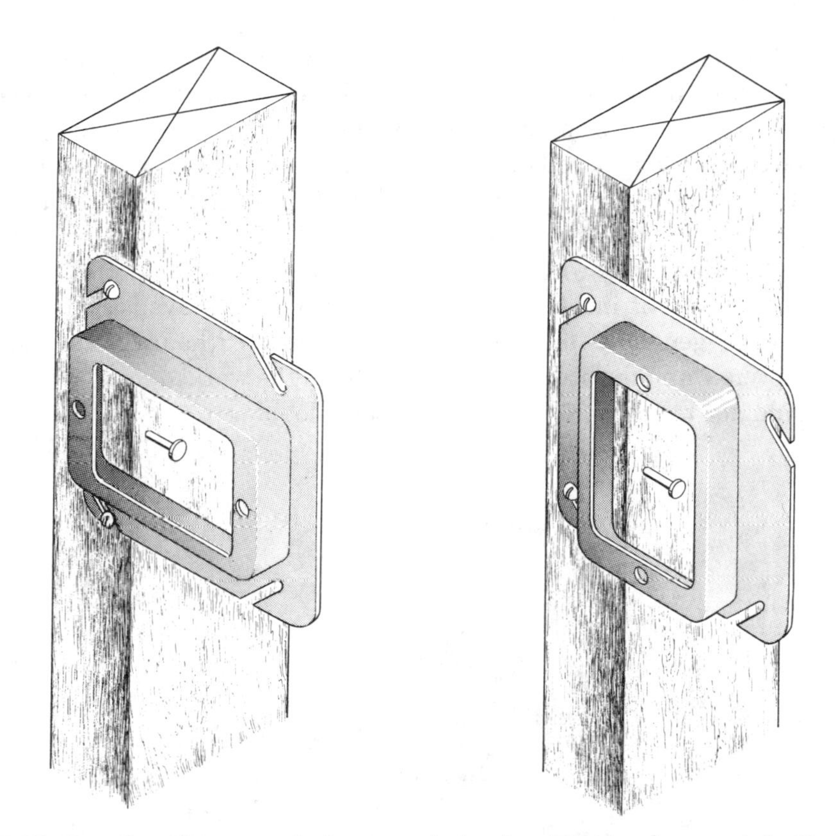

Fig. 8-18. Mounting 4" square, single gang, plaster ring. This can be mounted either horizontally or vertically, depending on correct position of switch to be used.

see that they do not have some minor variations. The relays, however, should not be attached until the plastering has been completed. That is, all wiring is completed beforehand, except installation of relays and switches. These items are withheld to prevent damage during the plastering operation. They are installed with "finish work" when receptacle outlets and fixtures are installed.

Installing Low Voltage Wire

When all the switch positions have been located and plaster rings installed, holes are drilled for low voltage wiring the same as for conventional wiring. Where wires run at right angles to the joist, the joist can be drilled where necessary or the wire may be run on joist attic surfaces with protective wood guard stripping used; and where the wires run parallel, they should be held loosely with drive rings or insulated staples. Tightly driven staples create problems. The spools of wire should be supported on a rod or rack as shown in Fig. 8-19 and placed below the relay gang box for easy pulling in all directions. The wire should be minimum 300-volt insulation capacity rating, generally of plastic. Cotton covered annunciator or bell wire should not be used.

If a relay is to be operated from three switch points, the wire should be fed through the attic space, using the drilled holes, to the farthest one of the three. It should be attached to the nail (8-penny) at the switch ring (see Fig. 8-18) so it will not pull loose and will be available for switch connection. Without cutting the wire it should be worked back toward the spool, looping down to the second

Fig. 8-19. Low voltage wire spool rack for easy pulling.

Fig. 8-20. Running multi-conductor cable through joists.

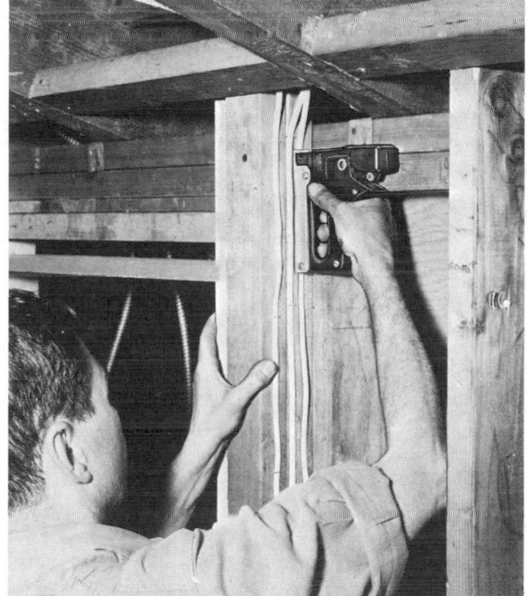

Fig. 8-21. Stapling cable to studding with staple gun. (General Electric Co.)

switch location and then to the first in this group. This method uses a little more wire but avoids splices. It also brings but one supply lead for the switches back to the transformer-rectifier for connection. If these switches are to operate the living room light, for example, the identification chart should be checked for the proper relay and it should be connected at this time. This procedure should be continued until all switch loops are connected to their proper relay. The low voltage wiring should then be ready for stapling.

Each wire should be stapled separately with an insulated staple so there is less chance of insulation damage and a short circuit occurring.

Color or number coding can be used, or two- and three-wire cable, which has an identifying feature, may be employed in remote control wiring. These systems vary somewhat. In one instance, black and white are used for the transformer secondary conductors. The white conductor leads to the common point on the relays (X, or No. 1), and the black conductor leads to the common point on the push buttons (X, or No. 1). Blue and red conductors are used between push buttons and relays. The red connects to the ON

Fig. 8-22. Low voltage wire should be stapled loosely. If for any reason the wires must be pulled tight, insulated staples should be used.

terminal of the relay (C, or No. 2) and the ON terminal of the push button (C, or No. 2). The blue conductor connects the OFF terminal (O, or No. 3) of the relay and switch.

With proper notation, the identified two- and three-wire cable can be used in the same way. Three-wire cable will be found especially useful in connecting push buttons in parallel and in making runs to the selector switch. Two-wire cable is most often used for runs from the transformer. These conductors must be run through joists and studs, Fig. 8-20. Wires used for remote control should be supported at 4½ foot intervals, usually by means of staples, Fig. 8-21. When low voltage wire is stapled it is not pulled tight. See Fig. 8-22. If it is pulled tight, insulated staples should be used.

Rewiring

Each remodeling job offers individual problems, and no single plan will satisfy all conditions. However, close examination of remote control wiring will show many ways it can be used to advantage.

Surface mounted switches make the low voltage system advantageous for rewiring, because they can be mounted so easily. Furthermore, the small two- and three-wire cable can be run behind moldings or along baseboards, and through areas that would not accept armored cable.

If it is necessary to run cable across a plastered wall, it can be laid in a shallow groove and plastered over. A method of inserting an additional outlet box to handle the relay is shown in Fig. 8-23. This procedure is used where the original box will not hold the relay, which is frequently the case with older wiring equipment. When it is desirable to replace a single wall switch with several distributed push buttons, the low voltage leads from the new push button locations are connected to the operating coil of a relay. The 120-volt leads that went to the old switch are connected to the high voltage contacts of the relay. Chapter 15, "Remodeling Wiring," may further assist with methods.

Further examination of this low voltage, remote control wiring system will show that it is often superior to conventional wiring, but will not replace it in every case. Low voltage wiring, however, should be given consideration on new construction projects as well as in old work.

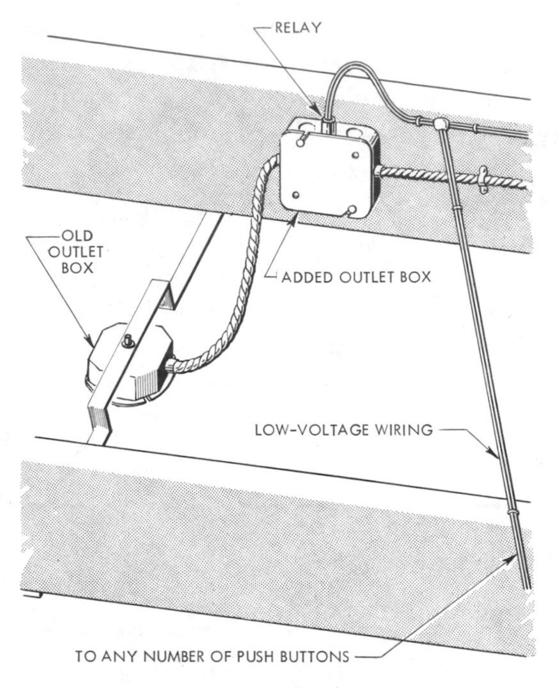

Fig. 8-23. Use of additional outlet box for mounting relay.

Troubleshooting

Sometimes a system may need trouble-shooting. If there is power available at the relays, they can generally be considered as operative until other parts of the system have been checked. Power consuming devices may be at fault and should be checked first (for example, bad light bulbs or poor contact at cord caps and receptacles). Painted switches may cause difficulty. Broken switch plates that allow low voltage wires to dangle freely are liable to short out. A continuous drain on the low voltage system will considerably reduce the normal expected life of the rectifier.

If a relay fails to operate it may be due to the relay making contact against a dead short circuit in the 120 volt wiring at the lighting fixture. Such a dead short can pass enough current before the circuit breaker trips, or a fuse blows, to weld the contacts in the relay together. If a relay failure is found, the cause should be eliminated before the relay is replaced.

If the low-voltage power unit is replaced and the cause of failure is not eliminated, naturally a reoccurrence of the trouble can be expected to happen again.

There is a two-wire control system also used, but it operates on a relay ratchet principle.

QUESTIONS FOR DISCUSSION OR SELF STUDY.

1. What method of electrical energy is used with remote control wiring systems?
2. How many low voltage leads are there on the relay discussed?
3. With outlet box mounted relays, what voltage is used inside the box?
4. How may two or more outlets be controlled by a single button?
5. What maximum distance can three or four relays be operated?
6. What is the approximate maximum distance for operation of a single relay?
7. Does current flow through the relay coil all the time the light is turned on or only momentarily?
8. Which type of outlet box is used for the described transformer mounting?
9. How are low voltage and high voltage wiring separated?
10. How many wiring terminals do single control switches have?
11. What is the mounting method of low voltage switches?
12. What devices are used for central control points?
13. What kind of wire is used for switch loops?
14. How are three and four way switching methods used with this system?
15. Name the common relay locating methods.
16. How is relay noise averted?
17. List different transformer locations and mountings described.
18. Where is the most desirable location for a selector switch?
19. How are switch loop wires secured?
20. What must be done if the outlet boxes in old work or remodeling jobs are too small or impossible to mount relays?
21. List advantages of the remote-control low voltage control system.
22. Why should relays be identified with the circuits they control?

QUESTIONS ON THE NATIONAL ELECTRICAL CODE

1. Define from the code: remote control circuit.
2. Under which classification system voltage and current limitation is residential relay switching included?
3. What is the maximum overcurrent rating protecting the 115 volt primary of a control transformer?
4. What is the minimum separation installation of normal low voltage wiring from lighting circuits?
5. If the minimum separation must be reduced what added protection is necessary?
6. How must conductors of different systems be separated in raceways and boxes?
7. Where three or more control conduc-

tors are used, what is the recommendation for grouping or covering?

8. What is the minimum voltage limitation of insulation on 24 volt control wiring?

9. What reliance is placed on current limitation to stop dangerous control curents?

10. What limitations are placed on circuits extending beyond one building?

11. What are grounding requirements for remote control equipment?

12. How should remote control wiring be protected around attic crawl holes?

13. What is the maximum number of remote control switches that can be installed to operate one relay?

14. How do the 115 volt wiring restrictions differ from the restrictions of other systems?

15. What are requirements for running control wiring as open wiring?

16. What are requirements for splicing control wiring in wooden frame construction?

17. How is control wiring protected from physical damage?

18. What is the stapling or supporting requirement for control cable?

19. How are control cables terminated?

20. What are limitations of relay switching control systems with circuits employing a neutral conductor?

Large Appliances and Air Conditioning

Chapter

9

The electrician can often suggest to a homeowner electrical appliances that may not have been known or thought of when the building was planned. The more information the wireman has on these appliances the more advantages he can point out—especially if it is more convenient and economical to wire on new construction.

Large appliances tend to consume the most electrical energy. This requires larger wire sizes, conduit, circuit breakers or fuses. Most important, it may require quite a larger electrical service than originally planned. It is a distinct advantage to install a larger service on new construction to accommodate greater power demands for immediate needs or for the future.

The most commonly installed and popular large power consuming appliances are discussed in this chapter.

Electric Ranges

The first large electric appliance to gain popularity in the home was the electric range. Fig. 9-1 illustrates modern ranges. They are clean, efficient, and equipped with automatic timing devices that make cooking a comparatively simple chore.

A three-wire, 120-240 (or 115-230) volt supply is needed for this unit with ample wire ampacity (amperes capacity). Fig. 9-2 shows a simplified diagram of service entrance and circuit wiring for a range installation.

The wiring shown here is installed

Fig. 9-1. Modern electric ranges. (Left: General Electric Co. Right: Hotpoint Division of General Electric Co.)

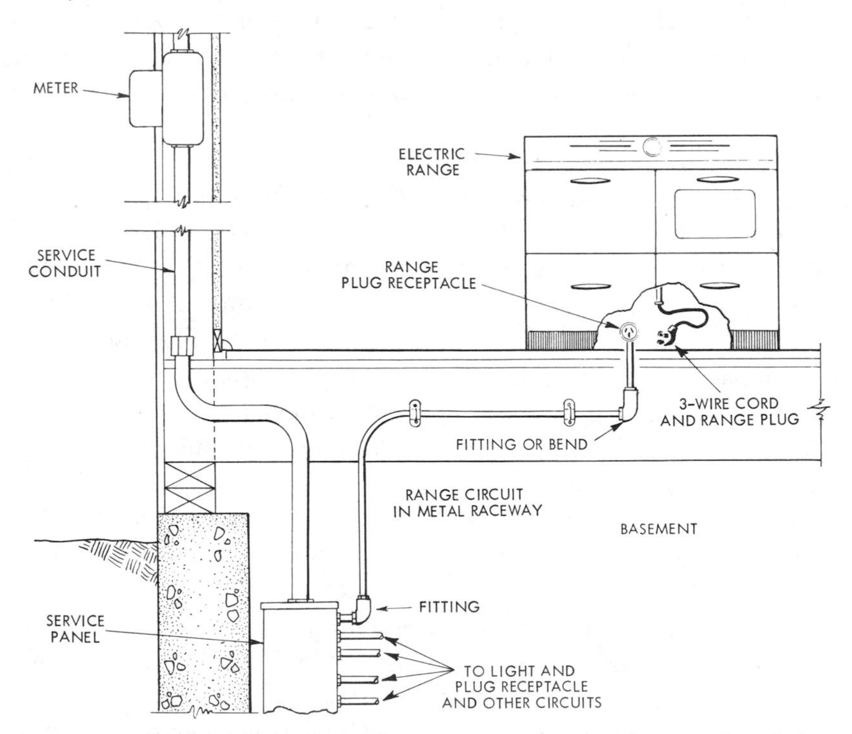

Fig. 9-2. An example of service and circuit arrangement for electric range installation.

in either rigid or thin-wall conduit. It could be done with armored cable, or even with service entrance cable, provided local authorities allow it. Nonmetallic sheathed cable may be used for the range circuit.

Normally, the range circuit terminates in a heavy-duty, 50-amp receptacle, and the range is equipped with a three- or four-wire cord and plug. In existing systems or installations (see Fig. 9-3) the frame is grounded to the neutral conductor of the cord (three-wire). This is also true for electric clothes dryers. When using four terminals, one is used solely for grounding the equipment. The receptacles, Fig. 9-3, are used for either flush (Fig. 9-3, right) or for surface mounting (Fig. 9-3, left). It is customary to use the flush type in new construction, and the surface type on alteration work. The surface outlet may be installed on the wall or floor under the range in remodeling work. A 4-inch square box must be used to mount the flush outlet. The cord and plug are the same in either case.

Fig. 9-4 shows the wiring materials needed for a flush mounting with wall finished installation for range or electric clothes dryer. Four inch square outlet boxes (Fig. 9-4) are used. Dryers and combination clothes-washer and dryer units normally require 30 ampere receptacles compared with 50 ampere receptacles for ranges.

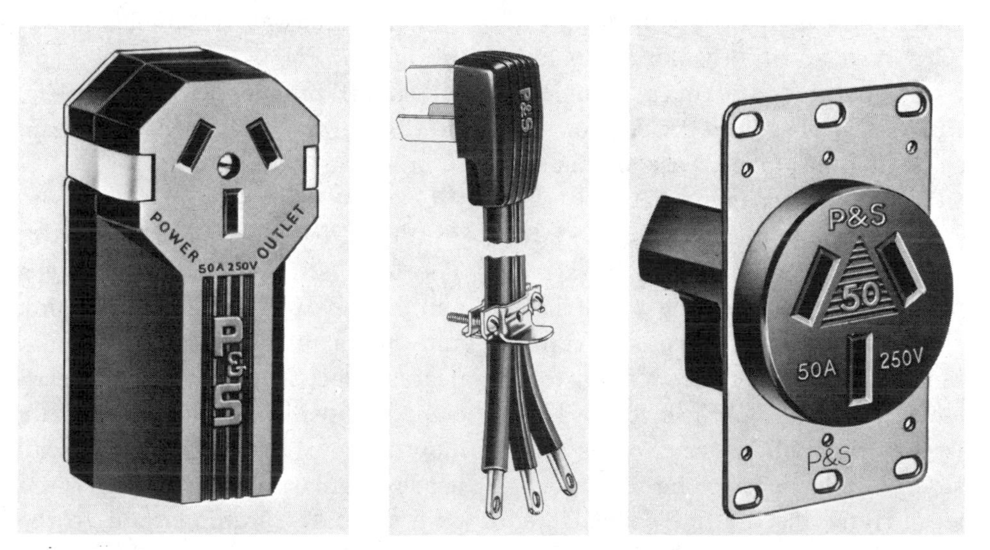

Fig. 9-3. Range plug and receptacles in existing systems. (Pass & Seymour, Inc.)

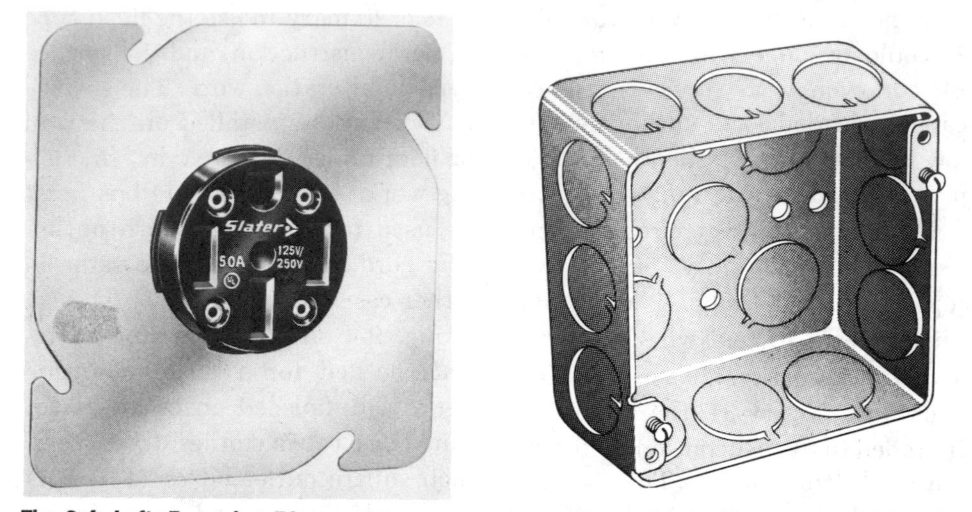

Fig. 9-4. Left: Four-wire, 50 ampere range receptacle to mount into 4″ square outlet box for flush into-the-wall installation. (Slater Electric, Inc.) Right: A 4″ square box for transformer mounting.

Built-In Units

The electric range of Fig. 9-1 has been supplanted, to a large extent, by equipment like that in Fig. 9-5. The oven and the cooking top are like separate parts of a complete range which has been divided for the sake of appearance, ease of operation, and kitchen convenience.

The two units are connected to individual circuits of lower rating than a range, depending upon their size. A common practice is to run a feeder from the service location to a small panelboard in the kitchen. This is generally done when there is considerable distance to the main service entrance distribution panel. Here, separate circuits, as shown in Fig. 9-6, are run to each heating device and to other fixed appliances such as garbage disposal and dishwasher.

A closer look at the panel is shown in Fig. 9-7. Fig. 9-7, left, shows typical circuit breaker wiring and Fig. 9-7, right, the flush door and trim to finish the installation.

Garbage disposal units are waste grinders driven by an electric motor. The unit is mounted, usually by a plumber, between the kitchen sink and the waste disposal line. It is connected electrically by the electrician. An operating switch is usually mounted on the sink splash wall nearby. This unit should be wired on a separate 20 amp circuit to the panel.

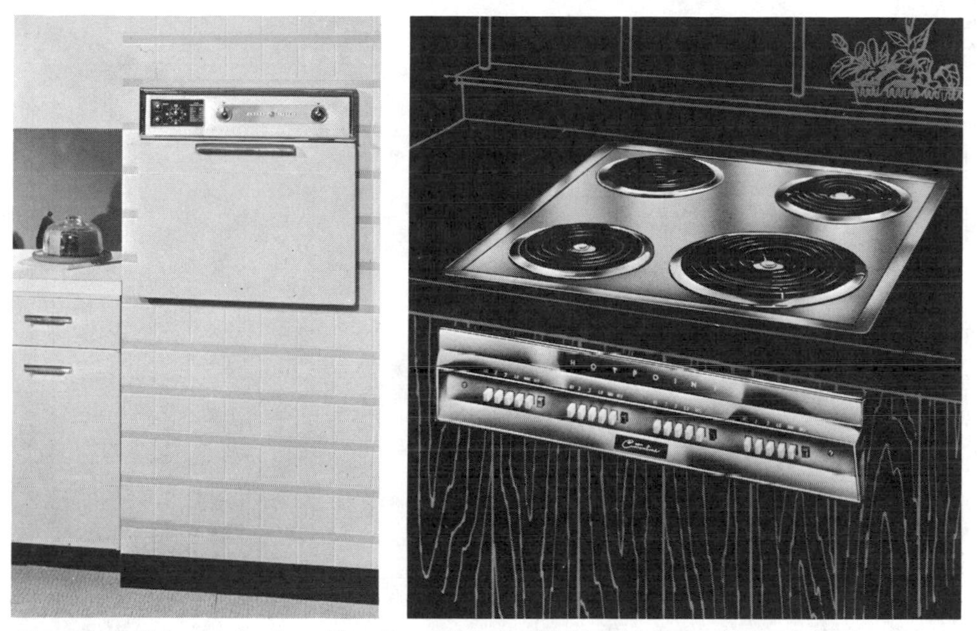

Fig. 9-5. Built-in oven and cooking top. (Hotpoint Division of General Electric Co.)

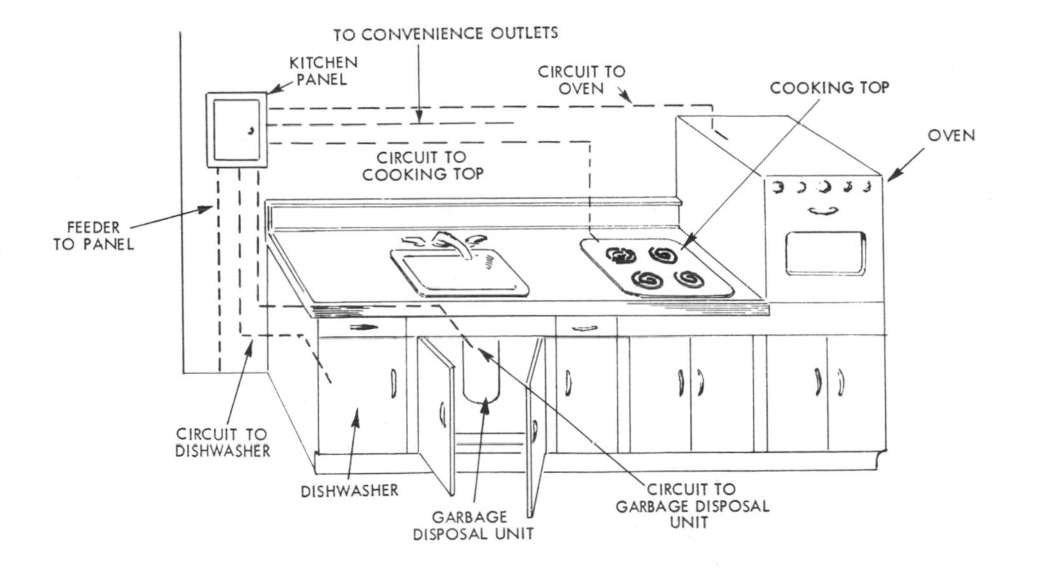

Fig. 9-6. Circuits to kitchen equipment.

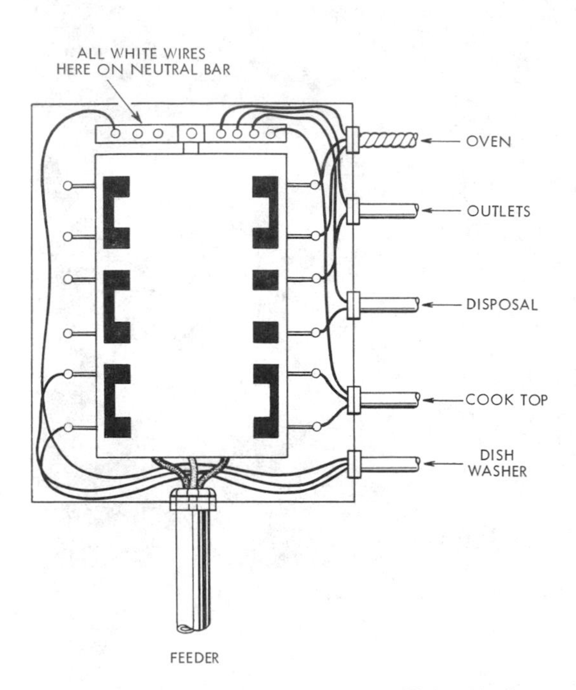

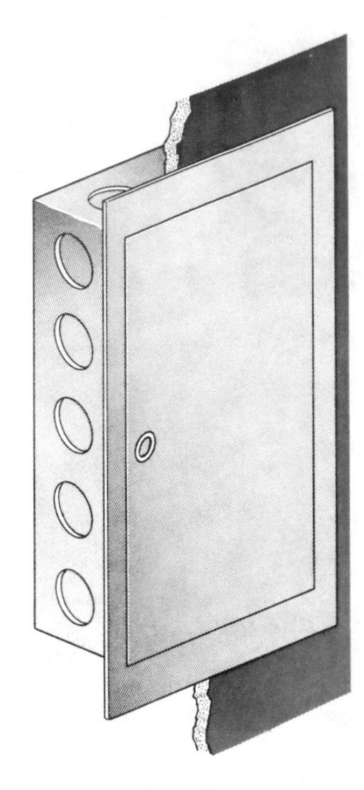

Fig. 9-7. Left: Typical panel wiring. Right: Flush cover panel door.

A dishwasher is a motor driven appliance also and should be connected to a separate 20 ampere circuit for satisfactory service and to meet code requirements.

A length of flexible conduit or armored cable is usually run from the panel or a nearby accessible junction box. The end for the appliance is nailed to a stud at the appliance location and is allowed to extend out of the wall a few feet for connection to the appliance terminal box after plastering and painting. Other desired methods of wiring should be discussed with the local department of building and safety.

The most amazing cooking advance of the century is electronic cooking. An electronic range or "built-in" cooks with high frequency energy called microwaves. These microwaves are absorbed by food causing the food molecules to vibrate against each other resulting in friction. This friction creates heat within the food and cooks it faster than any other method known. These units look like the conventional oven but require more power to operate. This effects an increase in all branch circuit component sizes and the service also. They are wired on separate circuits like a conventional built-in or range. Wiremen sometimes call electronic cooking units "radar ranges."

Kitchen Ventilators

One of the greatest boons to the family cook is the addition of ventilation equipment — adequate in size and properly installed. Fig. 9-8 shows a combination hood ventilator and food warmer. This versatile electrical appliance contains two 250 watt infra-red fixtures for keeping plates and foods hot. Extra range lighting is also contained. For ridding the kitchen of odors, heat, smoke, and moisture there is a multi-speed exhaust blower. All of these power consuming devices must be considered by the wireman when laying out his wiring distribution and load center.

Dryers

The automatic clothes dryer, Fig. 9-9, left, is another useful item in the home. The average size of the 230-volt unit is 5000 watts (5 kw). A

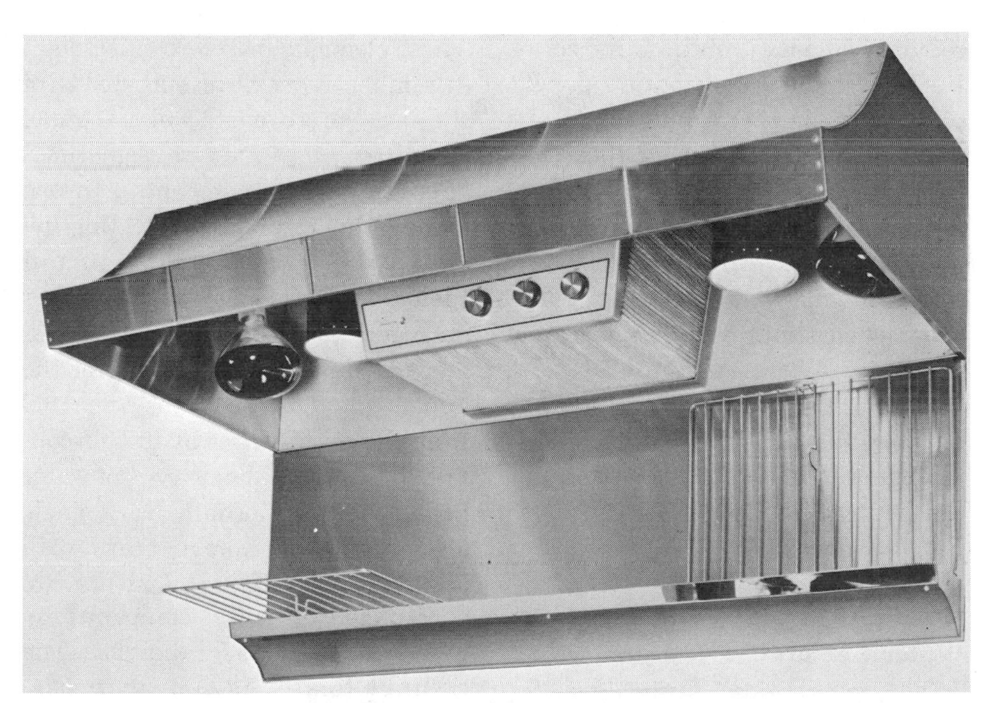

Fig. 9-8. Combination kitchen ventilator and food warmer. (Norris Thermador Corp.)

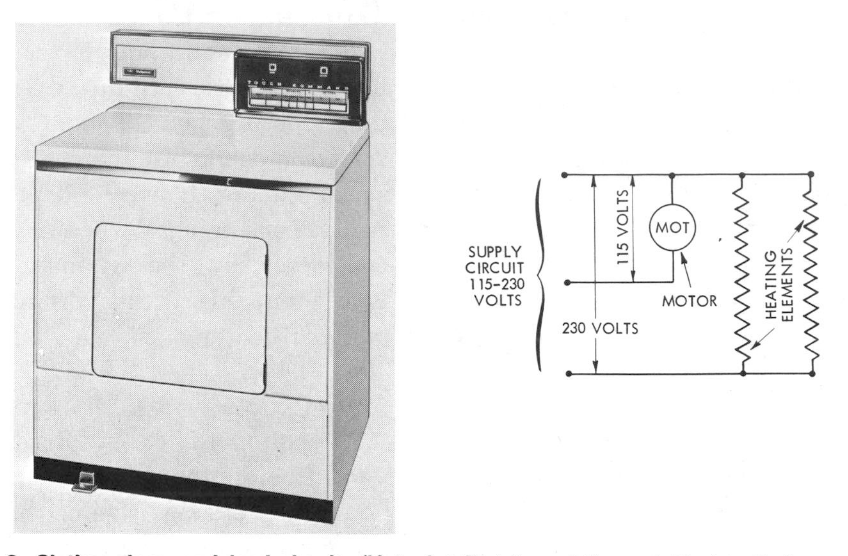

Fig. 9-9. Clothes dryer and load circuit. (Hotpoint Division of General Electric Co.)

separate circuit is run from the service location or a distribution panel, terminating in a plug and cord connection similar to that for the electric range. The receptacle and plug for dryer mostly are not larger than 30-amp. A three-wire, 115-230 volt circuit is employed here because the heating elements are designed for 230 volts, while the motor is rated at 115 volts. The heaters are connected between the 230-volt wires, the motor between one of these wires and the neutral conductor, as shown in Fig. 9-9, right.

Water Heaters

The round water heater, Fig. 9-10A, has two heating elements. These elements may be connected in a number of ways to suit needs of the user, or to enable him to take advantage of special rates sometimes offered by power companies. In one mode of operation, both heating devices come on at the same time if their thermostats so direct. Another arrangement permits the lower element to draw current only after the upper one has been disconnected from the circuit wires by its self-contained relay. The home type of water heater, however, usually has a single element. Large storage tanks with low recovery heating elements are encouraged by power companies by giving special rates for their use. The cabinet heater is shown in Fig. 9-10B.

In some parts of the country power companies give special heating rates if current is taken only during hours when the supply lines are not heavily loaded. Fig. 9-11 shows a simplified scheme for making use

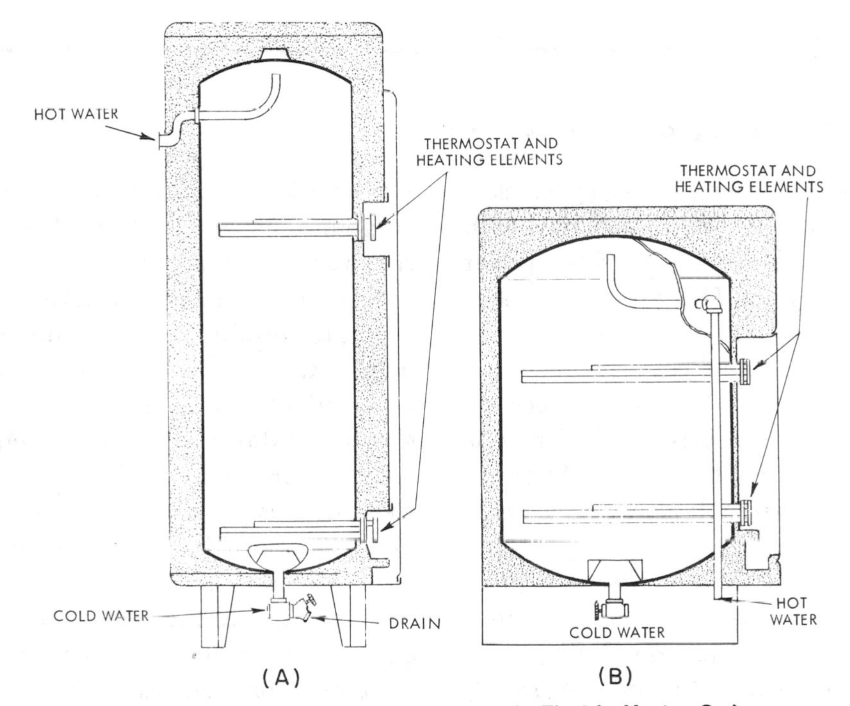

Fig. 9-10. Round and cabinet type waterheaters. (Wesix Electric Heater Co.)

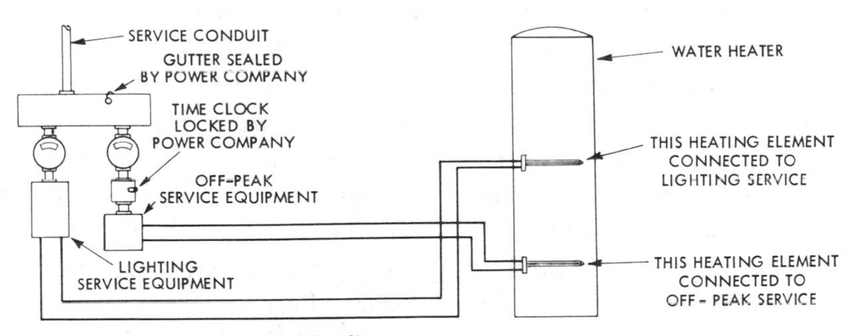

Fig. 9-11. Off-peak water heater circuit.

of this plan. One of the two elements of the water heater is connected to the lighting service meter, the other to an off-peak meter and a time clock which permits current to flow during certain prescribed hours. The time clock is sealed by the power company to prevent tampering with the mechanism. The element connected to the lighting meter is the smaller one, which maintains a desired temperature after the large element has created the proper degree of heat.

Space Heating and Air Conditioning

The increasing efficiency of electric power generation, coupled with greater use, has kept electric power rates very low. Domestic power rates have decreased during the past years. Other fuel costs, on the other hand, have increased appreciably over this same period. This trend is a forerunner of the future, when electrically heated homes will become commonplace.

The essential purpose of space heating is to replace heat lost through transmission from inside the premises to the outside. Determination of the amount of power required, expressed in watts, is a complicated process. It depends upon the difference in temperature between that desired inside the house and the prevailing temperature of the atmosphere. It requires a knowledge of average and maximum temperature variations likely to be encountered, and an understanding of heat-transmission constants of building materials. After these factors have been properly assessed, calculations may be performed to arrive at an approximate answer.

Heating engineers achieve more accurate results in new buildings whose exact design features are readily available to them. On existing structures, however, even the engineers run into trouble. Since there is too much risk of creating a dissatisfied customer, the matter of laying out a heating system is not a proper one for the average electrician or contractor without a considerable amount of previous experience.

After considerable experience with the heating of a particular type of structure, one may acquire a fair degree of skill. Until then, experts should be expected to assume risks connected with design of installations. It is within the province of the electrician, nevertheless, to recommend and to install some of the various types of unit heating devices treated here.

Heat Required. The heat required for each room must equal or exceed the heat lost through WALLS, CEILINGS, FLOORS, and GLASS by heat transfer to the colder outdoors, and by infiltration of air around windows and doors, and through the normal opening and closing of doors to the outside.

The heating requirement calculations coupled with heat loss calculations are quite complicated. A rule of thumb requirement for average residences, well insulated from the weather, is 1½ watts per cubic foot, or 10 to 12 watts per square foot of floor space.

Placement of Heating System Equipment. Logic should assist the installer in the proper placement of heaters. In some cases, however, the construction of the home, interior decor, and the personal tastes of the architect or home owner will require some compromise regarding original location by design engineer.

Fortunately, there is a wide selection of heaters from which to select to provide the most compatible system.

The following should be considered on every job:

1. *Avoid* placement of wall heaters on the inside (warm) walls.

2. *Avoid* excessive heat concentration in any one spot of any room. Generally, any room having a heat requirement *greater* than 4500 watts, should have two or more heat sources and possibly two or more control points. For best heat distribution, highest comfort standards, and most economical operation, rooms with lengths more than twice the width should have two or more heat sources (if its greatest dimension is 20 feet or more).

3. *Never* place an automatic heater in a central hall to provide heat for adjoining bedrooms.

4. *Never* install a manually operated heater in a home, except in auxiliary or service areas and in gamerooms, hobbyshop rooms, etc. Heating engineers, employed by the manufacturers of heating equipment, are also available for needed advice.

Central Heating

A central electrical heating system has a large heating chamber which contains resistors, a motor and blower, and a system of air ducts that lead to wall or floor outlets in various rooms. Flow of heat is governed by a thermostat (a subject which will be discussed later).

The center of a central heating system is the electric furnace.

Electric Furnaces. The exterior of many electric furnaces, as shown in Fig. 9-12, is well designed and pleasing in appearance. Electric furnaces heat *only* by convection (transfer of heat by circulation). The resistance heating elements (Fig. 9-13) are arranged in a series and operate only as heat is needed. The motor driven blower propels the air

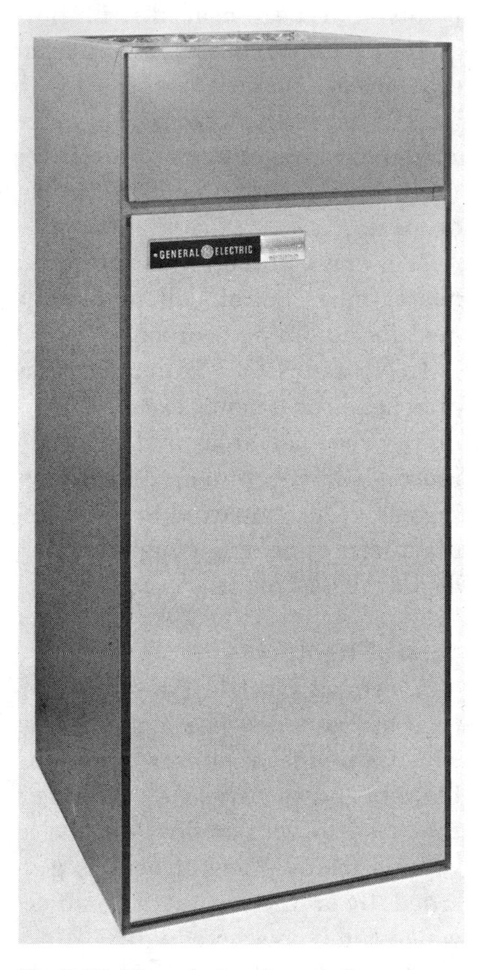

Fig. 9-12. The exterior of many furnaces are tastefully designed, as is this electric furnace. (General Electric Co.)

over the heating elements, through the ducts, and into the rooms. Electric furnaces are not limited to forced-air heating; they may also be used with a boiler to supply hot water for radiators, or convectors, and radiant panel heaters.

Some electric furnaces for residential heating are small enough to be placed in a closet or suspended from the basement ceiling. These are used to heat an entire house.

The wireman must know the total current required for the unit he is installing. This information is necessary to determine wire and conduit sizes. He must also know the location of the furnace terminal connection box so that he may provide a conduit in the proper place on rough wiring.

Duct Heating. A blower circulates air through the ducts, as shown in the illustration, Fig. 9-14. There is a heater in the main duct, near the blower, and an individual heater in each of the branch ducts. See Fig. 9-15. A master thermostat, connected to a magnetic switch, determines whether or not current shall be supplied to the circuit-breaker panel. Individual duct heaters are regulated by separate thermostats which are located in the several rooms connected to the branch ducts. Circuit breakers in the panel protect the duct heaters, the wires leading to them, and wires leading to the motor.

Supplemental or Unit Heating

Floor Furnaces. This type of furnace, Fig. 9-16, supplies heat to separate rooms, or acts as an addition to a central heating system when located in hallways, stairwells, below large windows, or other places

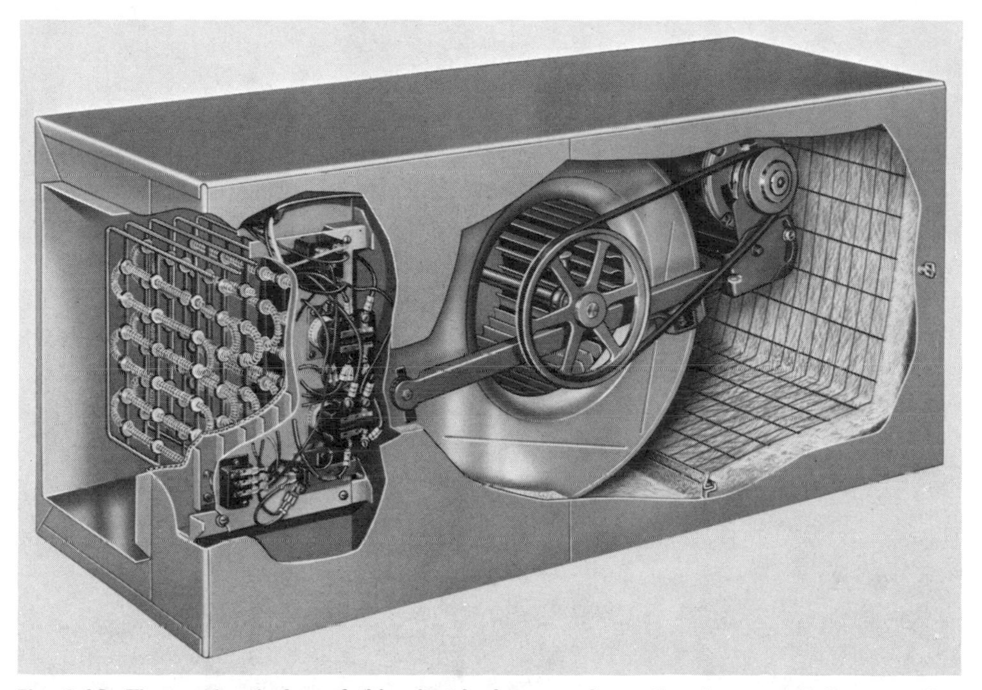

Fig. 9-13. The sectional view of this electric furnace shows the thermostatically controlled resistance elements, blower and filter. Many electric furnaces are sufficiently small to be concealed in a closet. (Lennox Industries, Inc.)

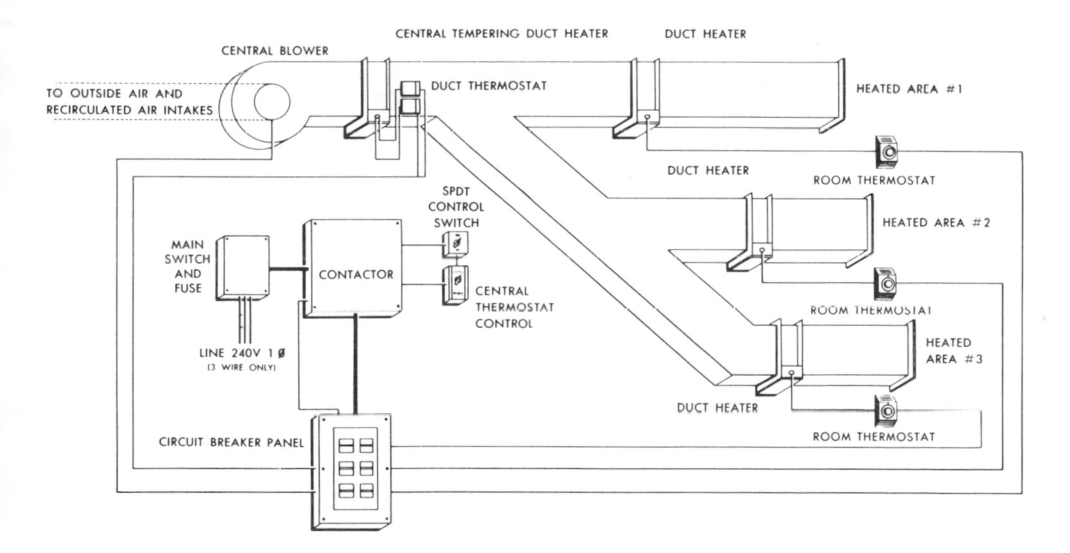

Fig. 9-14. Duct heating system. (Wesix Electric Heater Co.)

Fig. 9-15. The electric duct heater is inserted directly into the ductwork. (Norris Thermador Corp.)

Fig. 9-16. Floor furnaces. Floor insert heaters are particularly suited for installation below floor to ceiling windows or sliding glass doors. (Top: Wesix Electric Heater Co. Bottom: Electromode, Division Friden, Inc.)

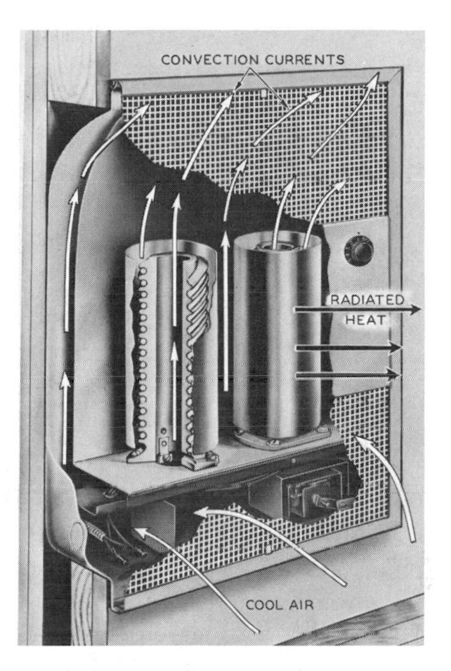

Fig. 9-17. Top: Wall heater. Bottom: Mounting box. (1) Mount box vertically between studs with flange flush with finished tile or plaster surface. Run branch circuit cable through knock-out on either side or bottom of box. (2) Connect heater to circuit. (3) Mount finish grille. The box, or can, is installed on rough construction; the heater, on finish. (Top: Wesix Electric Heater Co.)

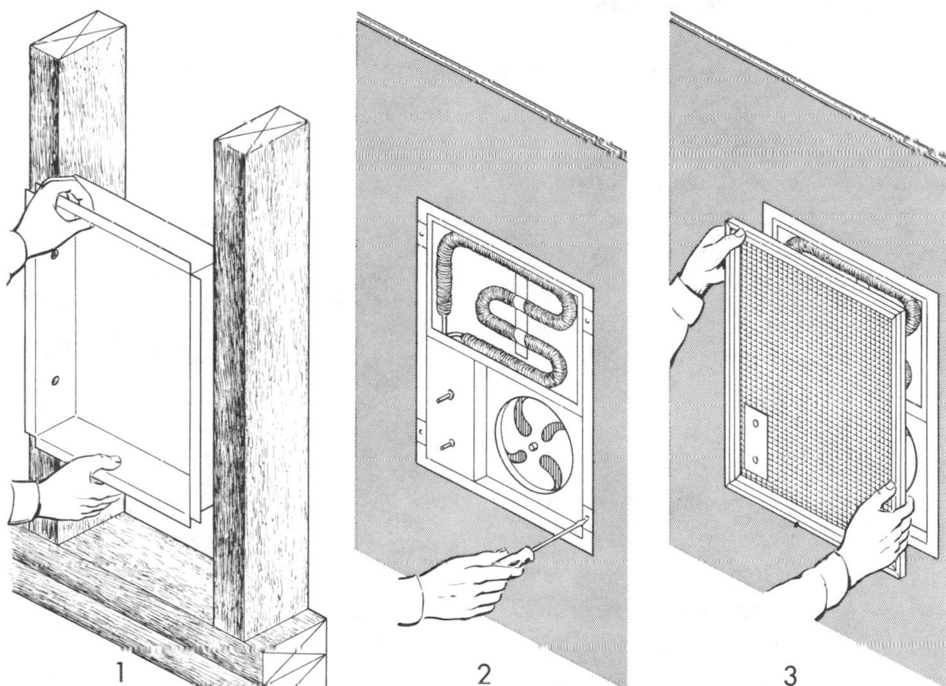

which are not adequately heated by the main furnace. Different sizes and shapes are available. Floor furnaces may be controlled automatically by thermostats, or manually by wall switches.

Wall Heaters and Portable Devices. A wall heater may be recessed, as in Fig. 9-17, top, or surface mounted. It offers a combination of radiation, convection, and fan-blower circulation heating. The term *radiation* applies to heat rays sent out at right angles to hot surfaces. The term *convection* refers to a circulation of air created in the room by action of the heater. Cool air near the floor rises through the heated interior of the unit to pass along the ceiling to the far wall, then downward to the floor, and along it to the heater again.

In addition to the wire-wound elements shown here, wall heating elements are made in the form of glass or ceramic panels which have high-resistance conductors embedded in the material of which they are made. In any case, they serve a like purpose.

Wall heaters are most often supplied by individual circuits, depending upon their capacity. Their output is controlled through built-in thermostats, built-in switches, wall switches or thermostats. Fig. 9-17, bottom, illustrates the installation of a wall air heater: (1) illustrates a steel heater box being mounted vertically between studs for rough wiring. The bottom of the box should be located approximately 10 inches from the floor. Ample length of wire should be allowed for final heater connection. The box should be set back enough to allow the flange to be flush with the finished wall surface. Branch circuit wires are connected on finish and the heater is mounted in the box or can. The grill is installed last (3).

Portable heaters, such as the one shown in Fig. 9-18, are available in various sizes. The portable unit operates in the same manner as the wall heater. It is equipped with a cord, and has a plug that can be inserted in a standard plug receptacle for heater sizes generally up to 1650 watts. Large portable heaters are generally 240 volts and must have

Fig. 9-18. Portable forced air heater. (Norris Thermador Corp.)

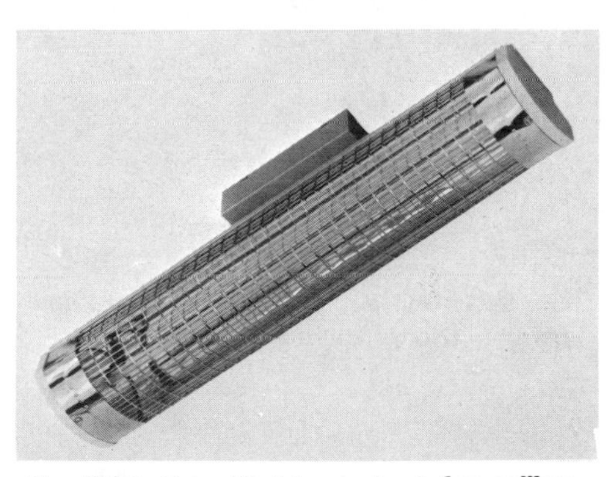

Fig. 9-19. Top: Resistance heat fan ceiling heater. Center: Infra-red heater. Bottom: Infra-red heater may be installed like fluorescent fixtures. (Top: Norris Thermador Corp. Center: Emerson Electric Co. Bottom: Electromode, Division Friden, Inc.)

polarized receptacles installed for 240 volt equipment only. However, adequate wire size and protection must be provided.

Ceiling Heaters. Ceiling heaters, Fig. 9-19, are often used in bathrooms. In general, three types are available: one having a resistance element, one with a resistance heater with motor driven fan circulator, and the other with infrared lamps. The resistance type is sometimes combined with a circular fluorescent lamp or with an incandescent lamp to provide light and heat at the same time. The infrared unit, which also performs both functions, utilizes one, two, or three infrared lamps. Here again heater capacities dictate wire sizes and if independent circuits should be used. Standard outlet boxes are used for wiring and mounting heaters.

Fig. 9-20, left and right, shows radiant ceiling heaters commonly used in bathrooms. They have no fans for forced-air movement. Similarly, unit forced-air heaters, suspended from the ceiling, are employed mostly in commercial and industrial buildings. One is shown in Fig. 9-21. These are sometimes used in home workshops or basements.

Baseboard Heaters. Baseboard heating strips may be either recessed on new construction or surface-mounted on existing buildings along outside walls of rooms. These strips can be employed as the sole heating

Fig. 9-20. Resistance wall heaters commonly used in bathrooms. (Norris Thermador Corp.)

Fig. 9-21. Unit air heater for overhead suspension. (Norris Thermador Corp.)

source, or as an addition to a central plant. They are especially valuable in rooms which have large window areas, Fig. 9-22A. Heat furnished by such units effectively counteracts a tendency for chilling drafts to sweep down over glass panes and across the floor.

Fig. 9-22B shows terminals and wiring space provided for connection

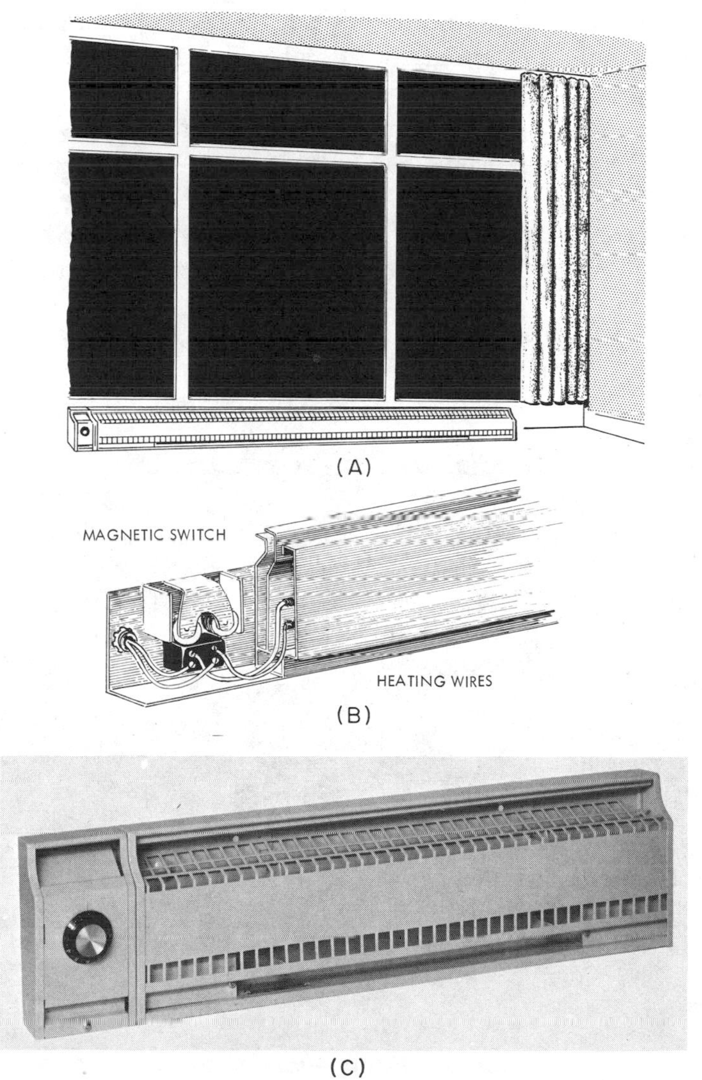

Fig. 9-22. Baseboard heating with thermostat built in. (Bottom: Norris Thermador Corp.)

of circuit wires, together with the magnetic switch, on some makes, that controls flow of current to the elements. A thermostat mounted on the wall of the room governs the magnetic switch to control the heavy load.

The particular unit illustrated here consumes 157 watts per linear foot (approximately 0.7 amp at 230 volts). The degree of heat in the connection box is so low that standard Type TW branch circuit wires may be used.

These branch circuit wires may be brought into the heater enclosure for connection in any system approved method. Connections from flex or cable are made as shown in Fig. 9-22B and wired to relay or thermostat. First, the end of the flexible conduit or cable must be allowed to protrude into the room until plastered around. When installing the mounting can, plaster is chipped around the flex or cable end to allow for a connector. The mounting can or enclosure is screwed through the plaster by the wireman. When the can is secured and wires connected, the cover grille is installed as shown in Fig. 9-22C.

A variety of heater lengths are

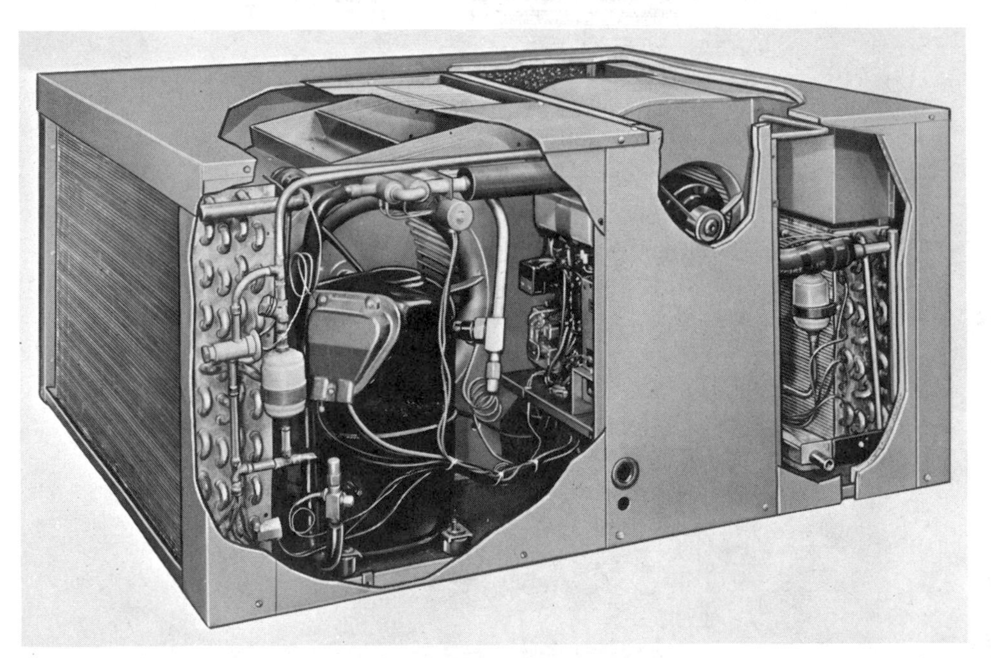

Fig. 9-23. Heat pumps may draw heat from air, earth or water. This is an air cooled heat pump. Ductwork is required. (Climatrol Division, Worthington Air Conditioning Co.)

available and some may be connected in series. However, increased branch circuit wire ampacities should be kept in mind with increased heater loads.

Heat Pumps

Perhaps the ultimate in electric heating systems is the heat pump which heats, cools, controls humidity and filters air. See Fig. 9-23.

Heat pumps are a combination of refrigerating coils and an electric motor. They depend on the principle of transferring heat from one point to another. The motor drives a compressor. In winter, liquid from the compressor is forced through a valve into expansion coils which may be buried in the ground, or placed in water or outside air. As the liquid expands and becomes a gas it draws heat out of the ground, water or air. This gas is forced through the compressor to the smaller coils inside the house, at which point it gives up its heat. In summer, the process is reversed, heat being drawn from inside and dissipated in the ground, air or water. In this case, the pump serves as an air conditioner year around.

Simplified diagram in Fig. 9-24 shows operation of standard heat pump. Heat is extracted from outside air (there's some even at low temperatures) then increased within

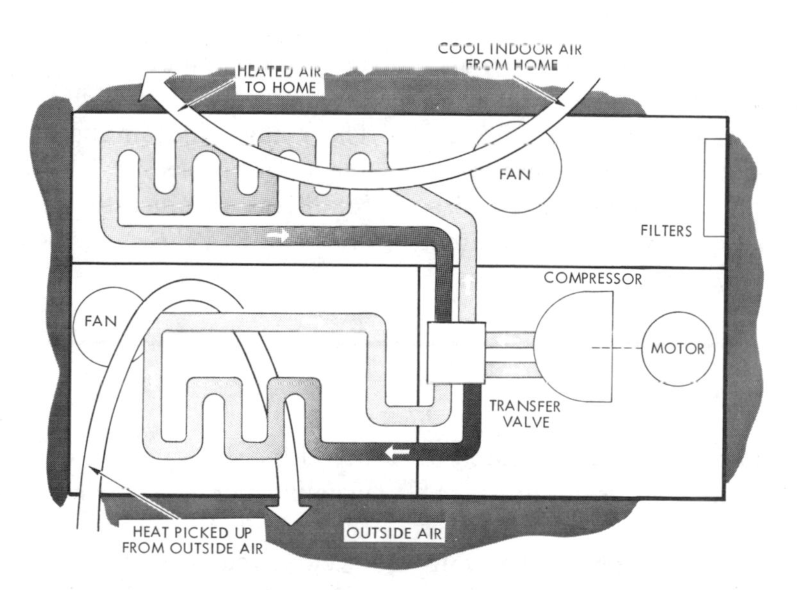

Fig. 9-24. A reversible air conditioner, the heat pump works all year long through a duct system.

the pump and blown, at desired temperatures, into the house. At the same time, cool inside air is blown into the pump for the same treatment. Reversed process cools air in summer. Heat pumps, thermostatically controlled, may be installed inside or outside the house. They are generally used in temperate climatic zones. They are generally used in a central, year-around air conditioning system. Each room is supplied with air through sheet metal duct work.

Air Conditioning

The electrician must work with the air conditioning contractor's men who generally install all air-duct work and often install a package unit containing heat-pump, controls and electrical terminal connection panel. Most air conditioning problems for the wiremen are concerned with the amount of electrical load; this is used in determining wire and conduit size for the power consuming equipment. Depending on the size, 115 or 230 volt (3 wire) is required. This affects the main electrical service size also. The wireman must also know where his conduit is to be located in respect to the terminal box on the air conditioning unit.

If he knows where the connection terminal for the unit is located exactly, including dimensions and measurements, he will better know how to terminate his conduit provisions on rough construction. It will also tell him how to proceed on final wiring. This information can be gained from the air-conditioning contractor or manufacturer of the equipment. He may also supply protective conduits for thermostat wiring also. This would be when units are located outside of the structure.

Window air-conditioners are usually an after thought and are related to remodeling wiring. Adequate voltage, wire, conduit and service sizes are the problems for wireman here. The name plate on the equipment and existing electrical circuits and conditions will dictate steps to follow in adding these extra loads.

Thermostats

Thermostats are manufactured in diverse forms, precision units are illustrated in Fig. 9-25. Contact points are built to handle low-voltage or line-voltage current, as the case may be. Some are designed to be used for either. Thermostat wiring is usually color coded or numbered by the manufacturer.

Two circuits are illustrated in Fig. 9-26. A single pole thermostat circuit is shown in Fig. 9-26A. (Fig. 9-25,

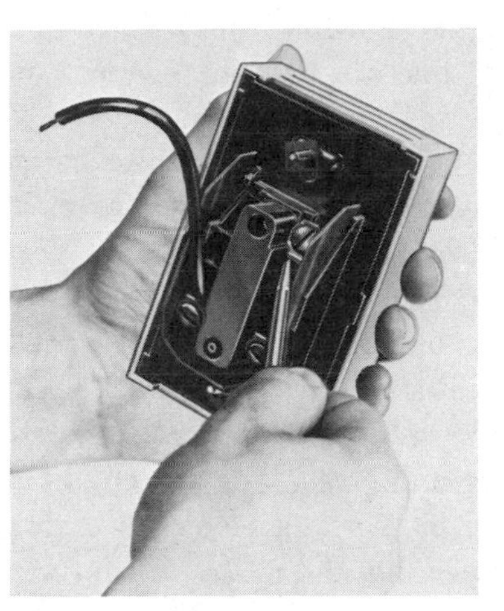

Fig. 9-25. Top left: Electric heat line-voltage thermostat, front view. Top right: Rear view of single pole line voltage thermostat. Screwdriver is pointing to incoming line terminal. Some have screw terminals and some have wire leads. Bottom: Outside view of thermostat. (Honeywell)

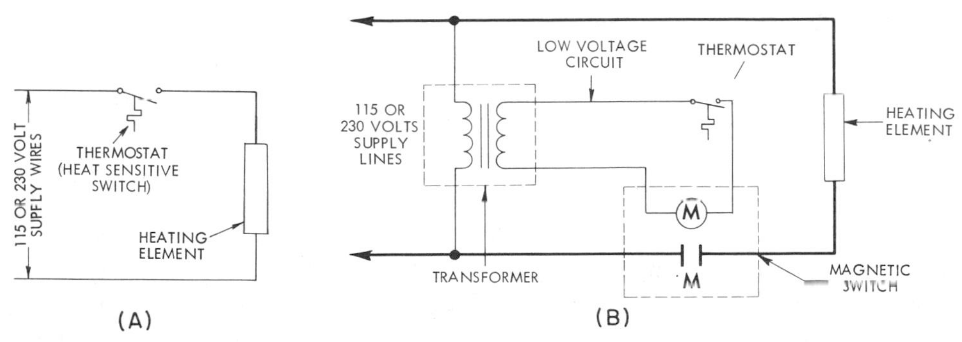

Fig. 9-26. Thermostat circuits.

top right, shows wiring.) Only one supply line to heating element is broken by line voltage thermostat. A double-pole line voltage thermostat switches both supply wires, as may be required by local codes. See Fig. 9-26A. The thermostat turns the heating unit on and off and carries the heater load current. In Fig. 9-26B, it causes a magnetic switch to open and close. The switch is for turning on heating elements or for disconnecting them. A transformer is employed in this illustration to reduce the voltage for the thermostat. Magnetic switches are used in conjunction with line-voltage units, such as this one in Fig. 9-26, left, if contact points of the thermostat are not heavy enough to carry load current. It may be noted that thermostats are employed also to operate electrical valves in other fuel systems. In Fig. 9-26, right, when the thermostat closes, the coil "M" of the magnetic switch is energized causing contact "M" to close supplying the heater load on the higher voltage.

Location of a thermostat is an important consideration in any type of heating. It should be placed where it can not be affected unduly by heat from the source. For example, a thermostat used in connection with a baseboard heating installation should not be installed directly above the strip. If possible, it should not be positioned where it is exposed to drafts, as between two doors. The thermostat would be situated within the moderate temperature range, that is, not too high on the wall or too close to the floor. The preferable height is four to five feet.

Many heaters are provided with built-in thermostats which are calibrated to provide a high standard of comfort with reasonable temperature tolerance. It is imperative that wall thermostats be located and installed where they will sample the average room conditions.

Placement of thermostats on the outside cold wall should be avoided; also avoid placement next to large glass areas where the sun's rays may fall directly on the thermostat. Do not locate thermostats on a wall whose other side butts up to heat generating appliances (ovens, cooktops, etc.). Hot or cold water pipes or air ducts in the same or adjacent stud space should be avoided.

In new installations, wiring the thermostat is usually a simple task; but in old work, certain procedures must be followed in order to make a neat installation. Fig. 9-27 shows how to drill holes for thermostat wires. After deciding on the location, remove the molding directly below this point and drill a small "marker" hole through the floor at an angle of about 45°. The molding can be replaced at once. Then drill vertically upward through the basement ceiling or under the house, using the first hole as a guide. Drill a ½″ hole

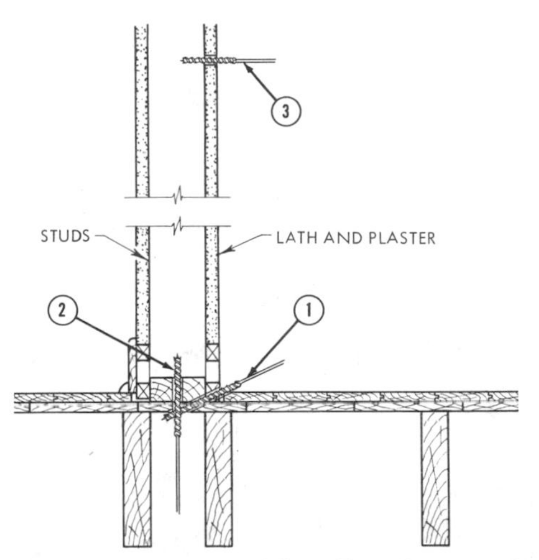

Fig. 9-27. Method of installing thermostat wires.

at the exact spot where the thermostat is to set. Partitions may be located for drilling by other means such as using a flashlight and finding the bottom 2 x 4 through diagonal sub flooring spaces, or driving a nail through the floor beside the partition, and measuring from the nail to the partition center for drilling under the house. The control wires are now fished in, allowing about 6″ for making connections. A switch box is necessary for line-voltage thermostats. See Chapter 15, "Remodeling Wiring."

Plug the hole around the wires with rock wool, or some other insulating material, to prevent drafts of air at the back of the thermostat. If this precaution is not taken, faulty operation may result. When mounting the thermostat on the wall, use a spirit level or a plumb bob to insure a vertical position. The calibration does not always hold true unless the unit is perfectly upright for some makes.

Line-Voltage Control Units

If the thermostat is a line-voltage type (one connected to the branch circuit without a transformer), the method of locating and pulling in the wires will be the same; but the line wires will have to be in armored or nonmetallic sheathed cable or flexible conduit and secured to an outlet box, as shown in Fig. 9-28. The wiring methods must be the same as for lighting outlets and receptacles in that particular house.

Regardless of the automatic control with which thermostats will be used, they should, of course, be installed as explained. If a line-voltage type thermostat is used, the transformer can be dispensed with and a junction box substituted for convenience in making connections and to meet code requirements.

Thermostat Installation Suggestions

Finally, here are some practical suggestions that experienced installers have learned the hard way and pass along.

1. Single pole thermostats rather than two pole thermostats should be

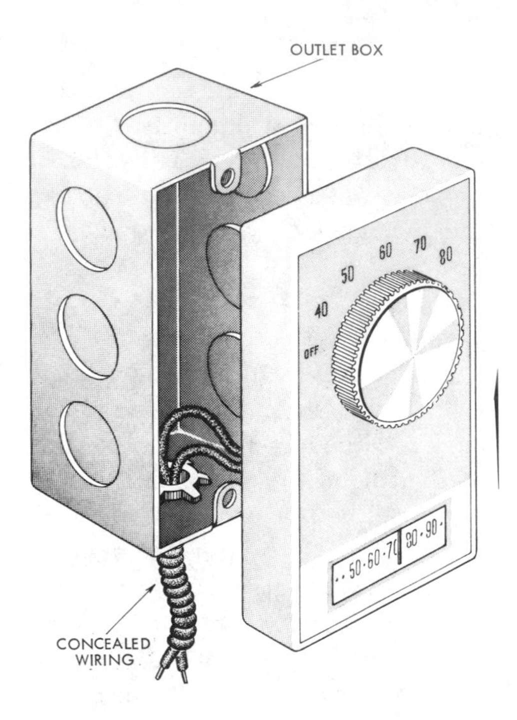

Fig. 9-28. Mounting line-voltage thermostat using concealed wiring.

used unless local codes require two pole. They are easier to install, cost less, and usually operate with less differential and inaccuracies (droop).

2. Installation time in new construction can be saved by installing 4″ square boxes with single device covers or plaster rings. The extra wiring space reduces labor cost of mounting—especially on two pole type thermostats.

3. Loading thermostats to maximum wattage rating should be avoided. The more load, the more inaccuracies and more sticking of contacts may result.

Wiring for Motors

Heat pumps, sump pumps for basements, and heating and cooling blowers in residential construction have motors which are sizeable enough to warrant some special considerations.

In order to comply with provisions of the code, the wireman should understand the basic components of a motor circuit. Its four essential elements are: (1) size of circuit disconnect switch, (2) size of branch circuit overcurrent protection device, (3) size of motor overcurrent protective device, and (4) size of the circuit conductors. The wiring process can best be illustrated by way of an actual example.

Connecting a ¼ HP, 115-Volt, Single-Phase Motor

The motor in Fig. 9-29A, rated at 5.8 amps, drives a water sump pump in the basement of a residence. A diagram applicable to most motor installations is shown in Fig. 9-29B.

Fig. 9-30A shows an additional use of a selector switch so that basement seeping or flooding water can be pumped either in the "hand" position or "automatic." In the automatic position the sump pump goes on when the water level reaches a certain height. The float closes a switch starting the pump motor. As the water is pumped out the float is lowered, opening the switch and stopping the motor automatically. Fig. 9-30B illustrates the selector switch.

The maximum rating of a branch-circuit protective device (fuse) for a motor which is started without series resistors or voltage-reducing apparatus should be 300 percent of name plate current. The name plate current is 5.8 amps, so that the largest permissible circuit protector is equal to 3 x 5.8 amps or 17.4 amps.

Most codes permit the next larger standard overcurrent device to be used, so that a 20-amp fuse may be acceptable here. The circuit disconnect switch must be able to carry the 17.4 amp starting current, the nearest fusible switch being of 30-amp

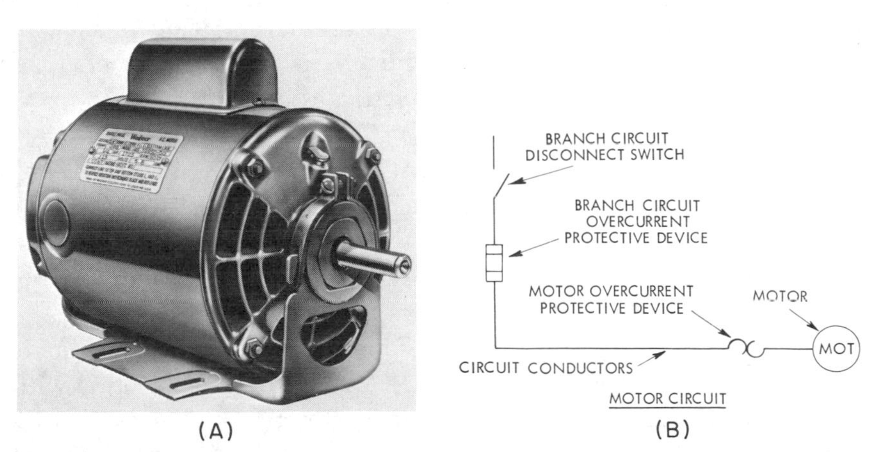

Fig. 9-29. Small motor and circuit diagram. (Wagner Electric Corp.)

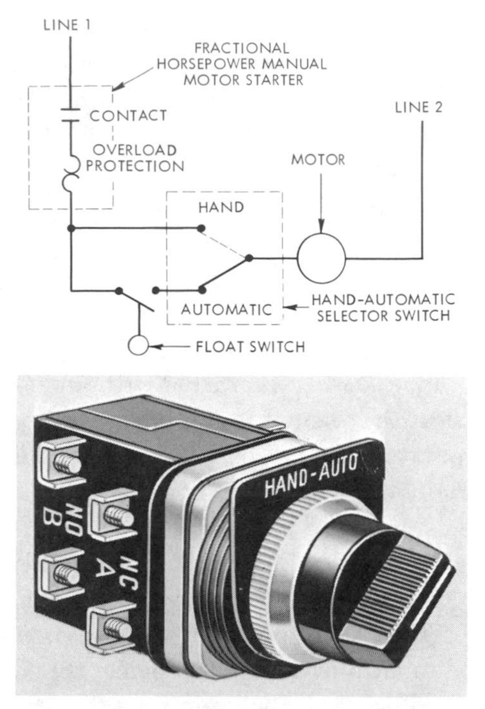

Fig. 9-30. Top: Schematic diagram showing motor control wiring. Bottom: Selector switch, open unit. (Bottom: Allen-Bradley Co.)

rating. In the present instance, the motor is fixed, and it is automatically started, so that running (overload) protection is required for two reasons.

The maximum size of the running protective device should not exceed, but can be less, 125 percent of name plate current rating for a 40-degree motor, or 115 percent for other motors. The term "40-degree" means that the motor windings, under normal operating conditions, never reach a temperature more than 40° (Centigrade) higher than that of the surrounding air. The term "other motors" includes those rated: "50-degrees," "Short-time Duty," and "Intermittent." Markings of this general nature are stamped on the name plates.

When running protection is necessary, rating of the device should not be greater than 125 percent of motor current rating, but can be less. Here, the maximum value is equal to 1.25 x 5.8 amps, or 7.25 amps. The motor should be checked to see if there is a built-in thermal protective device. If so, additional running protection is not required, because the thermal switch in the motor cuts off flow of current when motor windings become too hot. On some motors, current flow will not be resumed until the windings have cooled. On others, there is a button, usually red in color, projecting from one of the endbells. After a shut down from overheating, the motor cannot operate until this button is pressed. Some are automatically reset.

The size of wire needed to supply the motor must be determined. Most codes provide that the circuit conductors shall have a continuous current rating not less than 125 percent of motor name plate current rating. This value has already been found as 7.25 amps in connection with the rating of the overcurrent unit. According to NEC tables, the allowable current-carrying capacity of AWG No. 14 copper conductor is 15 amps. Since No. 14 is the smallest permissible size of conductor for general wiring, it can be used here. However, some codes may require a mini- of #12 for this circuit because of high starting currents.

Finally, the motor should be grounded. If supplied by metal-clad wiring (that is, flexible conduit or armored cable) the Code requires that it be adequately grounded. If the wiring is non-metallic sheathed cable, grounding is by a different means. It is best at all times to ground the unit. Metal-clad wiring provides its own grounding medium. In the case of nonmetallic installations, where the cable does not have a separate grounding wire, a piece of No. 14 AWG copper wire, either bare or insulated, may be run from the motor to a cold water pipe.

The name plates on motors should be read for exact currents, because charts and tables do not always match the motor.

These are brief insights into problems encountered regarding code applications. References should be made to the National Electrical Code book regarding all problems of the code. The back pages of the NEC contain helpful examples, similar to the one described, with section references to explain and clarify.

Radiant Heat

Radiant energy travels in straight lines until intercepted or absorbed by some body or object and converted to heat. Radiant heat passes through certain materials such as glass without perceptibly heating them, and it is reflected by various other materials. It will also pass through air regardless of its temperature without heating it to any appreciable extent.

A good example of the effect of radiant heat is the difference of noticeable warmth to a person while standing in the sun or under the shade of a tree. While under the tree the person feels colder than when in the rays of the sun, even though the temperature is the same in both places.

Electric Radiant Heating Panels

Fabricated panels, available in a number of sizes, are designed for flush or surface mounting. They rely on heat-producing wires embedded in plasterboard or tempered glass, or aluminum circuitry sandwiched between flat sheets of durable polyester to provide radiant heat. These

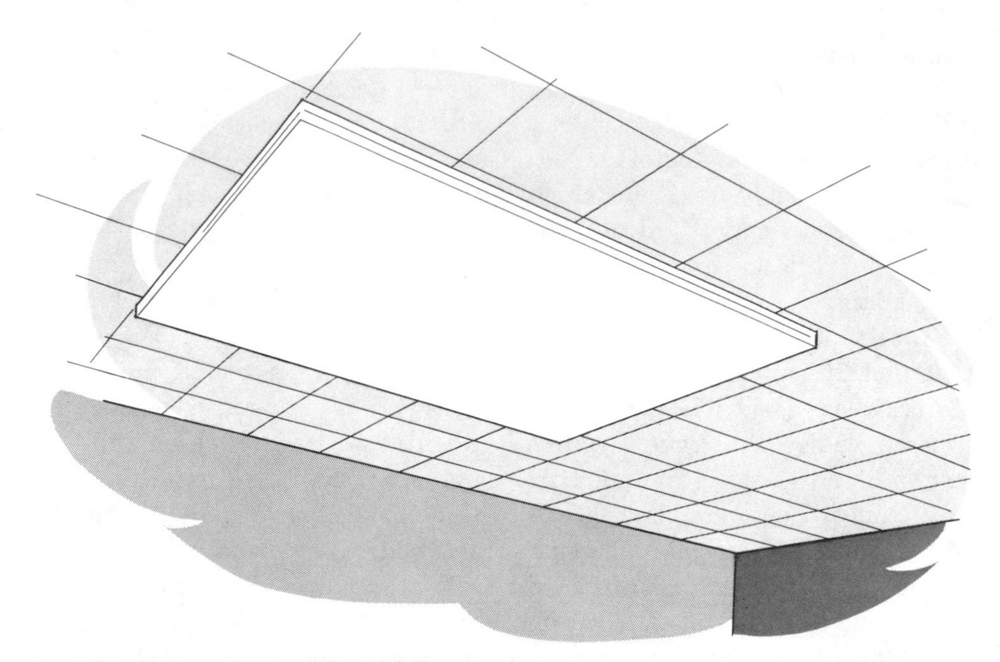

Fig. 9-31. Surface electric radiant heating panel.

are ideal over large window areas, in entryways, vestibules and other places where concentrated heat is needed. When surfaces are painted to blend with surrounding ceilings, panels are not noticeable and blend with overall decorative scheme. See Fig. 9-31.

An accessible attic space is required in order that connections may be made between boxes and circuit conductors of each heating panel.

Electric Radiant Heating Cable

For completely invisible heating, cable can be embedded in plaster or between two sheets of wallboard. Each room has temperature control, permitting room-to-room variations for individual needs. Ceiling electric cable radiates heat evenly to rooms below. Proper installation is of utmost importance in deriving full efficiency from the system.

Installation of the cable is permanent and care should be exercised to avoid any abnormal condition which would eventually cause damage to the cable or circuit overload. Electric radiant heating cable may be installed in the ceiling or in the floor; however, the ceiling installation provides greatest efficiency if properly insulated.

Thermal Insulation Recommendations

Insulation is essential to satisfactory performance. Some manufacturers recommend a minimum insulation standard of 6 inches in the ceiling, 3⅝ inches in the wall, and 2 inches in the floor. All the insulation that space will permit should be used for satisfactory heating.

Lath used in plaster construction should be of a fire-resistant material such as plaster and gypsum board. Cable cannot be attached to metal lath or other conducting surfaces. (Refer to the National Electrical Code, "Fixed Electric Space Heating Equipment" for complete code requirements.)

Radiant Heating Cable: Plaster Ceiling Installation

Installing Thermostat Junction Box. The thermostat junction box should be located a minimum distance of 60 inches from the floor. See Fig. 9-32. Two holes in the ceiling 6 inches from the wall should be drilled directly above the thermostat junction box. Two holes should be drilled in the ceiling joist (if necessary) so that the non-heating preloomed (extra sleeve insulated) leads can be run between the thermostat location and ceiling holes. Fig. 9-33.

Ceiling Area Layout

Heating cable is factory assembled with non-heating leads attached and should not be cut. Lengths vary from 17 to 2,600 feet. The capacity averages 2¾ watts per foot. It can be embedded or left exposed, depending on use. Lines should be marked on the ceiling 6 inches away from side walls throughout the perimeter of the room, because the cable must be installed at least 6 inches away from the wall. See Fig. 9-34.

The room should be divided into two equal parts, so that the dividing line is parallel to the longest *outside* wall. The half adjacent to the outside wall is the "close" spacing section, while the remaining half of the room is the "open" spacing section. Refer to Fig. 9-34 and use "close" cable spacing in close spacing section and "open" cable spacing in open spacing section. With this method more cable, and thus more heat, will be distributed in the section adjoining the outside wall where the heat loss is greater.

To determine the cable spacing the steps below should be followed:

1. Calculate the usable ceiling area of the room in square feet.

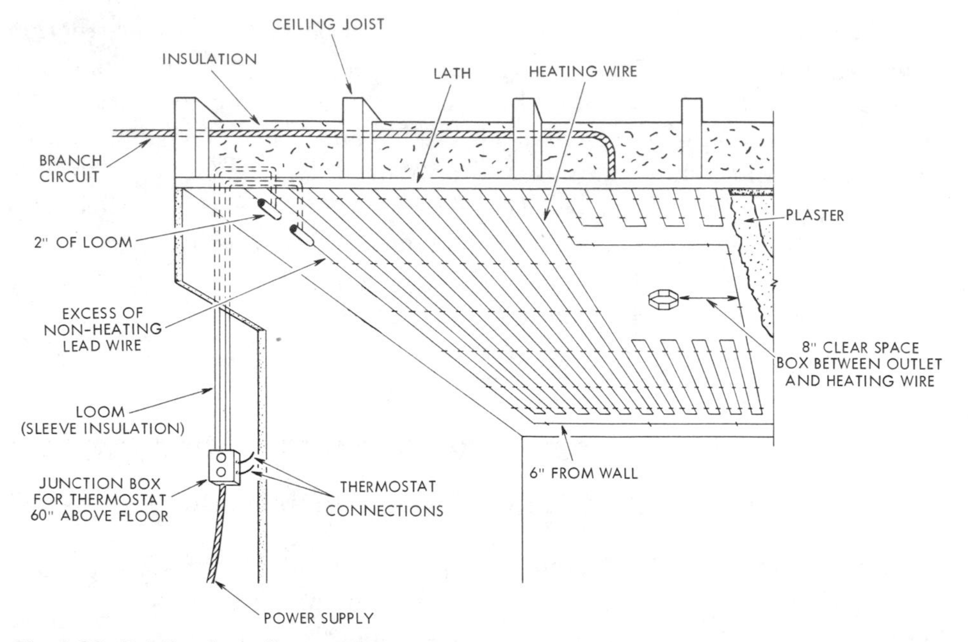

Fig. 9-32. Details of a radiant cable installation before plastering. (Emerson Electric Mfg. Co.)

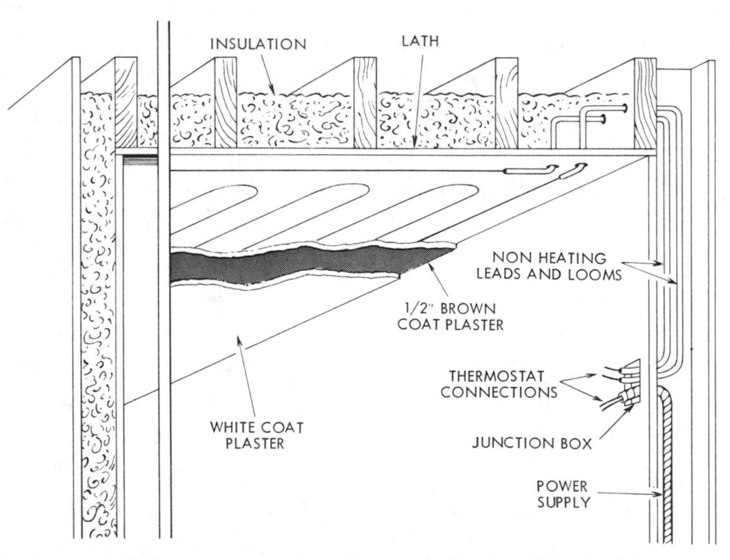

Fig. 9-33. Typical installation cross section. (Emerson Electric Mfg. Co.)

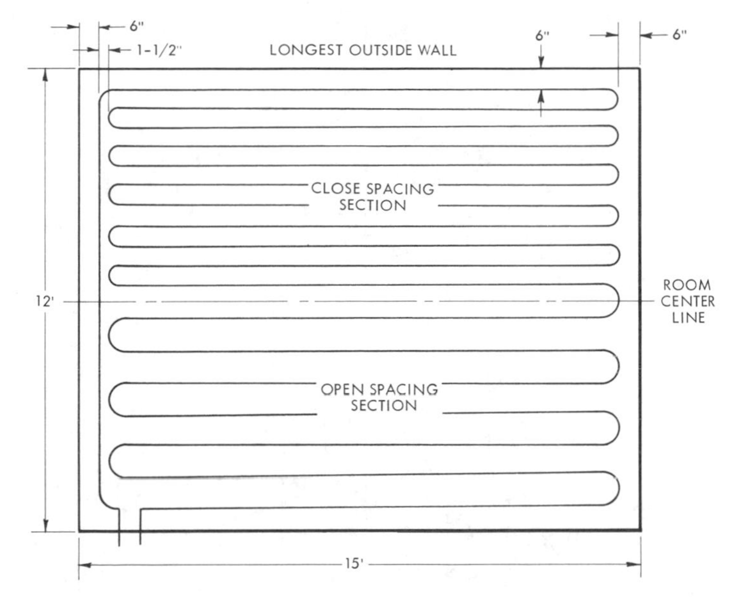

Fig. 9-34. Cable layout diagram plaster installation. (Emerson Electric Mfg. Co.)

2. Determine the heat loss of the room if possible (refer to electric heat and cooling manuals), then select the correct cable or estimate 10 watts per square foot. Watts per square foot of ceiling area should be multiplied to find cable wattage. The nearest logical cable wattage should be selected.
3. Divide the wattage of the cable by the usable ceiling area to obtain the required watts per square foot.

$$\frac{\text{Cable Wattage}}{\text{Usable Ceiling Area (sq. ft.)}}$$

= Watts per square foot.

4. Locate the watts per square foot (obtained in step 3) on the bottom scale of the Cable of the Cable Spacing Chart, Fig. 9-35. Follow this number up to the point where it intersects the curve (see "arrowed" example on the chart). Then move across to the left to obtain the cable spacing for close spacing section. Always "round off" to the next *lowest* ⅛-inch spacing.
5. To determine the spacing in the open spacing section, follow the same procedure except move from the point of intersection to the right of the chart and read the spacing to the nearest ⅛-inch. *Note:* Maintain this spacing as accurately

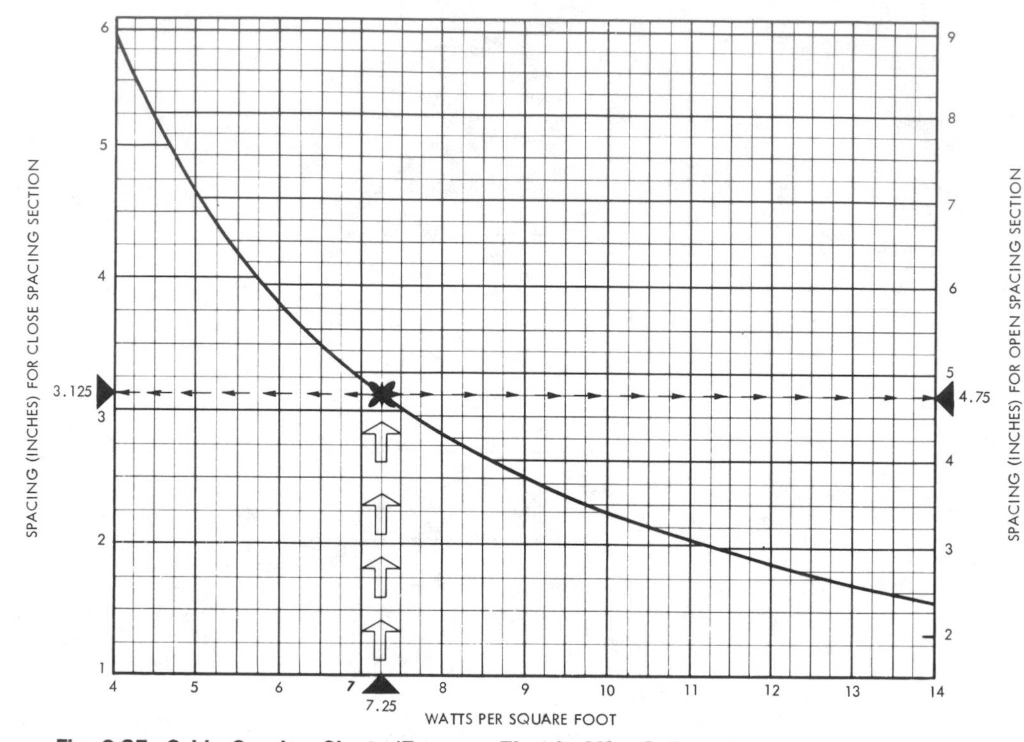

Fig. 9-35. Cable Spacing Chart. (Emerson Electric Mfg. Co.)

as possible except where it is necessary to deviate in order to clear nail rows, light fixtures, etc.

6. Marks should be placed along the ceiling for each run, making sure that the marks are spaced precisely according to results of above calculation. A convenient method of accomplishing this is to notch two wood strips, or two yardsticks, with the proper spacing. The two notched strips are held by operators on opposite ends of the room. A chalk line is stretched between the notched strips and snapped. This method places the marks completely across the room and thus insures accuracy of spacing.

Caution: Cable shall not be installed in closets, over cabinets which extend to the ceiling, under walls or partitions, or over walls or partitions, which extend to the ceiling. *Exception:* Single runs of cable may pass over partitions where they are imbedded.

Installing Cable

The cable spool should be placed on a spike or peg driven into the wall, so that the cable can be unwound easily without kinking as it is strung into position.

Run the non-heating, pre-loomed leads through one of the holes in the ceiling, then through the hole in the joist (if required), down inside the wall and out through the hole into the thermostat junction box. Excess pre-loomed, non-heating lead can be pulled up and stapled to the ceiling. (See Fig. 9-36.)

Caution: The identification tags should not be removed nor should the non-heating leads be cut or shortened. Cable must not be located less than 8 inches from any electrical fixtures, outlets, junction boxes, etc.

After stapling the unused portion of the pre-loomed, non-heating lead to the ceiling, continue along the inside wall (6 inches from the wall) to the close spacing section.

Staple the cable at least every 6 inches on straight runs. Also staple once at center of loop, then 3 inches and 6 inches from the loop, as shown in Fig. 9-37.

Caution: Cable is always positioned a minimum of 2 inches away from any metal reinforcements or metal lath. The cable should not be spaced closer than 1½ inches in plaster installations. Do not tamper with the cable splice; do not alter, cut or adjust the cable length.

Crisscross both sections of the room (as shown in Fig. 9-34) using the close spacing in close spacing

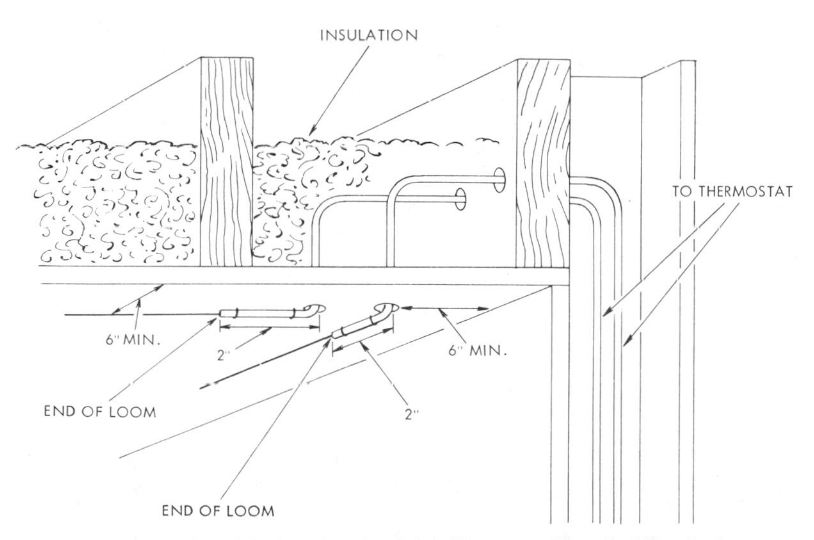

Fig. 9-36. Installing loom insulated ends of cable. (Emerson Electric Mfg. Co.)

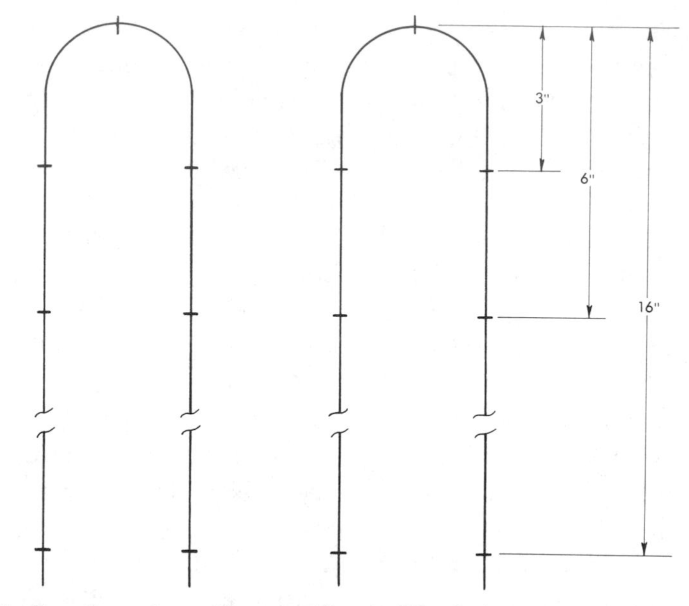

Fig. 9-37. Stapling dimensions. (Emerson Electric Mfg. Co.)

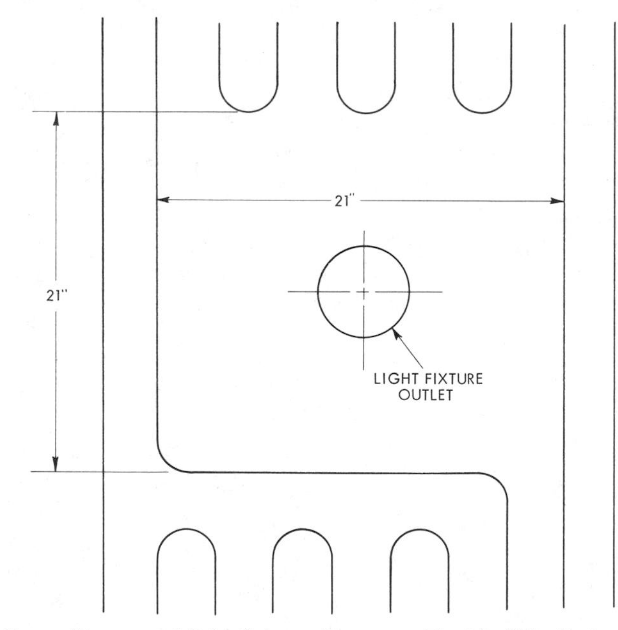

Fig. 9-38. Locating cable around light fixture. (Emerson Electric Mfg. Co.)

section and open spacing in open spacing section. Fig. 9-38 illustrates how to run the cable around a lighting fixture.

Note: The cable is marked at the half-way point. When this mark is reached, approximately three-fourths of the close spacing section should be completed.

This method is specifically designed so that the cable will end up short which means that a small portion of the ceiling will not be covered by cable. By following the instructions the uncovered portion will be adjacent to the inside wall.

The return non-heating lead should be installed in the same manner as described previously. Make sure the same length of cable (with identification tag) extends into the thermostat box. Surplus non-heating lead may be crisscrossed, stapled to the ceiling and plastered over.

If it is desired to use uniform cable spacing, the following formula to determine the distance between the wires can be used:

$$S = \frac{12\ (L\text{-}1)\ (W\text{-}1)}{CL\text{-}(2W + L)}$$

Where:

S = Cable spacing in inches.
L = Ceiling dimension parallel to cable runs in feet.
W = Ceiling dimension perpendicular to cable runs in feet.
CL = Cable length in feet.

Sample Problem

Cable to be installed in a room 15 ft. x 12 ft. with cable running parallel to the 15 ft. dimension. The heat loss is 2500 watts and the cable selected has a length of 910 ft.

$$S = \frac{12\ (L\text{-}1)\ (W\text{-}1)}{CL\text{-}(2W + L)}$$

$$S = \frac{12\ (15\text{-}1)\ (12\text{-}1)}{910\text{-}(2 \times 12 + 15)}$$

$$S = \frac{12 \times 14 \times 11}{910\text{-}(24 + 15)}$$

$$S = \frac{1848}{910\text{-}39} = \frac{1848}{871}$$

$$S = 2.12 \text{ inches or } 2\tfrac{1}{8} \text{ in.}$$

Testing for Continuity

The color or numbering of the non-heating leads should be observed in order to determine the voltage to be connected to them. For example, red indicates 240 volts, and yellow indicates 120 volts. The voltage is also printed on the spool package and identification tags attached to the non-heating leads.

Connect an ohmmeter to the two leads to make certain that no breaks are present in the circuit. With ohmmeter connected and a consistent reading obtained, use a broom and "sweep" the cable runs. If this produces a fluctuation of the ohmmeter needle it is an indication that a loose connection or break exists which must be spliced before plastering.

Another method of testing for continuity is to use a battery and

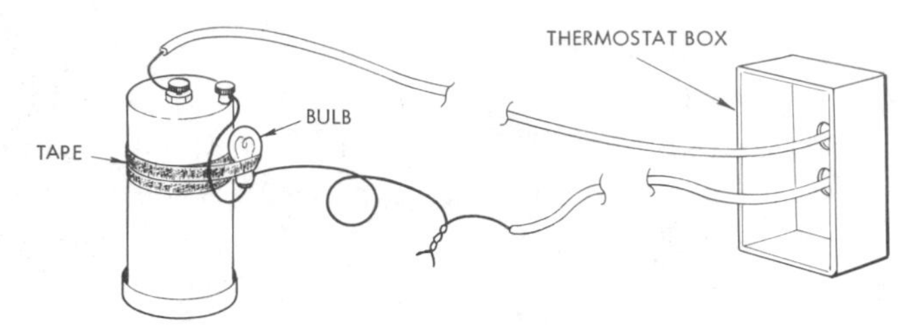

Fig. 9-39. Diagram for continuity test lamp. (Emerson Electric Mfg. Co.)

light bulb. Because of the wide range of resistances encountered in the various wattage ratings of cables, it is important to use the fresh 6-volt batteries and 6 volt bulbs. Connect the cable, battery, and bulb in series as shown in Fig. 9-39.

If the cable wattage rating is more than 1500 watts, use only one 6-volt battery. If the wattage rating is less than 1500 watts, this cable circuit has more resistance, therefore, use two 6-volt batteries connected in series. If only one battery is used on a cable of less than 1500 watts, the bulb will not glow due to increased circuit resistance. If two batteries are used on a cable of more than 1500 watts, the bulb will burn out due to less resistance of higher wattage cable.

If the bulb lights, when connected as described above, and does not flicker when swept as described earlier, the cable has no breaks.

When checking high wattage cable, the bulb will glow very dimly. This is normal and should not be mistaken for trouble in the cable.

Installing of Splice

Locating a Break in Ceiling Cable. If a break occurs in the cable, either before or after plastering, it may be located by connecting the secondary windings of a high-voltage transformer to the cable lead wires. When the primary side of the transformer is connected to an appropriate power source, a high-voltage discharge, or spark, will occur at the break, causing an audible buzzing sound.

A hearing aid can be used to locate the break more precisely by moving the microphone along the ceiling until the loudest sound is observed.

If the ceiling is already plastered, the break can be repaired from the attic space by carefully digging down into the plaster to expose the wire.

Repairing the Break. Use a standard 14 AWG compression con-

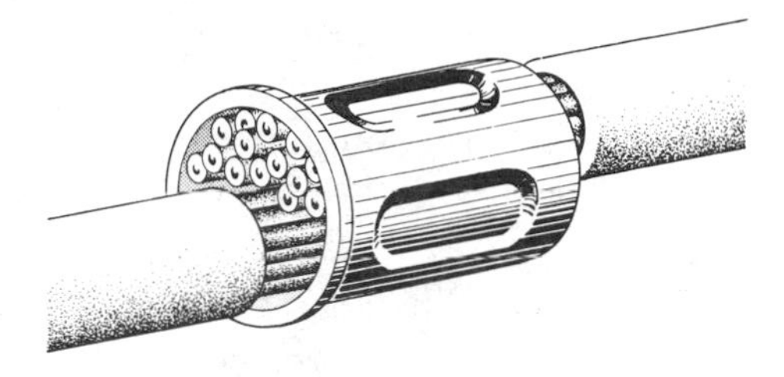

Fig. 9-40. Compression connector cable splice.

nector and splice the break in the cable as shown in Fig. 9-40. Cover the splice with tape to a thickness of 3/64 inch. The tape should extend to at least one inch on each side of splice.

Be sure to use a thermosetting tape listed by Underwriters Laboratories for a temperature of 80° C.

Test for Heating

Temporarily, the two exposed cable leads should be connected to the proper voltage for a period of at least five minutes in order to determine if heating occurs. If an interruption in heating the cable is experienced, recheck for continuity with the ohmmeter or test lamp.

Call in the local electrical inspector for approval before concealing the cable in non-insulating plaster. The plaster should be allowed to dry before energizing the heat cable.

Radiant Heating Cable: Dry Wall Ceiling Installation

For dry wall ceiling installation the location of each ceiling joist should be marked on the walls before ceiling is installed so that nailing or stapling of cable may be facilitated. The thermostat junction box should be located a minimum distance of 60 inches from the floor.

Lines must be marked on the ceiling 6 inches away from side walls throughout the perimeter of the room because the cable must be installed at least 6 inches away from the wall.

The room is divided into two equal parts, so that the dividing line is parallel to the longest *outside* wall. The half adjacent to the outside wall is the "close" spacing section, while the remaining half of the room is the "open" spacing section. Refer to Fig. 9-34 and use "close" cable spacing in close spacing section and "open" cable spacing in open spacing section. With this method more cable, and thus more heat, will be distributed in the section adjoining the outside wall where the heat loss is greater.

The cable is installed in the same manner as for a plastered ceiling, except as follows. (See Fig. 9-41.)

In dry-wall construction, the cables must be installed parallel to the ceiling joists in order to provide proper clearance at joist locations so there will be no interference with nails.

Run the cable from the thermostat junction box along the inside wall to the close spacing section (Fig. 9-41). Do not start the first run of cable along an outside wall parallel to the room center line (See Fig. 9-41).

The cable should be spaced two inches at each joist (one inch on each side). When a cable run is one inch from the center of the joist, cross over the joist keeping the one inch spacing on the other side of the joist. Care must be taken to eliminate the possibility of driving a nail into the cable, when installing the finish dry-wall. The dry-wall installer should be cautioned.

Complete Section I (Fig. 9-41) using close spacing and then continue to Section II using open spacing. Enough cable must be left in Section II to return to the thermostat junction box. This method is specifically designed so that the cable ends up a proper length.

Mark on the wall a space three

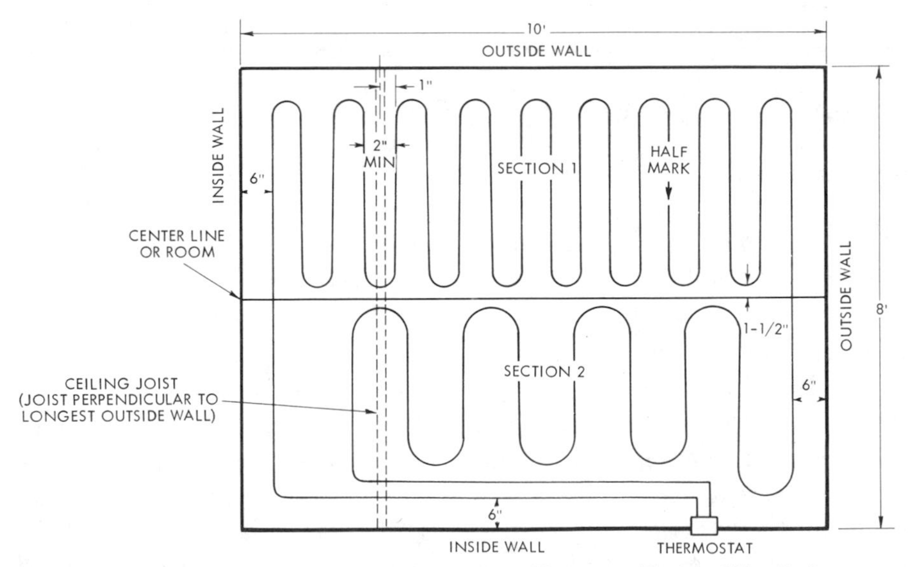

Fig. 9-41. Cable layout diagram, drywall installation. (Emerson Electric Mfg. Co.)

TABLE I. CURRENT CARRYING CAPACITIES*

AMPS.	TYPE RH WIRE SIZE	TYPE TW WIRE SIZE
15	12	10
20	10	8
30	8	6
50	4	2

*50°C AMBIENT TEMPERATURE

inches on each side of the center line, so that no nail will be driven into the cable where it crosses the joist near the center of the room.

Extreme care should be used when nailing where cables cross the joists. No nails should be hammered in the space four inches to eight inches away from the wall or where the thermostat lead wires are stapled to the ceiling.

All branch circuit wiring should be installed above the insulation. When the wire is installed in the insulation it must be at least two inches above the heated ceiling and should be considered as operating at an ambient temperature, 50° C (122° F), which is within code limits of building wire.

The current carrying capacities in 50° C ambient temperatures are shown in Table I, as recommendations. If it is desired to use uniform cable spacing, refer back to "Plaster Ceiling Installation" for calculating the cable spacing.

Radiant Heating Cable: Concrete Slab Floor Installation

On concrete slab floors, one inch of rigid waterproof perimeter insulation between the edge of the slab and the foundation is required. Two inches if insulation to a depth of at least two feet should be set in between the foundation and the slab-floor gravel fill, for best operating results. See Fig. 9-42.

Basic Procedure. The thermostat junction box is located on the inside wall 60 inches from the floor. Follow the method described earlier in this section on "Plaster Ceiling Installation."

A piece of rigid conduit should be installed to house the non-heating lead wires between the floor and switch box. The lower end of the conduit should be embedded approximately 6 inches in the concrete; and where the non-heating leads emerge from the conduit, a smooth porcelain bushing should be installed for pro-

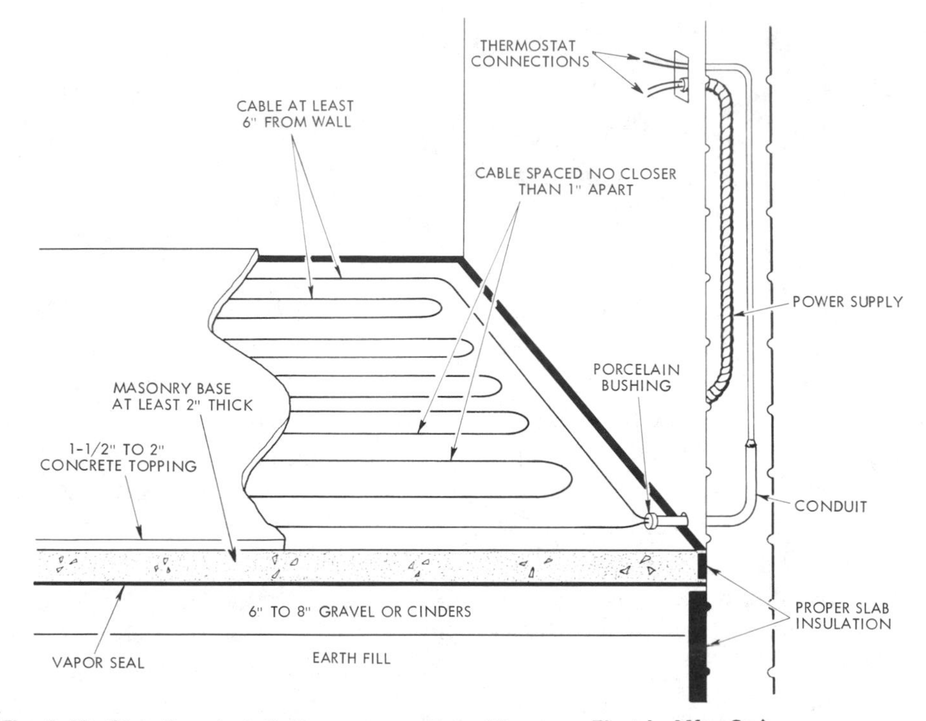

Fig. 9-42. Slab floor installation cross section. (Emerson Electric Mfg. Co.)

tection. It may be necessary to tape the bushing and leads to avoid seepage of fresh cement into the conduit when the slab is poured.

Unwind the heating element as described for the ceiling installation with the non-heating, pre-loomed lead wire installed through the conduit to the thermostat junction box. Leave the 6 inch length of non-heating, pre-loomed lead wire extending out of the thermostat junction box in order to expose identification tags and provide ease of connections. Excess non-heating leads should be embedded in the concrete in the same manner as the heating cable element.

Either of the two methods for determining the spacing can be used; mark the cable locations on the floor. Refer to "Ceiling Area Layout," or "Installing Cable." Do not space floor cable less than one inch apart.

Run the cable along the floor 6 inches from side wall to the opposite outside or exposed wall. Fasten cables to the floor with staples while the concrete is green enough to permit the staples to penetrate it.

Fasten the remaining portion of the cable to the floor on one inch (or wider) spacing. Run the return non-heating, pre-loomed lead wire through the conduit to the thermostat junction box.

Electric cable heating will give satisfactory results if the general builder provides for necessary insulation. To protect cable insulation make certain that only rubber tired wheelbarrows are used for crossing cables in order to pour concrete. Also make certain that traffic is kept to a minimum.

Radiant Heating Cable: Hallways

Where there are three exposed walls or an irregular shaped room, such as a long narrow hallway, use the following formula to determine the spacing:

$$S = \frac{12\ (L-1)\ (W-1)}{CL}$$

Where:

S = Cable spacing in inches.
L = Room dimension parallel to cable runs, in feet.
W = Room dimension perpendicular to cable runs, in feet.
CL = Cable length, in feet.

An example of a typical electric radiant heating cable selection chart is shown in Table II. Note wattage selections, volts, amps and cable lengths.

TABLE II. ELECTRIC RADIANT HEATING CABLE SELECTION CHART

Cat. No.	Watts	Volts	Color Non-Ht. Lead	Lgth. (ft.)	Amps.	Cat. No.	Watts	Volts	Color Non-Ht. Lead	Lgth. (ft.)	Amps.
C21	200	120	Yellow	72.5	1.7	C528	520	208	Blue	189	2.5
C31	300	120	Yellow	109	2.5	C708	700	208	Blue	254	3.4
C41	400	120	Yellow	146	3.3	C1048	1040	208	Blue	378	5.0
C51	500	120	Yellow	181	4.2	C1398	1390	208	Blue	505	6.7
C61	600	120	Yellow	218	5.0	C1738	1730	208	Blue	630	8.4
C71	700	120	Yellow	255	5.8	C1918	1910	208	Blue	694	9.2
C751	750	120	Yellow	273	6.3	C2178	2170	208	Blue	790	10.5
C81	800	120	Yellow	291	6.7	C2608	2600	208	Blue	946	12.5
C91	900	120	Yellow	327	7.5	C3118	3110	208	Blue	1136	15.0
C101	1000	120	Yellow	364	8.3	C4008	4000	208	Blue	1450	19.3
C111	1100	120	Yellow	400	9.2						
C1251	1250	120	Yellow	455	10.4						
C141	1400	120	Yellow	509	11.7						
C151	1500	120	Yellow	545	12.5						
C161	1600	120	Yellow	551	13.3						
C1751	1750	120	Yellow	636	14.6						

TABLE II (CON'T)

Cat. No.	Watts	Volts	Color Non-Ht. Lead	Lgth. (ft.)	Amps.
C0252	250	240	Blue	92	1.1
C42	400	240	Blue	145	1.7
C62	600	240	Blue	218	2.5
C82	800	240	Blue	292	3.4
C102	1000	240	Blue	362	4.2
C122	1200	240	Blue	436	5.0
C142	1400	240	Blue	509	5.9
C152	1500	240	Blue	545	6.3
C162	1600	240	Blue	582	6.7
C182	1800	240	Blue	654	7.5
C202	2000	240	Blue	728	8.3
C222	2200	240	Blue	800	9.2
C252	2500	240	Blue	910	10.4
C282	2800	240	Blue	1018	11.8
C302	3000	240	Blue	1090	12.5
C322	3200	240	Blue	1162	13.3
C352	3500	240	Blue	1272	14.6
C382	3800	240	Blue	1380	15.8
C402	4000	240	Blue	1452	16.7
C462	4600	240	Blue	1672	19.2
C502	5000	240	Blue	1820	20.9

EMERSON ELECTRIC, BUILDER PRODUCTS DIV.

QUESTIONS FOR DISCUSSION OR SELF STUDY

1. What voltages are necessary for electric range circuit wiring?
2. Besides voltage requirements, what additional factor warrants consideration for range wiring?
3. How is an electric range circuit terminated?
4. List the purposes of the four terminals of a range receptacle.
5. What is the difference between a central heating system and a duct-heating installation?
6. What is a low-recovery electric water heater?
7. How are branch duct heaters controlled in a duct-heating system?
8. What type of insulated wire is used to connect perimeter or baseboard heaters?
9. Describe the sequence of operation in the wire diagram shown in Fig. 9-30.
10. Why is closer spacing of radiant heating cable recommended near outer walls and windows?
11. Why is it important not to shorten heating cable by cutting?
12. What is the average recommended watts per square foot for heating calculations?
13. Why must heating cable not be located less than eight inches near electrical outlets, fixtures, or junction boxes?
14. How may heating cable breaks, or opens, be located after the ceiling is plastered?
15. When the half-way mark of the heating cable is reached upon installation, approximately how much of the installation should be completed?
16. A room 18′ x 21′ would require how many watts to heat, using 12 watts per square foot?
17. What is the recommended elevation above the floor that thermostats should be located?
18. Normally, on which walls should thermostats be located?
19. Which described thermostat must pass full current of the connected heater?
20. List the essential components of a motor circuit which must be installed in accordance with the NEC.

QUESTIONS ON THE NATIONAL ELECTRICAL CODE

1. Define: Branch Circuit.
2. Define: Branch Circuit—Appliance.
3. Define: Ampacity.
4. Describe the differences between: a fixed appliance, portable appliance, and stationary appliance.
5. What is the minimum size conductor for a 9 K.W. range?
6. How are ranges and dryers grounded?
7. How are frames of wall-mounted ovens and counter-mounted cooking units grounded?
8. What per cent of a fixed resistance heater load is the ampacity of branch circuit conductors rated?
9. What are code limitations on locations of fixed space heating equipment?
10. What code requirements for disconnecting means of fixed space heating equipment are there?
11. What code limitations exist for line thermostat controllers for fixed electric space heating?
12. Heating cables are required to be secured or stapled at what maximum interval or spacing?
13. What are the color codes for these heating cable voltages: 120 volts, 240 volts?
14. What are the wiring clearances above heated ceilings and within thermal insulation?
15. How are wire ampacities affected in interior walls or partitions around heated surfaces?
16. What are the area restrictions for installing ceiling radiant heating cable?
17. What are the clearances from objects and openings for the installation of ceiling heating cable?
18. What code requirements exist for splicing heating cable?
19. What is the minimum spacing of heating cable in dry board and plaster installations?
20. What is the maximum spacing for heating cable supports?
21. What size fuse protection should an air conditioning motor system require that draws 22 amperes at 230 volts?
22. If a circuit breaker were used in place of a fuse in Question #21, what would be the required size?
23. What size wire and conduit should be used in Question #21?
24. What size motor overload protection is required?
25. What is a controller and how is it used?

Chapter

10 Residential Furnace Controls

Besides installing electrical heating devices, the electrician is called upon to wire for electrical accessories to coal-, gas-, or oil-heating systems, particularly in colder climates of the country. The most common automatic control unit applicable to all types of heating is the thermostat which has been discussed in the preceding chapter. There are a number of other electrical control devices, but most of them apply to only one system.

Aquastat. Used in connection with hot-water heating plants, the aquastat is a temperature-sensitive switch that governs a control circuit to shut off the heating system as the water in the boiler reaches a set temperature. There are two general types of these controls; namely, the immersion type, Fig. 10-1 and the surface type, Fig. 10-2. The *immersion* unit has a tube which screws into the top or side of the boiler, or into a fitting in the hot-water riser located as near the boiler as possible.

The *surface* or *clamp-on type*

Fig. 10-1. Immersion type of aquastat. (Minneapolis-Honeywell Regulator Co.)

Fig. 10-2. Clamp-on aquastat. (Minneapolis-Honeywell Regulator Co.)

Fig. 10-3. Pressuretrol. (Minneapolis-Honeywell Regulator Co.)

aquastat is strapped to the hot-water riser and depends upon transfer of heat from the pipe to its operating element. It is used when it is impossible to drill a hole in the boiler or riser. Because of air currents and resultant temperature variations around the boiler this device is not as satisfactory as the immersion type aquastat.

Pressurestat. The pressurestat or pressuretrol, Fig. 10-3, performs the same duty with respect to the steam boiler as the aquastat does with respect to the hot-water boiler. It opens the control circuit to shut off the heating equipment, when the desired pressure has been reached. This pressure, in a residential installation, is usually about five pounds. The device must be screwed into the boiler so that its temperature-actuating bulb is in direct contact with the steam.

Airstat. This unit provides the high-limit protection essential to the operation of a warm-air furnace. Bolted to the dome of the furnace, with the operating tube inside the bonnet, the airstat opens the control circuit when air temperature inside the furnace reaches a predetermined value.

Furnacestat. This device is used with a warm-air furnace that has a circulating fan in the duct system.

Fig. 10-4 shows a furnacestat used to prevent circulation of cold air through the rooms. It maintains an open fan circuit until the temperature of the air in the bonnet reaches a preset minimum value. Typical wiring is shown in Fig. 10-5. Since this device, like the airstat, is located in the furnace bonnet, it is possible to obtain a combined unit known as a *combination furnace controller*, Fig. 10-6.

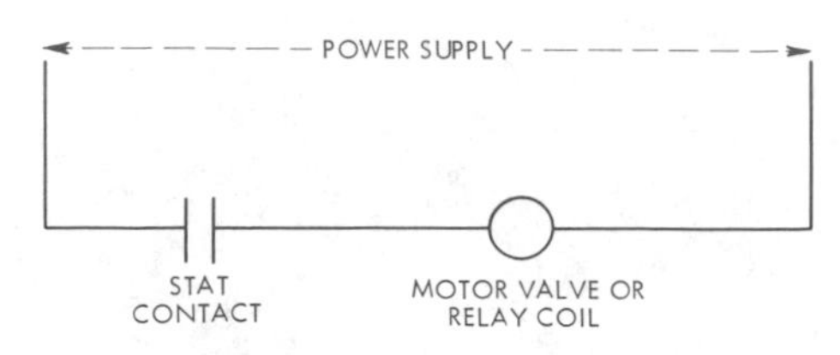

Fig. 10-5. Electrical schematic common to automatic furnace controls. Contact may be normally open or closed. Open contact is shown.

Fig. 10-4. Furnacestat. (Minneapolis-Honeywell Regulator Co.)

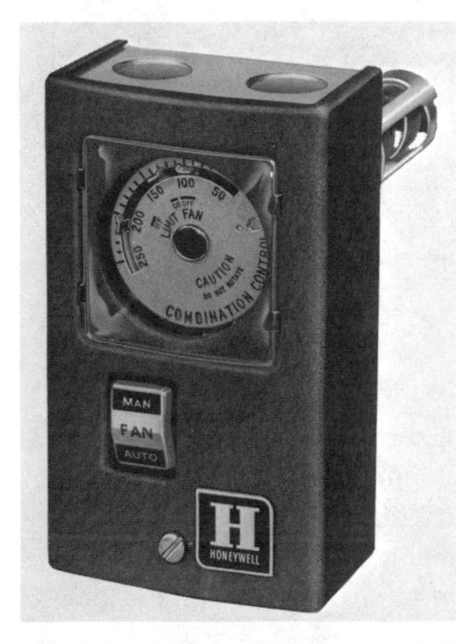

Fig. 10-6. Combination furnace control. (Minneapolis-Honeywell Regulator Co.)

Installing Heating System Controls

Motor-Operated Draft Control

Now that most of the important heat-control equipment has been briefly explained, the installation of draft control on handfired boilers will be considered. A draft-control system includes a motor which rotates a half-turn in either direction to open or close the dampers according to temperature demands. Fig. 10-7 shows a damper control installation for a hot-water or steam boiler;

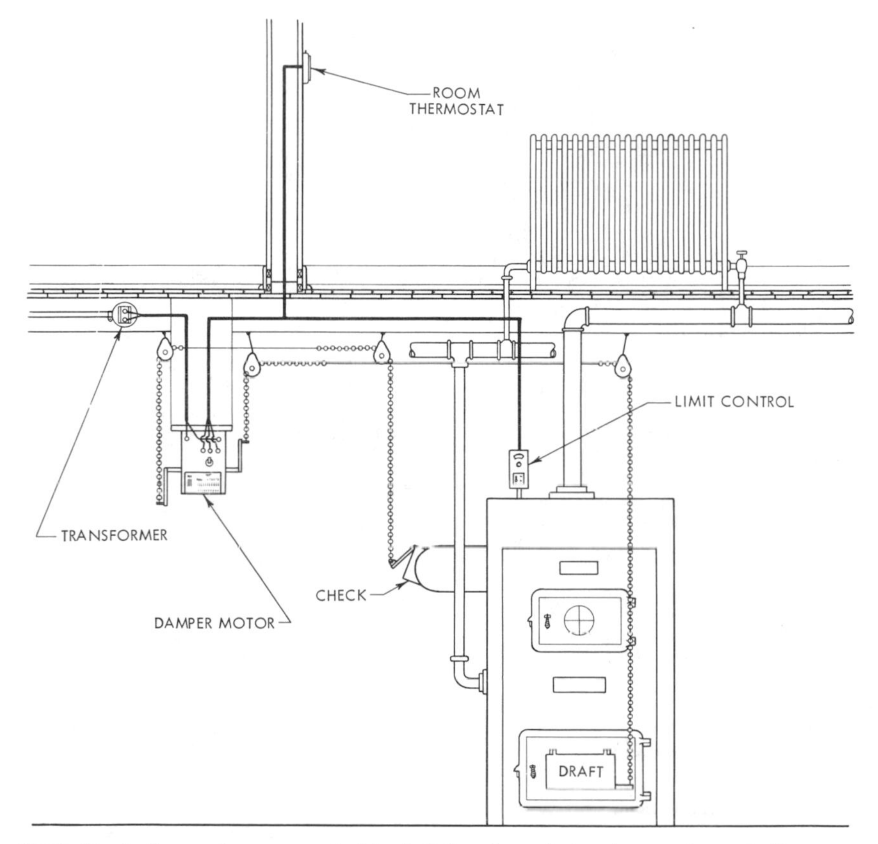

Fig. 10-7. Illustrating a damper-control installation for a hot-water or steam boiler.

and Fig. 10-8 shows the same type of installation on a warm-air furnace. Both connections in Fig. 10-7 and Fig. 10-8 are alike except for differences in the limit controls. Limit controls may be high temperature controls for air and water, or they may be for control of stack temperature and electrically connected into the control circuit to prevent excessive temperatures in the heating system. They are also used to prevent excessive steam pressure. A low limit control switch is used to prevent firing of a boiler when the water supply is low.

A high limit control switch is shown in Fig. 10-10, upper part, (see page 250 with suggested alternatives of application: that is, to prevent excessive air or water temperatures or excessive steam

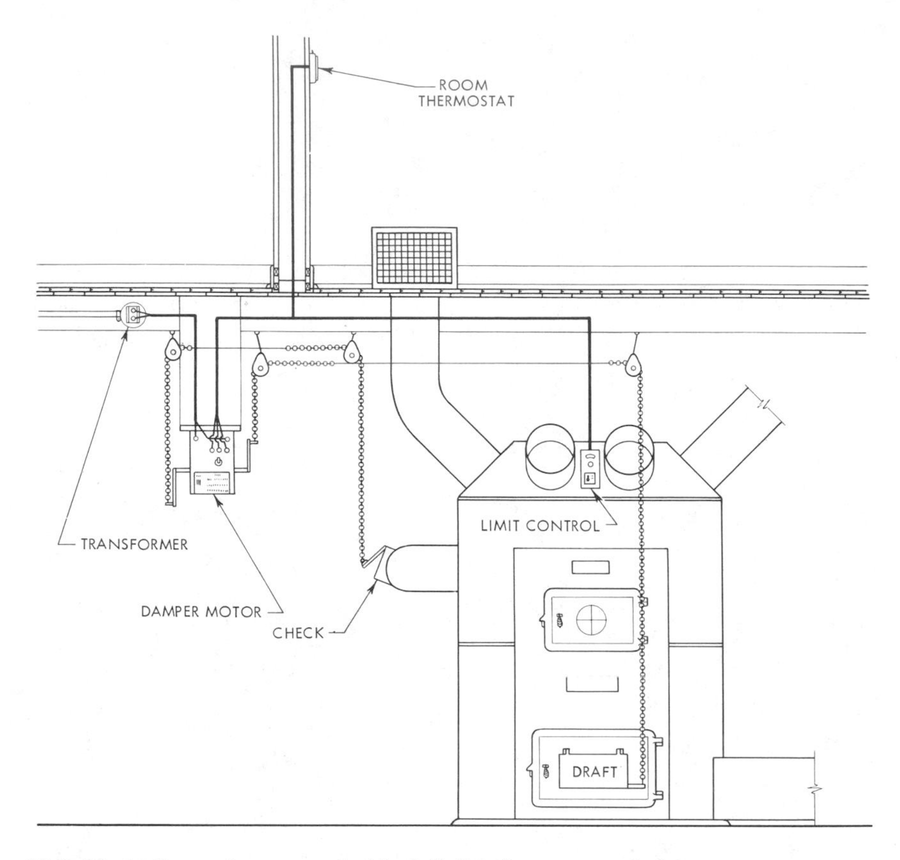

Fig. 10-8. Illustrating a damper-control installation for a warm-air furnace.

pressure. A normally closed contact opens a heat-firing system when excessive temperatures occur. This circuit is similar to the one shown in Fig. 10-5, except a normally closed contact would be used.

On some installations, a 115-volt motor operates the dampers, while a 24-volt transformer supplies the control circuit.

Care should be taken to locate the motor in a dry, clean place, and in such a position that the chains between motor arms and dampers move freely on the pulleys. The connection diagram for a damper-control low-voltage installation is shown in Fig. 10-9. The letters *R*, *B*, and *W* refer to colors of insulation on the wires—red, black and white.

For automatic heat-control wiring connected to the 120-volt serv-

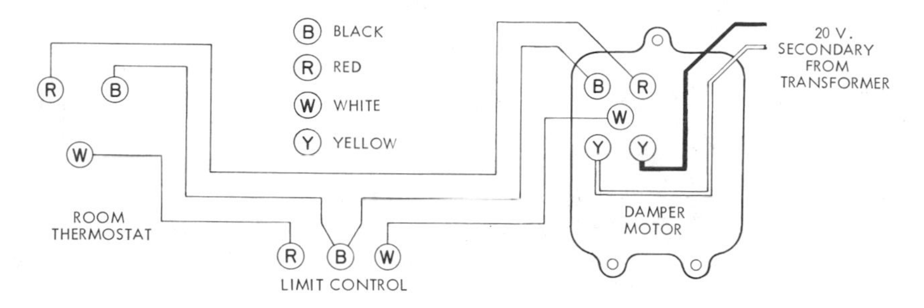

Fig. 10-9. Diagram of connections for a damper-control low-voltage installation.

ice, No. 14 wire is adequate; for 24-volt control wiring, No. 18 is the minimum size wire to be used.

Stoker Controls

In the foregoing discussion of damper controls, the only concern was to open and close the dampers in accordance with demand for heat. The fire had to be maintained clinker-free if the damper controls were to function properly. With a stoker, furnace coal feed and forced draft are two additional automatic features which must be handled by the controls. Fig. 10-10 shows the equipment layout and wiring diagram for a stoker installation. This scheme can be used for warm air, hot water, or steam heat, if the correct form of high-limit control is chosen.

Coal will not burn long in the absence of oxygen. Since the only supply of air is by way of the forced draft unit, a stoker relay, or timer, is employed. This device causes the fan to send a blast of air through the combustion chamber at spaced intervals. Such action enables the fire to continue burning during comparatively warm periods when the house thermostat may not call for heat.

From the diagram shown in Fig. 10-10, it is evident that line voltage is used for the high-limit control, the stoker relay, and the stoker motor. Low-voltage wiring runs from the thermostat to the stoker relay. The dotted lines indicate a low-limit control switch. This relay is used to maintain boiler water at a temperature sufficient for domestic use in the nonheating season.

Oil-Burner Controls

The installation of oil-burner controls is complicated by the addition of a stack-relay switch. This switch or *protector relay*, as it is sometimes called, is installed in the flue pipe. It functions to shut off the burner in case the stack does not come up to a certain predetermined temperature within 45 seconds after the motor

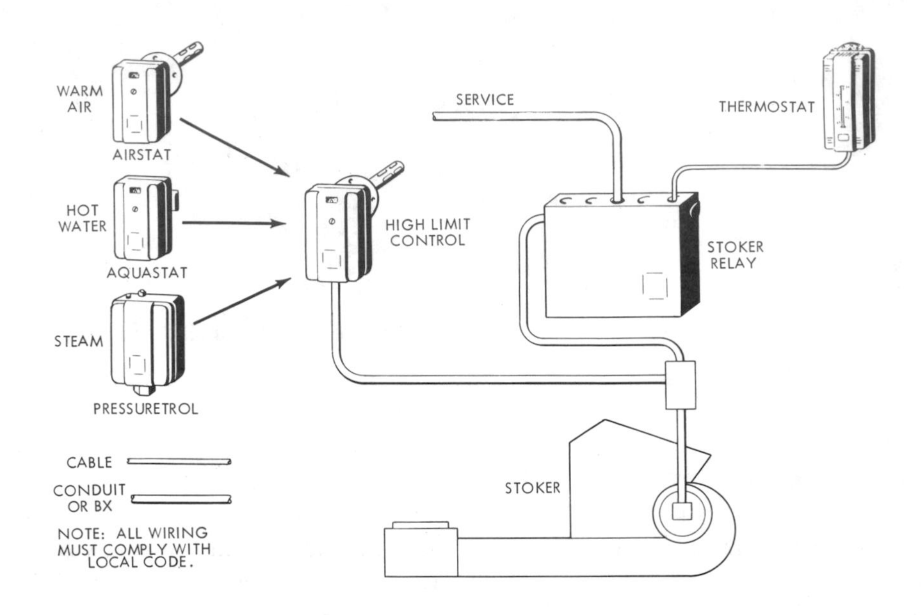

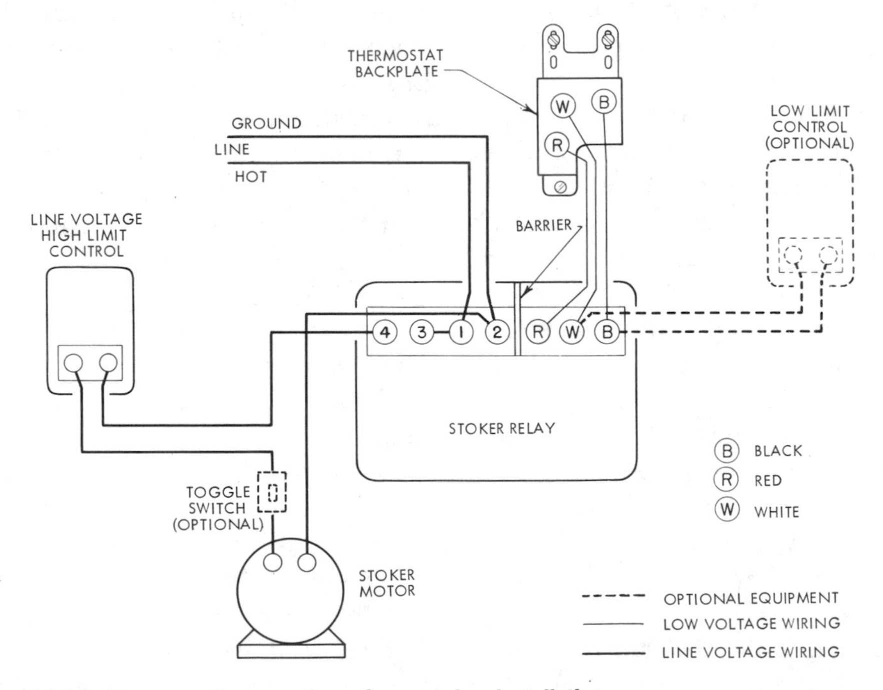

Fig. 10-10. Diagram of connections for a stoker installation.

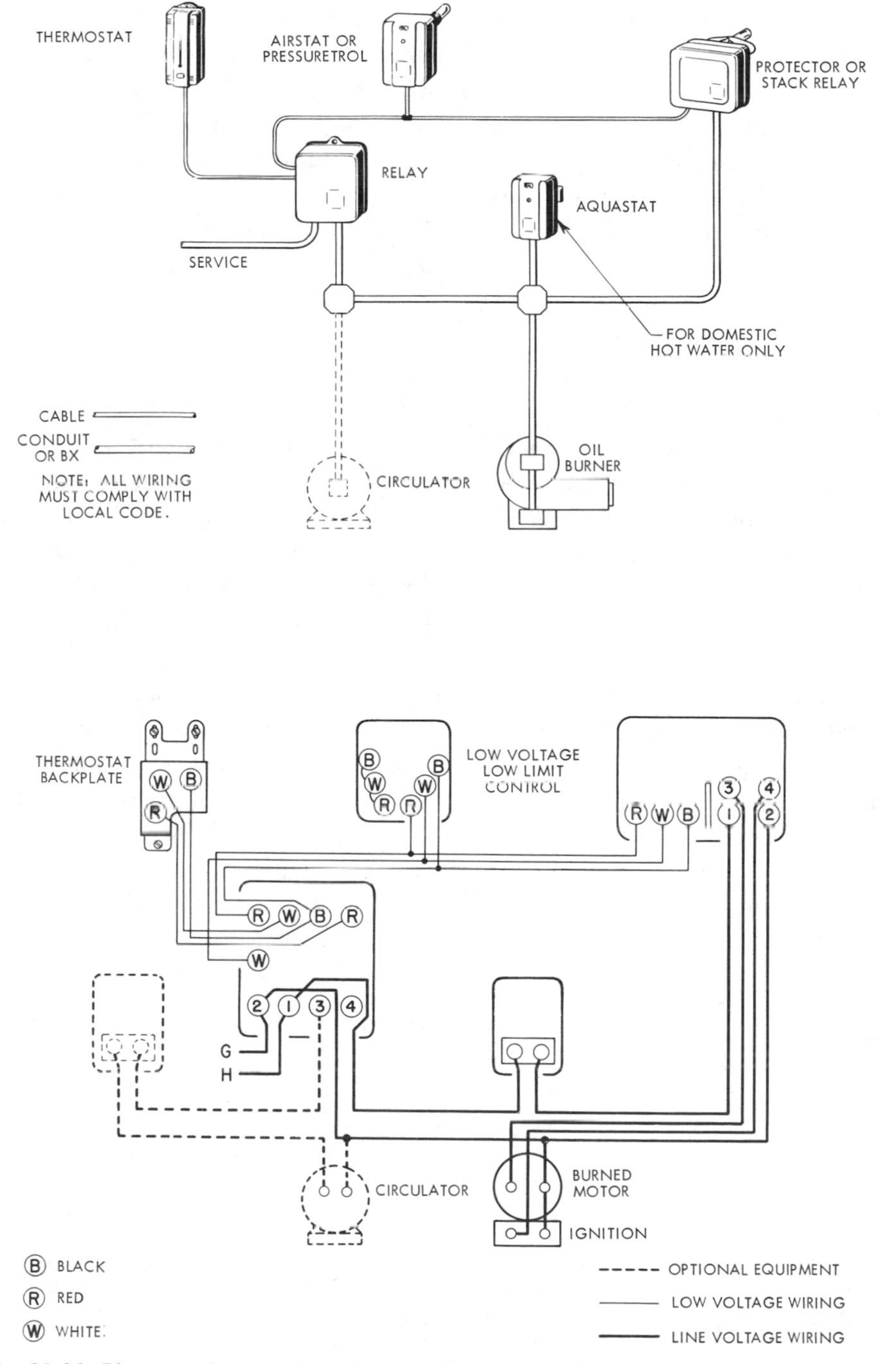

Fig. 10-11. Diagram of connections for an oil-burner installation.

starts. The relay prevents flooding of the basement with oil in case the ignition system fails.

Ignition of the oil is accomplished with the aid of a spark gap which is connected to the high-voltage terminals of a transformer installed on or near the burner.

The spark may be continuous, or it can be made to cease when the oil starts to burn. Certain factors will determine the correct method. The continuous spark system is generally advised when down drafts are common in the particular location. With a continuous spark, it is necessary to replace the electrodes more often than with the intermittent spark.

After the spark has been shut off, the oil continues to burn because the temperature of the firebrick lining is high enough to ignite the oil.

The conduit layout and diagram of connections for an oil-burner installation is shown in Fig. 10-11. Note that supply wires are connected directly to the motor-starting relay housing which also contains the low-voltage transformer. The low-temperature control and circulator, indicated by dotted lines, should be used on a forced-feed hot-water system only, and not with a steam or hot-water system.

Gas-Burner Controls

The main factor to bear in mind when installing automatic controls for gas burners is that a dangerous quantity of gas must not be permitted to accumulate in the combustion chamber. There is a variety of safety devices for dealing with this hazard. In general, there are two means of feeding gas to the boiler when the room thermostat calls for heat: by means of a diaphragm valve, or by means of a solenoid-operated valve. The diaphragm valve is pressure actuated. A solenoid coil opens a small

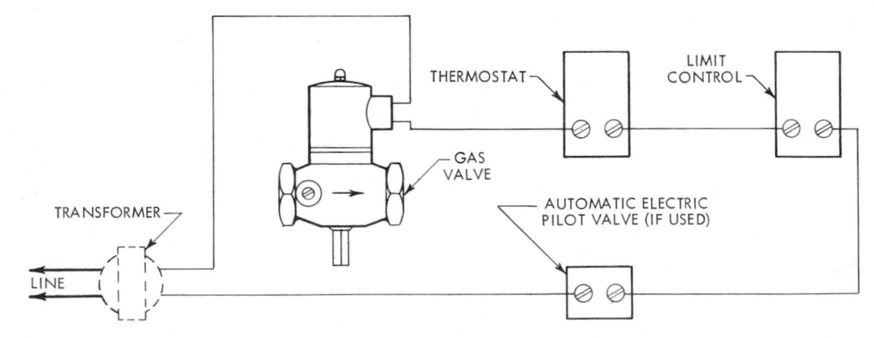

Fig. 10-12. Gas-burner controls.

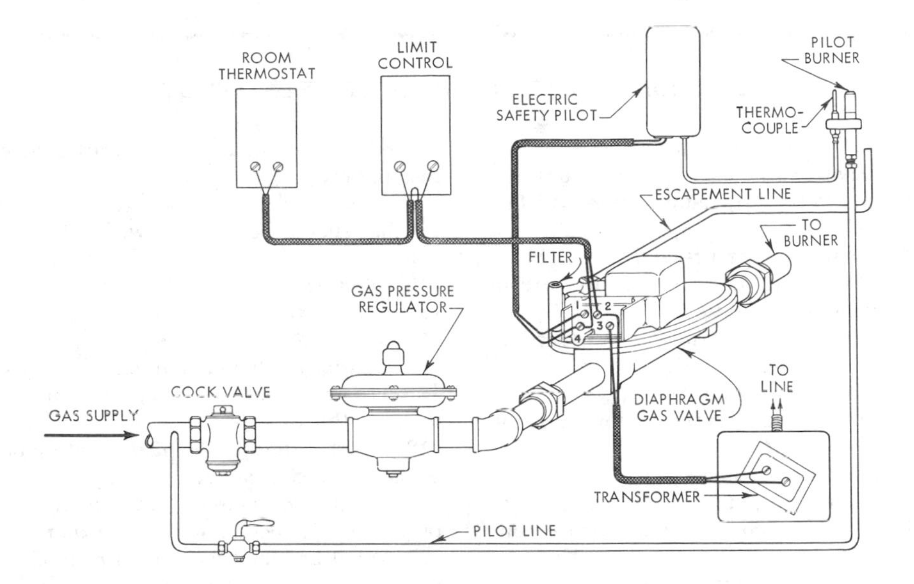

Fig. 10-13. Illustrating a gas-burner installation.

valve which permits gas to escape from the top of the diaphragm. When this occurs, normal pressure underneath raises the diaphragm, and gas flows into the boiler. When a solenoid type main valve is used as in Fig. 10-12, completion of the operating circuit energizes the solenoid coil and pulls up a plunger which in turn opens the valve.

One difference between gas-burner and stoker or oil-burner controls is that all the gas-burner devices operate on low voltage. This is possible because no motor-driven fuel pump is required for a gas unit.

Fig. 10-13 shows a layout of the relative positions of equipment needed for a gas-burner installation. Note that the electric safety pilot is wired to a thermocouple. A thermocouple has two dissimilar metals joined at a point and when subjected to heat produces electricity. If the pilot burner is extinguished, no current will be generated in the thermocouple element, thus causing the circuit for the diaphragm- or solenoid-operated valve to remain open.

All protective devices should be so designed as to terminate operation of a unit without use of electricity. Hence in the event of current failure the fuel supply is automatically cut off.

QUESTIONS FOR DISCUSSION OR SELF STUDY

1. For what is an aquastat used?
2. What are the two types of aquastat?
3. Which type aquastat is more satisfactory? Why?
4. What is a pressurestat?
5. What is the usual steam pressure in a residential installation?
6. How is the pressuretrol fastened to the steam boiler?
7. What is an airstat?
8. At what point on the equipment is the airstat installed?
9. What is a furnacestat?
10. Where is it placed on the equipment?
11. What is a combination furnace controller?
12. How are automatic draft dampers usually operated?
13. What voltage is used customarily for control units of an automatic draft-damper installation?
14. What is the smallest size of conductor that should be used for wiring 120-volt valves?
15. What is the smallest size of conductor suitable for wiring 24-volt controls?
16. What is a stoker relay?
17. In Fig. 10-12 which automatic control switches must be closed before gas will flow through the valve?
18. What is the specific purpose of an oil-burner protector relay?
19. What two types of ignition are employed in connection with oil burners?
20. What are the two general methods of feeding gas to a boiler when heat is called for by the thermostat?

Installing Service Equipment

Chapter **11**

All electrical energy supplied to power consuming devices and appliances within a building must pass through the electrical service entrance equipment where it is metered, protected and distributed through branch circuits. The size is determined by the amount of power consuming devices connected, such as those discussed in Chapter 9.

Power from the local power company is delivered into the house through lead-in cables called a *service drop* if it is overhead. A *watt-hour meter*, connected between the service drop and the main power switch or *service disconnect*, measures the amount of power used by the house circuits. See Fig. 11-1. A main control center, also called a *distribution panel*, *service panel* or *panelboard*, which contains the *fuse box* or *circuit breakers*, connects to the service wires and delivers current to the various outlets. Fig. 11-2 graphically describes the path of the current from the pole to service head, etc. Some communities or areas within a community have underground services that enter the house below grade.

In larger residential construction the connection from the single panelboard to the circuit outlets is modified. This method employs a main control or panelboard to which several *branch circuit control centers* are connected. See Fig. 11-2. The number of load centers is dependent somewhat upon the number and type of circuits required and the distance of electrical loads from the main service entrance. These panels are located throughout the house as close to the point of usage as possible. The number of individual load

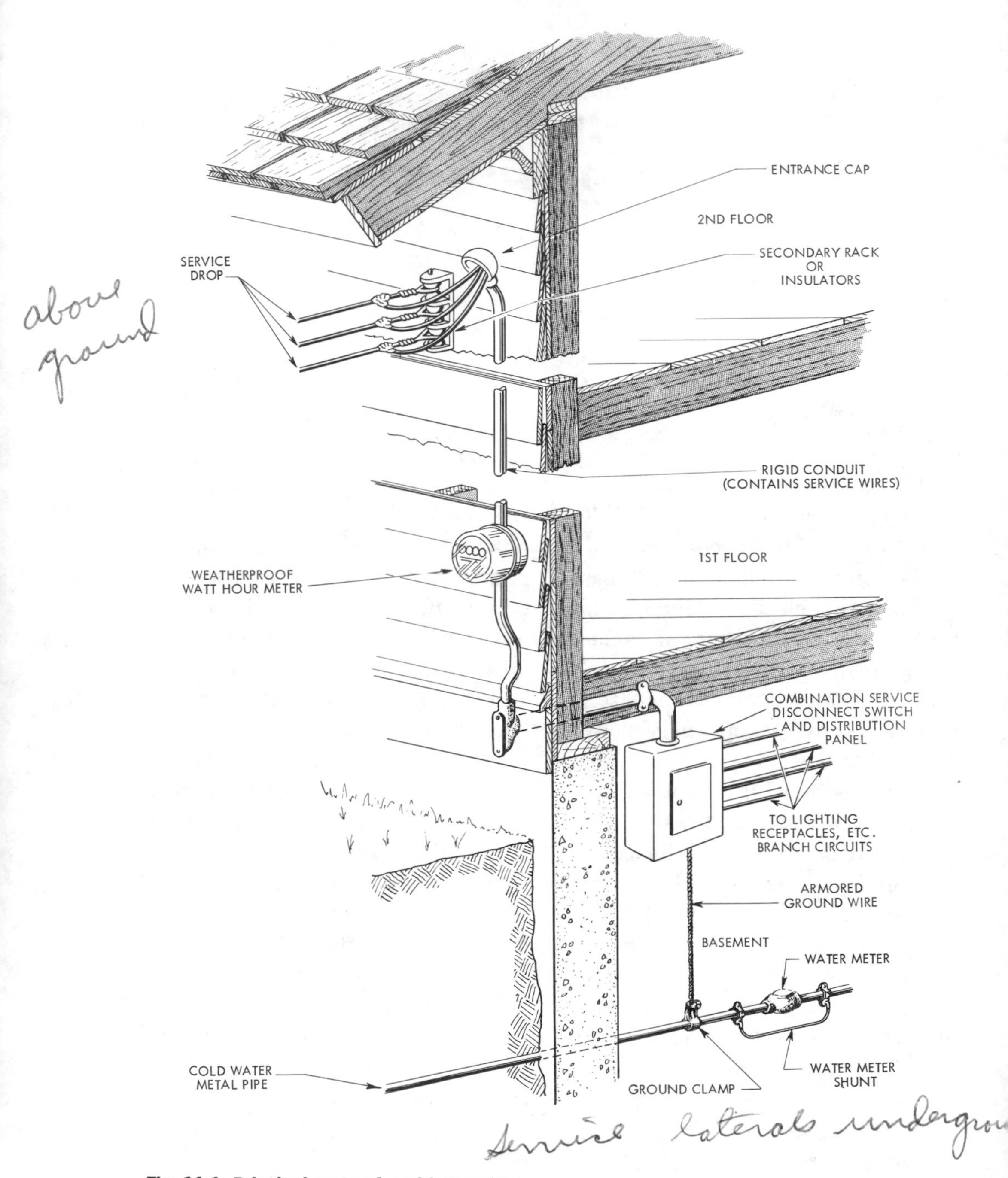

Fig. 11-1. Principal parts of a wiring system.

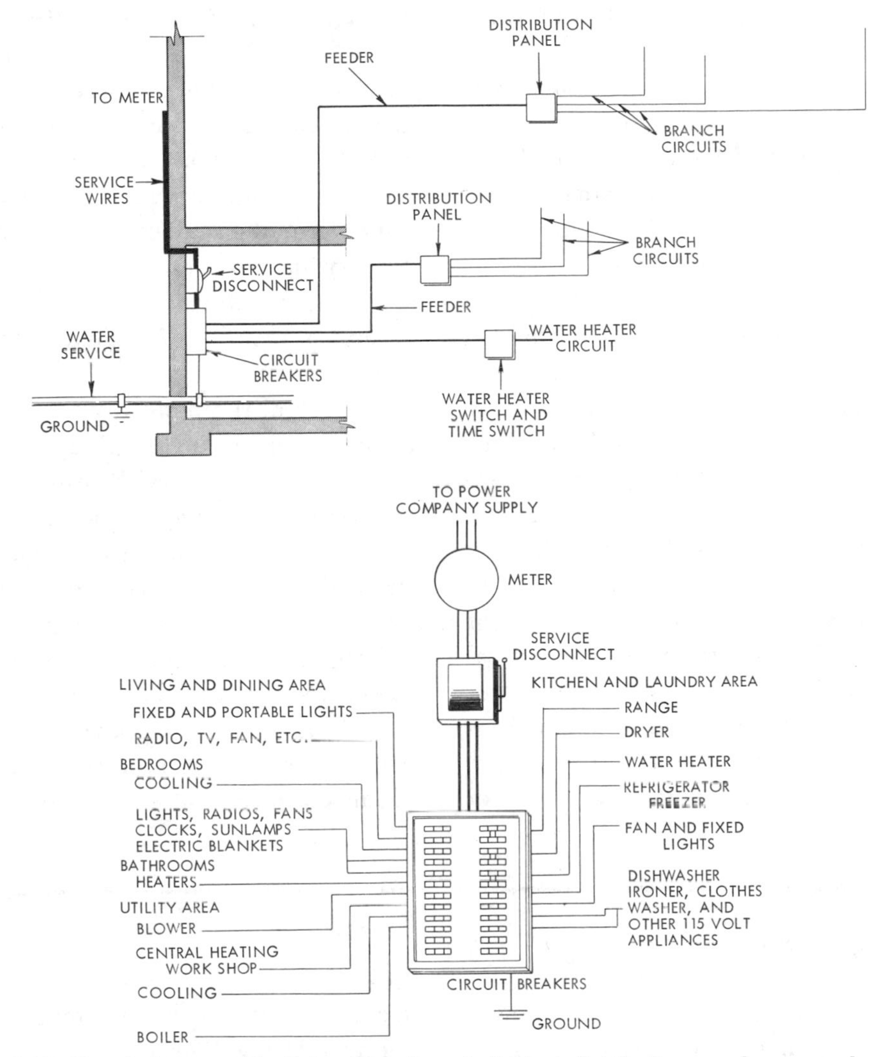

Fig. 11-2. Top: In large residential construction, individual distribution panels are used. Bottom: Relationship of branch circuits and service equipment.

centers is dependent upon the needs of the home. The distribution panels may be conveniently located in the kitchen, utility room or closets. By placing the load centers close to the loads, the possibility of a voltage drop is greatly reduced by increased feeder wire size. All electrical services are required to be grounded to earth for protection of life and property.

(Grounding is explained in Chapter 12.)

The *size, methods*, and *amount* of service equipment is determined by the amount of power consumed and by the local electrical codes. Power utility companies also impose requirements of their own.

The electric wiring system in a residential building may be divided into separate sections or classifications, each of which has its specific purpose. In Fig. 11-1, relationships between the various sections are shown, and a manner in which they may be connected. A coordinated study of this figure with the following descriptions is recommended for a better understanding of electrical services. In Fig. 11-1 the meter socket and distribution panel is mounted and connected together with conduit as shown. (See Chapter 13, "Methods of Conduit Wiring.") Wire is pulled into place and grounded where required. (See Chapter 12, "Grounding for Safety".) After all branch circuit wiring is brought to the service panel, circuit breakers are installed and wiring connected. This is generally done when all "finish" or "trim" is installed for final inspection.

As stated, the main divisions of a complete wiring installation may be listed as follows: service drop, service wires, metering, service switch, branch circuits and grounding. In succeeding paragraphs the service drop, services wires and service switch, along with conductor electrical protection, are discussed in detail. Branch circuits have already been discussed; grounding will be discussed in the following chapter.

Service Drop

The feed wires from the power company lines to the secondary rack on the customer's building are known collectively as the *service drop*.

The service drop, shown in Fig. 11-1, is actually not part of the wiring system in a building. It is the connection provided by the power company from area distribution lines to the building. It is discussed here because failure to observe important points relating to the service drop, such as point of attachment and driveway clearances may result in delay and expense. Service head locations, service drop point of attachment, and clearances over driveways, walkways and swimming pools is determined by the department of building and safety and the local power servicing agency.

Service Conductors

Service conductors connect the power company service drop to the service switch.

The wires may be carried to the building either underground or overhead, depending upon the location of the power company distribution lines. The electrical wireman should know: (1) whether underground or overhead service will be used; (2) the voltage and amperage; (3) whether single- or three-phase; (4) whether two, three, or four wires will be employed; (5) the type of watt-hour meter, and where it will be located; (6) and the point at which these wires are to enter the building.

Underground Service Connections

These connections are of three general types: (1) from the service switch directly to a utility company manhole in the street; (2) from the service switch to a utility company sidewalk handhole; and (3) from the service switch to a pole riser.

When the service originates in the manhole, it is necessary to make arrangements with the company for joint installation of conduit and wire because no one except a utility company workman is permitted to enter the chamber.

When the service originates in the sidewalk handhole, Fig. 11-3, the

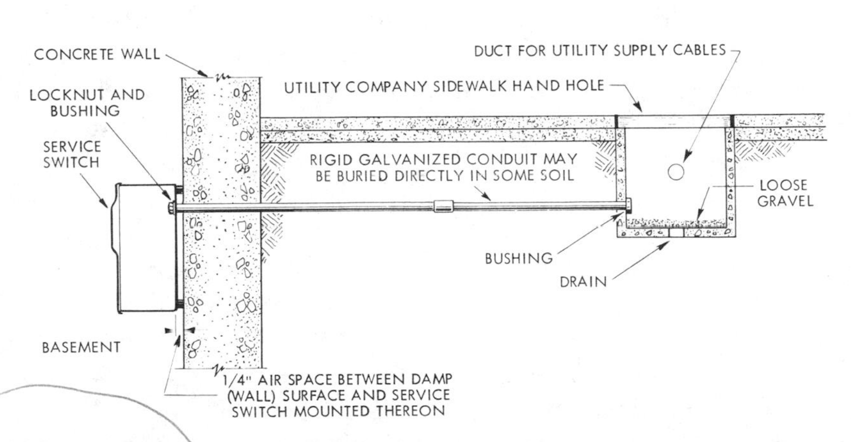

Fig. 11-3. Underground service conduit to sidewalk handhole; underground service conductors are called a "service lateral."

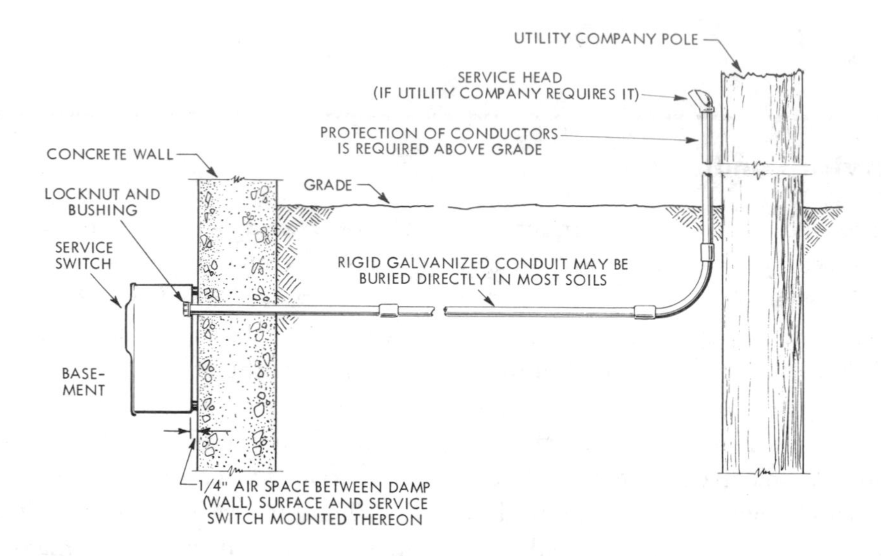

Fig. 11-4. Underground service conduit and pole riser.

electrical contractor is usually permitted to run his conduit and wire into it. Utility personnel connect the wires to service cables which pass through ducts from one handhole to another along the street.

A pole riser service, Fig. 11-4, is also a joint enterprise between electrical contractor and the utility company. In many cases, the electrical contractor simply installs conduit to a point at least 8 feet above grade, a minimum height for such protection, and pulls in enough wire to reach the crossarm at the top of the pole.

Utility linemen extend the wires up the pole, and may cover them with approved molding. In some cases, a short piece of conduit and a service head are used to terminate the run at the crossarm, the contractor supplying these items to the company. In other cases, the contractor must furnish a service head and enough conduit to reach the desired point.

In new construction, entry through the wall is made before the concrete is poured, a single length of conduit being secured in place as forms are assembled. Another way is a sleeve between the forms large enough for passage of the service conduit.

A method of feed to a meter located above grade level is shown in Fig. 11-5. Note the meter socket can be mounted flush into the wall making it more architecturally pleasing than when mounted on the wall surface.

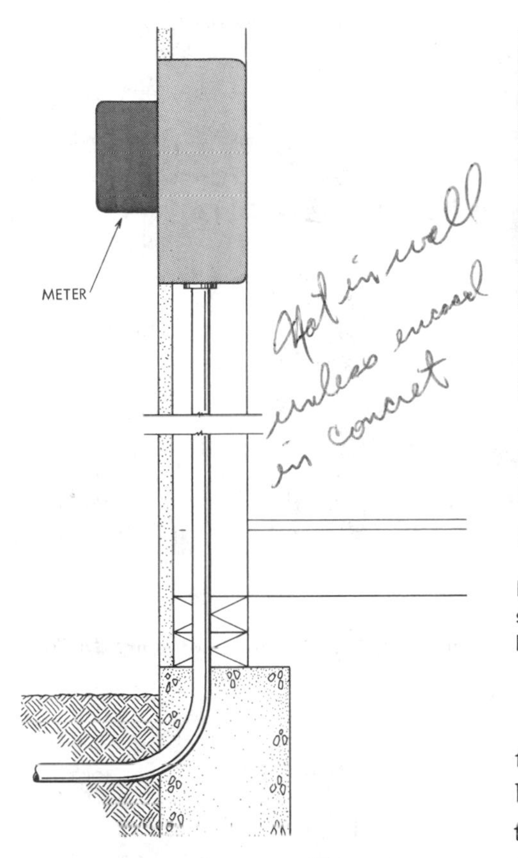

Fig. 11-5. Underground electrical service to flush meter can and distribution panel; this is more architecturally pleasing.

Fig. 11-6. Installing pull box for underground service conductors for a large, all electric home. (Boron Electric, Inc.)

Pull boxes for installing underground service conductors are sometimes necessary or required. Such a pull box is shown in Fig. 11-6. Conduits and conductors are installed to metering equipment from the pull box. Note iron pull wire installed in conduit. Years ago, lead-covered wires had to be used for underground service. With the advent of different types of insulations, however, lead has in most cases been replaced by type RH, RHW or TW conductors, and others. Specifications of these conductors and others are found in the NEC.

Overhead Service Connections

The overhead *service drop* is usually installed and connected by the power company. It is secured to the building by appropriate means at a point where the line gang can obtain a direct pull from the nearest service pole. Fig. 11-1 shows service insulators for supporting the power company's service drop. Single insulators with a simple screw point for mounting in wood are available for single installation; they also come mounted in sets of three should they be required. The wires must be high enough to provide proper clearance above grade, and should not come within 3 feet of any door, window opening, or fire exit. The exact location of the pull-off should be learned from the power company. It is always better to install a service head above the splicing point to the service drop. This will prevent water from entering the conduits and service equipment should the service wire driploops ever be accidentally removed.

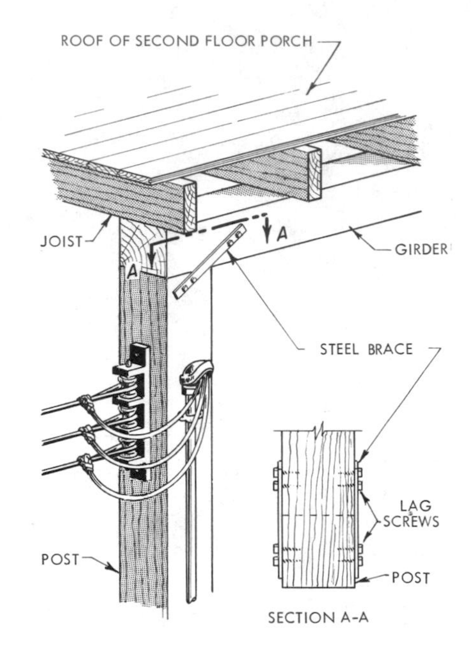

Fig. 11-7. Pull-off from a two story dwelling.

Reinforcing Building

Many service drops are long and heavy; therefore, the pull-off structure must be sturdy enough to withstand strain imposed under existing weather conditions. If there is any doubt on this point, reinforcement should be provided in a manner best suited to meet these requirements. Several illustrations are given showing how this can be done in situations commonly encountered.

Fig. 11-7 shows the service drop pull-off from the porch of a two-story dwelling or apartment building. Two steel braces are bolted or lag-screwed to the post and to the wood girder, as shown.

Fig. 11-8 shows a similar arrangement for a three-story building; the pull-off is made at the third-floor level.

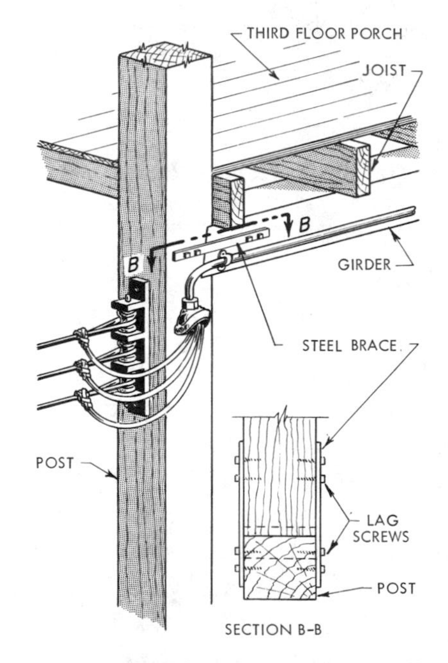

Fig. 11-8. Pull-off from a three story dwelling.

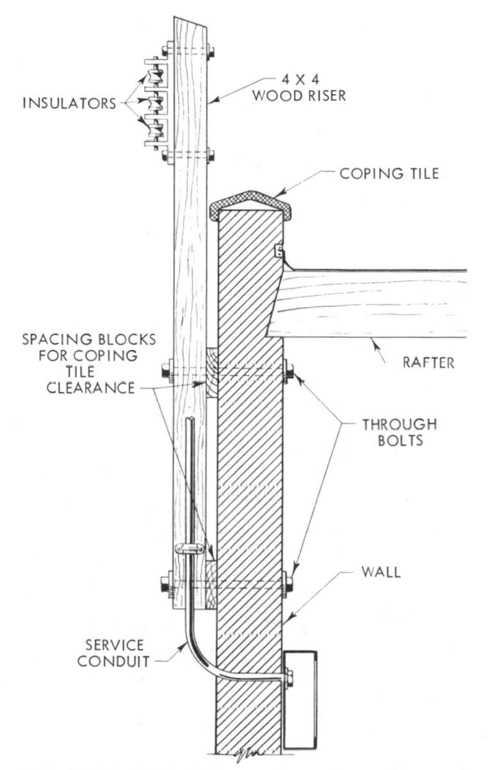

Fig. 11-9. Pull-off from riser above the roof.

If the building is not high enough to provide proper clearance for the service drop, a wooden riser 4″ x 4″ or larger, Fig. 11-9, is employed to give the necessary height. The riser is located as near as possible to the corner of the building or to a traverse masonry wall (interior wall running at right angle to outside wall) in order to get additional support. It must be kept in mind, however, that a building with light walls cannot support a heavy pull. It is safer and better to erect a pole alongside the building in such a case.

Fig. 11-10, top, shows a service

Fig. 11-10. Top: Typical service head or cap, sometimes called a weather-head. This has three piece design for ease of installation and wire pulling. Bottom: Service drop attached directly to service rigid conduit on single family dwelling. Note power company using three conductor, twisted service drop here. (Top: All Steel Equipment Inc.)

head and Fig. 11-10, bottom, shows how it is installed with the service drop attached directly to the rigid service conduit. The conduit shown is 1½ inch size. In this installation the power company used a 3 conductor, twisted service drop instead of three single conductor wires. This is a common practice. Note weatherproofing roof-jack at conduit base. The installation of three supporting insulators was unnecessary here. The power company servicing agency or the local department of building and safety should be consulted regarding service requirements. It is generally accepted that the supporting insulators (secondary rack) have 18 inch minimum roof clearance. However local codes or power companies should be consulted.

Conductor and Conduit Sizes

The proper service conductor size and conduit sizes can be determined from examples explained in the back of the National Electrical Code book. The calculated load for appliances, etc., in amperes is given for examples of residences. For example, if the code requires service conductors for 100 amperes, the conductor ampacity table should be consulted to determine what size wire will carry 100 amperes. The index and table of contents in code books are very helpful in locating information on points in question. Care should be taken in selecting the proper tables and columns. For example, larger aluminum wire sizes are required than copper for the same current. Different types of wire insulations and designations are rated for different ampacities. Type TW is different from type RHW. Finally, tables of conductor conduit fill or maximum number of conductors in trade sizes of conduit or tubing should be consulted to determine the proper size conduit for the size and type of wire used, and number of wires to be used. Three wires is the normal number of service conductors. Care should be exercised again to consult the proper table in the code book. Tables for new and old work are different. Columns are different for different wire types.

Service Entrance Cable

Special service-entrance cable is also commonly used instead of separate conductors pulled into conduit. This is shown in the oval shaped cable in Fig. 11-11. Note how the bare spiral wound wire is used as a neutral conductor by the wireman twisting the strands together at each end for terminal connections. The spiral winding prevents tampering with conductors ahead of the meter. Some entrance cables have additional steel armor wrappings.

Fig. 11-12 (top left) shows a service entrance cap for use with oval

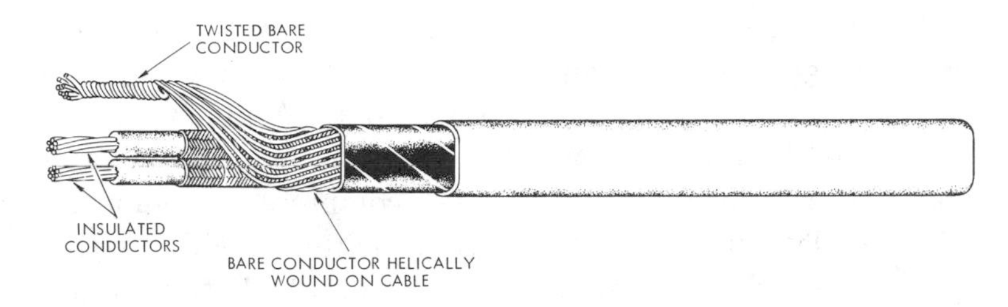

Fig. 11-11. Service entrance concentric cable with spirally wound bare conductor which is twisted at ends by the wireman and used as a neutral wire.

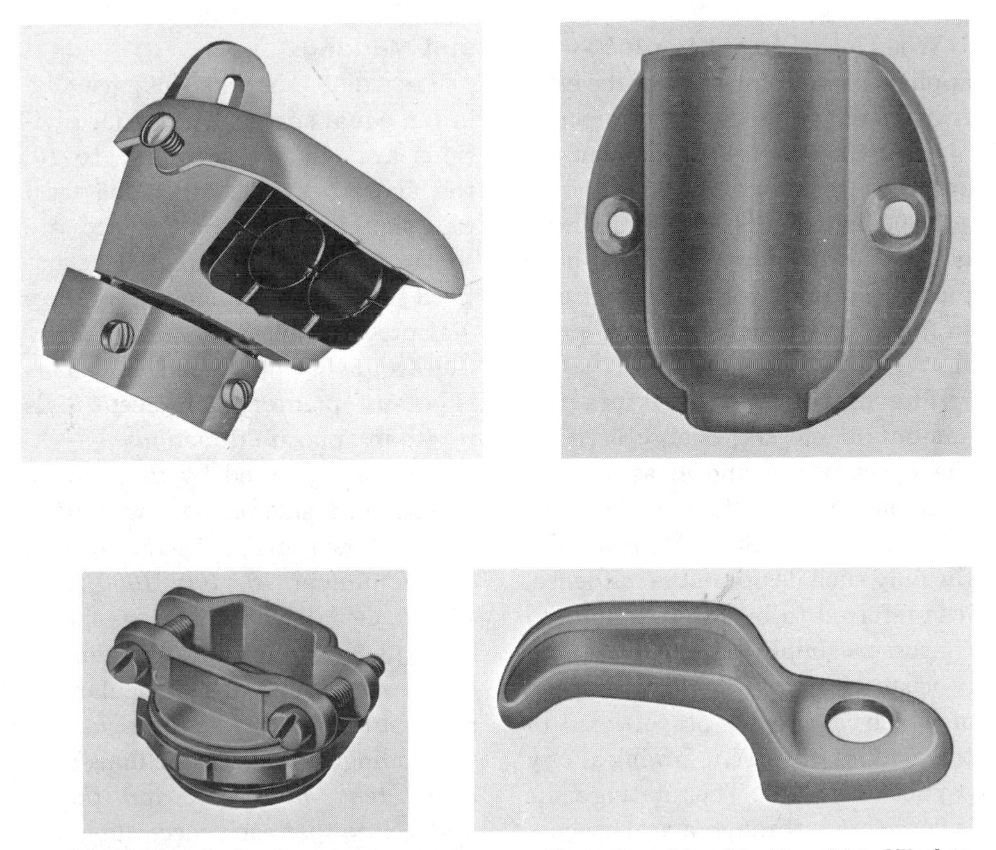

Fig. 11-12. Top left: Service entrance cap for use with oval service cable. Top right: Sill plate used with service entrance cable. Bottom left: Oval service cable connector. Bottom right: Oval service cable strap. (All Steel Equipment Inc.)

service cable from sizes 8/2 (#8,2 conductors) through 2/3 (#2,3 conductors).

A sill plate used with service entrance cable is shown in Fig. 11-12 (top right). The plate provides service cable protection where it enters the building. Sealing compound is used to keep out water.

Fig. 11-12 (bottom left) shows a clamp type oval service connector for use where cable enters cabinet knock-outs. Watertight connectors are available with a rubber grommet that compresses around the cable when tightened with a nut.

Fig. 11-12 (bottom right) illustrates a mounting strap for use with oval service cable.

Watt-Hour Meter

When a consumer has a 1000-watt appliance and connects it to the electric circuit for one hour, the amount of energy used is 1000 watt-hours or one watt consumed for 1,000 hours is 1,000 watt-hours. A kilowatt-hour is 1000 watt-hours. The watt-hour unit is too small for convenient use, so the kilowatt-hour is used in practice.

The meter that registers the amount of electric energy used by the consumer is known as a *watt-hour meter*, even though it records *kilowatt hours*. Since it is usually the only such device in the residence, it is referred to by wiremen and the consumer simply as the *meter*. It consists of a small motor the speed of which is directly proportional to the amount of current flowing at any particular instant. The rotating shaft is connected, through a gear mechanism, to dials which record the total number of kilowatt-hours consumed.

Dial Markings

Each dial is marked off, usually, in ten equal spaces. The major divisions are numbered from *1* to *10*, the figure *0* at the top being used to designate *10*. Two meter faces are shown in Fig. 11-13. It will be noticed that the numbering of any dial proceeds in the opposite direction from that of its neighbor. This is because pointers of adjacent dials rotate in opposite directions.

Dials are marked by *units, tens,* and so on, inside of the ring numerals, as shown in Fig. 11-12, bottom, or by number *10, 100, 1000,* and *10,000,* just above the circle as shown in 11-13, top, to indicate the number of kilowatt-hours each particular one represents. Starting at the *right* and proceeding to the *left*, the dials read *units, tens, hundreds,* and *thousands*. Note that each succeeding dial has ten times the value of its right-hand neighbor. On the right-hand

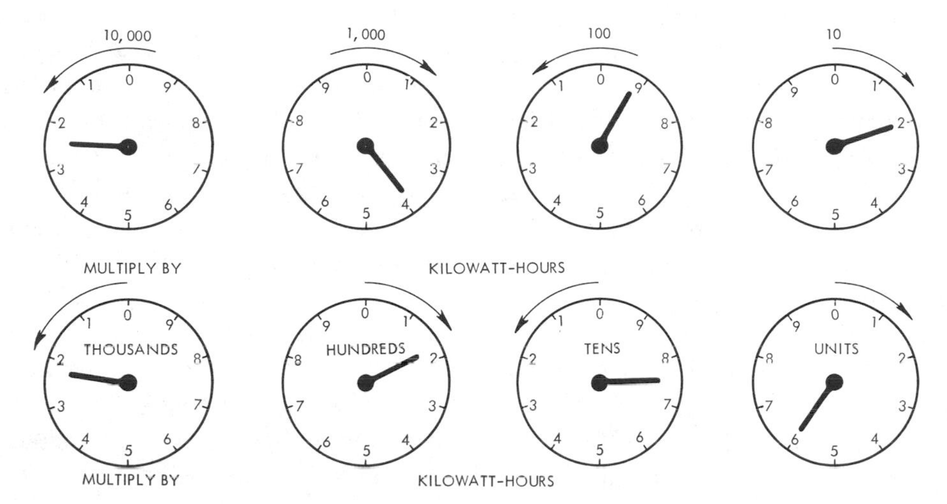

Fig. 11-13. How to read dial markings on a kilowatt-hour meter.

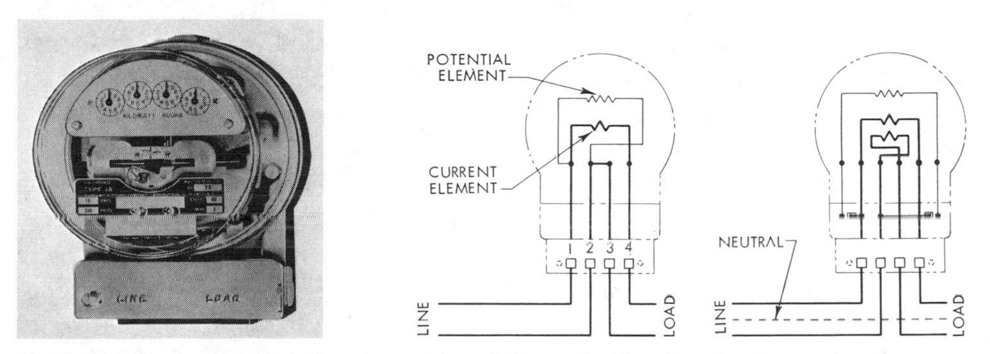

Fig. 11-14. Bottom connected meter and how it is used with a 2 or 3 wire supply. (Sangamo Electric Co.)

dial, each of the numbered divisions has a value of one kilowatt-hour, each complete revolution of the pointer representing ten kilowatt-hours. Most residential meters appear without word markings, only numbers. This may be seen in Fig. 11-14.

Reading a Meter. When reading a meter start at the right hand dial and proceed to the left. Jot down the smaller of the two numbers between which the right-hand (units) pointer lies. To the left of the first number, mark down the reading of the 10-dial, and continue with the remaining ones. Thus the readings in Fig. 11-13 bottom are, respectively, *six, seven, one, two,* the figure being written as: . . . . *6*, then . . . . *76*, then . . *176*, with the final reading *2176*. Suppose this value represents last month's reading. Proceeding in the same way, (assuming the upper dial number readings are this month's figures of the same meter) the reading obtained from the upper set of dials is *2392*. Subtracting the former from the latter, $2392 - 2176 = 216$, the consumption during this period was *216* kilowatt-hours.

Care must be taken when a pointer seems to be directly opposite one of the figures, such as the *4* on the upper 1000-dial. The number must be recorded as *3*, because the pointer on the 100-dial at the right has reached only *9*, not having completed the full revolution to move the 1000-pointer to *4*.

With certain meters, the dial reading must be multiplied by a number called the *meter constant,* in order to obtain actual kilowatt-hours. In such devices, there is a notation such as: "Multiply by 10" on the dial or in some other conspicuous place on the meter. Thus, if the dials shown in Fig. 11-13 were marked *multiply by 10*, the consumption would be 216×10, or *2160* kilowatt-hours.

Meter Wiring

The number of wires brought out for the meter and their arrangement depend on the character of the service, the type of meter, and the kind of load, as well as on standard metering practice set up by the power company. Full information on these points should be obtained from the company before the work is done. Typical *bottom connected* residential meter diagrams applicable to various types of underground or overhead electric services are shown in Fig. 11-14. The unit illustrated here is a bottom-connected watthour meter having four terminals. The two left-hand ones are connected to the incoming line; the two right-hand ones to the customer's load. The metal cover plate has the words *Line* and *Load* stamped on it.

Internal Meter Diagrams. In Fig. 11-14 current elements in the meter diagrams are represented by the heavy lines, potential elements by the lighter lines.

In the 2-wire meter, a supply wire enters at terminal No. 1 and connects to both potential and series windings. The other end of the series coil goes to terminal No. 4, and thence to the load. The remaining supply, where it connects to the potential coil before continuing, enters terminal No. 2, goes through a short connection to terminal No. 3, and out to the load.

In a 3-wire meter the two outside wires pass into and out of the meter, the potential winding bridging them, as shown. The neutral wire goes directly to the load.

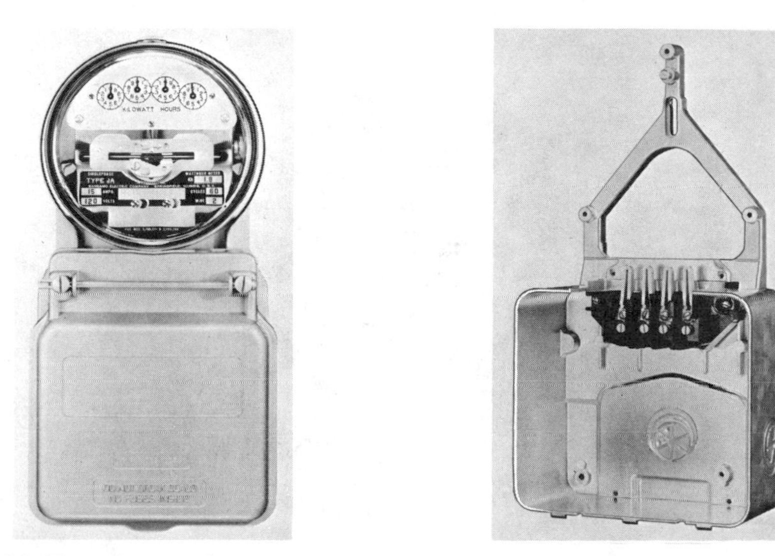

Fig. 11-15. Meter fitting for bottom connected meter. (Sangamo Electric Co.)

Meter Installation

Bottom Connected Meter. The simplest kind of meter installation is a bottom-connected unit, Fig. 11-15, mounted just above the service switch. An opening is provided in the top of the switch to fit the bottom of the meter. After connections have been made, a metal cover is placed over the meter terminals and sealed by the power company.

Socket-type Meter. Power companies required meters that could be quickly installed or removed. The socket-type meter, Fig. 11-16, was developed to meet this need. The internal construction and diagram are the same as for other types. The terminals are in the form of blades which fit into clips, as shown. The meter-socket fitting is often supplied by the power company. In some parts of the country the electrical contractor must supply the meter socket. The neutral conductor may be grounded in the meter and in the service panel, local power company requirements prevailing. Meter sockets are sometimes combined with service disconnect switches and circuit breakers in the same enclosure. This is called a combination service panel.

The wireman must fasten the meter socket to the building, install the service conduit (or service-entrance cable) and wires, and then attach them to the clips in the fitting.

Supply wires are connected to the upper terminals, and load wires to the bottom. In a 3-wire meter the neutral wire usually passes through

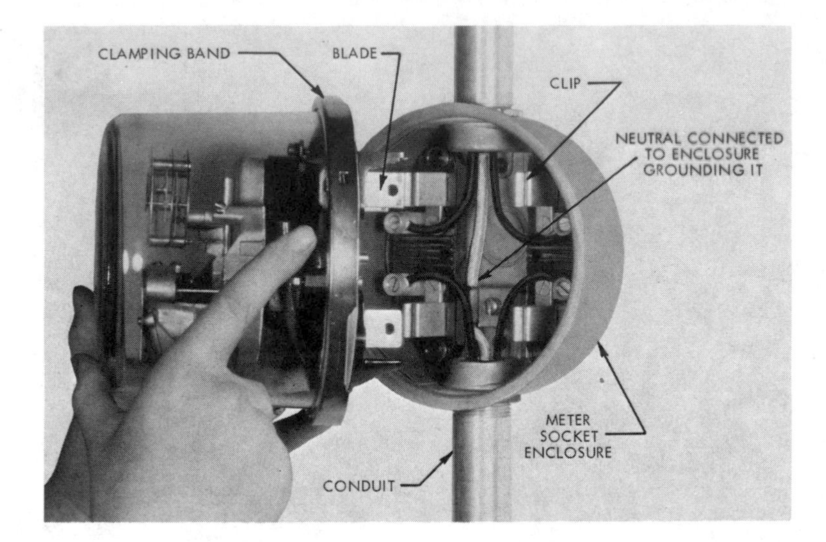

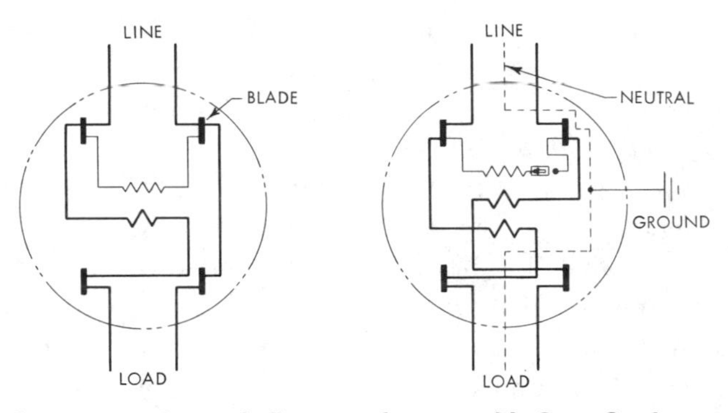

Fig. 11-16. Socket type meter and diagrams for use with 2 or 3 wire supply. (Sangamo Electric Co.)

the fitting without making contact. Sometimes, however, it is attached to the enclosure in order to ground the fitting without use of a separate grounding conductor. Note threaded conduit provisions for the connecting enclosure. (Grounding details are given in the next chapter.)

Special Water-Heater Meter

In some parts of the country the power companies give special heating rates if current is taken only during hours when the supply lines are not heavily loaded. Fig. 11-17 shows a simplified scheme for making use of this plan. One of the two elements

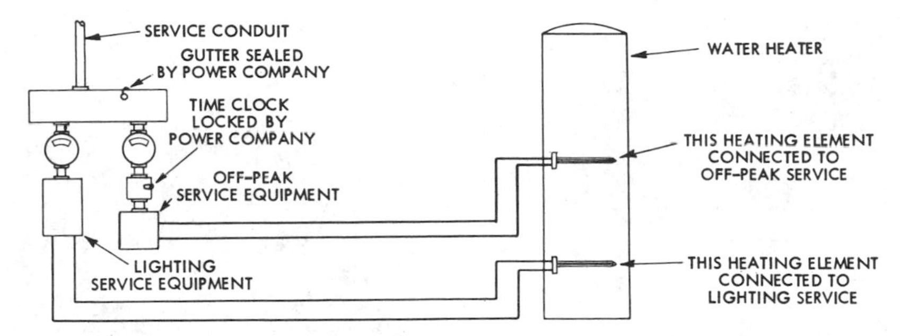

Fig. 11-17. Off peak water heater circuit.

of the water heater is connected to the lighting service meter, the other to an off-peak meter and a time clock which permits current to flow during certain prescribed hours. The time clock is sealed by the power company to prevent tampering with the mechanism. The element connected to the lighting meter is the smaller one, which maintains a desired temperature after the large element has created the proper degree of heat.

Service Switch and Distribution Panel

The service switch is the main disconnect switch. Means must be provided for disconnecting conductors in the building from the service entrance conductors.

The service switch, Fig. 11-18, left, is in many installations a fusible type, the capacity and design of which is suited to the particular application. Thus, it provides both overcurrent protection and disconnecting means. In some cases, the service switch incorporates a meter-fitting and a branch-circuit cabinet. In other installations, meter, service main switch and overcurrent protection for all conductors are within a common enclosure.

Each ungrounded service-entrance conductor and branch circuit wire requires overcurrent protection. The ampere ratings of overcurrent protection can be determined when service and branch-circuit wire sizes are determined by following examples in the back of the National Electrical Code book.

The codes list certain requirements for a service switch. Examples are: (1) it must be fully enclosed; (2) the operating lever or handle must be on the outside of the enclo-

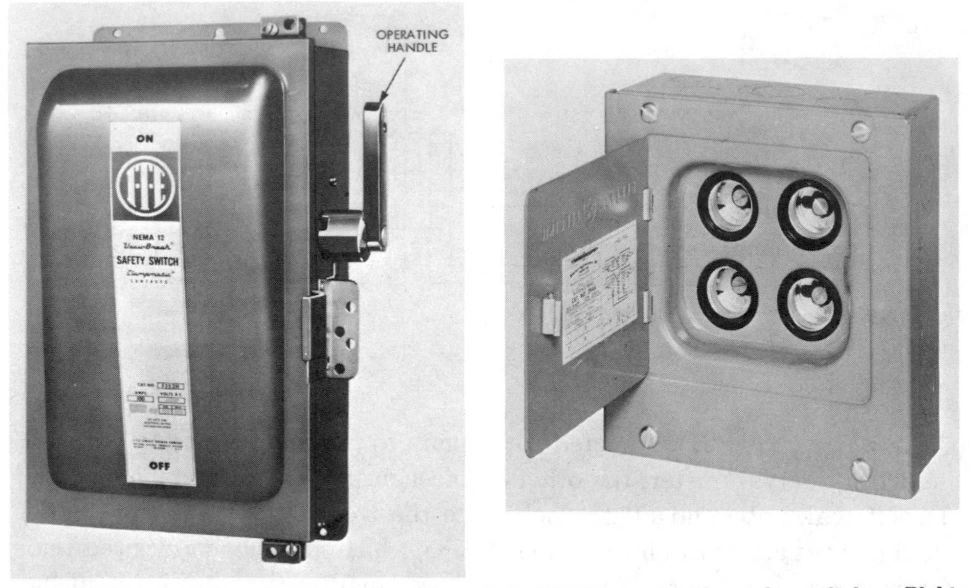

Fig. 11-18. Left: Externally operated safety switch (E-X-O) used with service switches. Right: 4 circuit fuse panel. (Left: I-T-E Imperial Corp. Right: General Electric Co.)

sure in plain view and within reach of a person standing on the floor; (3) there must be some means for disconnecting the supply conductors from their source of electric energy; (4) there must be a plain indication of whether the switch is in closed or open position; (5) provision for overcurrent protection is necessary; (6) there must be provision for grounding the neutral conductor; (7) in damp or wet locations, a 1/4 inch air space between the switch enclosure and the mounting surface.

Distribution Panel

The distribution panel is supplied by the main disconnecting switch. It distributes electrical energy to branch circuits through its many circuit breakers or fuses that protect the wires. The distribution "panel" is equally referred to as "service panel," "load center," "panelboard" and sometimes "main control center." These names are used interchangeably in the text to assist the learner in proper electrical terminology. A small circuit "panel" is shown in Fig. 11-18 (right).

The service panel, Fig. 11-19, is equipped with a main pullout type disconnect switch and eight plug fuses. The fuses protect eight lighting and utility circuits. If an electric range were to be included in the

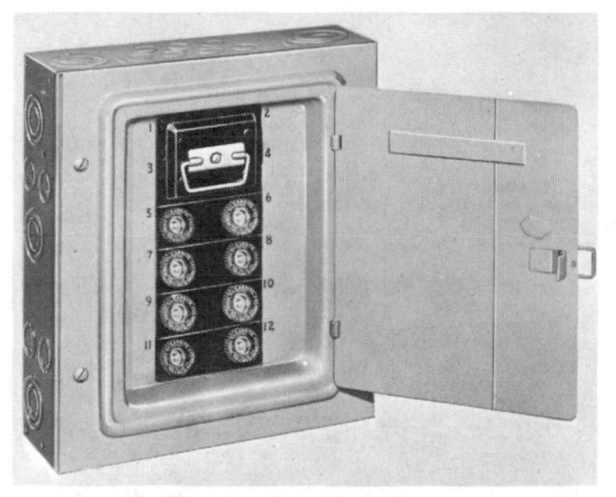

Fig. 11-19. Service panel.

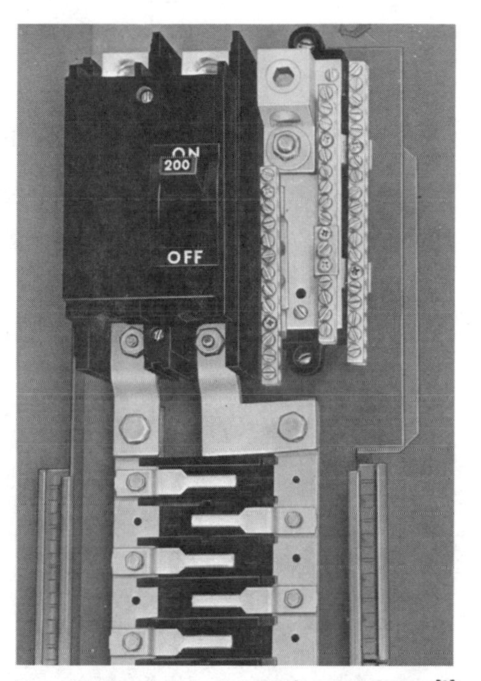

Fig. 11-20. Load distribution center with main circuit breaker and neutral terminals at the same end. Note bus bar arrangement at bottom of main breaker awaiting installation of branch circuit breakers. (Square D Co.)

installation, a second pull-out switch would be needed. It would be found, usually, directly below or to the right of the main pull-out. Note here that a main fuse and disconnect are in the same enclosure. Circuit breaker panels like this are available also.

A 200 amp load distribution center is shown in Fig. 11-20. Note a main disconnecting circuit breaker, top left. All supply wiring is connected to the top of this panel. Other convenient arrangements are available. Fig. 11-20 also shows the copper bus bar arrangement from the bottom of the circuit breaker for mounting circuit breakers for branch circuits. Different panels have different methods of mounting circuit breakers for branch circuit protection. Some clip into space provided and some screw into tapped holes of the vertical pieces of copper conductors. Some are a combination of the two.

An individual load center is shown in Fig. 11-21. This panel is ideally suited for residential electrical distribution. It is surface mounted and suitable for commercial installations also. Note blank spacing for installation of circuit breakers for future electrical expansion. Cabinets are mounted on properly secured wood backing before conduits and wire are installed.

Fig. 11-22 illustrates a combination service entrance containing a

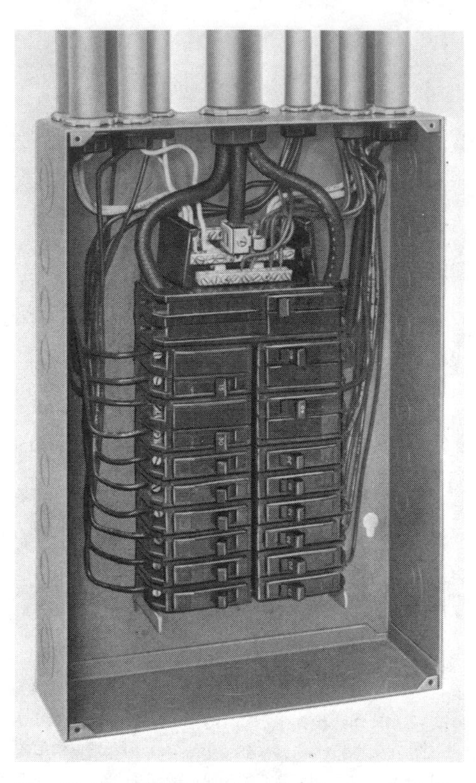

Fig. 11-21. Wired load center before installation of trim. (Square D Co.)

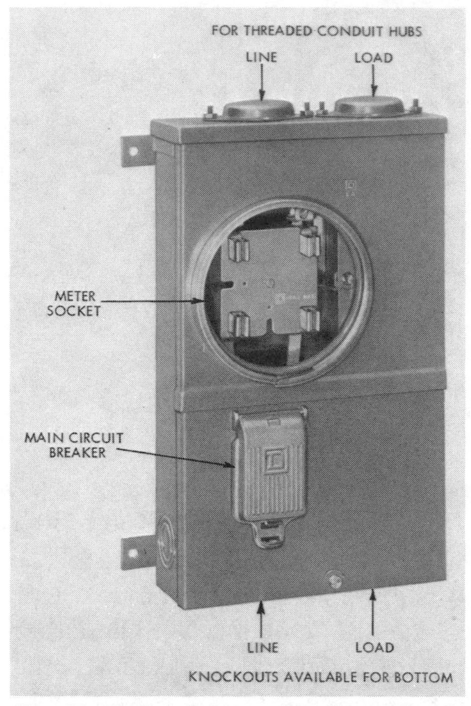

Fig. 11-22. Raintight combination main circuit breaker and service entrance. (Square D Co.)

meter socket and main circuit breaker. This panel is sometimes used where the electrical contractor must supply the meter socket instead of the power company. The main breaker in this combination device could feed the load center shown in Fig. 11-21 at a remote inside location. The load center in Fig. 11-21 is not of weatherproof construction and therefore must be installed inside. Fig. 11-22 is of raintight construction for outdoor installation. Wiring and conduit options may be at the top, bottom or lower sides through knock-outs. If the top is used, threaded hubs are necessary with mounting gaskets to maintain a weatherproof condition. Conduits are then screwed into the hub. Fig. 11-23 shows typical arrangements of Fig. 11-22, combination service entrance, with meter socket and main circuit breaker. Fig. 11-24, top, illustrates a raintight combination meter socket, main cir-

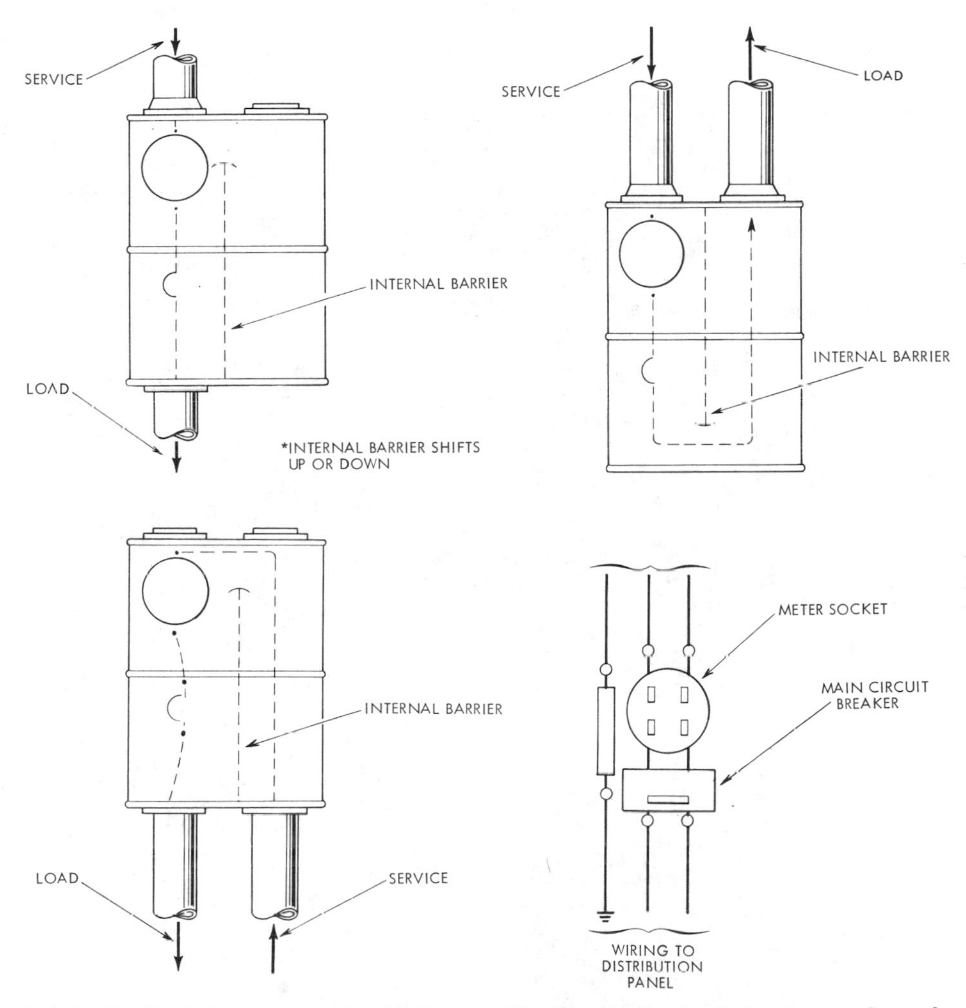

Fig. 11-23. Typical arrangements of the combination main circuit breaker and service entrance shown in Fig. 11-22. Note that the internal barrier shifts up and down. Top left: Overhead service with bottom load center feed. Top right: Overhead service, top load center feed. Bottom left: Underground service, bottom load feed. Bottom right: Wiring diagram for meter socket, main circuit breaker and neutral.

cuit breaker and distribution circuit breaker, service entrance. This is for an overhead service approach. Proper size threaded conduit hubs are to be installed for adequate conduit size and rain protection. The entrance can comes from the factory with a metal hole protection cover as

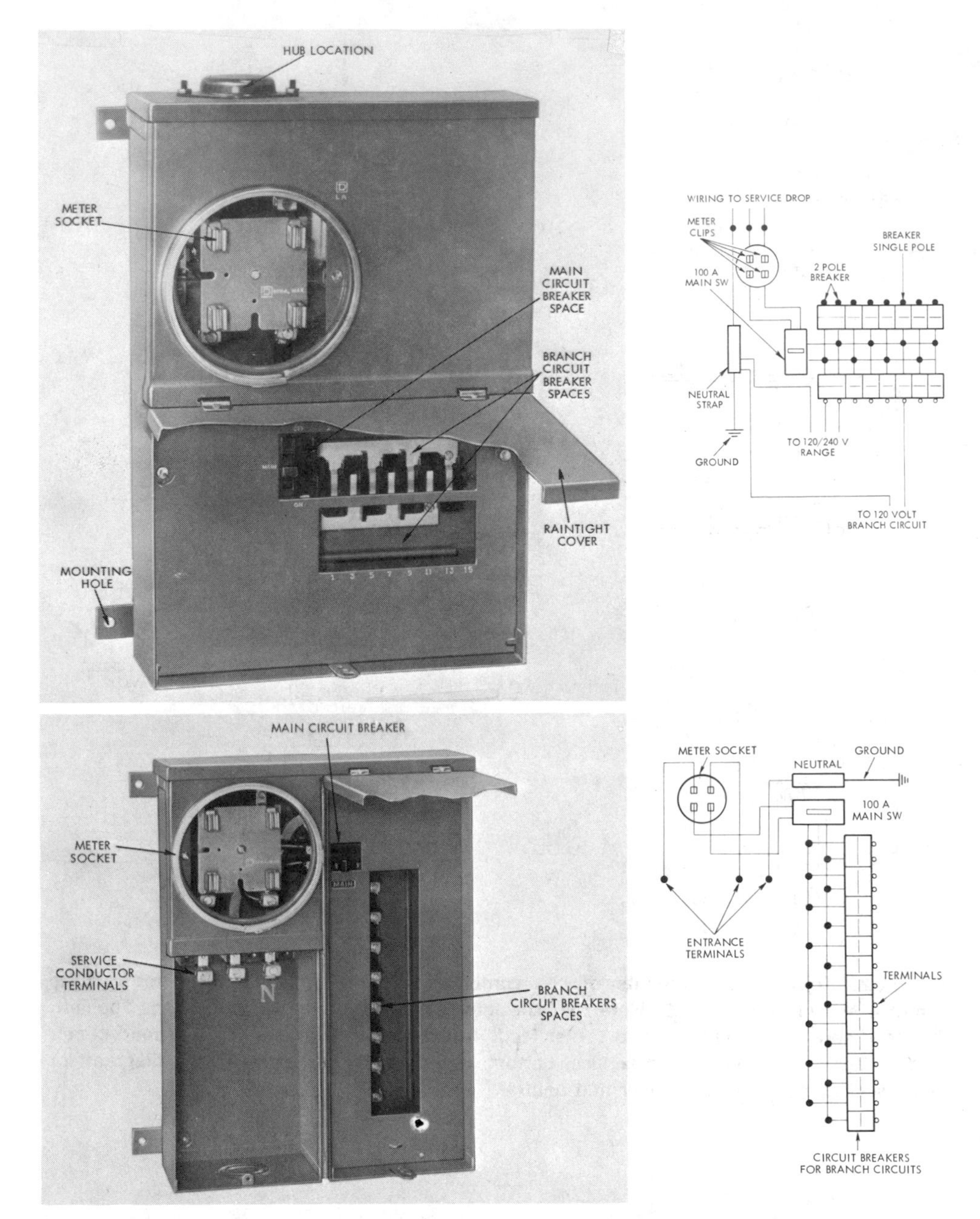

Fig. 11-24. Top: Raintight combination service entrance with branch circuit breakers. A diagram (to the right) gives the internal wiring and some idea of the external wiring. Bottom: Raintight combination underground service, surface mounting. A diagram (to the right) is given of the internal factory wiring of the meter socket and circuit breakers of the panel. (Square D Co.)

seen in Fig. 11-24, top left. Fig. 11-24, bottom, is similar to the combination at the top, except for an underground service approach.

In Fig. 11-24, bottom, service conductor terminals may be seen below the meter socket. A main circuit breaker is located top of right hand compartment, and branch circuit breaker spaces are provided below the main circuit breaker. A steel barrier separates the metered conductors from the nonmetered. These will be pulled in by the wireman. Both cabinets are designed for mounting on the surface, such as a brick or block structure.

Note in the wiring diagrams in Fig. 11-24 that the neutral wire is not switched but connects to a "neutral strap" and to earth ground. There are enough screws on this neutral strap for each 115 volt circuit, one colored wire from a circuit breaker, and a white wire from the neutral. The meter is plugged into the meter socket, by the power company after inspection approval, completing the electrical circuit to the main circuit breaker. Copper bus bars feed the individual branch circuit breakers from the main breaker. Two-pole circuit breakers are for branch circuits supplying appliances requiring 240 volts, such as a range or water and space heaters. Lighting and convenience outlet circuits are 115 (or 120) volt circuits and require only one breaker or fuse for the "hot" wire only, the other being the neutral also from the panel.

Panel Load Balancing

Most dwellings are supplied with a three-wire 115/230 volt single phase service. (Note that 120/240 volts are used interchangeably with 115/230 volts due to different servicing agencies.) The reasons why this system is used are not entirely a matter of supplying appliances and other equipment with the proper voltages. Their advantages are pointed out in terms of resistance, amperage and voltage drop of the total electrical system.

Codes require that each connected load be as evenly divided between the conductors that supply it as practical conditions will permit. Not only should the loads be divided equally between the supplying mains or feeders but the further divisions of load should be divided equally among branch circuits.

Referring to Fig. 11-25, following the arrows, we note a circuit from Line 2 (L-2) through closed circuit breaker #1, through a convenience outlet branch circuit, connected to an equally loaded circuit, back through breaker #2, to Line 1 (L 1). If these circuits were unbalanced, the neutral would have to carry the difference where it is connected between both circuit #1 and #2. If circuit #1 had 20 amps connected load, and circuit #2 had 10 amps,

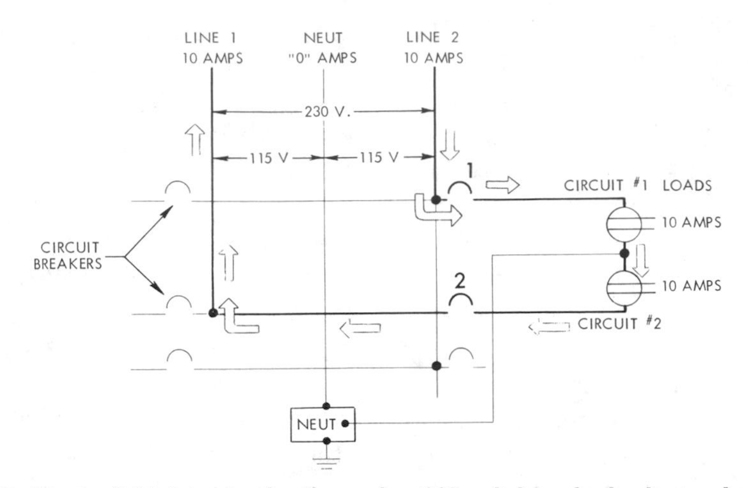

Fig. 11-25. Evenly divided loads of a three wire, 115 volt branch circuit, evenly divided on service distribution panel and supply conductors.

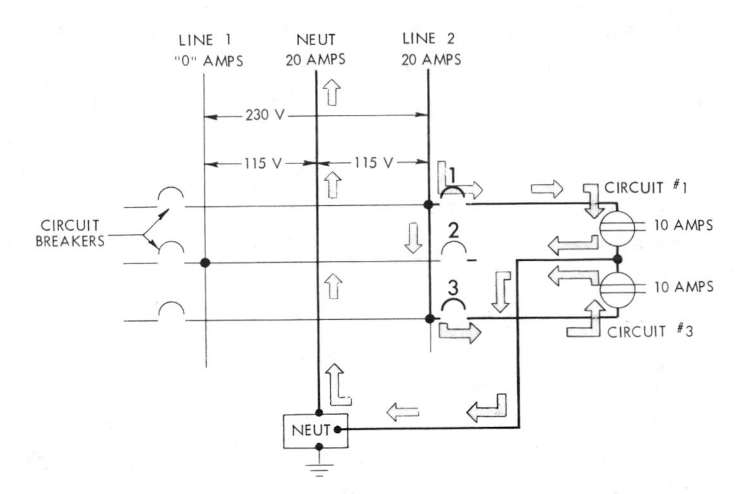

Fig. 11-26. Undesirable, unbalanced distribution panel and three wire branch circuit.

the neutral would carry the difference, that is, 10 amps.

This is not serious for this small example, but with a heavily loaded panel with many circuits, imbalance of the load connections could prove objectionable.

Fig. 11-26 illustrates a simplified undesirable, unbalanced distribution panel and 3-wire branch circuits #1 and #3. Following the arrows we note both branch circuits #1 and #3 are connected to Line 2 (circuit breaker #2 is connected to L 1). It can be seen here the neutral wire must carry the load of both branch circuits and that L1 has no current flow. Further, Fig. 11-26 utilizes two wires at 115 volts with a load of 20 amps. The more desirable connection in Fig. 11-25 with a perfectly balanced load connects both branch circuits #1 and #2 in series across 230 volts with one-half the current of 10 amperes. Power consumption is the same in branch circuits.

A neutral conductor carries the unbalanced current from other conductors.

It can be seen in comparing Fig. 11-25 and Fig. 11-26 that if the load is not evenly distributed between the outside conductors in a three-wire system at a service or distribution center, the possibilities of utilizing power available through conductors of the system supplying the service drops is largely a haphazard chance.

The manner in which a badly unbalanced load on a three-wire circuit results in undervoltage, due to voltage drop, on the heavily loaded side and an overvoltage on the lightly loaded side can be seen in Figs. 11-25 and 11-26.

A study of these conditions will also serve to show why a fuse or circuit breaker is not permitted by codes in a neutral conductor. With an increased neutral wire load, a protective device could open the neutral wire circuit. This disconnects the electrical system grounding wire and also causes a voltage imbalance in the branch circuits that are not balanced causing burned out lamps and other equipment. The neutral conductor maintains 115 volts for a 3 wire branch circuit system as shown in Fig. 11-26 and carries only the unbalanced current from other conductors.

Load balancing at the distribution panel and branch circuits helps to eliminate the undesirable voltage drop that causes a power loss (explained in Chapter 1). Most newly constructed residences are adequately wired, even at minimum code wiring standards, and are within code tolerances regarding voltage drop of branch circuits and panel feeders.

However, it may be safer to calculate questionable branch circuits, such as distant floodlighting, and large residences which are wired to the minimum code standards.

Two Wire System

Some very small residences may be served with two wires, 115 volts only. If this occurs, there is no choice or problem in balancing the loads because there is only two wires. These electric services are being discouraged, however.

Conductor Electrical Protection

Wire is the heart of any electrical system. The electrical system will fail without proper conductor insulation. Without proper protection to safeguard wire insulations, conductors themselves may melt or start fires due to extreme overheating caused by overloading or short-circuits within the system or attached loads. Today, conductors are now insulated better than ever. Insulation is protected mechanically by conduits, raceways, armor and cable braids, or outer protective coverings. Conductor insulations must also be properly protected from overheating or burning from within.

A wire will safely pass only the amount of amperage for which it is designed. Therefore, a "gage" is required to allow only permissible maximum current flow. This gage, or electrical protective safety valve, comes in two common forms: the *circuit breaker* and the *fuse*.

Fig. 11-27. "Finish" in process of being installed. Only those openings for breakers are removed. Note branch circuit directory card on door to be filled out with the proper entry. (Los Angeles Trade-Technical College)

Circuit Breakers

Fig. 11-27 illustrates circuit breakers mounted in a distribution panel. A circuit breaker is defined by the NEC as a device that can open and close a circuit by non-automatic means, and which opens the circuit automatically on a predetermined overload of current without injury to itself (when properly applied within its rating). Circuit breakers look like switches, as shown in Fig. 11-27, and are generally group located in this manner. They are sometimes used to disconnect the branch circuit from the line but they are principally designed to interrupt the

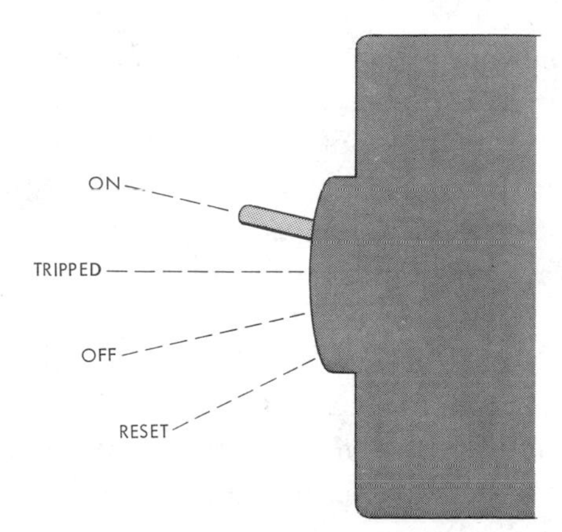

Fig. 11-28. Typical positions on a circuit breaker for the handle. To restore energy from a tripped overload, the handle must be moved to "reset" then to "on."

circuit upon overcurrents, thereby protecting the wire. Once a circuit breaker has been "tripped," automatically disconnecting the circuit from the line, the handle moves from the "on" position to a center location as shown in Fig. 11-28.

To restore energy to the circuit after the overcurrent is corrected and breaker cooled, the handle must be moved to the "reset" position, then to "on."

Most circuit breakers installed in residences contain thermal bi-metal elements for slow overloads and a magnetic arrangement for instantaneous trips in case of short circuits.

Bi-metal elements are similar in action to some thermostats, with the exception that the heat is generated within by the overload current. The bi-metal (two dissimilar metals fused together) bends due to greater expansion of one metal and trips a mechanism. Breakers are carefully calibrated at the factory and are not field adjustable.

Circuit breakers are rated in amperes to correspond to wire ampacities. This capacity is marked on the handle of the breaker. The "on" and "off" positions are marked on the breaker. Because circuit breakers are connected into hot wires only, one single-pole breaker protects a 120 volt branch circuit while a double pole (2 pole) protects a 240 volt circuit. Panel mounting methods of circuit breakers vary with different manufacturers. Some clip into place and some must be held with screws.

When trim cabinet covers are installed, only those openings for breakers should be removed. This is done by rapping sharply with a screwdriver at extreme corners away from the metal ties. This can be seen in Fig. 11-29.

Filler plates are available to close unused open holes. Referring back to Fig. 11-27, note that one spare breaker opening remains (at bottom) and one opening in the center remains to be closed with a filler plate.

The directory card on the inside of the door should be filled in indicating which number breaker controls what branch circuit.

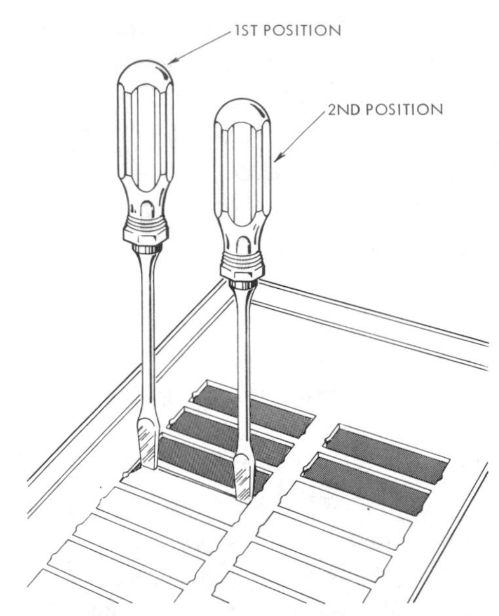

Fig. 11-29. Only those twistouts should be removed which match the location of projected breakers. This is done by wrapping sharply with a screwdriver at extreme corners away from the tie, as shown in the two sequential positions.

Fuses

The fuse is similar in purpose to a circuit breaker, that is, it protects the wire electrically. A fuse is made of a line of metal alloy properly sized to melt at a certain point, thereby interrupting the flow of current before the wire is damaged.

These links come in different forms and device mountings. Some are renewable, some are not.

The common plug fuse of the Edison-medium base (same as a common lamp base) is shown in Fig. 11-30. The base screw-thread contact surface is the same as that of a household lamp base.

Plug fuses of this type are classified at not over 125 volts and 0 to 30 amperes. Fifteen amperes or less are distinguished from those of a larger rating by a hexagonal window opening in the cap or by some other prominent hexagonal feature, such as the form of the top or cap itself. The shape is regulated by code (NEC).

Through the window one can see if the fuse is good or "blown."

Although all fuses are required to be marked with their rated amperage it can be seen at a glance if a hexagonal shaped fuse has been mistakenly replaced with a larger rated, round shaped fuse. The physical identification is not designed to be an aid in changing a fuse in the dark. It is extremely hazardous to work with any live electrical equipment without proper illumination.

Non-Inter-Changeable Type "S" Fuses. It can be seen that medium, Edison base fuses can be dangerously interchanged. That is, a 30 amp fuse can be inserted in place of a 15 amp fuse which was intended to protect a wire with a capacity of 15 amperes. To prevent such an occurrence of panel tampering, an adapter that screws in like a fuse and locks into place should be installed. See Fig. 11-31.

Fuses with special thread sizes to match the adapter are then used in-

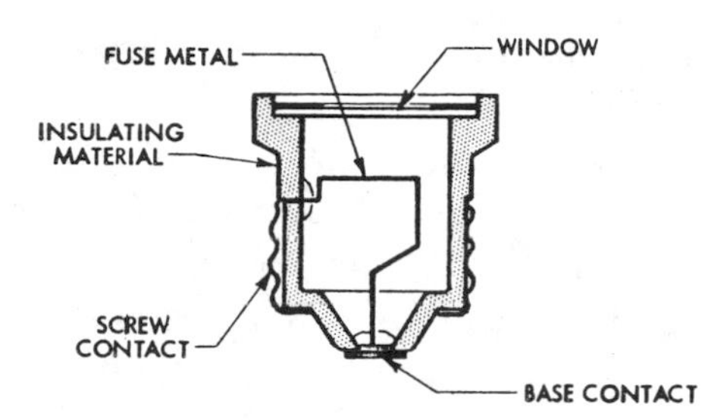

Fig. 11-30. Left: Plug fuse designed to pass safely 15 amps. Note fuse link in hexagonal shaped window. Right: Cross section of plug fuse. (Bussman Mfg. Co.)

Fig. 11-31. Adapter to accept special fuse sizes. Barb on lower front prevents easy removal of adapter. (Bussman Mfg. Co.)

Fig. 11-32. Type "S" fustat for use with special adapter and certain amperage ratings. (Bussman Mfg. Co.)

stead of ordinary fuses. These plug fuses are called type "S" fuses and "fustats." See Fig. 11-32. With the adapter installed the fustat fuse can then be inserted or removed in the same manner as the ordinary plug fuse. This method prevents people ignorant of the important of proper fuse protection from using too large a size or using pennies or other materials to bridge the fuse, thereby eliminating all protection.

Time-Lag Fuses. The ordinary plug fuse has very little time delay action. If an ordinary plug fuse is carrying a slight or medium load and a motor driven appliance on the same circuit becomes energized, the sudden inrush of amperes of the motor starting current will very often blow the fuse thereby causing an entirely unnecessary interruption of service. A dual-element fuse is available that offers "short circuit" and "momentary overload" protection with no damage to the wire insulation. One form is shown in Fig. 11-32.

The ordinary fuse link for short circuits may be seen, through the window, on the right. The other end of the spring seen on the left is attached to the bottom end of the normal fuse element joined in a "solder-pot" called a *thermal cutout*. On an overload the heat generated is transmitted to the solder in the thermal cutout and when the solder softens or melts from the heat of an overload, the fuse link is pulled out of contact with the thermal cutout thereby opening the circuit. See Fig. 11-33.

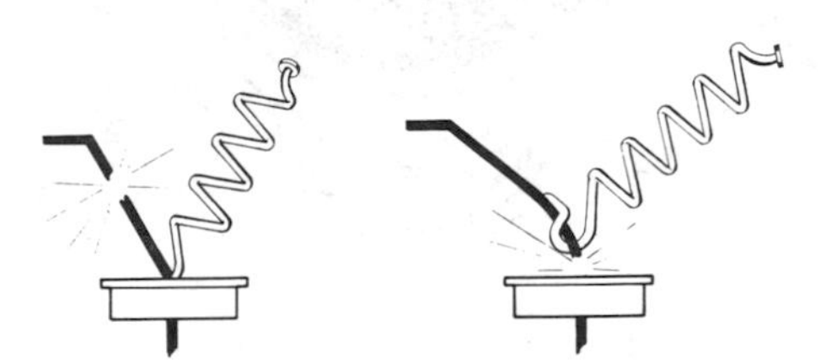

Fig. 11-33. Left: On a short circuit, the fuse link blows just as in other types of fuses. Right: On a slow overload, the heat generated is transmitted to the solder in the thermal cutout. When the solder softens from an overload, the fuse link is pulled out of contact with the thermal circuit.

The fuse link of a dual element fuse is made of metal with a higher melting point than ordinary single element fuses. This is why it will not blow on overloads as ordinary fuses do.

Cartridge Fuses

For amperages above 30 amps and for voltages of 250 and 600 maximum, tubular cartridge fuses are used. These come in two distinctly different shapes, depending on their rating. See Fig. 11-34. Cartridge fuses are available for one-time service only, or with renewable fuse links, and with the dual element "fusetron."

The fusetron dual-element type contains a fuse element that handles short-circuits and heavy overloads, plus a high time delay thermal cutout element that handles light overloads. See cutaway view in Fig. 11-35. The center spring portions are time-lag elements to hold the circuit energized on harmless overloads to prevent needless power outages. End fuse links are designed for short-circuits and heavy overloads.

The blades in Fig. 11-35, left, fit into fuse clips of a size and shape to

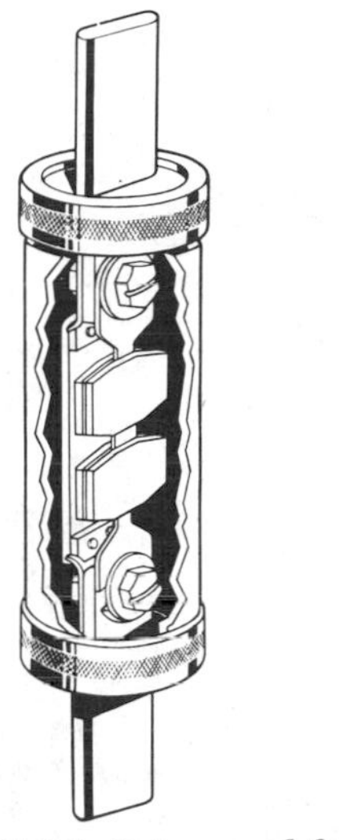

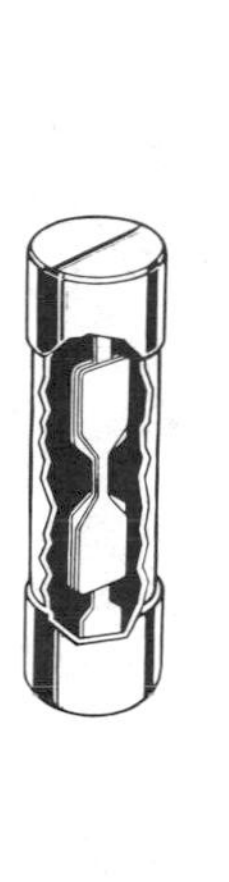

Fig. 11-34. Cutaway of fuse types available in renewable and non-renewable links. Left: Blade-contact fuse available in sizes larger than 60 amps. Right: Furrule-contact cartridge fuse available in sizes to 60 amps. Both fuses shown are renewable type. Fuse links may be replaced. (Bussman Mfg. Co.)

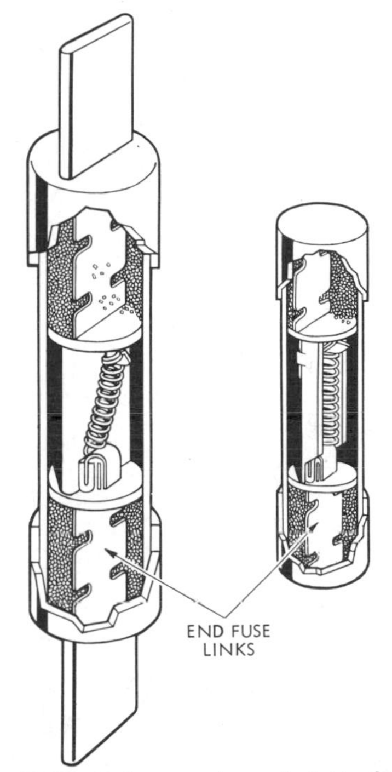

Fig. 11-35. Cutaway view of dual element fuses (non-renewable). Both sizes are of the time-delay type. (Bussman Mfg. Co.)

match, usually within an enclosure of an externally operated disconnect switch. It contains a disconnect switch and fuse holder clips. The same is true of Fig. 11-35, right. Fuseholder sizes, such as those contained in externally operated disconnect (E-X-O) safety switches (Fig. 11-18), are classed by the same maximum current ratings and sizes as the fuses used in them.

Cartridge fuses come in several group sizes designed to fit into certain switches as indicated in Table I. The same method is used for both 250 volt and 600 volt groups with the exception of the increased length and slightly increased diameter of the 600 volt series (600 volt fuses are industrial applications not residential). See Table I.

Selecting a switch size of 30 amps as an example, means any available fuse protection can be installed from 0 to 30 amperes—as shown in Table I under group amps.

An externally operated disconnect switch is commonly called "E-X-O". Normally an E-X-O switch contains

240-61

TABLE 1 E-X-O SWITCH AND FUSE SIZES
250 AND 600 VOLT SIZES

SWITCH SIZE	GROUP AMPS	DESCRIPTION	DIA. & LGT.	STANDARD FUSE AMP. RATINGS
30A	0-30	SMALL FERRULE	9/16"-2" *13/16"-5"	1-3-6-10-15-20-25-30
60A	31-60	LARGE FERRULE	13/16"-3" *1 1/16"-5 1/2"	35-40-45-50-60
100A	61-100	SMALL BLADE	1"-5 7/8" *1 1/4"-7 7/8"	70-80-90-100
200A	101-200	MEDIUM BLADE	1 1/2"-7 1/8" 1 3/4"-9 5/8"	110-125-150-175-200
400A	201-400	LARGE BLADE	2"-8 5/8" *2 1/2"-11 5/8"	225-250-300-350-400
600A	401-600	GIANT BLADE	2 1/2"-10 3/8" *2 1/2"-13 3/8"	450-500-600

*600 VOLT RATING

a disconnect switch above fuse holders within the same enclosure. See Fig. 11-36. It contains wire protection (fuses when installed) and circuit disconnecting means. This would be similar to a circuit breaker, that is, it affords a disconnecting means and wire electrical protection. From Table I we find that a 30 amp switch size is required for a 15 amp fuse. If wires were brought into a switch and were to be protected for a size equal to 110 amps, a 200 amp switch size is necessary because 110 amps is between 101-200 group amps on the table. Maximum fuse sizes are equal to switch sizes or capacities.

Fig. 11-19 contains eight plug fuses, and mounted above (concealed) are two cartridge fuses (above 30 amps). A handle can be seen to pull out these fuses. This also serves to disconnect the panel

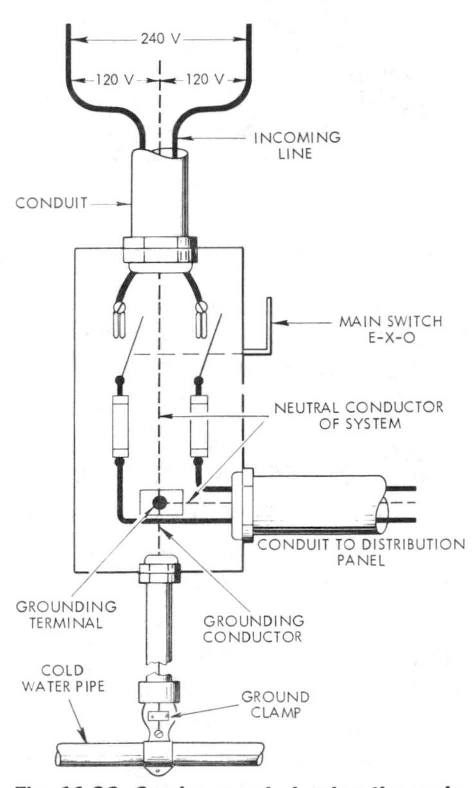

Fig. 11-36. Service panel showing the main switch E-X-O.

from the line. These are called "pull-out-fuses."

It would be well to memorize ampacities of wire sizes in raceway, starting with #14, for purposes of selecting proper over-current protection and matching wires with electrical loads to situations that prevail. For example, if a conductor and its type of insulation is designed to carry no more than 15 amperes, a 15 amp fuse connected in series with the wire will prevent excessive current (amperes) flow from overheating and damaging the wire. If the insulated wire used is copper, be sure to look at the table allowable ampacities of insulated copper conductors — not aluminum. Read code table headings carefully in order to match with situations in question.

Current information is available in the latest NEC. The index should be consulted for latest location.

Cartridge Fuse Pullers. Fig. 11-37 illustrates some cartridge fuse pullers. They come in different sizes for different size fuses and are a great help in changing fuses. Take caution, however: fuses should never be pulled under load or with switch blades closed. Fuses are de-energized by operating the switch handle to the "off" position thereby removing electrical energy from the supply side. Fuses should be de-energized before removal to prevent arcing at contact surfaces under load, should a good fuse be removed. De-energizing a fuse prevents accidental contact with live parts when the fuse is removed. *All* electrical equipment and circuits should be de-energized if possible before any work is attempted. This is necessary to prevent serious injury or death by electrocution and/or damage to property.

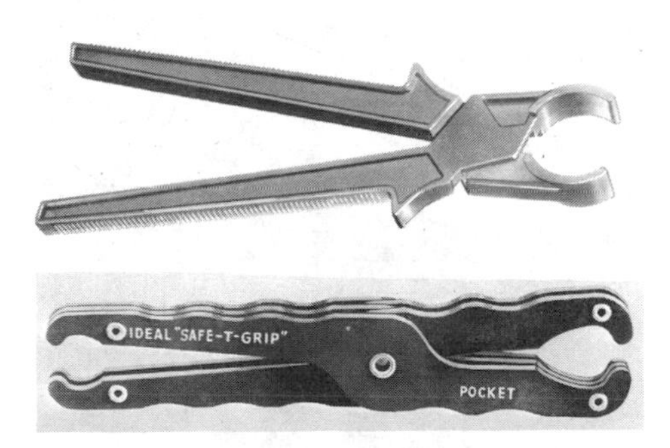

Fig. 11-37. Top: Nylon fuse puller. Bottom: Fiber cartridge fuse puller. Pullers come in different sizes. (Ideal Industries, Inc.)

Locating a Faulty Fuse

Before fuses are replaced in a switch they should be tested to determine if it is necessary to replace them. Fig. 11-38 illustrates a common voltage tester used to test the fuses. Fig. 11-39 shows a step-by-step procedure for locating a defective fuse. A test should be made first to determine if power is available and the test instrument is in working order. This is shown in Fig. 11-39A. If an available indication is recorded, the second suggested procedure is

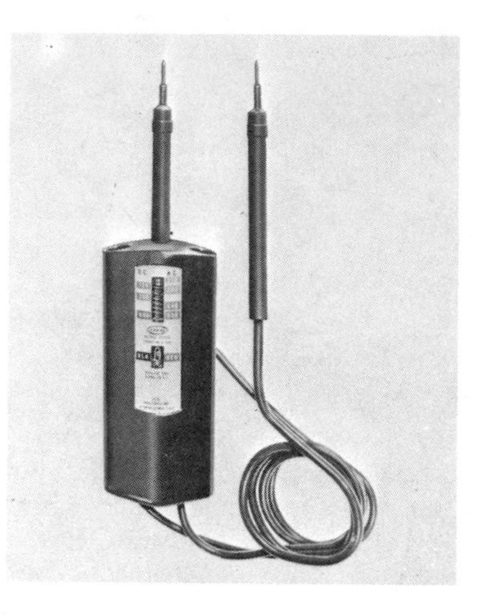

Fig. 11-38. Voltage tester commonly used by wireman. (Ideal Industries, Inc.)

shown in Fig. 11-39B. No indication of a present voltage tells us one or two fuses are blown or faulty.

If an indication of voltage is present at the tester as shown in Fig. 11-39C, it is obvious that the left fuse is good thereby passing voltage to the voltage tester. Should no voltage be present when testing as shown in Fig. 11-39D, the indication is that the right hand fuse is open, thereby passing no voltage to the tester. This fuse should then be replaced with the proper amperage rating after the fault has been cleared. Do not replace fuse until it has been de-energized by opening the switch.

Installing a new fuse without removing the cause of the first trouble will only mean another blown fuse and another useless interruption of service. If a short-circuit blew a fuse, the line must be cleared (the cause of the short-circuit removed) before a new fuse is installed.

If the circuit is overloaded, obviously the power overload will have to be relieved. If it is only a temporary overload, however, a new fuse may be installed almost immediately. A temporary overload may be a current surge such as a motor driven appliance. Examples may be starting an overloaded clothes washing machine or a waste disposal. The load should be lessened before fuse renewal.

External heating of fuses reduces their current carrying capacity. Heat from the sun on a closed service panel has caused many power failures. Simply planting a bush or tree for shade is in the realm of the wireman's recommendations.

Poor contact on the fuse, near the fuse or in the fuse, also causes use-

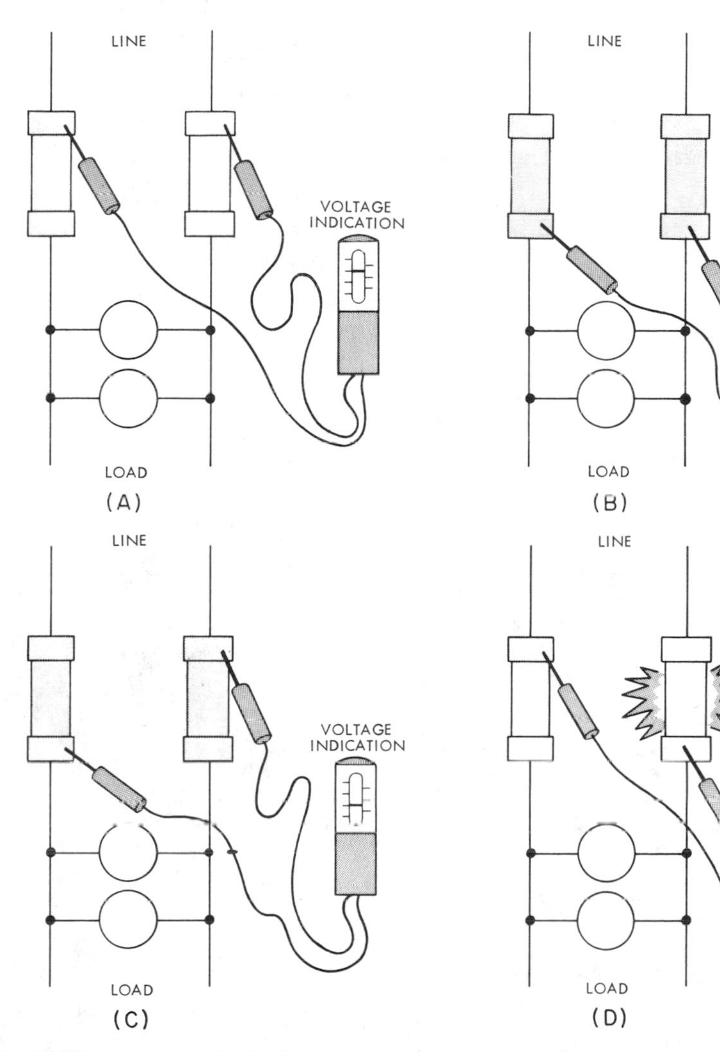

Fig. 11-39. Locating a faulty fuse.

less power failures. A fuse is a contact making device. If the surfaces that make contact in the fuse clips are discolored or corroded, the fuse has been making, or will be making, poor contact in the clips. The fuses should be de-energized and the fuse or fuses removed for cleaning along with the clips. The cleaned fuse may be re-installed for satisfactory service. Fuse and clip contact surfaces may be cleaned with fine sandpaper.

Service Appearance and Convenience

Fig. 11-40 illustrates a well-planned electrical service being installed. A well-planned job costs no more than other installations. Better insulated conductors permit smaller size conduits. Smaller or minimum

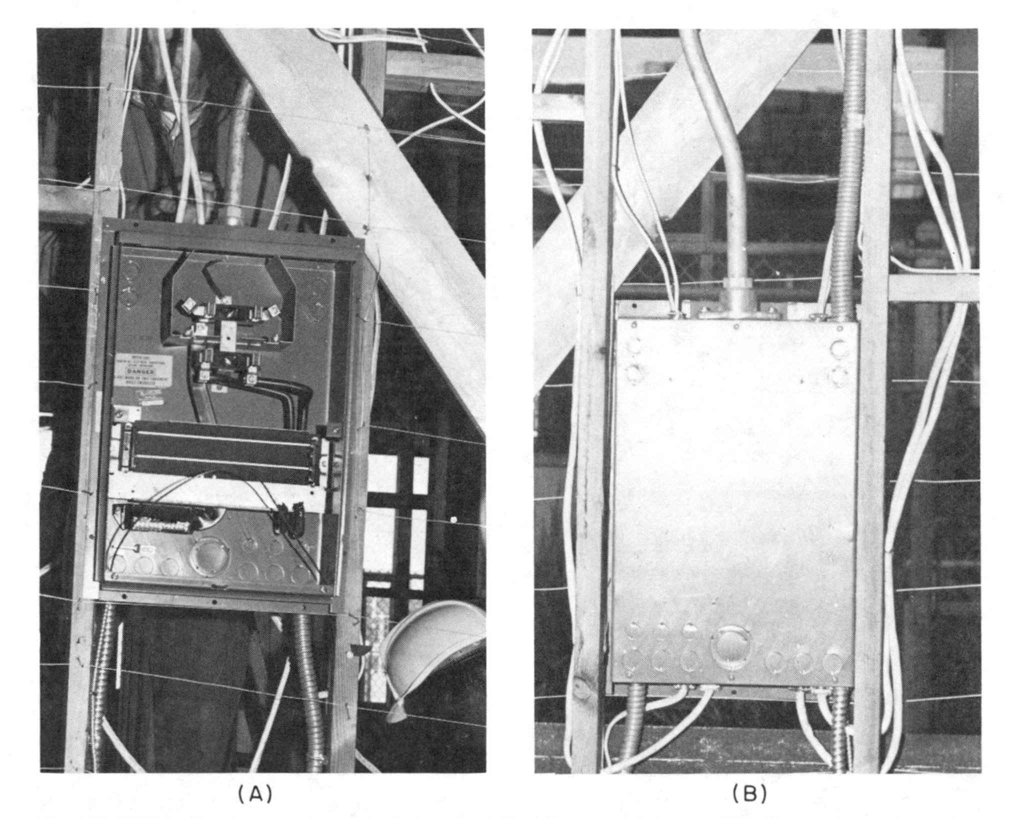

(A) (B)

Fig. 11-40(A). Service equipment being installed by electricians. (B). Rear view of service equipment. (Los Angeles Trade-Technical College)

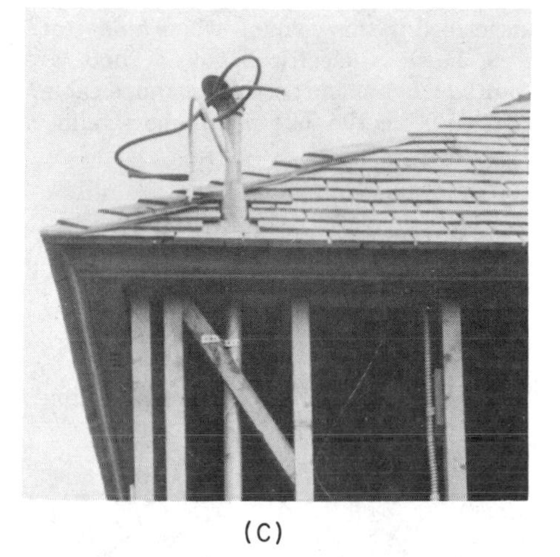

Fig. 11-40 (continued) (C). Typical and adequate wiring installation which is architecturally pleasing and convenient. (Los Angeles Trade-Technical College)

size conduits for services do not allow for much future expansion of electrical capacity though. This method is simple to install because it is nailed between the studding in frame construction. It also improves the appearance of the exterior of the home. Fig. 11-40(A) illustrates service "can" installed before wiring, circuit breakers, meter and trim are installed. Fig. 11-40(B) shows the back of the service which is inside the house.

Note method of extending service head to above the lower roof line in Fig. 11-40(C). Power company requirements and local safety codes should be checked and followed on location of insulators for supporting the power company's service drop, location of meter, for service head location, service drop clearance above walk-ways and driveways, and service requirements that may be questionable.

QUESTIONS FOR DISCUSSION OR SELF STUDY

1. What are service drop wires?
2. Who decides where the service drop shall be located?
3. Who generally connects and installs the service drop, wireman or power company lineman?
4. Who connects the service wires in a sidewalk handhole?
5. How is proper clearance for the service drop obtained in buildings of low height?
6. Who usually runs the underground service?
7. How many principal parts are there to a common service entrance cap? What is its purpose?
8. How does the service entrance cable

described in this chapter utilize a neutral?

9. How is electrical power theft prevented when using service entrance cable?

10. What is the following meter reading?

11. What is the following meter reading?

12. How is the following meter socket connected by the wireman for three wire service?

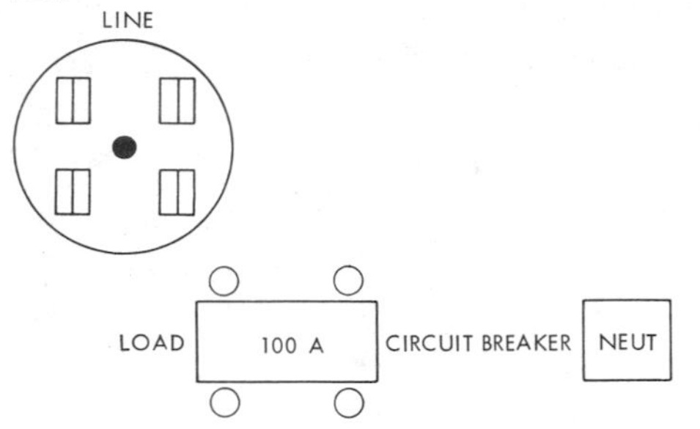

13. What provisions are made to disconnect all ungrounded conductors?

14. How many screws on the neutral bus strap should be available?

15. Why should panel wiring be arranged to evenly divide the load on a 3-wire service?

16. What are the two types of wire overcurrent protection?

17. When circuit breakers trip due to a short circuit or overload, how is power restored?

18. Why are plug fusestats virtually foolproof?

19. Why are time lag fuses desireable?

20. What switch size is required for a 20 amp fuse?

21. What size fuses change from ferrule to blades?

22. List standard E-X-O safety switch sizes.

23. Why should fuses be de-energized before removal?

24. Describe the procedure for locating a faulty fuse.

25. How are fuses and clip contact surfaces cleaned?

26. Why is cleaning sometimes necessary?

27. At what time in the installation procedure is the circuit breakers installed?

28. What are end fuse links designed to protect against?

29. What is the center spring in a fuse designed to protect against?

QUESTIONS ON THE NATIONAL ELECTRICAL CODE

1. In general what is the maximum number of service conductor sets? What are exceptions?

2. What size of service conductors must be used?

3. What is the recommended minimum amperage rating for 3 wire, individual residences?

4. What is the minimum size service entrance conductors? Copper? Aluminum?

5. What is the minimum size service drop conductors to be used?

6. Service drop point of attachment on elevation shall be such as to maintain what minimum clearances over the following:

    Sidewalks?
    Driveways?
    Parking lots?
    Alleys?
    Roofs?

Finished grade?
Public streets?
Truck traffic ways?

7. What is the minimum height of a point of attachment over finished grade?

8. What means of service drop attachment must be provided?

9. How must underground service conductors be protected from physical damage: In the ground? On poles? When entering a building?

10. A single family dwelling has a floor area of 1800 square feet, 14 KW range, 2.5 KW water heater, 1.2 KW dishwasher, 4.5 KW clothes dryer, 9 KW space heating, one 6 amp, 230 volt room air conditioner and three 10 amp 115 volt room air conditioning units.

PROBLEM #1

Find the following watts for a computed load:
General Lighting?
Small Appliance Load?
Laundry?
Range?
Net Computed load so far?
Size of copper service conductors if other loads were omitted?
Size of service conduit?
Number of single and double pole branch circuit breakers?
Size of main switch?
Size of grounding wire?

PROBLEM #2

Including the remaining electrical appliances find:
Size of copper service conductors?
Size of service conduit?
Size of main switch?
Number of single pole circuit breakers?
Number of double pole circuit breakers?
Size of grounding conductor?

11. What is the minimum spacing of service wires by mounting insulators?

12. What is the maximum distance of supporting intervals for service cable?

13. Where it is impractical to locate the service head above the point of attachment what code limitations prevail?

14. How is service equipment grounded and bonded?

15. When a service rigid metal raceway is interrupted by flexible conduit what are the bonding requirements?

16. Where should the service disconnect switch be located?

17. What are approved disconnect switches?

18. What are minimum rated service disconnecting means in amperes?

19. Which method of terminal connection is prohibited in service equipment?

20. Which conductors require overcurrent protection?

21. What is the purpose of overcurrent protection?

22. What is the maximum voltage rating to ground for which a plug fuse may be used?

23. When are cartridge fuses used?

24. What is the maximum amperage rating of a plug fuse?

25. What standard size fuse is required for #12 copper wire?

26. To which side of an externally operated disconnect switch is the power supply connected?

27. What is the largest standard size fuse used to protect a 250 volt circuit?

28. What is the standard voltage and amperage ratings of type "S" fuses?

29. How are 0 to 15 ampere rating plug fuses distinguished from those of larger ratings?

30. What are standard E-X-O switch sizes?

31. A 100 ampere service requires what size grounding wire?

32. When 4 to 6 branch circuit wires are pulled into the same conduit, the reduction factor of ampacity is what?

33. What is the largest size solid conductor that may be pulled into a conduit?

34. What is the requirements of small appliance branch circuits?

35. How many small appliance branch circuits are required in a kitchen?

36. What is the maximum allowable voltage drop for the following: Branch circuits?

Feeders and branch circuits?

37. What is the permissible load of a 20 ampere branch circuit?

38. How is the branch circuit wire size determined for a one horsepower motor driven load?

39. What are the code requirements regarding the mounting of electrical pull-boxes and cabinets on damp surfaces such as basement masonry?

40. What is the maximum number of branch circuit overcurrent devices allowed in a panel?

Grounding for Safety

Chapter

12

A wiring system must afford protection to life and property against faults caused by electrical disturbances, lightning, failure of electrical equipment which is part of the wiring system, or failure of equipment and appliances that are connected to the system.

For this reason all metal enclosures of the wiring system, as well as the noncurrent-carrying, or neutral, conductors, should be tied together and reduced to a common earth potential.

The following explanation gives the reasons for grounding and how to provide for it. There are two distinct divisions of the grounding problem: (1) system grounding and (2) equipment grounding.

System Grounding

System grounding means the connection of the neutral conductor of the wiring system to the earth. Its purpose is to drain off any excessively high voltages that may accidentally entered the system as a result of lightning, an insulation breakdown in the supply transformer, or an accidental contact between the service wires and nearby high tension wires.

High voltages from such sources would probably be in the neighborhood of 2300 volts or more, which would cause the insulation on the ordinary 600 volt wire to break down. In that case, if the system is not grounded, current will flow from these points of insulation failure over the nearest available path to the earth. The resistance of the material through which the current flows will tend to generate heat and, where the surrounding material is inflammable, a fire could start. If the current finds no way to reach ground potential,

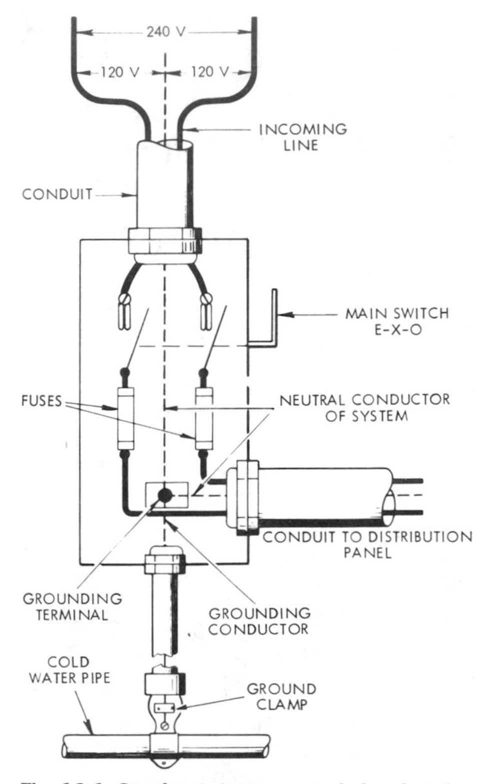

Fig. 12-1. Service entrance panel showing the system ground. Note that the grounded conductor is connected directly to the power supply. It is neither switched nor fused. The best ground connection is a continuous underground water metallic pipe system, but other grounding means are possible.

the high voltage may remain on the line and produce a serious shock hazard.

On the other hand, if the system is properly grounded by means of a low resistance conductor of sufficient capacity, such currents will be carried off to the earth immediately with a minimum danger of fire or shock. This is shown in Fig. 12-1. Size of grounding conductor is determined by the size of service conductors. This is given in the NEC. The grounding conductor joins the neutral of the incoming service wires to the grounding electrode at the grounding terminal or neutral block. Fig. 12-1 also shows the voltages of a 3-wire service as supplied from the power company's transformer. (Note that 115/230 volt references are also commonly used in place of 120/240 volts. These are used interchangeably throughout the text.)

In a grounded system, an accidental grounding of one of the current carrying conductors will result in a short circuit, and will cause the fuse or circuit breaker to open thereby disconnecting the live conductors.

Equipment Grounding

Equipment grounding means the connection of all exposed non-current carrying metal parts of the wiring system to the earth. This includes the steel raceway itself, boxes and similar component parts, as well as the metal enclosures of the apparatus and equipment.

The purpose of this grounding is to prevent a voltage higher than earth potential on the enclosures or equipment. It reduces the danger of shock in case a live conductor comes in contact with these conductive parts.

Fig. 12-2, top, shows equipment

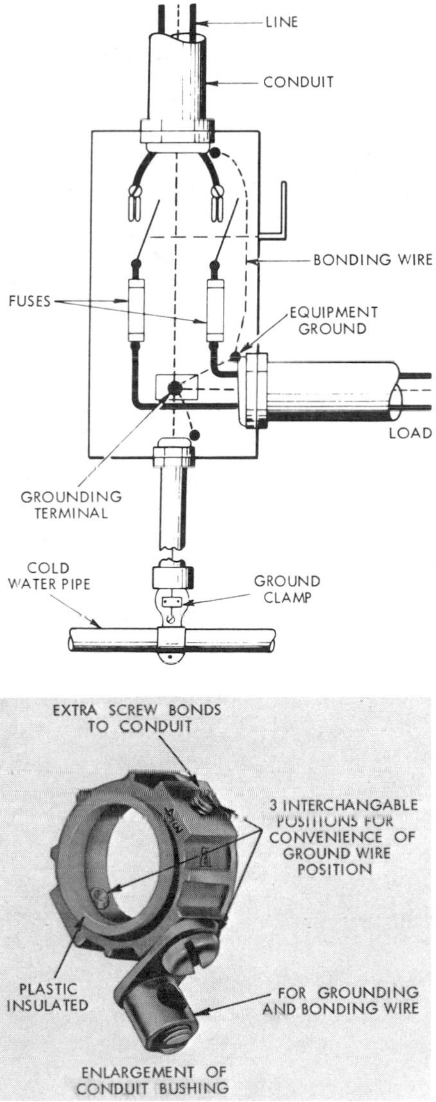

Fig. 12-2. Panel showing equipment grounding in addition to system ground. Note that the incoming and outgoing steel raceways are bonded together, and are also connected to the grounding terminal in the box. In this way, the same ground connection is used for grounding both the wiring system and the equipment. Bottom: Enlargement of conduit bushing with grounding lug and bonding screw. This bushing also contains plastic insulated throat (yellow). Any of 3 tapped grounding positions may be used here. (Bottom: All Steel Equipment Inc.)

grounding in which the incoming and outgoing steel conduits are connected together and then connected to a grounding terminal. This way the equipment is grounded (through the grounding terminal) to the system grounding. The enlargement in Fig. 12-2 (bottom) shows how and where the bonding screw may be connected to the conduit bushing.

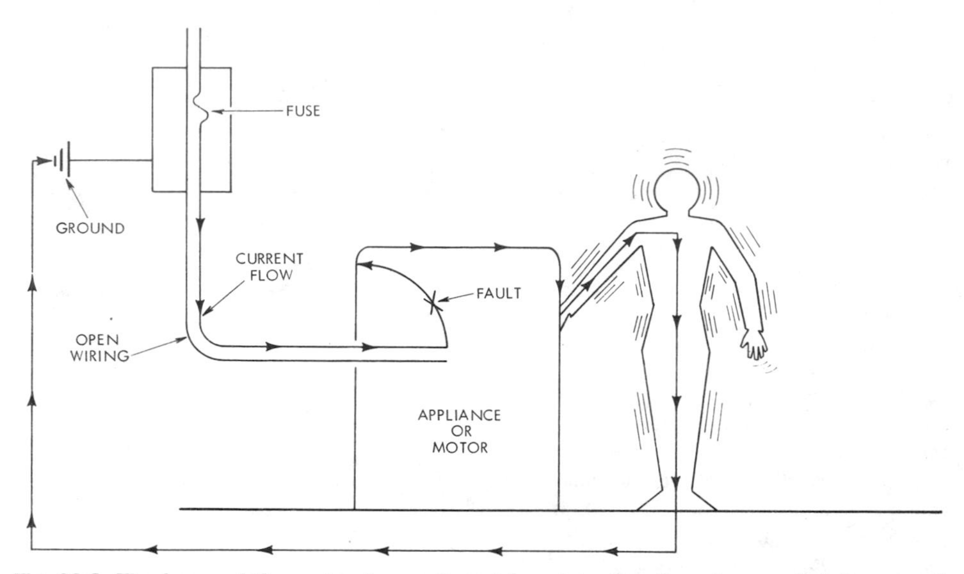

Fig. 12-3. The frame of the motor (or appliance) is unintentionally not grounded. As a result, any contact between the motor winding and the frame causes the current to flow through the person's body to the ground. Such a shock can cause serious harm or prove fatal.

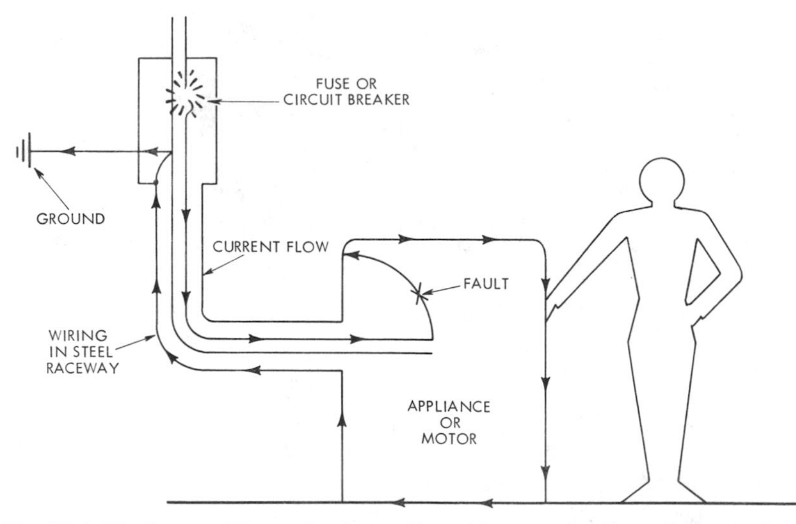

Fig. 12-4. The frame of the motor (or appliance) is grounded through the steel conduit. A similar fault current therefore causes the overcurrent protective device to open. The person is safe.

The ground bushing shown is better than the average ground bushing in that it has the extra features of a plastic insulated throat for added wire insulation protection. Also the extra screws, when tightened into conduit threads, give added bonding insurance. Three positions are available for the convenience of accepting the ground wire more easily. Most ground bushings contain only a screw and washer.

Fig. 12-3 illustrates the danger of ungrounded equipment. The simple solution to this dangerous condition is shown in Fig. 12-4. For the safety of people and for the protection of property, equipment must *always* be grounded.

Method of Grounding

The main requirements for an effective ground are continuity, ample capacity and permanence.

A continuous underground metallic water system is generally acknowledged to be the best electrical ground. Other suitable methods of grounding include continuous underground metallic steam or gas piping systems, or an artificial electrode such as a driven galvanized steel pipe or copper rod, or a buried iron or copper plate.

The system and equipment ground connections should be made to the same electrode at the service entrance. Equipment must be tied in solidly with the system ground. It is also important that wherever multiple grounds are used that they be tied together in order to avoid any difference of potential between the various parts of the system. Where steel conduit is used, it automatically performs these necessary functions in a simple and economical way. Complete rules for grounding are contained in the National Electrical Code and should be studied thoroughly.

The best grounding electrode is the iron pipe coming into the building from an underground cold water-supply system, not because of water contained in it, but by reason of the large surface area in contact with the earth. Similarly, a metal deep-well casing is good for this purpose. The underground metallic piping system should be used as the grounding electrode where such a piping system is available. There should be at least 10 feet or more of buried pipe. The wireman should be assured that the metallic pipe has no insulated couplings or changes to non-metallic

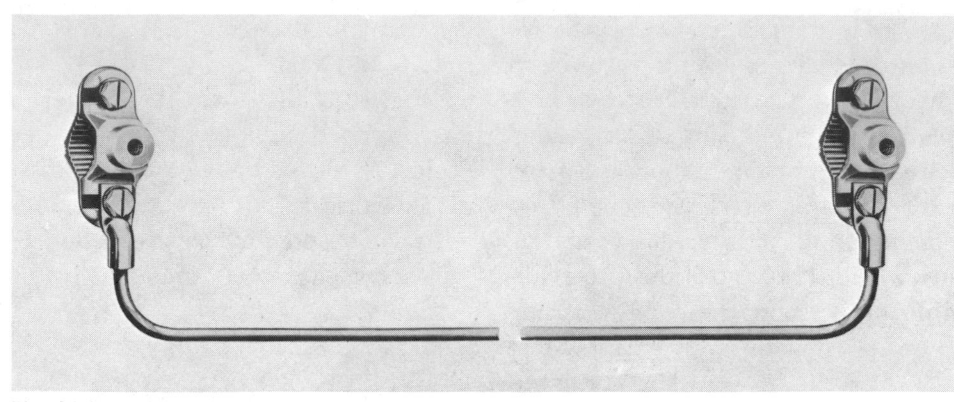

Fig. 12-5. Meter shunt.

piping underground. This is very important to proper grounding. Plastic underground piping systems are not conductors and are *not* to be used for grounding.

An underground gas-piping system, because of its large contact area with permanently moist earth is also used as a grounding electrode. A possible source of danger here is that a poorly installed ground connection may arc if the current flowing through it is strong enough. The arc, which is an electric discharge across a gap in a circuit, is extremely dangerous when coupled with a leaky gas pipe. Of course, the connection to the ground electrode (no matter what kind it may be) should be made with such approved fittings and in such a manner that *arcing* will not occur under any circumstances.

Connection of the grounding conductor to the underground metallic piping system should be done as near to the latter's point of entry as possible, on the supply side of whatever meter or valve may be located at this point. Where this cannot be done, a bonding jumper, or meter shunt, Fig. 12-5, should be installed so the electrical connection will not be interrupted if either meter or valve is taken out. This shunt is placed across the water meter. (See Fig. 11-1, Chapter 11, for an example of this type of installation.)

Ground Rods. When no satisfactory grounding electrode is readily available, the common practice is to drive one or more rods (connected in parallel) at such locations and to such depth as to provide a connection having a resistance not higher than 25 ohms to earth.

Where shale, hardpan, or a general rock formation makes it impossible to drive rods, an excavation to permanently moist soil should be made and the electrode buried there. When

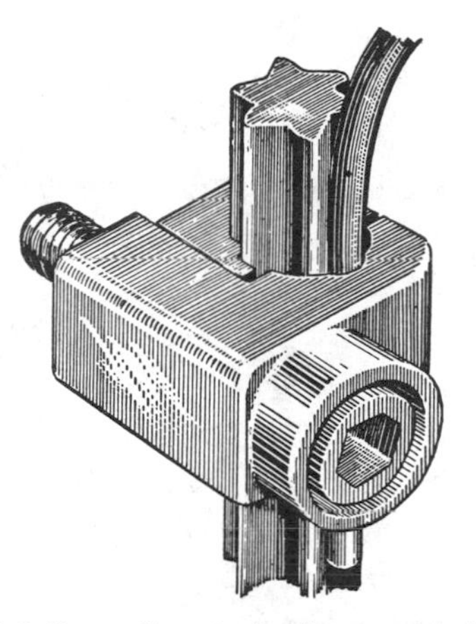

Fig. 12-6. Grounding electrode, ground clamp and ground connector. (Anaconda Wire and Cable Co.)

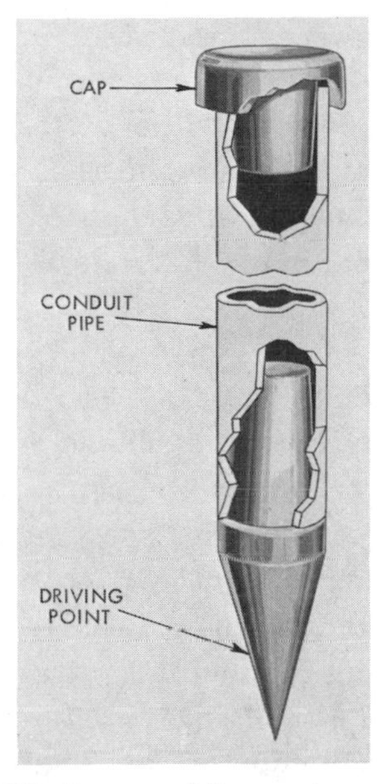

Fig. 12-7. How conduit may be used as a grounding electrode.

the excavation for foundation footings is used for this purpose, the electrode should be covered with several inches of concrete rather than embedding it in the earth. The N.E.C. should be consulted if this method is necessary.

A ground rod having a star-shaped cross section, together with its clamp, is shown in Fig. 12-6. This rod is not only stiffer than a round one of given cross-sectional area, but also has a greater contact surface with earth. This additional contact surface is desirable in localities where dryness or other soil condition tends to produce high resistivity, especially where the capacity of the installation is large or where motors are to be operated at more than 250 volts.

Where soil conditions favor the driving of grounding electrodes, it is permissible to use galvanized-rigid conduit or galvanized-steel pipe for this purpose. Minimum size should be ¾ inches. However, the lower end should be equipped with a driving point and the upper end with a cap, Fig. 12-7, to protect against crushing under the impact of a sledgehammer during installation. Galvanized conduit which has been given an outside layer of paraffin, lacquer, or other protective substance should not be used, because of the increase in contact resistance resulting from this coating.

Near the seacoast or in other localities where ground moisture is likely to be salty or acid, only rods made of copper or other corrosion-resistant material should be used. The wearing quality of galvanized steel is too uncertain to permit its use under these conditions.

The ground rod should not be driven all the way into or below the surface; a few inches should protrude above the soil so the cap is visible and accessible for inspection of the grounding conductor.

Grounding Conductor. The grounding conductor is attached to the pipe with fittings suitable for the type of conductor, armored cable, conduit, or wire. Fig. 12-8 (left) shows a clamp used with conduit. Fig. 12-8 (right) shows a ground clamp for either bare or armored ground wire. Bare or insulated wire, either copper or other corrosion-resistant material, solid or stranded, flat or round bus bar, is used for the system grounding conductor. Its carrying capacity must be proportional to the size of the installation, as indicated in the Code schedule. The grounding electrode should be thoroughly cleaned at the point where the ground clamp is to set, and the clamp should be tightly fastened.

At the service switch the grounding conductor is connected to the same terminal as the incoming *identified conductor*. The latter must also be fastened to the switch enclosure by means of a lug, a grounding wedge or a grounding bushing. Soldered connections should not be used.

Service switches, as well as service-branch-circuit combinations, may have the neutral block mounted directly on the enclosure, instead of being insulated therefrom. The identified service conductor is bolted directly to the neutral block, so that it is not disconnected from the source by opening of the switch.

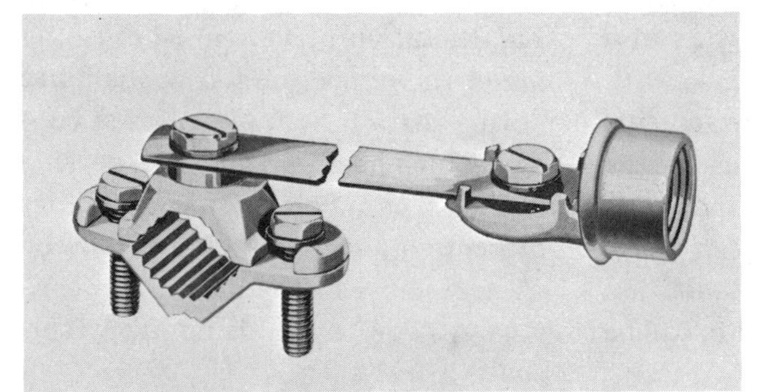

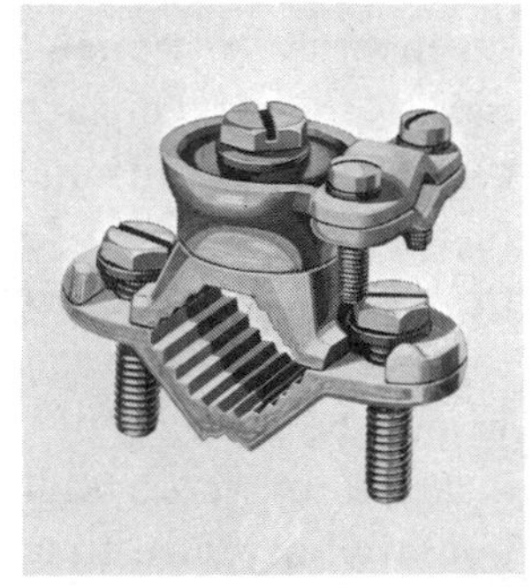

Fig. 12-8. Ground clamps for use with conduit and wire.

Equipment Ground

The equipment ground helps to eliminate electrical fire hazards as well as to remove the danger of serious injury from electric shock. Such danger arises when a person contacts the metal frame of electrical apparatus that has become *live* due to insulation failure of current-carrying parts. Figs. 12-3 and 12-4 illustrated this condition.

This ground connection should be applied to the metal frame or enclosure of all electrical apparatus and to conduit or other metallic raceways, regardless of whether a system ground is connected to the wiring at the service switch or not. Wiring systems that are completely metal-clad from service equipment to operating units provide such grounding automatically by way of the metal enclosure.

Codes require that secondary windings of transformers supplying interior wiring systems shall be grounded so as to insure that the maximum voltage to ground shall not exceed 150 volts. They recommend that this practice be followed even though the voltage to ground may go as high as 300 volts. This grounding connection is made by the power company. Hence, when electrical contact occurs between a current-carrying member and an ungrounded metal frame or enclosure the latter becomes live.

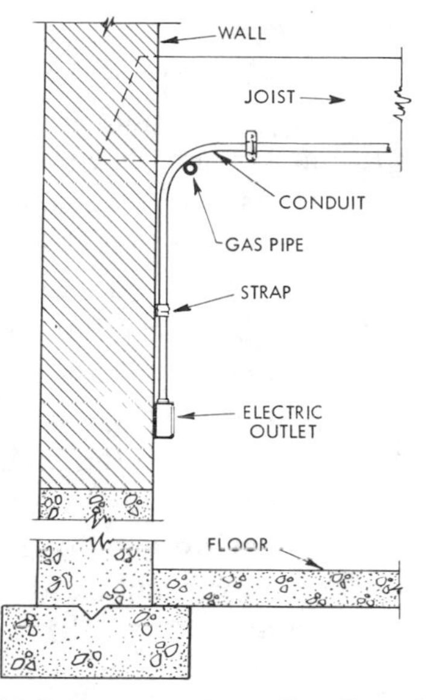

Fig. 12-9. How current can flow through the ungrounded conduit to gas pipe or other grounded object.

Inadequate Ground. Consider, for example, an electrical installation in the unfinished basement of a dwelling, where structural materials of the building normally insulate the conduit from contact with the ground. Fig. 12-9 shows a conduit running along the side of a joist, near its lower edge, and down to an outlet on the wall. A gas pipe runs across the bottom of the joists at right angles to the conduit, touching it lightly. If a wire becomes grounded to the raceway at some point in the building, it creates a live conduit. At the point of contact, current flows from the conduit to the gas pipe, thence to ground, and to the source of supply.

How Some Fires Occur. The flow of current, even though not sufficient to blow the circuit fuse, may be strong enough to set up an arc between the pipes where they touch. In time, the arc may burn a hole in the pipe, resulting in fire or explosion. If pressure between conduit and gas pipe is such as to maintain good electrical contact so that no arc develops, hazard of another sort may still exist. Farther along the gas pipe, say at the meter, electrical resistance may be high because of corroded threads. Resulting temperature rise at this point introduces a fire hazard.

Grounding Conduit System

Grounding of conductors to the conduit system may arise through breakdown of insulation. As pointed out earlier, this creates a live conduit and is a fire hazard. Proper grounding of the conduit system averts this hazard, causing the circuit fuse to blow and thus eliminating the possibility of an appreciable difference of potential occurring between conduit and ground. It should be noted that the majority of codes permit use of a single conductor for grounding both the wiring system and the service equipment. See Fig. 12-10.

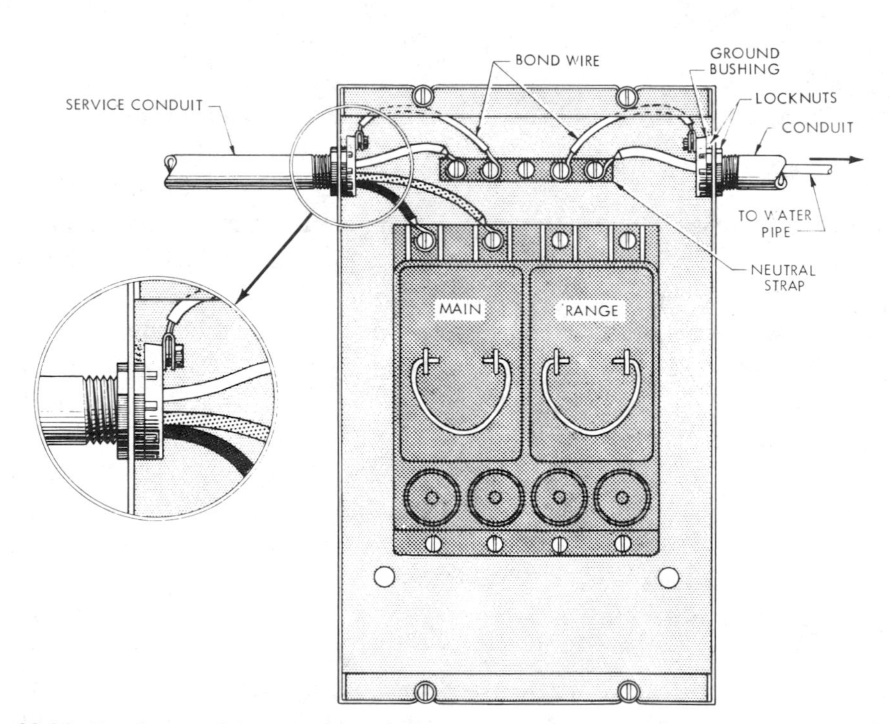

Fig. 12-10. Bonding conduit at service switch.

In order to ensure satisfactory electrical and mechanical connection throughout the entire conduit system, it is essential that every length be screwed all the way into its coupling or the threaded hub of its fitting. Where compression-type, setscrew, or intended types of thin-wall conduit couplings or connectors are used, it is necessary that each be thoroughly tightened. Where boxes of the knockout (K.O.) type are entered, a locknut is screwed onto the conduit before inserting it through the knockout. Another locknut is installed inside the box, holding the conduit tightly to the box and making a good connection. A ground bushing is then installed as shown in Fig. 12-10 if in service equipment. Regular bushings are used in other boxes. Care should be taken when installing locknuts so that they are installed to "bite" into the metal box making solid contact. Note that a different kind of grounding bushing is used in Fig. 12-10 than in Fig. 12-2. Some bushings like Fig. 12-2 have two screws, one for a grounded wire and one for screwing tightly into the cabinet making a good box to conduit bond.

Since only flat sides provide good seats for bushings and locknuts, conduits should leave outlet boxes squarely and not at an angle. Galvanized boxes offer a better ground than do those having an enamel finish (if enamel boxes are available) because the enamel, unless completely removed, greatly reduces the metallic contact surface between the box and the faces of bushings and locknuts. Two locknuts should be employed in the case of rigid conduit, one inside and one outside the box, if the voltage to ground exceeds 250 volts. Some require two lock nuts on every box with rigid conduit.

Bonding Conduit. Electrical codes require that conduits shall be bonded together at the service switch, Fig. 12-10, instead of depending entirely upon the bushing and locknut connection for the grounding contact. Bonding may be accomplished by the use of a copper wire or copper strip. Grounding bushings, Fig. 12-10, are also used for the purpose. This bushing has a large screw for securing the ground wire, and a setscrew (not shown) for locking the bushing to the conduit threads. Therefore, it will not become loose as a result of vibration. Fig. 12-11 shows service equipment with threaded hubs that are commonly used in the weather. Care should be taken not to cut too long a thread on conduits entering threaded hubs. Tightly threaded connections should be maintained for good ground continuity from conduit to enclosures. No grounding bushings are used here.

When extensive metal in or on buildings can become energized and is subject to personal contact, proper

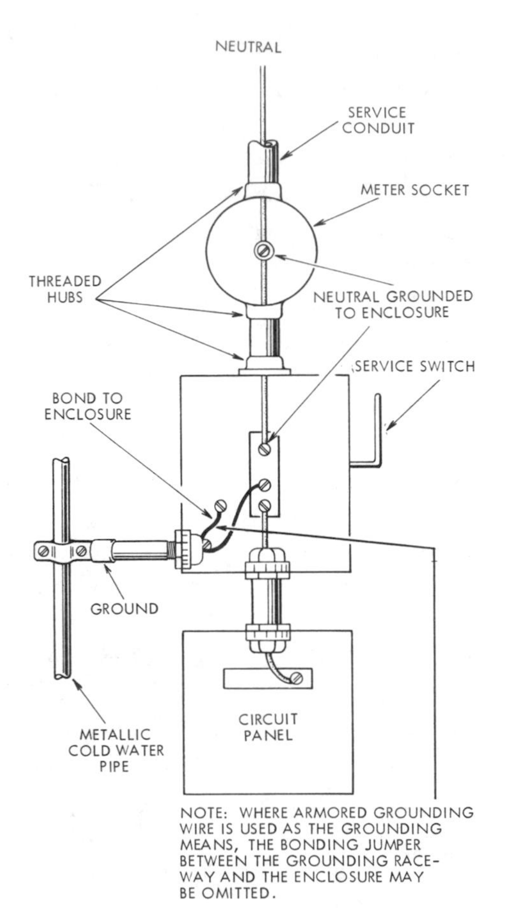

Fig. 12-11. Grounding of service equipment having threaded hubs on meter socket and main switch. Details vary in different localities.

bonding and grounding provide additional safety. This should be done to all water and gas piping systems.

Knockouts. Enclosures for service switches and other devices often have what are known as *combination* or *concentric knockouts* to accommodate any of several sizes of conduit without the necessity of punching or reaming out (that is, enlarging a hole or cutting a new one). The arrangement consists of placing the several partially pre-punched knockouts one within the other, as shown in Fig. 12-12, in the section of wiring trough. With these, the wireman removes only those knockouts which provide the size of hole needed for conduit to be used. In this case, unless the entering conduit requires the removal of the largest knockout, a bonding connection must be installed from the conduit to the enclosure in order to restore integrity of the equipment ground, because these knockout rings, even though left in place, may impair the electrical connection to ground.

Care should be taken when removing cabinet or gutter knockouts. Should larger knockouts be removed than required, reducing washers must be employed to accommodate a specific conduit size with an oversized hole. This may further complicate grounding problems. Knockouts should be removed one at a time, by alternately bending them inward and outward, back and forth, until the metal tab breaks. Referring to Fig. 12-13, top, the center knockout should be knocked inward: the screwdriver blade should be placed against the point farthest from the metal tie and struck inward. The metal is then bent back and forth to

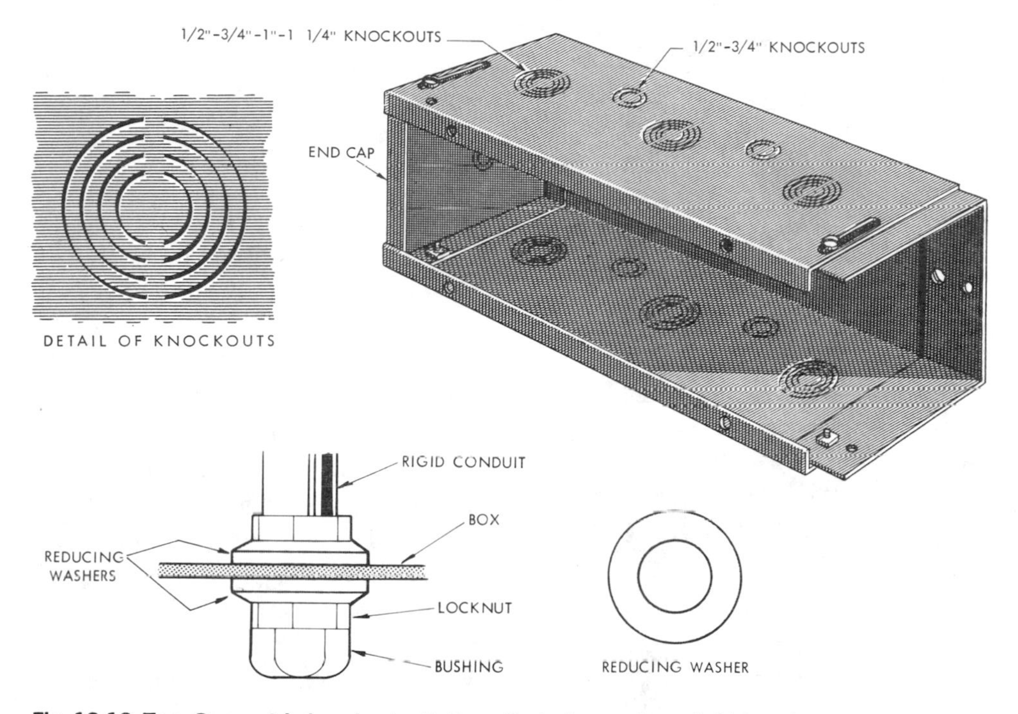

Fig. 12-12. Top: Concentric knockouts. Bottom: Reducing washers (left) installed and (right) top and bottom view.

break the tie. The rings should then be removed one at a time without straining the remaining rings. The ring should first be pried outward, as shown in Fig. 12-13, middle, with a screwdriver midway between ties using pliers flat against the box under the screwdriver. The rings should be bent outward with the pliers then back and forth to break the ties as in Fig. 12-13, bottom. The second ring (complete circle) should be removed, if desired, by striking the ring inward with the screwdriver blade against the point midway between the ties. The ring section is then bent inward and back and forth to break the ties.

Grounding Bushing. To make a good connection to ground, a grounding bushing should be installed on each conduit which enters the box through a concentric knockout. Connect each of the bushings to the enclosure with a piece of copper wire, either bare or insulated. The size of the wire to be used, as given in the Code schedule, depends in each case on the overcurrent or fusing protection of the wires entering through the particular concentric knockout. On a large conduit, use a lug for this

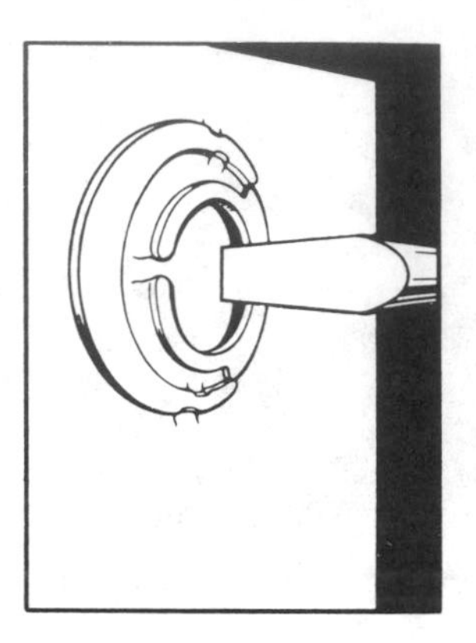

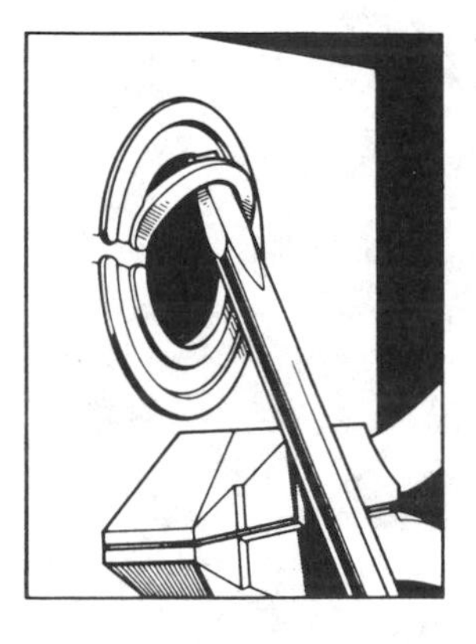

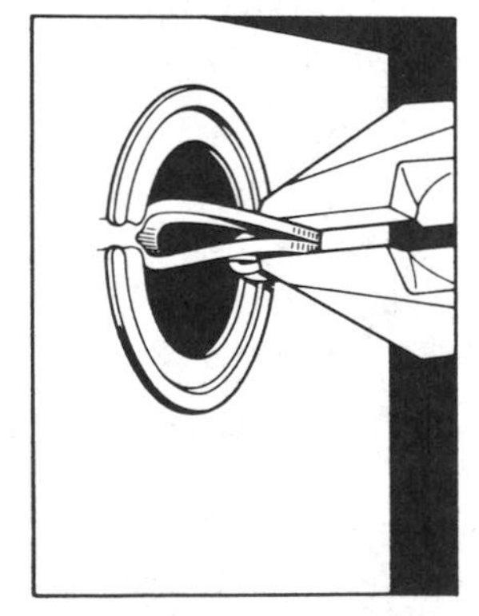

Fig. 12-13. Removing knockouts.

purpose, installed as previously described. Remove paint or enamel from the box at the point where the grounding connection is to be made.

Grounding Armored Cable

The basic grounding technique for metal clad cable was covered in Chapter 5.

Some enforcement agencies interpret that metal clad armored bushed cable is to be treated the same as flexible conduit. This ignores the bonding wire which flex does not contain, and calls upon the connectors to complete the grounding connection between outlet enclosure and metallic armor on the cable. When in doubt, the local code enforcement agency should be consulted. Some electrical inspectors now count the ground wire as another conductor utilizing outlet box capacity. Incorrectly some contractors are directing their wiremen to cut the ground off and ground the boxes externally to the nearest cold water pipe. Labor expenses are not any greater but material expenses may be lower due to the use of smaller outlet boxes.

Grounding Nonmetallic Sheathed Cable

On nonmetallic sheathed cablework, the grounding of boxes and motor frames or other enclosures of current-carrying devices may be done through individual grounding conductors or by means of a ground-

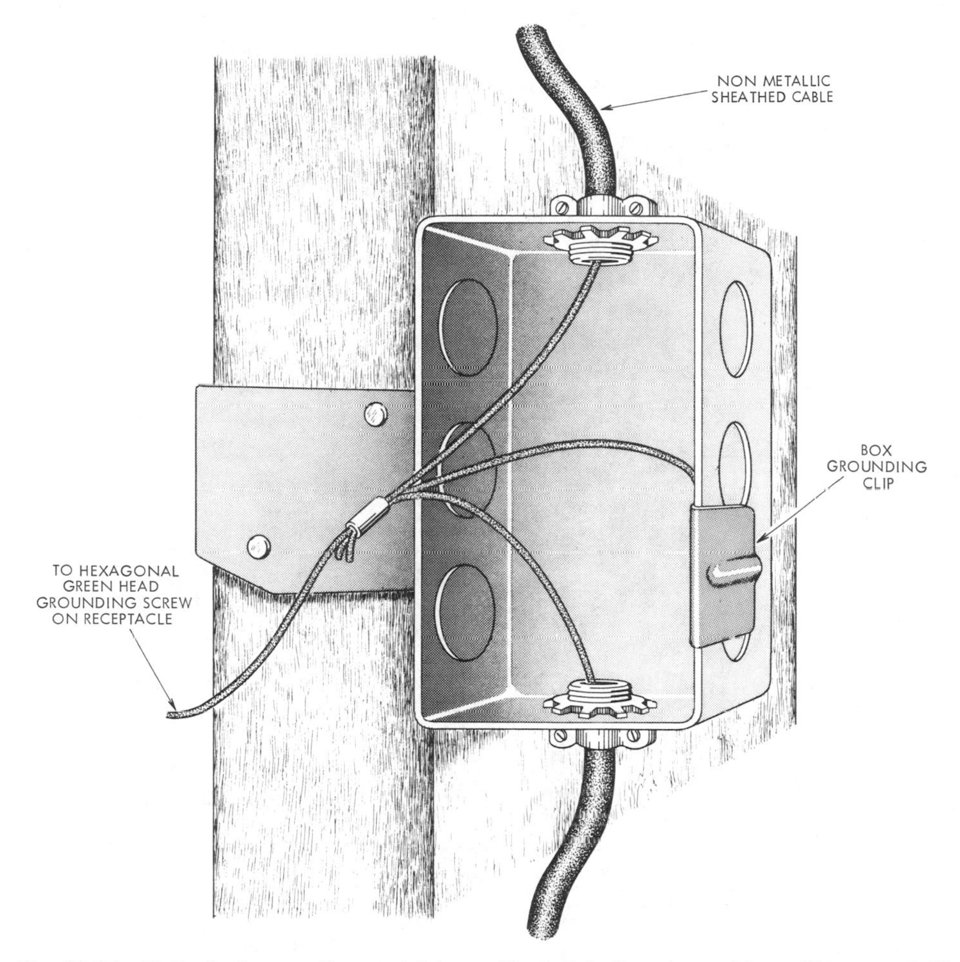

Fig. 12-14. Method of grounding metal box with metal clip when wiring with non-metallic sheathed cable. Receptacle grounding method is the same for armored cable.

ing conductor which is part of the cable assembly.

It is usually safer to have a continuous grounding conductor extending all the way back to the grounding electrode at the service location, than to ground each metal enclosure separately to the nearest water pipe. Fig. 12-14 shows a good grounding practice. Note how the receptacle can be removed without disturbing the grounding of other

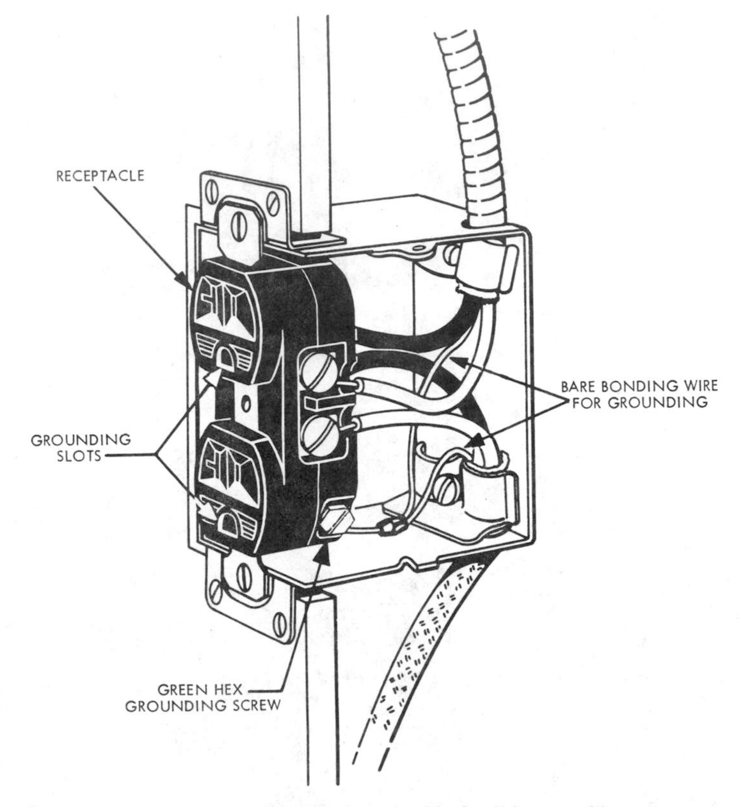

Fig. 12-15. When a grounding receptacle is used with flexible metallic or non-metallic cable, the bare bonding or grounding wire from each cable entering the box should be run to the green hex screw. Box may be grounded externally with metal wire clip or separate screw in the box. (Harvey Hubbell, Inc.)

equipment beyond this point. (Current carrying conductors are not shown in Fig. 12-14.)

Fig. 12-15 shows all wiring connected to proper terminals of a duplex receptacle. The box may be grounded separately under a screw placed in a box hole and to a cold water pipe, or a box grounding clip may be used in place of a screw should the metallic armor of the cable not be approved as a grounding conductor on remodeling jobs.

With a properly grounded metallic system (conduit or metal clad cable) receptacles are grounded through the system and, if it is felt necessary, may be grounded through an additional wire to the green hex screw as shown in Fig. 12-16. The metal yoke on the receptacle may be acceptable as an equipment ground in construction on wall surfaces as shown.

Fig. 12-17 shows relative wiring of a convenience outlet branch circuit to service panel. The grounding circuit is illustrated from outlet to ground through cable armor or conduit, to switch enclosure, to ground. Separate grounding conductors are also used.

Remember: Grounding is too important to be skimped or slighted.

Local safety and electrical codes may be consulted for additional information on grounding nonmetallic sheathed cable.

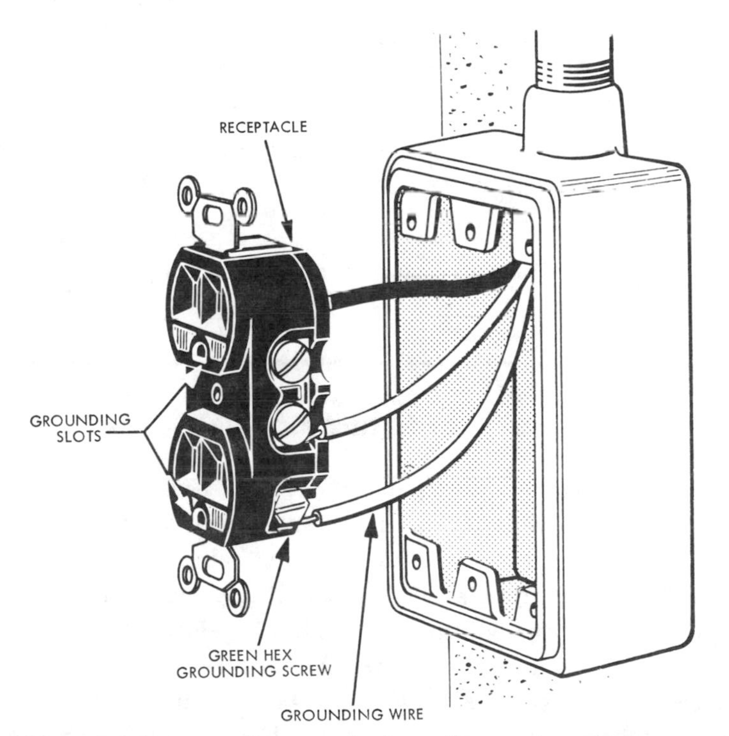

Fig. 12-16. Additional outlet grounding security by pulling and connecting a ground wire. (Harvey Hubbell, Inc.)

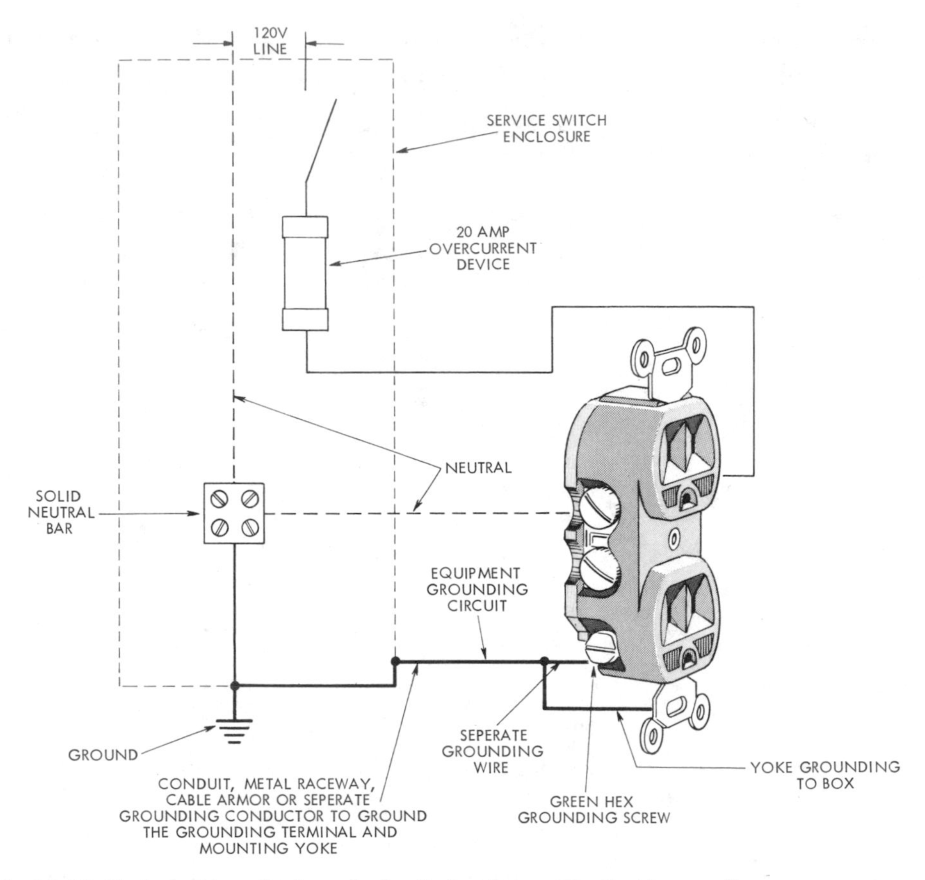

Fig. 12-17. Typical wiring of a branch circuit showing continuity of grounding.

Grounding Cord Caps, Cords and Cables

For the grounding of equipment operating from a flexible cord, a male-attachment plug cap is used which has one more contact blade than is required for carrying line current. This is the ground connection made to the grounded terminal of the receptacle as shown in Fig. 12-18.

This grounding blade is longer than the others. Hence, when the cap is being plugged into a receptacle, the ground connection is established before the current-carrying blades make contact. When the cap is being removed, the current-carrying blades are disconnected first, before the grounding connection is

Fig. 12-18. Because cord cap "U" grounding blade "GR" is longer than the current carrying blades, it insures a grounding connection while cord cap is being inserted or removed. Green hex screw is grounding blade's terminal. (Harvey Hubbell, Inc.)

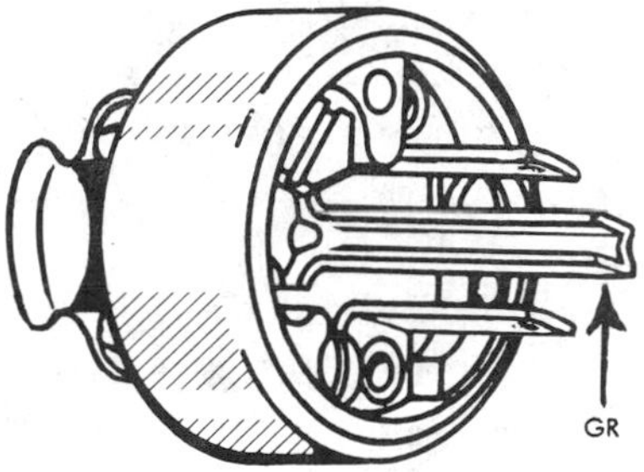

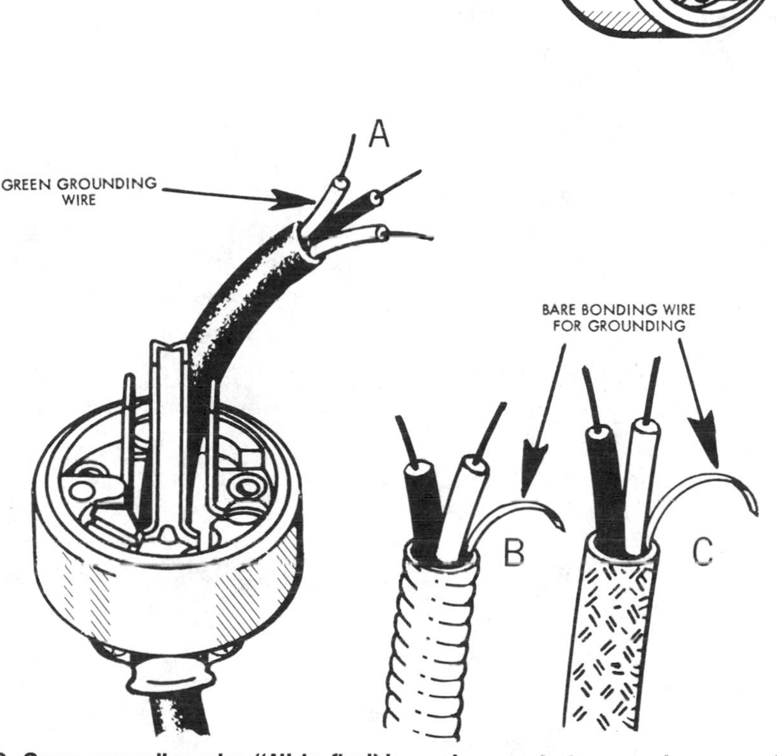

Fig. 12-19. Green grounding wire "A" in flexible cord connects to green hex screw in plug cap. Bare grounding wire "B" and metal covering grounds metal clad armored cable. Nonmetallic sheathed cable also has a bare bonding wire "C" for grounding purposes. (Harvey Hubbell, Inc.)

interrupted, thereby assuring continuous grounding protection.

The proper flexible cord for use with a grounding cap has, in addition to its current-carrying wires, a green wire for grounding only, as shown in Fig. 12-19A. One end of this green wire connects to the cap's green terminal being the "U" shaped ground blade. The other end is fastened to the metal frame or housing of the appliance or equipment to be grounded.

Flexible metal clad, Fig. 12-19B and non-metallic sheathed cables, Fig. 12-19C come with a bare grounding wire just inside the outer covering. This is the bonding wire that helps protect the unsuspecting homeowner from severe electric shock by carrying dangerous "fault currents" back to the main service and grounding electrode.

The purpose of all these grounding precautions is to carry electricity harmlessly away if it leaks into uninsulated metal because of misuse or failure of insulating materials or energized conductors. Current leaking in this manner is called "fault" current.

Fault currents can occur in any electrical system. But they are more common in a system in which flexible rather than rigid conduit are used, and in which the equipment is portable or semi-portable.

Wiring devices for use with flexible cords and portable equipment are designed with protective features which, when the devices are properly installed, conduct fault currents harmlessly to the ground, thereby reducing the hazardous effects of short circuits due to defective insulation or other causes.

Equipment rated for one voltage should be prevented from being connected accidentally to a circuit supplying a different voltage.

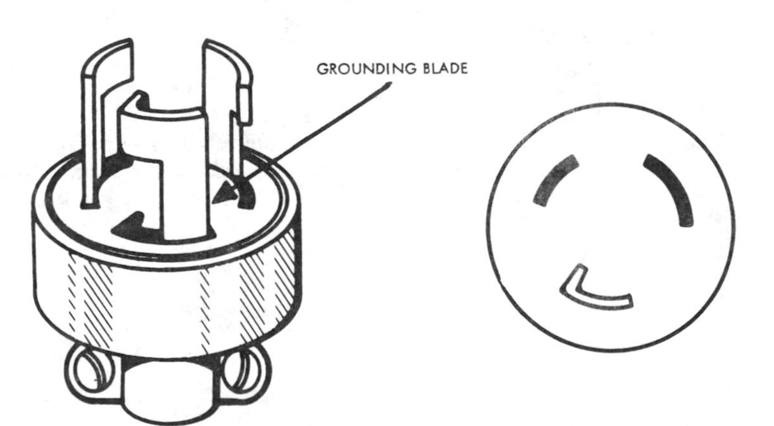

Fig. 12-20. "Twist lock" devices with more than two blades or slots are designated as grounding devices only when one of the blades or slots is elbow shaped as shown. (Harvey Hubbel, Inc.)

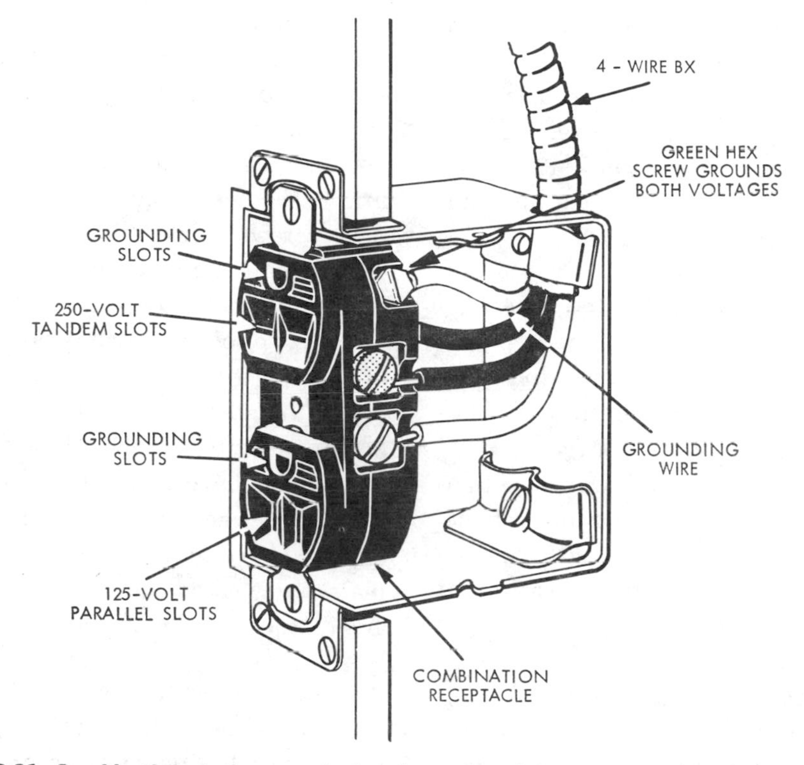

Fig. 12-21. Combination duplex receptacles shown here have common grounding terminals. Note slots are different for different voltages. (Harvey Hubbell, Inc.)

For the grounding of equipment operating from a flexible cord, a male attachment-plug cap is used which has one more contact blade than is required for carrying line current. See Figs. 12-18, 12-19, and 12-20. These are the longer blades, "U" or "elbow" shaped, for first and last connections.

In conventional wiring devices, the grounding blade is round or "U" shaped. In locking devices designed to prevent accidental disconnections, such as the "twist-lock" device (see Fig. 12-20), the grounding blade is elbow shaped. The slot accepts the elbow-shaped blade of the cap, which is rotated clockwise to lock the connection. Different physical sizes and configurations are for different amperages and voltages, none of which are interchangeable. Different arrangements of slots may be seen in Fig. 12-21. Note that different slot arrangements are for different voltages. Since keyed blade and slot configurations prevent mismatching of connectors, one configuration in a duplex grounding receptacle can supply 125 volts while a second configuration supplies 250 volts. Many other pairings of voltages and amperages are available.

Pigtail Grounding Connections

Fig. 12-22 shows a green pigtail grounding wire (A) often found on the cords of older power tools, but now banned by many codes.

One end of this wire is permanently connected to the motor housing. The user is supposed to fasten the other end beneath the screw that holds the cover plate to the grounded outlet box.

Many people neglect to make this connection. Therefore, cord pigtails should be replaced with standard grounding caps that will provide an automatically grounded connection. A temporary conversion from a 2-pole ungrounded receptacle to a 3-pole grounded type can be made by using a grounding adapter like the one shown at B in Fig. 12-22. The grounding slot in this adapter is connected to a pigtail which can be firmly fastened beneath the wall-plate screw. This grounds the adapter through the electrical system.

Pigtails, whether on cords or adapters, offer little or no assurance of proper grounding. Of the two types, the adapter is better because, once connected to the plate screw, it does not have to be disconnected in order to remove the cap. Pigtails on cords have to be connected to the plate screw when the cap is inserted and disconnected when it is removed.

Grounding Receptacles

A receptacle with a grounding slot has a clearly identified grounding terminal, Fig. 12-23, which may be connected to the buried grounding electrode at the service in one of several ways:

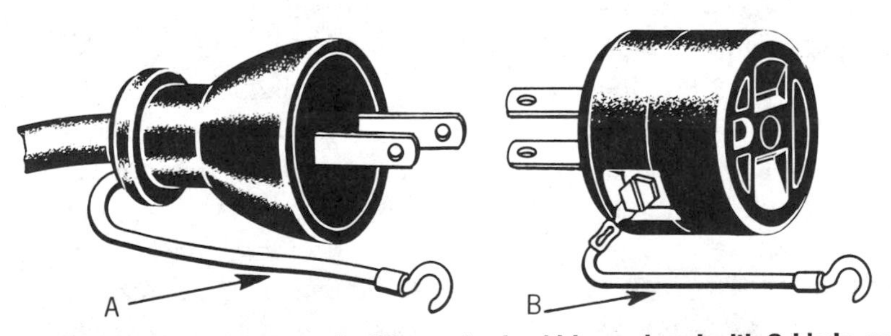

Fig. 12-22. Grounding pigtail "A" on flexible cords should be replaced with 3 blade grounding cap. Temporary grounding adapter grounds appliances through 2 slot receptacle if pigtail "B" is attached to metallic outlet box, through the center plate screw. (Harvey Hubbell, Inc.)

Fig. 12-23. Typical grounding receptacle has green hex screw grounding terminals on sides of mounting strap for the connection of green grounding wire. Grounding receptacles by metal to metal contact of receptacle mounting strap (yoke) is acceptable in some appliances and areas. (Harvey Hubbell, Inc.)

1. By a separate wire from the terminal to the electrode.
2. By an extra wire in the cables, conduits, or raceways that house the current-carrying wires.

To this extra wire are connected the grounding terminals and the housings of all wiring devices back to the service entrance switch. This switch is directly connected to the buried electrode.

3. By the metal enclosures of the wiring system itself. Flexible or rigid metal conduit properly installed forms a conductive contact with all metal wallboxes, junction boxes, switchboxes, and fuse or circuit-breaker panels back to the permanently grounded service entrance switch.

Also, the grounding terminal of the receptacle is internally connected to the metal parts by which the device is fastened to its housing, as in Fig. 12-23. If this housing is a metallic box already properly grounded, the receptacle itself will automatically be grounded through it in some installations.

It is generally recognized that grounding portable equipment as described reduces, but does not wholly prevent, electrical accidents. Only by properly insulating *all* exposed metal parts and housings is optimum protection achieved. Unfortunately, this is not always feasible because certain types of equipment would be too bulky to handle conveniently if they were protected with the ideal amount of insulation. Hence the need for wiring devices with built-in protective features.

Ground Fault Interrupter (GFI)

A GFI is a duplex-U-ground receptacle designed to protect people who come in contact with a "hot" line and to provide a path to ground through any part of the body.

Contact with a "hot line" can occur either through equipment malfunction or accidental contact with the live line voltage. The GFI will respond as soon as the current exceeds 5 milliamperes and disconnects the power within 1/40 of a second or less. (A 25 watt light bulb uses about 200 milliamperes.)

Installation is intended for such areas as receptacle outlets outdoors, bathrooms, temporary construction job site wiring, around swimming pools, and fountain equipment. The local department of building and safety should be consulted regarding local requirements and installation requirements.

QUESTIONS FOR DISCUSSION OR SELF STUDY.

1. What are the two divisions of grounding?
2. The connection of the neutral conductor of the wiring system to earth is in which division of grounding?
3. Which division of grounding is the connection of all exposed noncurrent-carrying metal parts of the wiring system to earth?
4. With everything properly grounded, what would the approximate voltage reading be between the service switch enclosure and a live line terminal?
5. Explain why a person would not become electrocuted if he were in contact with a properly grounded appliance housing should it have an insulation failure and become alive electrically?

6. Where should the two different divisions of grounding be actually connected?
7. What is the most desirable electrode for grounding?
8. Where required, what is the purpose of a meter bonding jumper or meter shunt?
9. What necessary precautions should an electrician take when selecting an underground water piping system as a grounding electrode?
10. When underground waterpiping systems are not available, what is the next best electrode?
11. What determines ground wire sizes and how can the size be calculated?
12. When grounding boxes with conductors inside, what considerations are necessary?
13. Which terminal on a duplex receptacle is the neutral and which is the ground?
14. Which color wire is designated as the neutral and which is the ground?
15. On a cord cap, which terminal is the equipment ground?
16. How is the ground connection made first when inserting a portable cord cap into a convenience outlet?
17. What is the purpose of different physical size and configurations of cord caps?
18. Why the need for wiring devices with built-in protective features?

QUESTIONS ON THE NATIONAL ELECTRICAL CODE

1. What is the maximum resistance permitted between a grounding electrode and earth?
2. Which conductor of the following services must be grounded:
   (a) Single-phase, 2 wire?
   (b) Single-phase, 3 wire?
3. What metal wire enclosures must be grounded?
4. How is equipment which is connected by cord and plug grounded?
5. What size service wire is required for a 200 ampre service?
6. What are code requirements for grounding of underground service conduit?
7. What are code requirements for grounding of underground service cable?
8. How are frames of electric ranges and electric clothes dryers grounded?
9. How is the electrical continuity of the grounding circuit maintained through the service equipment enclosures?
10. When a water pipe less than 10 feet long is used for grounding an electrical system what supplements are required?
11. What material and size of bonding jumpers are required?
12. What are other available grounding electrodes permitted other than an underground water piping system?
13. When the point of attachment of a ground clamp to water pipe is not on the street side of the water meter what requirements must be met?
14. What is the requirement for continuity and attachment of branch circuit grounding conductors to boxes?
15. What is the code requirements for grounding metallic boxes?
16. What is the requirement for ground attachment to electrodes?
17. What are the requirements for the installation of ground fault interrupters (GFI's) for residential occupancies and construction sites?
18. What are installation requirements for GFI's supplying lighting fixtures and lighting outlets and receptacles around pools?
19. What are GFI requirements around storable swimming pools?
20. What are GFI installation requirements for power supplies feeding fountain equipment?

Chapter

13 Methods of Conduit Wiring

The primary function of metallic conduits in the wiring system is to provide mechanical protection and electrical safety to both persons and property, as well as convenient and accessible ducts for the conductors.

A well designed electrical raceway system is one which has adequate capacity for future expansion and is readily adapted to changing conditions.

Mechanical Protection

A steel conduit system provides mechanical protection against forces from the outside and from within the system itself. During the construction period the wiring may be subjected to damage from the impact of wheelbarrows and trucks, and from being cut by a carpenter's saw or chisel, or punctured by nails and fasteners. Likewise, throughout the life of the structure, it is subject to injury as a result of possible stresses, pressures, heat and cold, fire, water or abuse. Conduit, because of its strength and durability, provides the necessary protection to the wiring system against such outside forces.

Steel conduit in commercial and industrial use also affords excellent protection against the dangers of explosions that may occur within the system due to the presence of combustible gases or fumes. Explosions due to slow gas seepage and electric arcs are safely confined to the inside of the conduit and are prevented from spreading to the outside where they might ignite surrounding gases or inflammable material. Explosion-proof boxes and fittings, that are required in hazardous locations by electrical codes, also help prevent seepage of gas into conduit systems.

Conduit Types

Metallic conduit for general-purpose wiring is made of either steel or

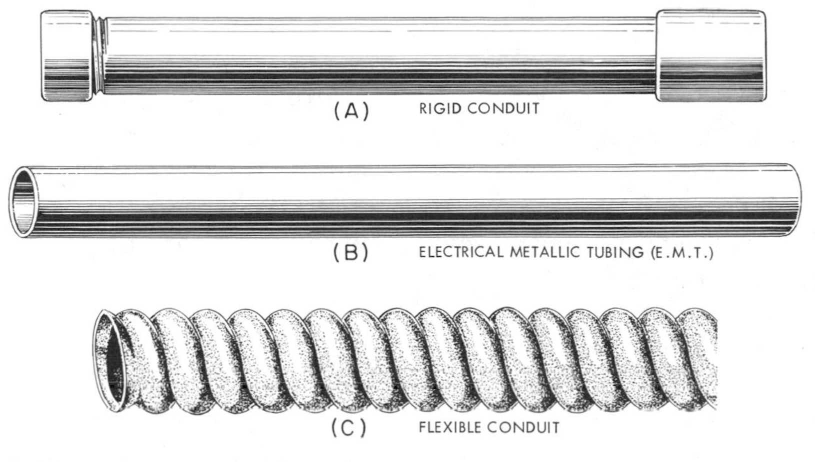

Fig. 13-1. Common types of metallic conduit.

aluminum. There are three common types: rigid conduit, electrical metallic tubing (E.M.T. or "EMT") and flexible conduit. See Fig. 13-1. The tubing is usually referred to as "thin-wall" conduit or EMT, and the terms will be employed interchangeably here. Rigid conduit is made to the same dimensions as standard pipe, the trade size of which is expressed, according to the nominal inside diameter, as ½", ¾", 1", and so on. Actually most inside measurements are greater than the designations.

Rigid Type. Rigid or non-flexible conduit, Fig. 13-1A, is supplied in 10-ft. lengths. It is treated to make the interior smooth, and coated outside to protect against corrosion. When plated with zinc by one of the galvanizing processes it is termed *galvanized conduit.* Black enamel conduit has very limited application, not being permitted if exposed to moisture or corrosion, and it is rarely used anymore in most parts of the country. Plastic covered conduits are also used in industry around highly corrosive fumes. When manufactured, rigid conduits must meet certain metallurgical specifications for easy bending qualities. After the raw pipe has been manufactured, it is cleaned, threaded on both ends with tapered threads, burr-reamed and finally it receives a protective coating if made of steel.

When the conduit leaves the factory each 10 foot length is equipped with a coupling on one end and a special thread-protector cap on the other. This is done to eliminate the possibility of injuring the threads during shipment, storage or other handling before installation. Approved conduit bears a U.L. (Underwriters Laboratories) sticker on each length.

Conduit systems should be completely installed before wiring is pulled into them. This means all cutting, threading, reaming, bending and strap-supporting should be completed before installing conductors. Conduit systems should be installed with the idea, constantly in mind, that wire must be pulled into them. Sharp edges should be carefully removed to avoid damage to wire insulation. A long sloping radius should be bent instead of a tight or small radius. This facilitates the pulling of wire later.

Table I shows the trade conduit sizes in inches and also inside measurements to the nearest 1/16 of an inch (the smallest division on an electrician's rule). A more precise dimension is given in decimals. Note that actual measurements are not the same as trade sizes but are slightly more in most cases. Note that thin wall is not made in all the trade sizes of rigid conduit. Also, outside diameters of thin wall and rigid are not the same. This means different size supporting straps and bending tools are required.

Plastic compound conduit is available for underground and special purpose applications — it is made without threads because couplings and conduit are cemented together. Plastic conduits are especially suited to commercial and industrial applications where there is an abundance of corrosive vapors and fumes.

EMT Type. The wall thickness of electrical metallic tubing (EMT) or "thin wall," Fig. 13-1B, is only about 40 percent of that of rigid conduit (see Table I) and is not to be threaded. It is easier to handle than its rigid counterpart, and may be installed more rapidly because of the

TABLE 1. CONDUIT DIMENSIONS

THIN WALL			TRADE SIZE INCHES	RIGID		
DIAMETER INCHES				DIAMETER INCHES		
INSIDE NEAREST 1/16"	INSIDE	OUTSIDE		OUTSIDE	INSIDE	INSIDE NEAREST 1/16"
5/8	0.622	0.706	1/2	0.840	0.622	5/8
13/16	0.824	0.922	3/4	1.050	0.824	13/16
1-1/16	1.049	1.163	1	1.315	1.049	1-1/16
1-3/8	1.380	1.508	1-1/4	1.660	1.380	1-3/8
1-5/8	1.610	1.738	1-1/2	1.900	1.610	1-5/8
2-1/16	2.067	2.195	2	2.375	2.067	2-1/16
2-3/4	2.731	2.875	2-1/2	2.875	2.469	2-1/2
3-3/8	3.356	3.500	3	3.500	3.068	3-1/16
	—	—	3-1/2	4.000	3.548	3-9/16
4-5/16	4.334	4.500	4	4.500	4.026	4
	—	—	5	5.563	5.047	5-1/16
	—	—	6	6.625	6.065	6-1/16

type of non-threaded fittings used with it. The largest obtainable size of electrical metallic tubing is 4 inches.

Flexible Type. Flexible metallic conduit, Fig. 13-1C, was first introduced under the name of "Greenfield." It is still sometimes referred to by this term throughout the trade. Because of its flexibility, this conduit is put up in coils instead of lengths. It may be used in lieu of the rigid type in most dry locations. The most common application, however, is for housing (residential wiring) and some commercial work, such as office and store spaces in wooden frame construction. The installation of flexible conduit requires the use of fittings for box connections.

Rigid Conduit: Installation Methods

Conduit Hickeys. The hickeys in Fig. 13-2 are for rigid conduit. On large jobs, where there are numerous duplicate bends or offsets, a bench type of tool, such as Fig. 13-3 is frequently called upon. This type is used for prefabrication of many similar bends. The hydraulic bender, Fig. 13-4 finds application where large sizes of conduits are involved. Fig. 13-4 shows the most rapid method of bending large size conduit. It utilizes an electric driven hydraulic pump driving a shoe that forms an elbow with a single ram stroke. This type is called a "one shot bender."

Conduit Bends

How to Make a Right-Angle Bend in Rigid Conduit. Suppose it becomes necessary to make a 90° bend in a piece of ½″ galvanized rigid conduit, so that it will fit into an

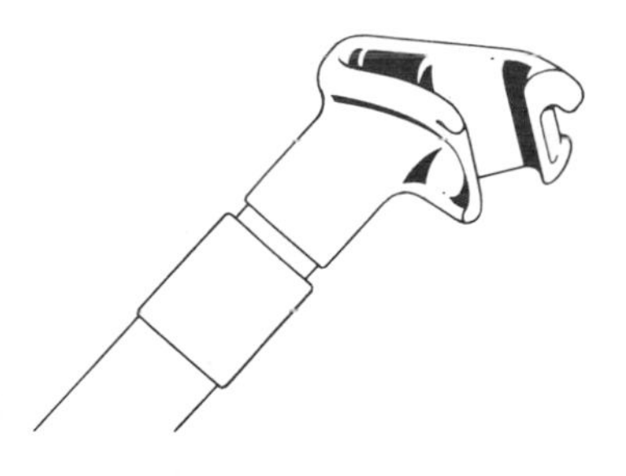

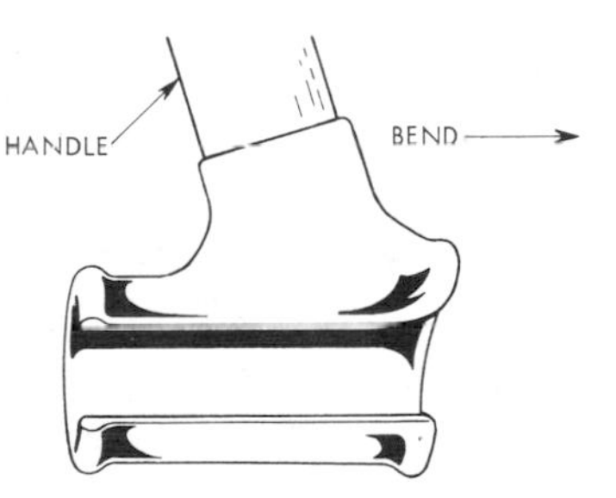

Fig. 13-2. Hickeys.

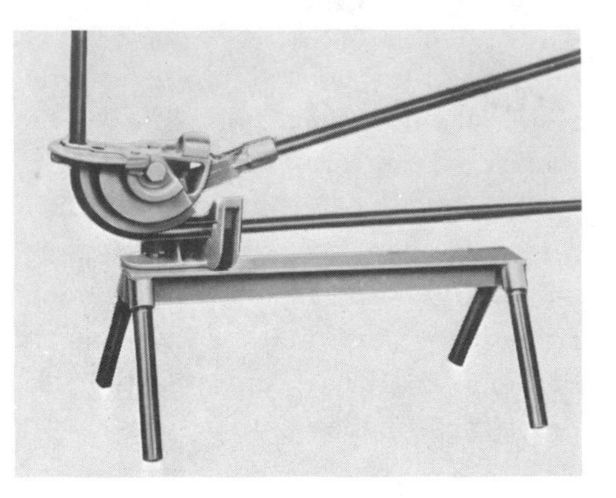

Fig. 13-3. Bench type conduit bender. (Greenlee Tool Co.)

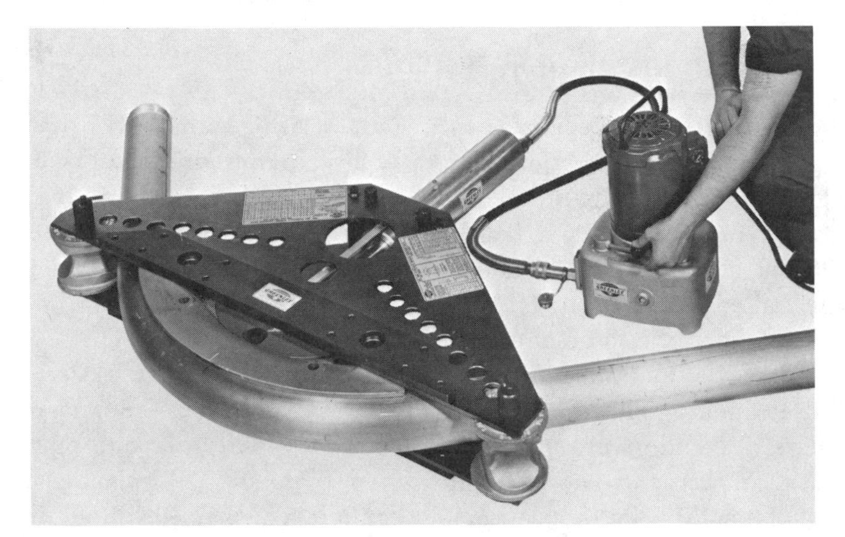

Fig. 13-4. A one shot electric-hydraulic bender forms a 90° angle in conduit with a single ram stroke. (Greenlee Tool Co.)

outlet box whose lower edge is 10⅜″ from the floor. Add ⅜″ for locknut and bushing inside the box, giving a total length of 10¾″. Mark this distance from the end of the conduit, and measure approximately 5″ in a backward direction to a second mark, Fig. 13-5A.

Lay the conduit on the floor, and against a wall or other obstacle to prevent back thrust (for safety of the learner), then slip the hickey

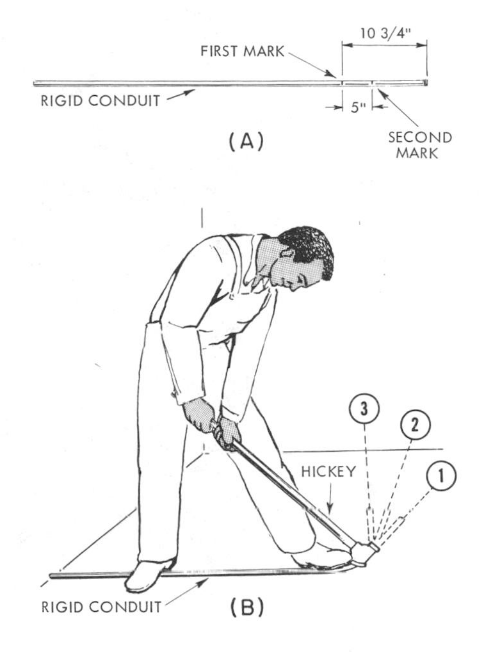

Fig. 13-5. Bending rigid conduit.

over the end. Place a foot on the pipe to hold it flat, and so that it will bend where desired (see Fig. 13-6) and adjust the back edge of the hickey to the second mark. Place the other foot on the conduit near the head of the hickey. This foot should be held tightly against the hickey to prevent a sloping, irregular 90° bend. Press down on the handle, meanwhile, until an angle of about 45° is made, as shown by dotted line *1* in Fig. 13-5B. Slide the hickey downward a short distance, and make another bend of about half the first angle, as indicated by dotted line *2* in the figure. Move the hickey downward about the same amount to finish the bend as shown by dotted line *3*.

Measure the completed bend. If a bit long, slide the hickey down to the neighborhood of its last position, and unbend a small amount. Then, raise the hickey to its first position, and take a rather sharp "bite" to form a right angle. If, on the other hand, the bend is too short, the process is reversed, first making a greater than 90° bend at the last or bottom mark, and then taking a backward "bite" at the top of the bend to make a right angle. Finally, check to see that the bend is exactly square.

The distance measured in the "backward" direction, stated here as 5″, Fig. 13-5A, varies according to size of conduit, material of which it is made, type of hickey, and precise method of performance. The dis-

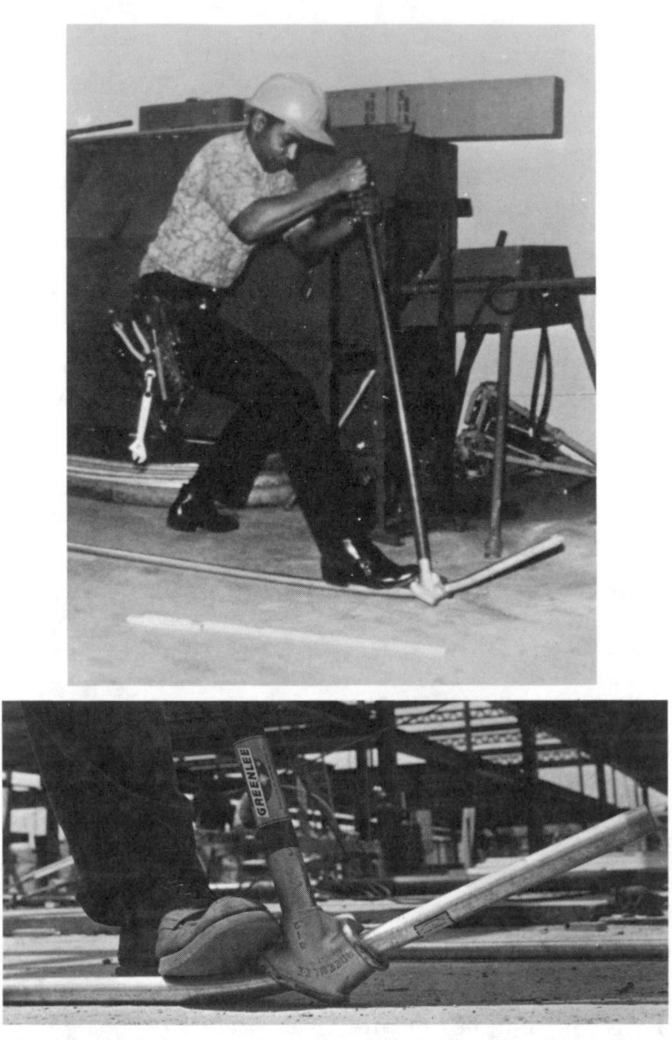

Fig. 13-6. Top: Proper position for bending of conduit. Bottom: Detail showing proper position of foot. Top: Photo by Charles S. Anderson, Electrical Industry Training Center, Fraser, Mich. Bottom: Greenlee Tool Co.

tance may be anywhere from 4½″ to 5½″ for ½″ rigid conduit, and as much as 8″ for 1″ conduit. Actual experience soon makes this operation simple and routine.

How to Make Offsets in Rigid Conduit. Offsets require the exercise of more skill than right-angle bends. The forming of complicated offsets and saddles can be learned

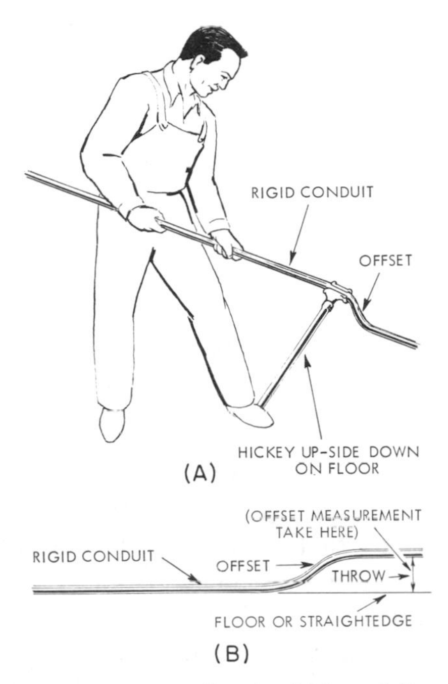

Fig. 13-7. Making offset in rigid conduit.

only through the experience of doing. Certain principles can be applied, however, and varied to suit material, tools, and personal methods of the workman. The general procedure may be explained with the aid of an example.

An offset is to be made in a length of ½″ rigid conduit, to pass along a 5″ wall projection. The distance from the threaded end of the conduit to the beginning of the offset is 12″. Mark this distance on the conduit and measure backward about 2½″. Stand the hickey upside down on the floor, Fig. 13-7A, with a foot against the lower end of the handle to steady it. Now insert the conduit into the hickey until the back edge of the hickey rests upon the second mark, and bend a 45° angle. Turn the conduit over, sliding it through the hickey only far enough to clear the curved portion, and bend a 45° angle, so that the short end and the long end are parallel.

Check the offset by holding it on the floor and measuring, Fig. 13-7B, to determine if the throw or offset is correct. If a straightedge is used here it will be easier and more accurate to read rule measurements by looking down upon the work.

For example, Fig. 13-7B is a top view if the offset is laid flat. The view is looking down on the offset and straightedge. This is an easy and accurate position for reading a rule. A straightedge is a straight line on the floor, a straight piece of wood stripping, or a straight hickey handle.

If a flat floor is being used for a straightedge, a side view is shown in Fig. 13-7B. This means that it is necessary to place an eye near the floor level to take a reading of the rule when determining measurements. This is more difficult and less accurate.

If the bend is too great, or if it is not great enough, proceed by taking a little bend out of each side if offset is too great. Making a little more bend in each end of the offset will make a larger throw or offset. Care

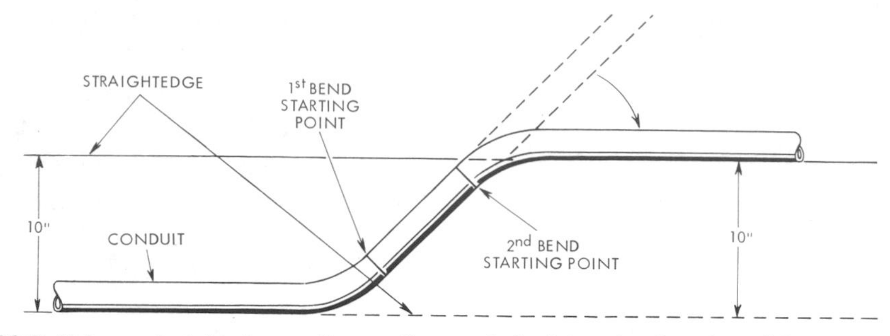

Fig. 13-8. Using a straightedge on line on floor to help determine location of 2nd bend on an offset.

should always be taken when working with hickeys not to bend too much in one position, thereby kinking or flattening the conduit. If the wall projects, say 10″, rather than 5″, it will be necessary to slide the conduit through the hickey a few inches after completing the first bend, before making the reverse one. Practice soon enables you to make correct distances and allowances for a given size offset. However, the proper use of a straightedge as shown in Fig. 13-8 will greatly assist in locating the second bending position for the offset. The second bend should begin inside the curve of the straightedge crossing (marked "2nd bend starting point" in Fig. 13-8). This is a relative position to the first bend, observing where the second bend should start, using the first as a pattern. The sharper or tighter the bend, the closer the bending point to the straightedge. The 10″ measurement, it should be remembered, was taken center to center of conduit or bottom to bottom, and is not an overall measurement. Also, bends should be parallel when finished, continuing in the same direction.

A useful offset is the box offset at outlet boxes in exposed work. Fig. 13-9, top, shows how conduit looks when *not* offset at the box. This is not good practice. Fig. 13-9, bottom, shows the neat appearance when the offset is formed at each side of the box. All conduits should leave outlet boxes at right angles as shown in Fig. 13-9, bottom.

How to Make Saddles. Saddles, like that in Fig. 13-10, are merely double offsets. One most common fault in making offsets or saddles is illustrated in Fig. 13-10B. When the conduit is not turned exactly halfway round after the first bend is made, the reverse bend is out of line with the first one. A good plan for avoiding this difficulty is to mark a line, such as (1) in Fig. 13-10C, the

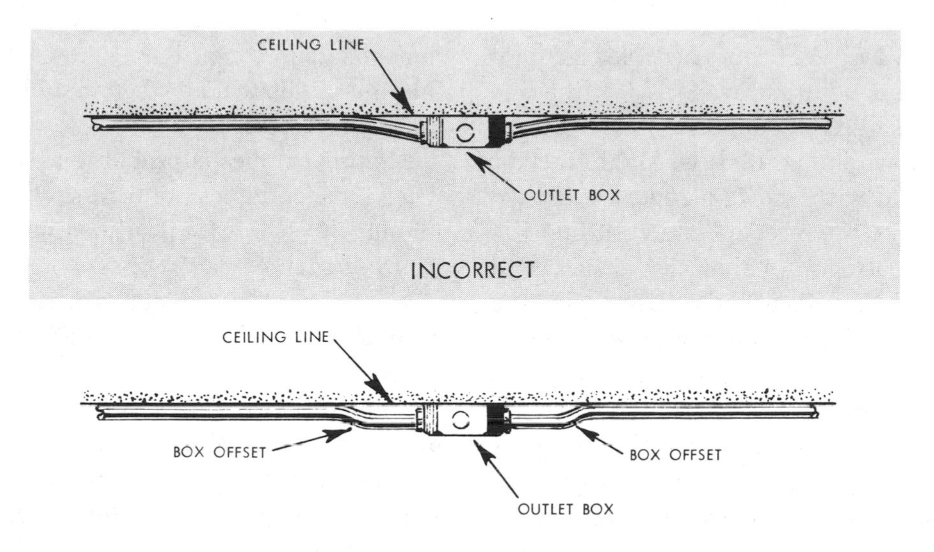

Fig. 13-9. The offset for the outlet box at *bottom* is CORRECT. (Top installation is wrong.)

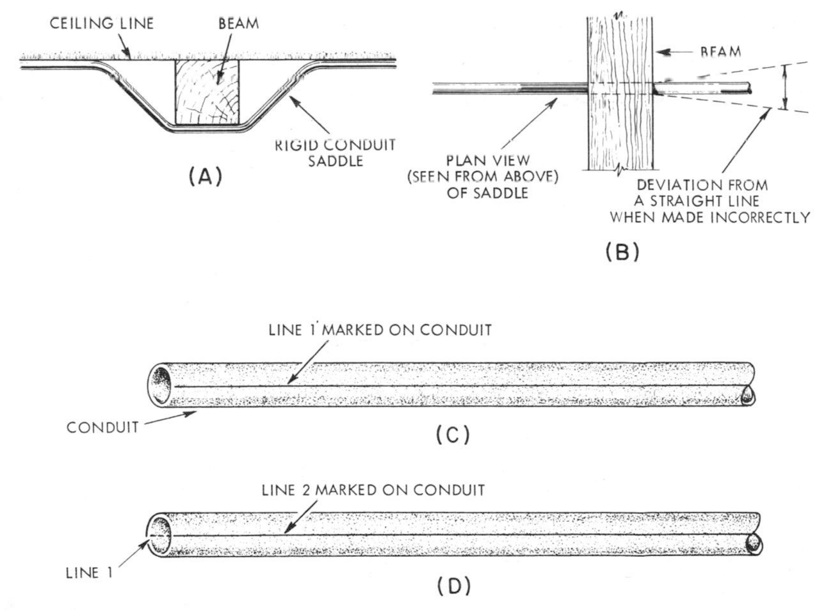

Fig. 13-10. Saddles and the markings for bending.

full length of the piece. Next, rotate the conduit halfway, and scribe line (2) directly opposite line (1), as shown in Fig. 13-10D. Make the first bend with line (1) uppermost, then make the reverse bend with line (2) uppermost. In this way either bend will be more exactly "true" with the other. With enough practice, and by sighting down the conduit while bending and being careful not to turn the conduit when the hickey is being moved for more bend, the turning-over maneuver becomes automatically accurate without the need for guide lines. Fig. 13-11 shows bending steps in making a saddle or a double offset. Care should be taken when making small saddles on bend number 3 so that it does not reduce bend number 1 or the amount of offset. It is better to strive for a certain measurement of clearance over the obstacle. The amount of clearance depends upon conduit sizes, how visible the saddle will be when installed and other factors.

Making saddles is probably easier when smaller angles are bent such as making 30° bends for positions 1 and 2 rather than 45° as in offsets.

Neat saddle and offsets are proof of an electrician's skill. The installation of rigid conduit and electrical metallic tubing requires considerably more mechanical skill than is needed for most of the other wiring systems. Bending conduit and tubing to exacting requirements is basically a manipulative skill that is acquired through practice.

Checking Bends with a Level

The true accuracy of finished bends may be checked with a pocket level—a valuable tool for a skilled craftsman. This is more important when bending larger size conduits

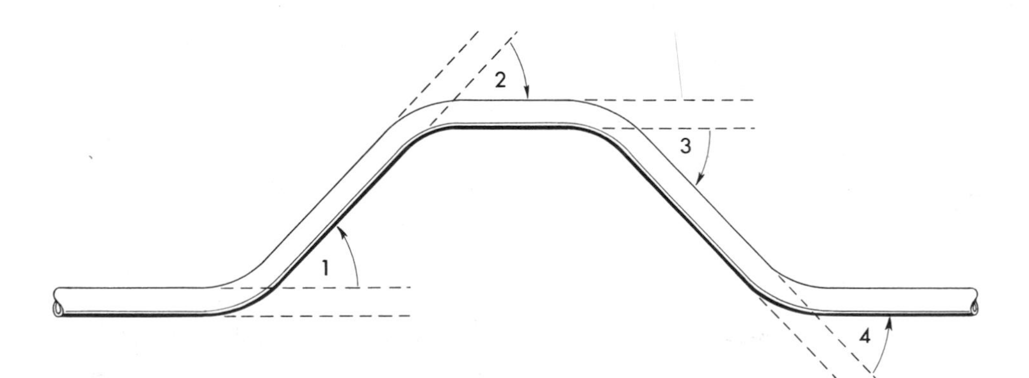

Fig. 13-11. Bending steps in making a saddle.

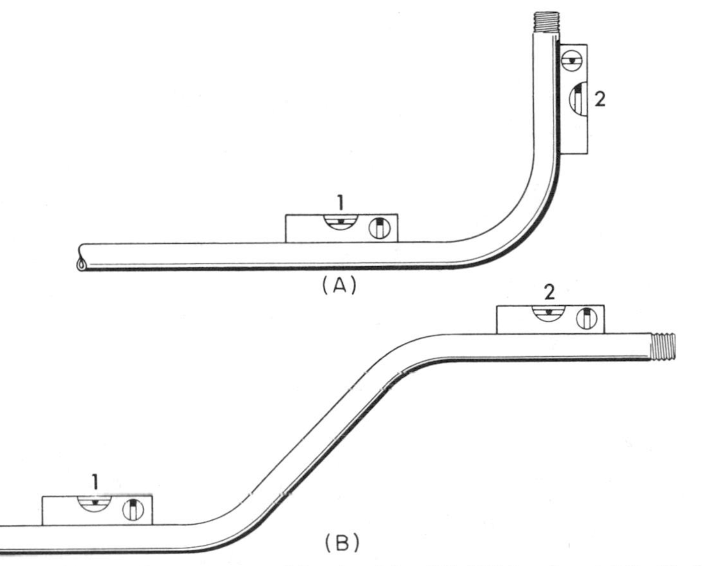

Fig. 13-12. Checking bends for accuracy with a level for (A) 90° bend and (B) offset.

because they cannot be forced into place if they are not bent accurately. Looking at Fig. 13-12A, the bubble on horizontal position 1 should be exactly the same as in position 2. It need not be level, merely in the same relative position. For example, if the bubble on the level in position 2 is exactly level, between the two lines, position 1 should be identical. However, if the position of the bubble is cut in half by the right hand line on the level in position 1 it should be cut by the right hand line in the vertical position 2. All this is also true for checking the bending accuracy of offsets, as in Fig. 13-12B. The bubble in positions 1 and 2 of Fig. 13-12B must also be in the same relative position.

Factory Conduit Elbows

While rigid conduit follows closely the regular dimensions of standard pipe, there is a marked difference in the bends or elbows, usually called ells. Water-pipe ells are castings with short radius, for 90° or 45° bends, but factory formed conduit ells are standard only for 90° bends. They are made from conduit tube bent to a long radius to facilitate the *pulling-in* of the wires. This radius varies from nearly six diameters (inside) for 1″ ells to four diameters for 4″ ells. Note that in Fig. 13-13 the

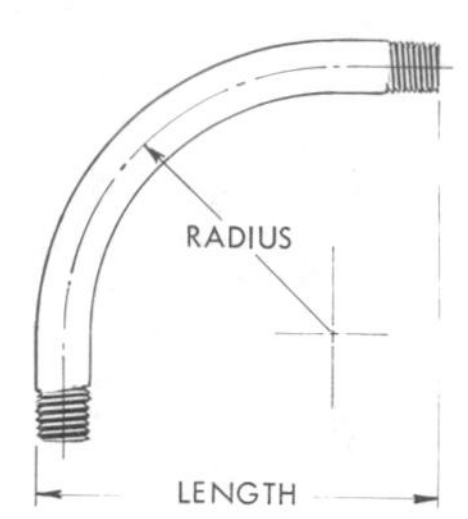

Fig. 13-13. Diagram showing length and radius of conduit elbow.

length is greater than the radius because beyond the completion of the 90° arc there is, at each end, an added straight length, roughly equal to 2″ in the smaller sizes increasing to nearly 5″ in the largest size. This length must be allowed for in determining the length of conduit which is to be used in a run which includes an ell. Since the lengths are not always the same for ells of a specific conduit size, it is wise to have the ell on the job before taking final measurements. Also it is advisable to check the ell to make sure that its angle is a true 90°, especially if two or more are to be installed side by side.

The ½″ and ¾″ ells are usually bent on the job from a length of conduit, with either an elbow former or a hickey, or by hand in a pipe vise with a bending mold. No conduit elbow should be bent to a shorter radius than six times the inside diameter of the conduit. They should be as large as possible to permit easier conductor installation.

Thin-wall ells are stocked by electrical supply houses also. They are not generally prefabricated in sizes smaller than 1″.

In addition to standard ells, bends of longer radii, commonly called *sweeps*, are used especially where multiconductor cables are to be installed. These sweeps are not carried in stock by supply houses, but must be ordered from the factory. They can also be formed on the job more cheaply by a skilled craftsman using a hydraulic bender.

Couplings

Conduit couplings are similar to standard pipe couplings except that they are finished just like conduit. A coupling is a larger sleeve-like device threaded inside to accommodate conduit (as seen in Fig. 13-14, right side). They are used to connect lengths of conduit together. When one conduit end cannot be screwed

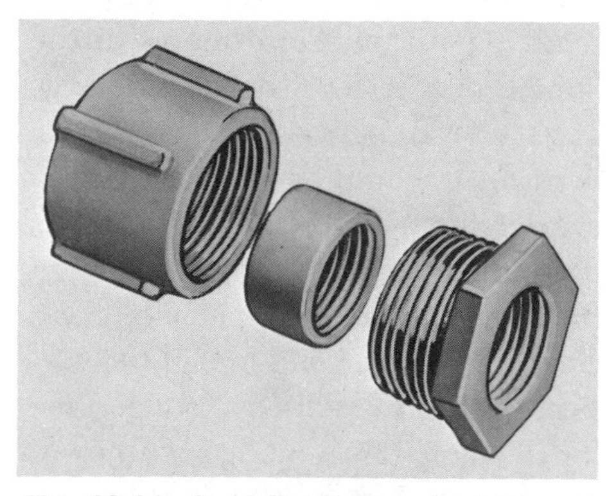

Fig. 13-14. Conduit union—a 3 piece coupling. (All Steel Equipment Inc.)

into a coupling due to a bend striking an obstruction, a three piece coupling is used, called a union. See Fig. 13-14. The left-hand piece is slipped over one conduit end before the center piece is screwed onto the other conduit end, then the left is coupled to the right, locking the center piece and conduits together. Neither conduit need be turned while coupling together.

Bushings and Locknuts

Where a conduit enters a box or other housing through a hole that is not threaded, the junction between the two is made firm and secure by means of a locknut, Fig. 13-15A, and a bushing, Fig. 13-15B. Bushings are made of metal and plastic. The locknut is screwed onto the threaded end of the conduit on the outside of the box and the conduit is then passed through the conduit hole in the box, and the bushing is then screwed onto the conduit as far as it will go and the locknut set up tight against the side of the box.

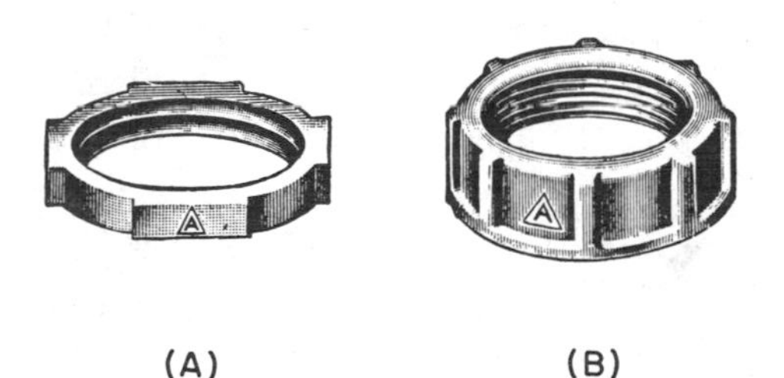

(A) (B)

Fig. 13-15. Locknut and conduit bushing. (Appleton Electric Co.)

Some localities require two locknuts, one inside the box in addition to the bushing, the other on the outside of the box. The National Electrical Code demands this double-locknut construction only where the voltage to ground exceeds 250 volts. When in doubt, consult the local code and the electrical inspector. Some local codes require two locknuts and a bushing on every installation.

How to Use Knock-out (K.O.) Punches

Quite often conduits must enter boxes and cabinets where no knockouts (K.O.) exist. This may occur most frequently around service equipment in residential construction. Fig. 13-16 shows cutters used in cutting a knock-out hole. Hole location should be laid out on box or cabinet as shown in Fig. 13-17A. The center lines A and B should extend well beyond size of hole to be punched. To make a ⅞″ hole needed to accommodate a ½″ conduit size, a 7/16″ hole should be drilled in the center of the layout, Fig. 13-17B. The die (top part) should be placed on the drive screw and inserted through the drilled hole, Fig. 13-17C. The punch (bottom part) is then threaded on until both die and punch are tight against the metal. The center lines for the hole are lined up with marks on the outside of the die. The drive screw is turned with a wrench

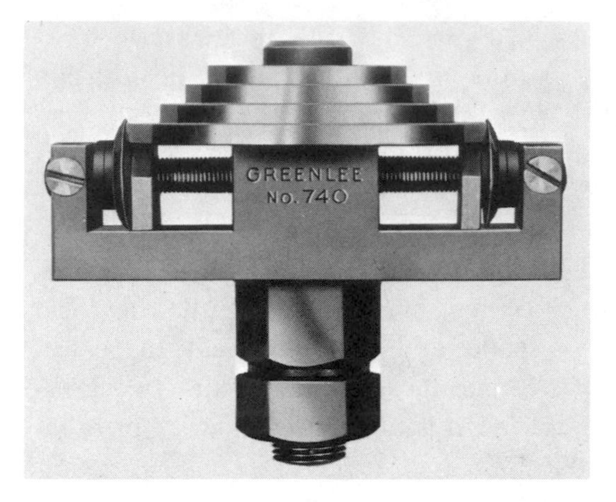

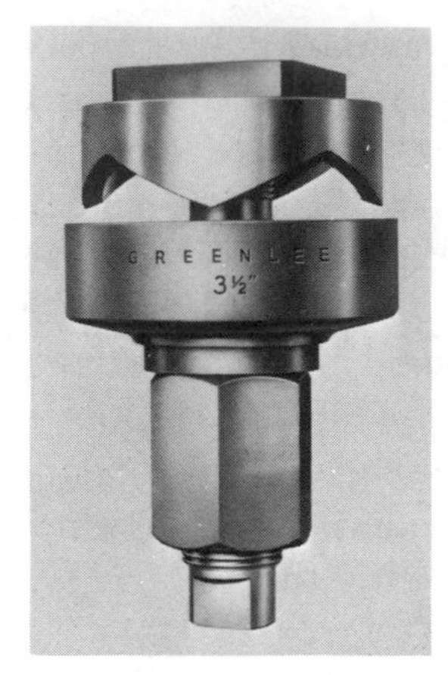

Fig. 13-16. Knockout cutters. (Greenlee Tool Co.)

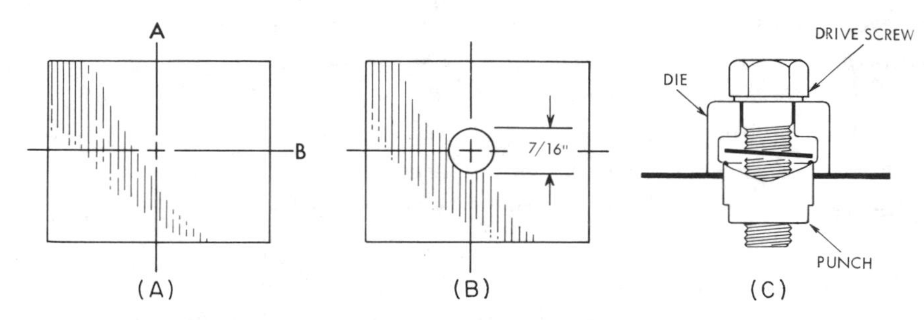

Fig. 13-17. Laying out and cutting a K. O. (Greenlee Tool Co.)

until the hole is punched. Continue a few turns in order to be sure slug drops free into the die. The punch is then removed from the drive screw to remove the slug from the die. Should the slug be curled slightly and be difficult to remove, flatten the slug with a hammer tap on a screwdriver so that it will fall out.

Care of Knock-out Punches

In sharpening a knock-out (K.O.) punch hold the punch so that the entire cutting face on one side of the punch is flat against the side of a medium-grit abrasive wheel. See Fig. 13-18. Repeat the procedure for the other cutting faces. Grind squarely upon angle of cutting edge, maintaining the two high-points equal

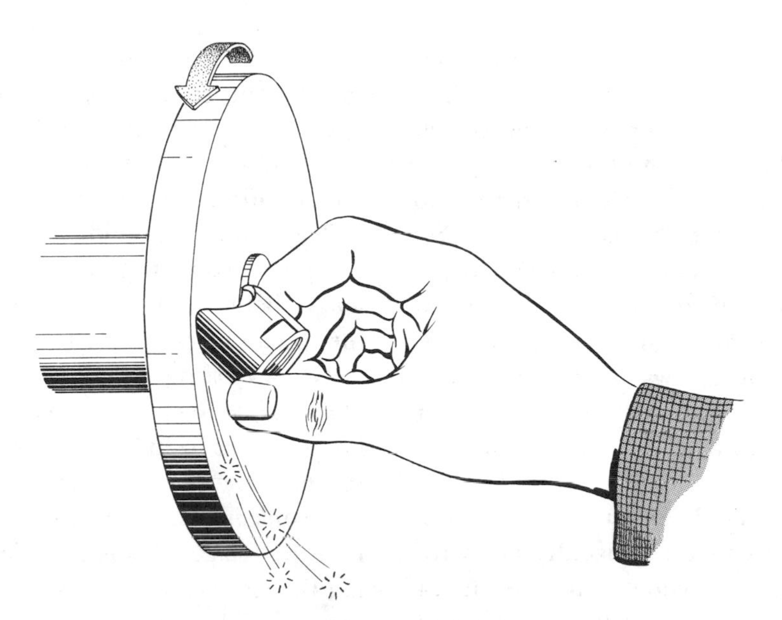

Fig. 13-18. Sharpening the knockout punch. Field sharpening the knockout punch with gentle pressure and extreme care on the flat side of an abrasive wheel. What is shown here is classified strictly as a job site emergency measure.

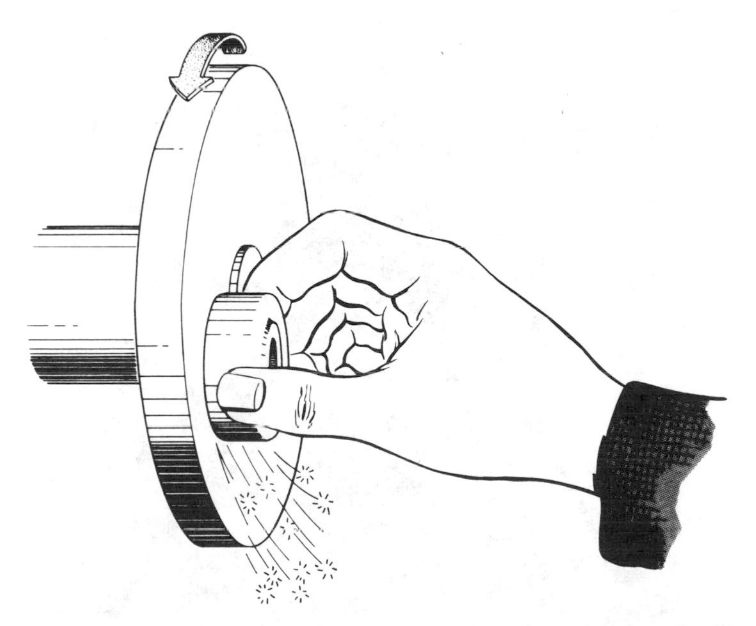

Fig. 13-19. Sharpening the die. Sharpening on the flat side of an abrasive wheel is classified strictly as a job site emergency measure.

and directly opposite from each other for the most efficient cutting. The outside diameter of the punch should *never* be ground.

To sharpen the die, grind the top surface of the die as shown in Fig. 13-19. Do not grind or file the inside diameter of the die or the metal shearing effect will be reduced or lost. The drive screws of all knock-out (K.O.) punches should be kept wet with lubricating oil.

Cutting Rigid Conduit

Rigid conduit is usually cut with a hacksaw. Blades with 18 or 24 teeth per inch are used. To prevent broken blades, conduits are cut by being held rigid in a pipe vise. A portable pipe vise is shown in Fig. 13-20. Hack saw blades are installed in the frame so that they cut pushing in a forward or downward position. This well-used portable pipe vise is ideally suited for this operation.

The conduit should be placed in the pipe vise so that there is ample length between the cut and vise jaws for threading purposes later.

When cutting threads on a portable vise, a downward motion of the handle must be maintained so as not to upset the portable vise. This vise also contains an adjustable screw on its top surface; this is used for making the vise stationary by tightening the adjustable screw, with the aid of a length of conduit, for extension to the ceiling.

Bends may be made of ½″ and

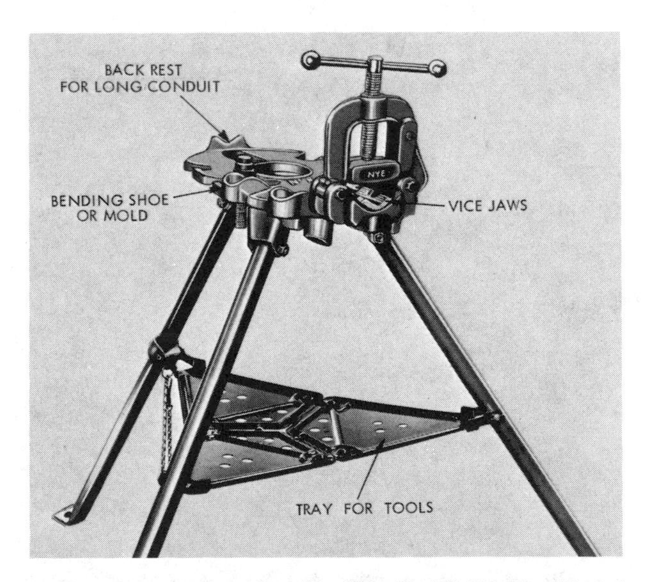

Fig. 13-20. Portable pipe vise stand and bender. (Nye Tool & Machine Works)

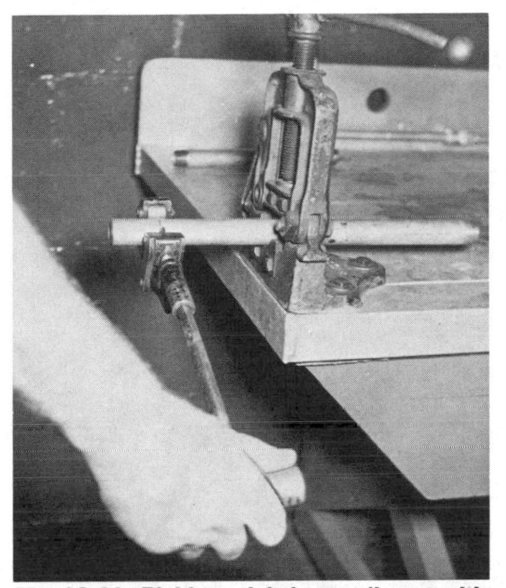

Fig. 13-21. Rigid conduit is usually cut with a hacksaw but pipe cutters, as shown, are sometimes furnished on the job instead of hacksaw blades.

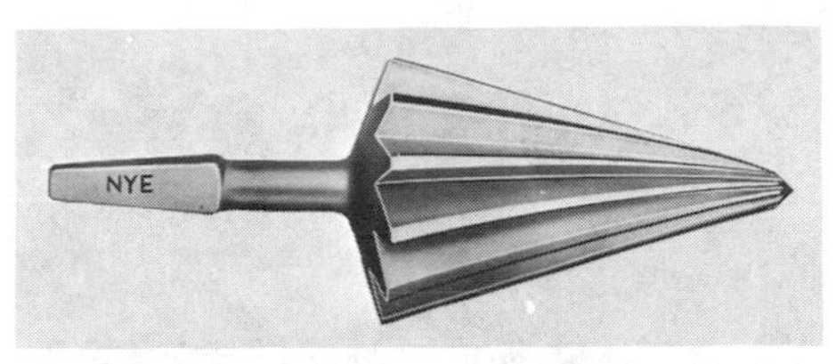

Fig. 13-22. Spiral flute reamer. (Nye Tool & Machine Works)

¾″ rigid conduit by inserting the pipe in the bending molds and pulling downward.

Conduit can be cut with a plumber's pipe cutter. See Fig. 13-21. This is sometimes furnished on the job instead of hacksaw blades. The pipe is held in a pipe vise and a small amount of cutting oil is placed on the area to be cut. Adjust the cutting wheel around the pipe at the place to be cut and revolve the pipe cutter around the pipe. Screw the handle in at each turn to tighten the cutting wheel.

In both cases, either using a pipe cutter or hacksaw, burrs will be left on the pipe. These, of course, must be removed or they will damage or cut the wire insulation when it is pulled through the conduit. The burrs or any sharp edges inside the mouth of the pipe may be removed with a reamer. See Fig. 13-22. The reamer is inserted in a brace for a few turns in reaming.

A pipe cutter unfortunately leaves a sharp, drawn-in edge that takes more reaming than the cut left by a hacksaw. When conduit is reamed extensive it reduces the wall thickness to a tapered, blade-like edge. When this thinned wall is coupled to another conduit in an extremely tight fit, the thin wall may curl inward. This of course can prove hazardous to wire insulation. The hack saw method is generally preferred because it leaves less burr to be reamed. Most wiremen carry reamers for conduit sizes under 1½″.

Fig. 13-23 shows how to use a reamer in a hand brace. Sharp inside edges and burrs must be removed on all conduits before installing and pulling wire through.

Fig. 13-23. The conduit must be reamed until all inside sharp edges that could damage wire insulation are removed.

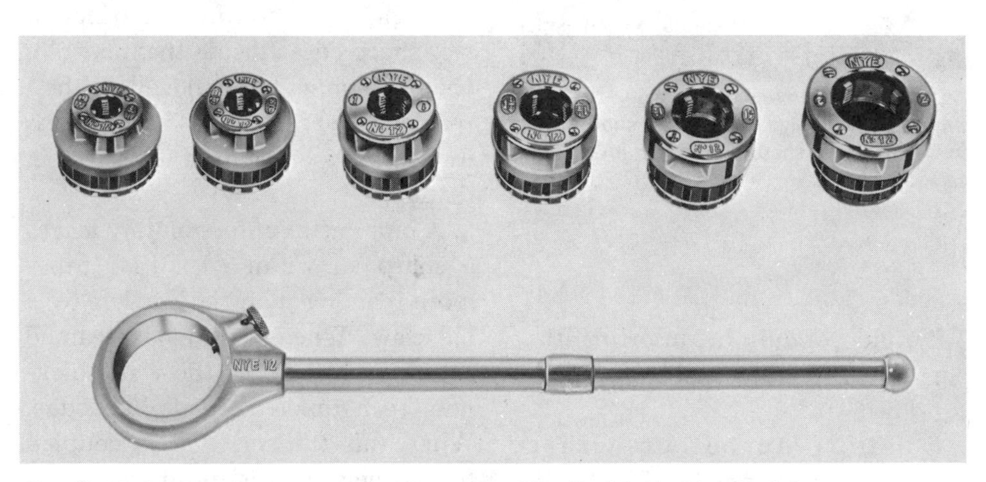

Fig. 13-24. Ratchet die stock with assorted dies. (Nye Tool & Machine Works)

Threading Rigid Conduit

Conduit threading equipment is the contractors' equipment. A ratchet die stock with different size pipe dies up to 1″ is shown in Fig. 13-24. The dies are easily inserted in or removed from the stock to accommodated different sizes of conduit within the range of the tool. When using these dies the handle should be pushed down with aid of body weight; the head should be held to one side for safety clearance. The pipe is held in a pipe vise while the threads are being turned.

For larger size conduits 1″ to 2″

Fig. 13-25. Adjustable die block. (Nye Tool & Machine Works)

Fig. 13-26. Die stock with adjustable guide. (Nye Tool & Machine Works)

an adjustable pipe die is used, Fig. 13-25. These types of dies are adjustable to two or more sizes up to 2″ and have a ratchet action handle. The largest conduit sizes to 6″ are threaded with a type shown in Fig. 13-26. The cutting dies are controlled with an adjustable guide, and the receding dies can be moved in or out to suit various diameters of conduit through manipulation of the adjustable sector handle. The lower bolts are locked onto the conduit before threading. The die is generally turned with a power unit that attaches to the square shaft.

Conduits must come close together in the couplings when installed and must find a firm seat in the shoulders of the threaded hubs of conduit fittings. It is important, therefore, to cut threads long enough. In other words, a full thread should be cut. The die should be run up on the conduit until the conduit just about comes through the die for about one-half or one full thread. This is the case for the dies shown in Fig. 13-24. This gives a thread length which is adequate for all purposes. If a thread is cut too long it will fit too loosely in a coupling or threaded hub and will corrode because threading removes zinc or galvanization.

Cutting oil should be applied to the pipe dies and threads as they are cut. Clean, sharp threads can be cut when the conduit is well oiled while threading. Clean, sharply cut threads make a better continuous grounding circuit and save unnecessary labor later.

EMT: Installation Methods

Roll type hand benders, as illustrated in Fig. 13-27, are generally used for bending EMT (electrical metallic tubing). They have high supporting sidewalls to prevent flattening or kinking of the tubing, and a long arc that permits the making of 90° bends in a single sweep without moving the bender to a new position along the tubing. There are different makes of benders with bending marks to follow for precision bending. These marks are molded into the shoe, with directions to add or subtract a dimension from a total bend figure.

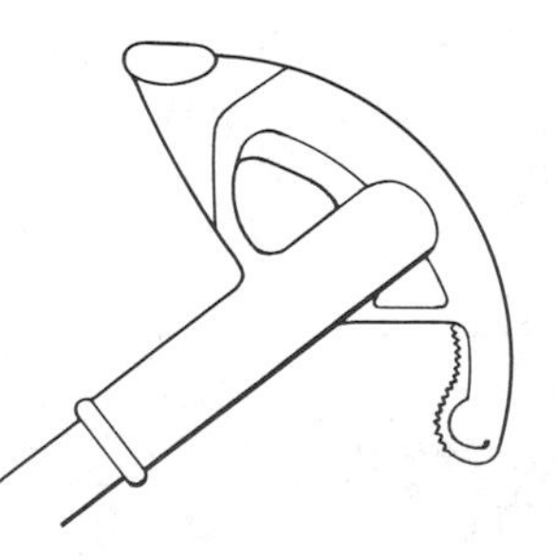

Fig. 13-27. Roll type hand bender used for bending tubing (EMT).

EMT Bends

How to Make a Right-Angle Bend in EMT. Generally, a 90° bend is made to a predetermined measurement. The tubing is marked (for example 10¾") from the end and the bender is located to the mark as shown. See Fig. 13-28A. This measurement in Fig. 13-28 is for a 10¾" stub-up, or riser. The back of the bender is lined up with the bending mark as shown in Fig. 13-28A. This is done visually. The accuracy of a finished bend (Fig. 13-28B) depends mostly on the accuracy of the electrician in lining up vertically the tubing mark with the back of the tubing form on the bender. With one foot on the back of the bender, the bend is made as shown in Fig. 13-29. When the handle is vertical, a 45° bend has been completed, a good

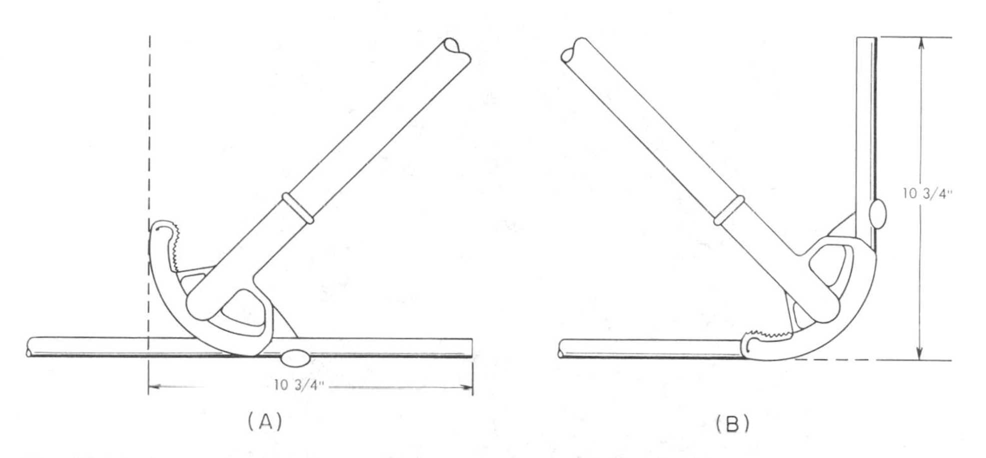

Fig. 13-28. A general method for bending a 90° bend to a predetermined measurement of 10¾".

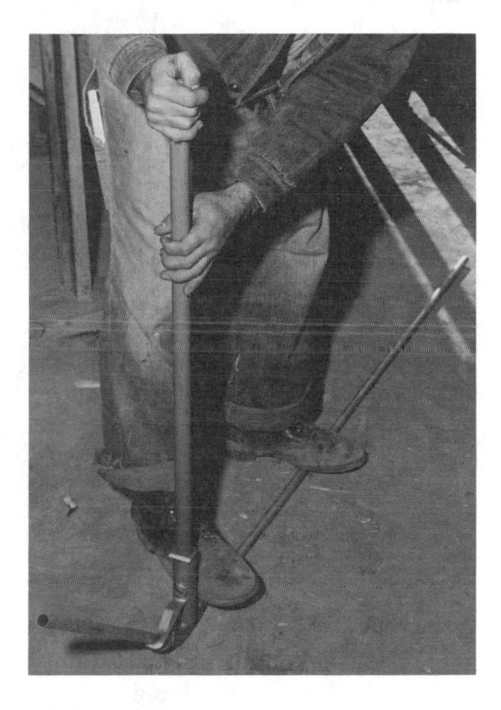

Fig. 13-29. Bending EMT on the floor. (Republic Steel Corp.)

start for an offset. Fig. 13-30 illustrates how to bend the tubing in an inverted position using the tubing as a lever.

The previously described method of making a 90° bend can be used with nearly any type of tube bender. It is a simple method to remember.

Fig. 13-30. Bending EMT with the bender in an inverted position using the tubing as a lever. (Republic Steel Corp.)

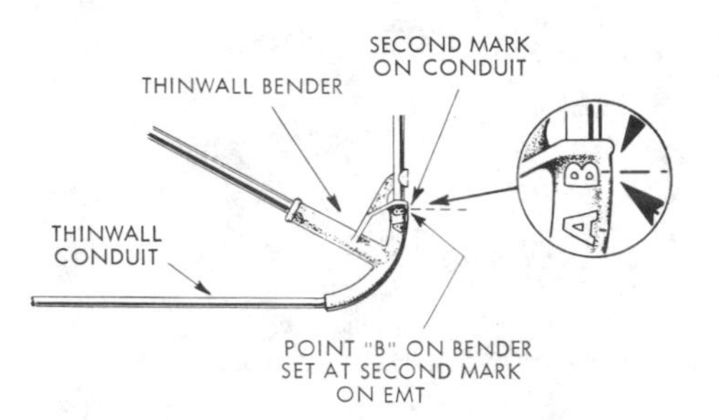

Fig. 13-31. Second bend of back-to-back bends.

A right angle bend may also be made using "A" and "B" markings on a bender. See Fig. 13-31. In order to make a 10¾″ bend in a length of ½″ thin-wall conduit, for example, a mark is written at a point 5¾″ from the end (10¾″ minus 5″). Point *B* on the tool is set at this mark, Fig. 13-31, and a 90° bend is made in a single movement of the handle. In using this method it will be noted that we are also doing the previous method in Fig. 13-28.

How to Make Back-to-Back Bends in EMT. For making "U" bends or "back-to-back" bends to connect outlet boxes, the head of the bending tool usually has a second mark, *A*, on the curved arc. The first bend, at the left in Fig. 13-32 is fashioned in the usual manner. To make the second one, measure the distance *D*,

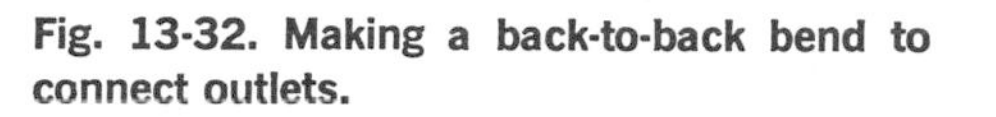
Fig. 13-32. Making a back-to-back bend to connect outlets.

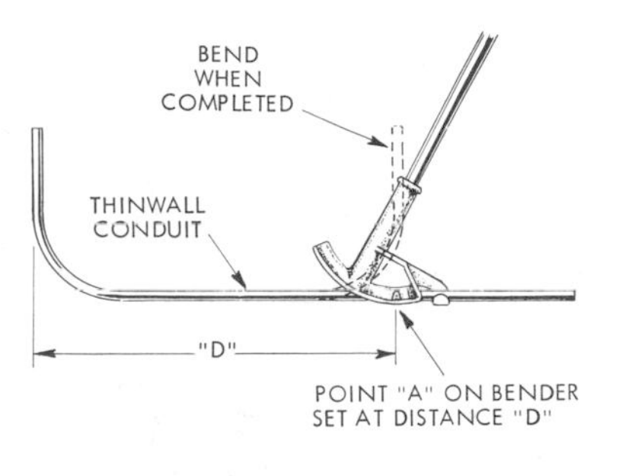

Fig. 13-33. Finished "back-to-back" bends and the bending position. (Republic Steel Corp.)

and mark the spot on the conduit. Reverse the tool bending direction or the EMT, setting its point *A* on the mark, and make a right-angle bend as indicated by the dotted line. Experience may produce other methods yielding the same results. Fig. 13-33 illustrates how the U bend or the back to back bend is completed. When the bender handle is in a 45° position, theoretically the EMT is bent into a 90° bend.

How to Make Offsets and Saddles in EMT. Saddles are somewhat

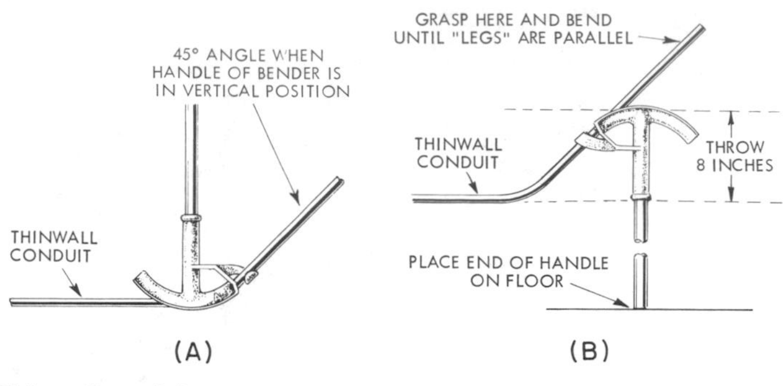

Fig. 13-34. Thin wall conduit.

easier to make with thin-wall than with rigid conduit. Suppose that an offset with an 8″ throw is required. In making the first bend, Fig. 13-34A, move the handle of the tool until it is in a vertical position, which means that a 45° angle is being produced. Reverse the bender, as in Fig. 23-34B, and slide along the conduit until the throw, as measured along the vertical handle, is equal to the desired amount. Some benders are marked in inches down the handle making it easier to measure or sight the 8 inches. Set the end of the handle on the floor and grasp the conduit, bending it until the two legs are parallel. The first bend of an offset is made at a point sufficiently far from the obstruction to allow clearance at the desired angle. The next bend, made at a point directly over or clearing the obstruction, is equal in angle and opposite in direction to the first bend.

In order to make this offset into a saddle, simply mark the saddle width on the conduit, and then proceed to make a "reverse" offset of the same dimensions.

An alternate method of making a saddle is by first bending the EMT 45° (Fig. 13-35A) at the point which will be directly over the obstruction. Next, with the aid of a straightedge (Fig. 13-35B), determine the length of the straight legs required to provide the necessary saddle rise. Complete the saddle by making a 22½° bend at the measured point on each straight leg. The 22½° bends are made in the direction opposite to the original 45° bend. A slight gain will be experienced when the last bends are made. Should the obstruction to cross be large in depth, the first bend should be moved to accommodate the additional tubing used in making all bends. If this is not done the EMT will be short for connection placing the saddle off center. Better tubing too long than too short.

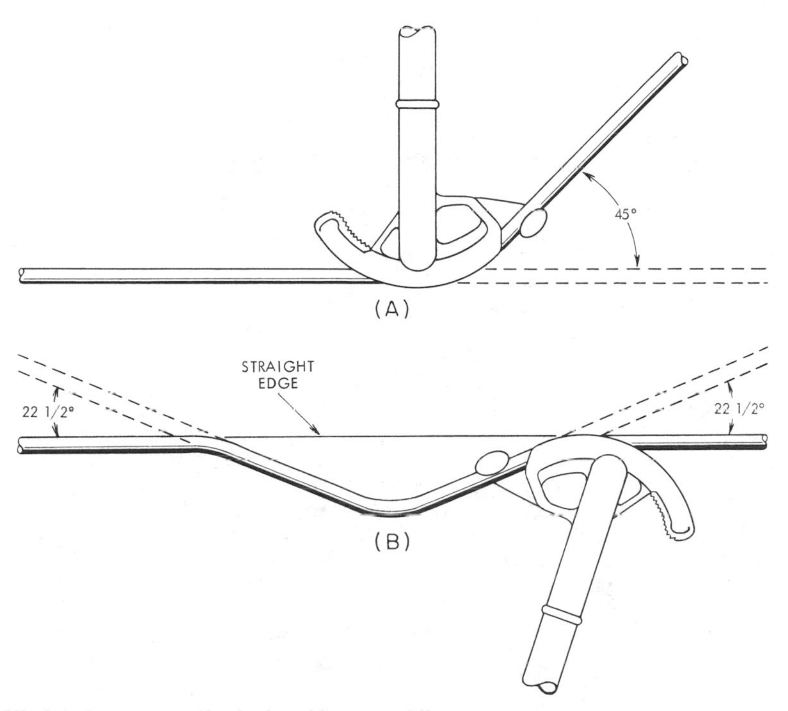

Fig. 13-35. An alternate method of making a saddle.

EMT Installation

The preliminary operations of spotting outlets and determining general circuit layout are the same for a thin-wall conduit job as for a cable installation. Hence, they need not be repeated here.

At one time the use of rigid conduit in partitions and ceilings was a laborious and time-consuming operation. Thin-wall conduit, largely because of the types of fittings specially adapted to it, makes an easier and far quicker job. Therefore, it has effectually replaced rigid conduit for this kind of work.

Fig. 13-36 shows two methods of running thin-wall conduit in these locations: boring timbers, and notching them. When boring, holes must be drilled large enough for the tubing to be inserted between the studs. The tubing is cut into rather short lengths, if necessary, calling for a multiplicity of couplings, and thus increasing the cost of material. However, tubing lengths are kept as long as possible. EMT can be bowed quite a bit while threading through holes in studs. Boring is the better of the two alternatives.

Because of its weakening effect

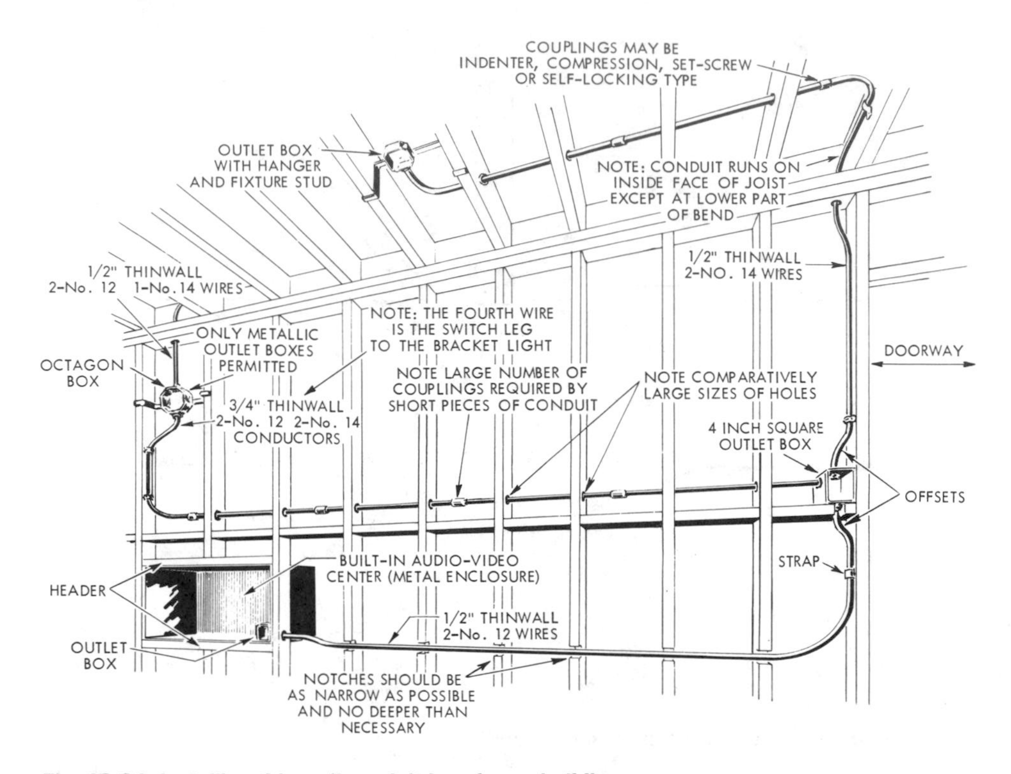

Fig. 13-36. Installing thin wall conduit in a frame building.

upon the structure, notching should be resorted to only where absolutely necessary. Notches should be as narrow as possible, and in no case deeper than 1/5 the stock of a bearing timber. A bearing timber supports floor joists or other weight.

Fig. 13-37 shows typical method of cutting small size EMT. Hacksaws with fine toothed blades should be used with thin wall. Blades should have 24 or 32 teeth per inch.

Fig. 13-38 illustrates the method of tightening compression coupling with two slip joint pliers. Note method of EMT entering outlet box from the top. This makes wire pulling and feeding easier.

Fig. 13-39 illustrates the method of approaching ceiling outbox with tubing in a building with a flat roof or a two story structure. Note how box bar hanger has been recessed into ceiling joist so that finished ceiling line will not be disrupted. This is not too important with plastered ceilings as it is with other types. Note also wiring has been pulled and

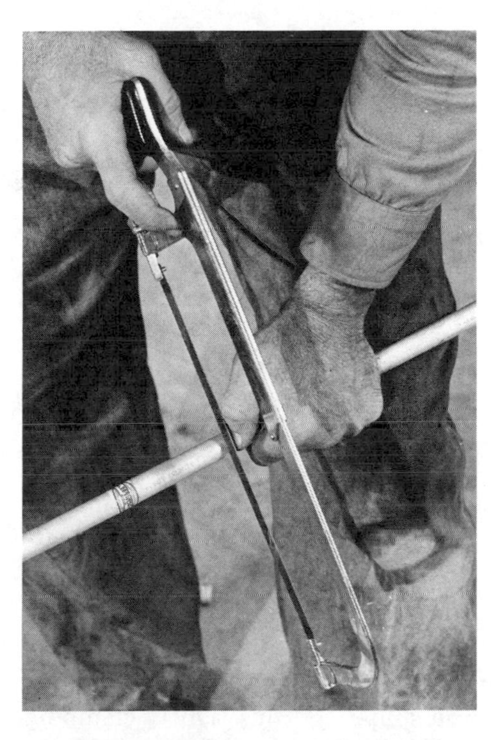

Fig. 13-37. Cutting EMT. (Republic Steel Corp.)

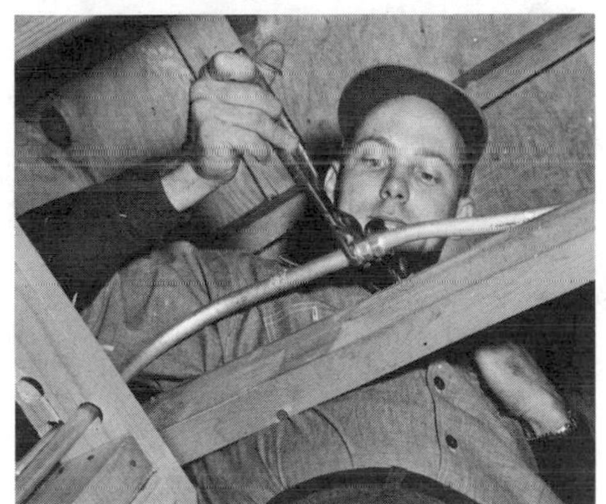

Fig. 13-38. Assembling and tightening compression type EMT coupling with two slip-joint pliers. (Republic Steel Co.)

Fig. 13-39. Method of feeding ceiling outlet box when there is a flat roof or two story building. (Republic Steel Corp.)

plaster ring has been installed, ready for covering of rough wiring by other crafts.

EMT Fittings

Thin-wall conduit couplings and box connectors are of two general types: Those for use anywhere that thin-wall conduit is permitted, commonly called *watertight;* and those permitted in dry places only.

Watertight Compression Couplings. Watertight couplings consist of a short body which slips on the tube, Fig. 13-40. Each end carries a male thread, and has a split compression ring in line with the axis of the bore. A gland screwed onto the threaded end of the body compresses the split ring, thus tightly clamping it to the tube.

Dry Location Couplings. Couplings for use in dry locations (which include those imbedded in concrete) are usually of the following types.

Setscrew Coupling and Box Connector. The setscrew connector is shown in Fig. 13-41. The tightened

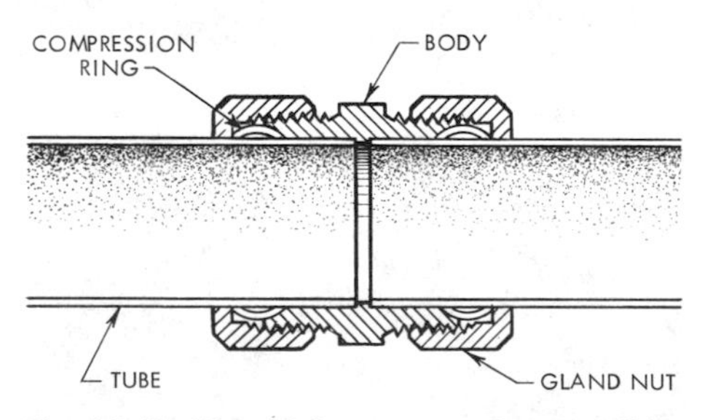

Fig. 13-40. Watertight compression coupling.

setscrew forms a slight dent in the tube. A coupling has two setscrews, one for each tube. Sizes 2½″ to 4″ generally have four set screws. The screw, since it is under pressure, forms a good metallic ground connection from one tube to another. Set screw connectors and couplings are used with rigid conduit also in many geographical locations.

Indenter Coupling. Indenter type couplings and connectors, Fig. 13-42, are slipped over the end of the tube, and a special tool is used to exert tremendous pressure on opposite sides of the connector and tube. If the indenter tool is moved a quarter turn, two more indentations are produced thus forming a tight connection between tube and fitting. An illustration of the indenting tool is shown in Fig. 13-43. Some indenter tools make four impressions at once.

Self-locking Coupling and Connector. Self-locking couplings and connectors are similar in construction to the watertight ones, Fig. 13-40, without the tightening nut. The compression ring on a self-locking coupling, however, is permanently assembled in a non-thread fitting. When connectors or couplings are tapped onto the end of the electrical metallic tubing the compression ring grips the tube securely. A conduit bushing should be put on the connector before tapping it on only to protect the threads.

Connectors. At one end connec-

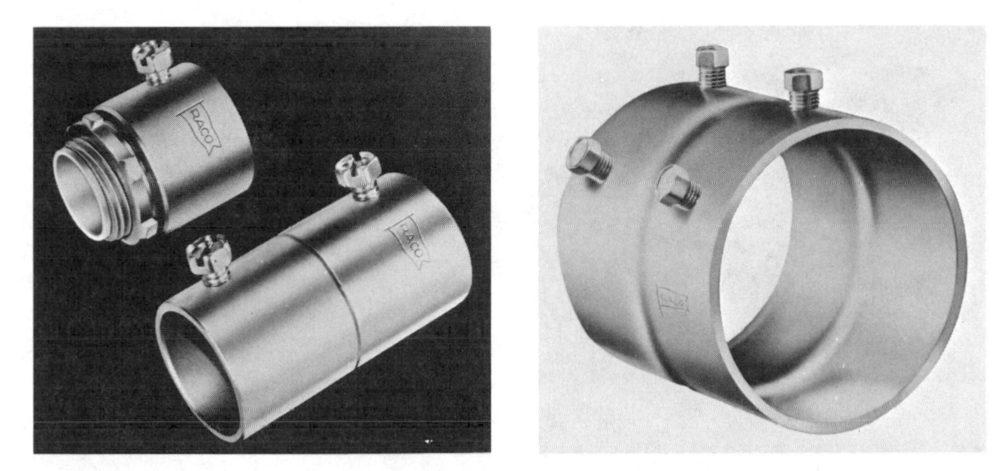

Fig. 13-41. Set screw connector and couplings. Left: EMT set screw box connector and EMT set screw coupling. Right: EMT set screw coupling for largest size tubing. (All Steel Equipment Co.)

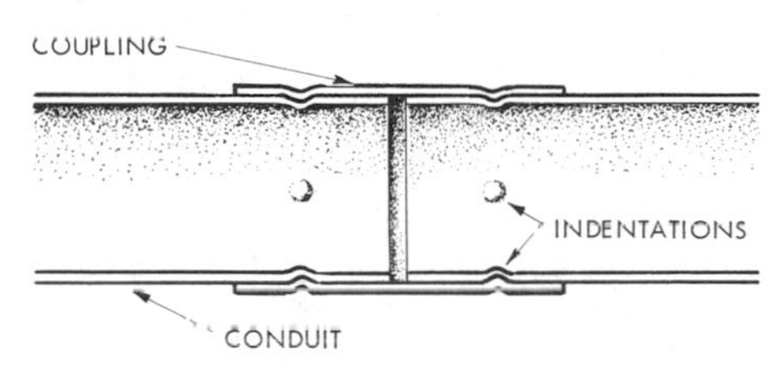

Fig. 13-42. Cross section of indented coupling.

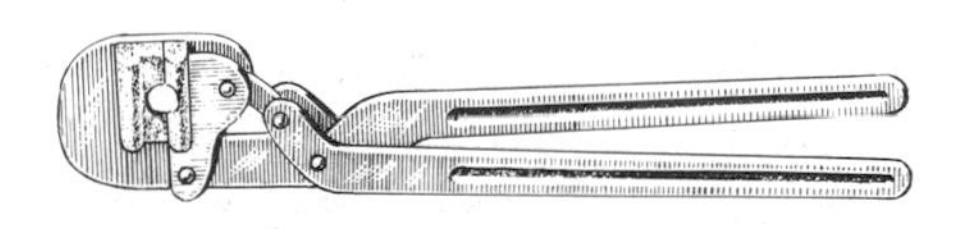

Fig. 13-43. Indenting tool. (Nye Tool & Machine Works)

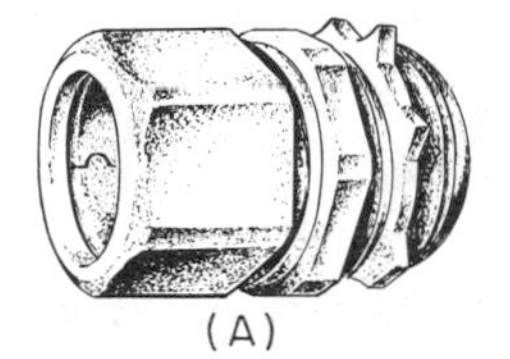

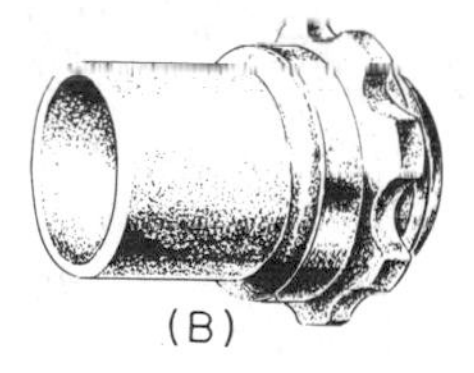

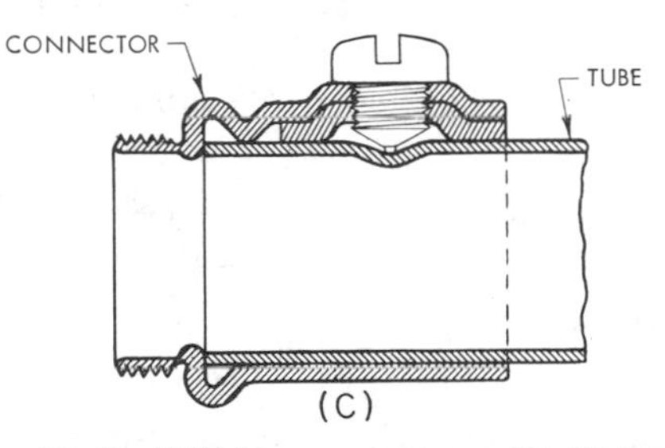

Fig. 13-44. EMT box connectors. (A) Watertight compression type, (B) indenter type, and (C) EMT set screw box connector.

Fig. 13-45. EMT compression connector mounting to 3 gang switch box. (All Steel Equipment Inc.)

tors have a body which slips over the thin-wall tubing, at the other end a short male-threaded extension, Fig. 13-44, for insertion into the outlet box. A locknut screwed onto this thread clamps box and connector together. Connectors are made in compression, setscrew, indenting, self-threading, or self-locking types —the same as couplings. Fig. 13-44B is indented with the EMT using the indenting tool. Fig. 13-44C shows how the setscrew locks the connector to the EMT. Fig. 13-45 shows an EMT compression connector mounted to basement, surface 3-gang switch box with 3-gang switch cover.

Connectors are also used when it is necessary to secure thin-wall conduit to a fitting which has a threaded hub, commonly used with rigid conduit.

Flexible Conduit: Installation Methods

From the electrical floor plans the *position* of all electrical outlets and switches are normally laid out on the studs, ceiling outlets on the floor. Locating a ceiling outlet on the floor with crayon informs the electrician that a ceiling outlet is required in this particular room and not a bracket light on a wall or a switched plug receptacle for table or floor lamps. It is easier to transfer this outlet location from the floor plan to the floor, rather than to a ceiling joist. It will be the responsibility of the electrician who mounts the outlet boxes to establish the exact location of the box. (See Chapter 4, "Methods of Wiring Non-Metallic

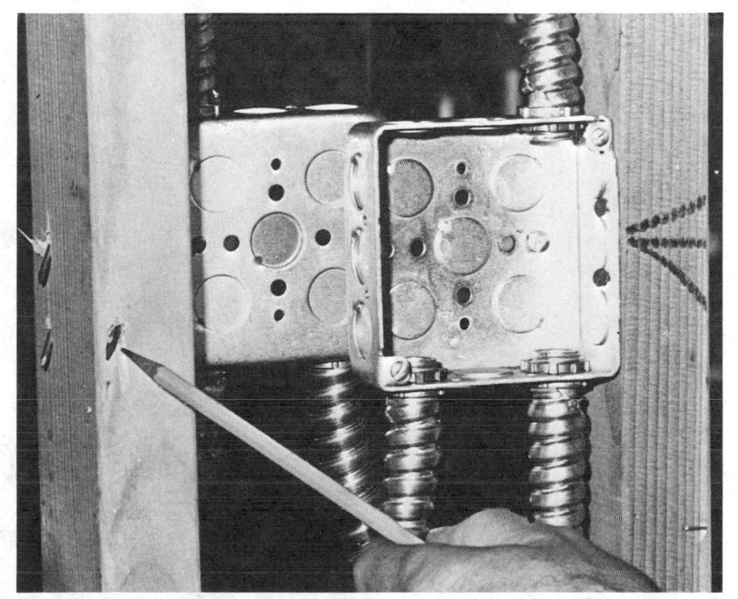

Fig. 13-46. Note crayon marked arrow on stud to the right of box. This locates the box. Note also the three nails bent over when installing 4 square boxes. (Los Angeles Trade-Technical College)

Sheathed Cable," for methods of locating room centers.) This process of mounting boxes is called "boxing." Boxing could be done by the "layout man" or by a different wireman, depending upon what production techniques are used.

Marks are generally put on with the lumber crayon so they can be easily seen. The wireman may have to measure the scale on the floor plan to transfer the location of a switch or outlet to the building frame. Note crayon marked arrow in Fig. 13-46, right side of the 4-square box on stud. Arrows are generally marked on studs pointing in the direction on which the side of the stud the box will be mounted, and normally the center of elevation of the box mounting is located. Sometimes a symbol very like the architectural symbol will be marked. For example, "SS" for two switches, or a "double cross" for convenience outlets.

Twelve inches, or hammer handle lengths near 12″, are used for convenience outlet elevations from the floor. Hammer handles are conveniently used for measuring height and spotting of convenience box before nailing to a stud, as seen in Fig. 13-47. Wooden handles, if too long, are cut for this gauging purpose.

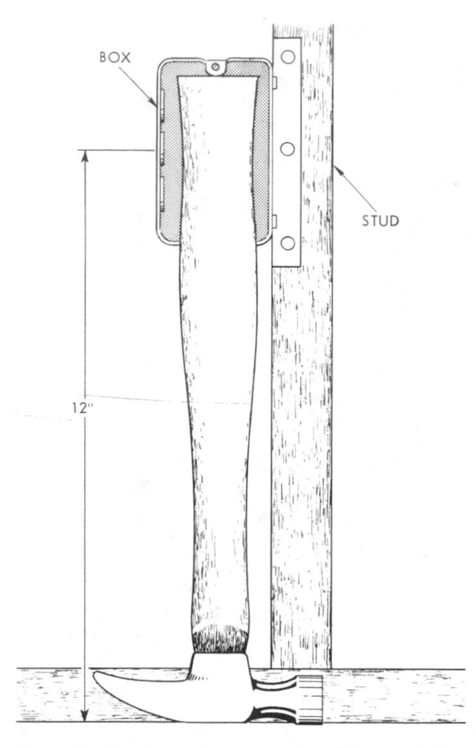

Fig. 13-47. Using hammer handle length for uniform installation of convenience outlets.

Fig. 13-48. Holes drilled so as not to weaken structural members. (Los Angeles Trade-Technical College)

The next step in the procedure is to mount the outlet boxes (drilling the frame could take place here, this is optional). It requires two eight penny nails driven in the side and one sixteen penny nail driven in diagonally. Boxes should be mounted flush with the face of the stud so normal half inch deep plaster rings will fit. Bracket boxes will be automatically set back. Care should be taken to clinch over the nails that protrude through the other side of the stud when mounting boxes (see Fig. 13-46). This is for safety and also to secure the box better. Distribution panel cabinets are also mounted at the same time as the outlet boxes (for branch circuits) are mounted. This, of course, is necessary in wiring, since it is important to know the height to cut the flexible conduit or raceway for panel feeders.

The next step would be to bore the holes through the studs in order to pull in the flexible conduit. When drilling structural members such as a joist it is necessary to drill as near the center as possible in order not to weaken them. See Fig. 13-48.

Going through drilled holes, 16 penny nails are usually driven at supporting intervals (specified by the code) into the stud and bent around the flexible conduit holding it tightly as shown in Fig. 13-49. Here we see that the flex is always kept in a smooth run when it is going from box to box or when entering the box. "S" bends and sharp kinks should be avoided. This will also make the pulling of the wire easier. It should be remembered that wire must be pulled into the flexible conduit; therefore, sweeping bends should be installed as much as possible. Even though the holes are drilled in a smooth workman-like manner, a wireman pulling flexible conduit could cross around a stud too sharply. It is also possible that he may break the flex with a driven

Fig. 13-49. Bending nails around flex at hole. (Los Angeles Trade-Technical College)

nail support. Care must be exercised.

The code requires that the flex be supported. Generally it should be supported within 12 inches on each side of every outlet box or fitting, and at intervals not exceeding 4½ feet on a run. Local codes should be checked for possible changes. *Nail straps* are commonly used. (See Fig. 13-55, page 358.) These are a combination of a nail and strap and are driven with a hammer on the nail side of the strap. These are different than the two hole straps that are used in many installations of rigid or electrical metallic tubing. Nails fit better here and are cheaper. Nail straps are used in places other than when passing through bored holes. For a change of direction, in flex at holes, and in straight runs, it is a good practice to use 16 penny nails driven over at an angle to secure the flex, rather than a nail strap.

When cutting the flex it is held at the top by either hand. A foot is

Fig. 13-50. Method of entering and leaving switch boxes. (Los Angeles Trade-Technical College)

placed on the loose end and the flex is then placed across the knee. The sawing is done with the free hand. Only *one* ribbon is sawed through and broken loose with both hands in a counter-clockwise manner similar to armored cable. (See Fig. 5-9, page 96.) The burr from the cut can be removed with plier jaws.

Note how the flexible conduit enters and leaves switch boxes. They normally connect to the top and the bottom, not on the sides or back. See Fig. 13-50. The connectors shown screw inside the flex spiral steel ribbon for locking. These have setscrews. There is no burr or sharp edge problem here. A flex connector, setscrew clamping type is shown in Fig. 13-51.

When the conduits are made up to the outlet boxes for safety, care should be taken in using a screwdriver to tighten the setscrew if these connectors (Fig. 13-51) are used. Many men, both right and left handed, have been injured by a slipping screwdriver when holding the flex or connector in the hand.

For surface-accessible or exposed-flex installations, a 90° box connector may be used for connecting to boxes. See Fig. 13-52. Wires must be pulled through flex and box *before* the 90° connector can be made up. Wires cannot be pulled through this connector. These are reasons for requiring the 90° connector exposed or accessible.

Fig. 13-53 illustrates how four inch square (4S) boxes are tied together when they are near back-to-back in the same bay stud partition. The flexible conduit is looped down from the bottom of one is properly secured, and is then ran up to the bottom of the other.

When the hole is not properly drilled, the flex may not go directly into the box. The sharp 90° turn or

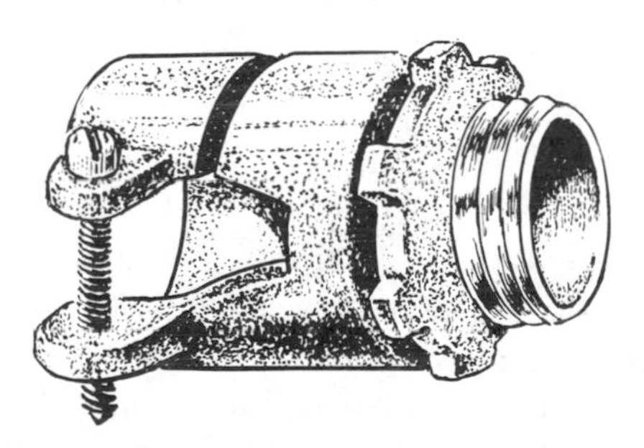

Fig. 13-51. Flexible conduit connector, setscrew clamping type.

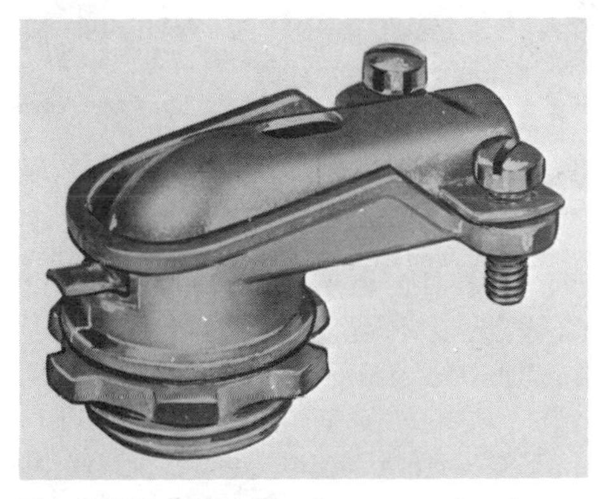

Fig. 13-52. A 90° flex box connector for exposed work. (All Steel Equipment Inc.)

Fig. 13-53. Method of tying boxes together. (Los Angeles Trade-Technical College)

bend of the flexible conduit going through a joist into an outlet box may cause problems. Note the looping approaches in Fig. 13-54. This makes wire pulling easier. Careful planning is the key word. Some outlets may have too many conduits entering the box for the number of wires they accommodate and may overload the box according to the code. Notice the different methods of approach to an outlet box and the connection of flexible conduit top and four sides of a 4″, square box,

Fig. 13-54. Multiple flex conduits entering an outlet box. (Los Angeles Trade-Technical College)

Fig. 13-54. It is much easier to correct any mistakes with a pencil on the plan before the outlet box is installed. The electrician should check the wire capacity of the outlet box in the Code book when in doubt.

Conduit Supports

Electrical codes have rigid rules regarding the supporting of all conduits. There are literally hundreds of wiring materials and devices and fittings used. The electrical apprentice, or learner, is urged to obtain different material catalogs from electrical supply houses for further study of identification and use, especially when going into industrial electrical construction. Fig. 13-55 illustrates the type of commonly used straps for nearly all conduits: (A) shows two hole strap, (B) one-hole malleable iron strap and (C) a nail strap.

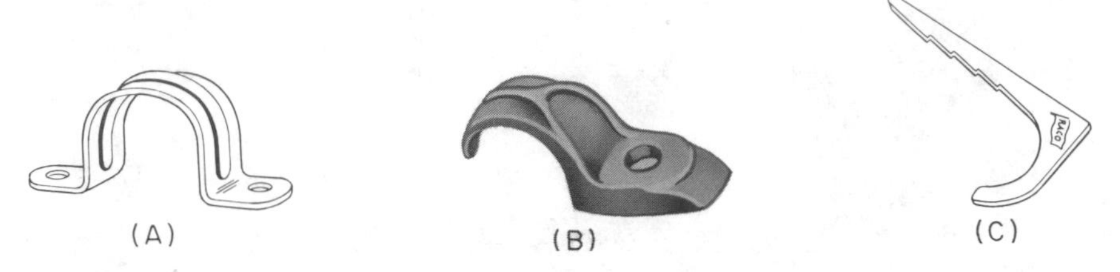

Fig. 13-55. Conduit straps. (A) Two hole, (B) one hole and (C) nail strap. (All Steel Equipment Inc.)

Fastening Devices

There are many devices on the market for securing objects to walls, ceilings, or floors of other than wood construction where nails and wood screws are used.

Toggle Bolts. *Toggle bolts* (Fig. 13-56) come in two general types but in many sizes. In one (Fig. 13-56, left) the toggle is hinged to the nut; in the other (Fig. 13-56, right) the toggle is hinged to the head of the bolt. The toggle arms are normally held in the outward position by springs within them. These devices can be used on hollow walls, ceilings, or similar locations. The first type, with toggle on the nut, is somewhat handier to use than the other one. To install it, drill or punch a hole large enough to pass the arms folded back as shown in Fig. 13-57A. Unscrew the toggle nut. Push the bolt through the hole in the base of the device to be mounted; then screw the nut on the bolt a few threads, so the arms of the toggle will fold back over the bolt. Next, push the toggle through the hole in the wall and screw up on the bolt until it is tight. Fig. 13-57B. It may be necessary to pull out on the bolt to hold the toggle arms against the inner surface of the wall, because that is the only way to keep it from turning.

If more than one toggle bolt is used to hold the device, all must be

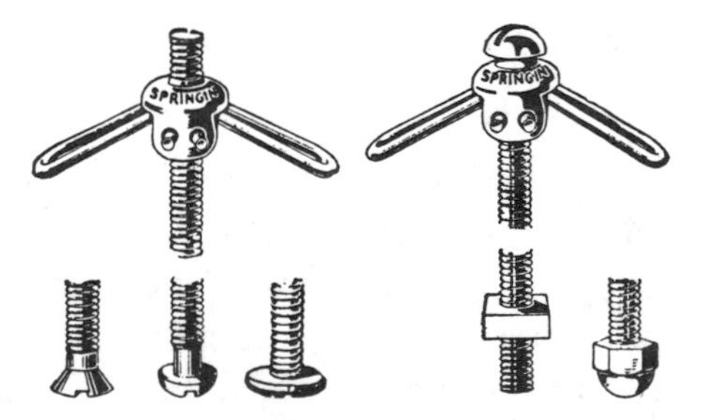

Fig. 13-56. Toggle bolts with different types of screw heads. (Star Expansion Bolt Co.)

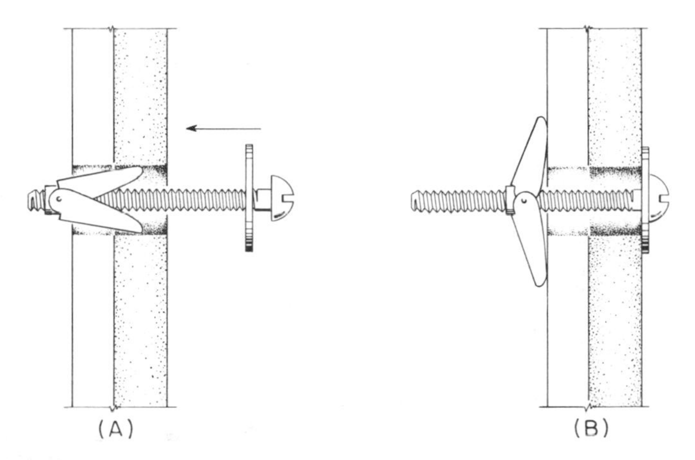

Fig. 13-57. Mounting on a lath and plaster wall using a toggle bolt.

inserted into their holes before the first is tightened. Greater care is required when installing the other type because the bolt may slide back into the wall and become lost.

Anchoring Devices

Where it is desired to mount an object on a dense surface instead of a hollow wall, some form of anchor must be employed. Anchoring devices are designed for application to a variety of surfaces, and experience will teach the best application of each.

Expansion Anchors. There are various types of expansion anchors. Fig. 13-58 shows one type anchor and calking tool and Fig. 13-59 shows another type expansion anchor. To install, drill a hole of the proper diameter and depth, insert the shell, and expand it by using a tool provided for this purpose, striking two or three moderate blows with a hammer. The smaller sizes are for machine screws only, but larger ones are made for lag screws and machine screws. Their holding power depends on the expansion of a lead sleeve. This expansion is caused by the wedge action of the conical nut, which is drawn into the sleeve by the threads of the screw.

Plug Types. The plug in Fig. 13-60 is made from jute, chemically treated and compressed, and is used with wood screws. The screw, the hole, and the plug to be used should all be the same size. The plug must be deep enough in the hole so the unthreaded shank of the screw does not reach it. To install, drill the hole, insert the plug, put in the screw and tighten it. The threads will force the jute fibers into small crevices in the

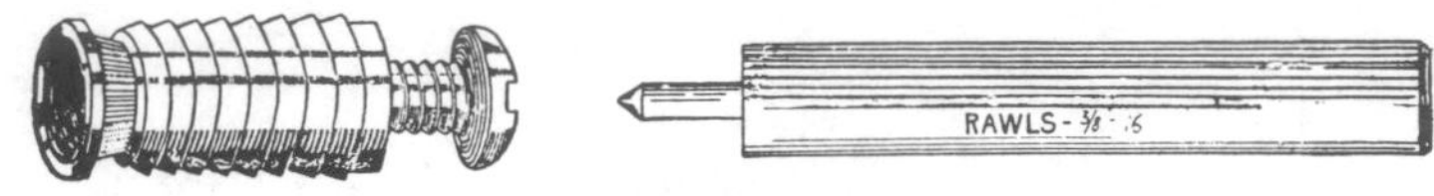

Fig. 13-58. Lead anchor and caulking tool. (Rawlplug Co.)

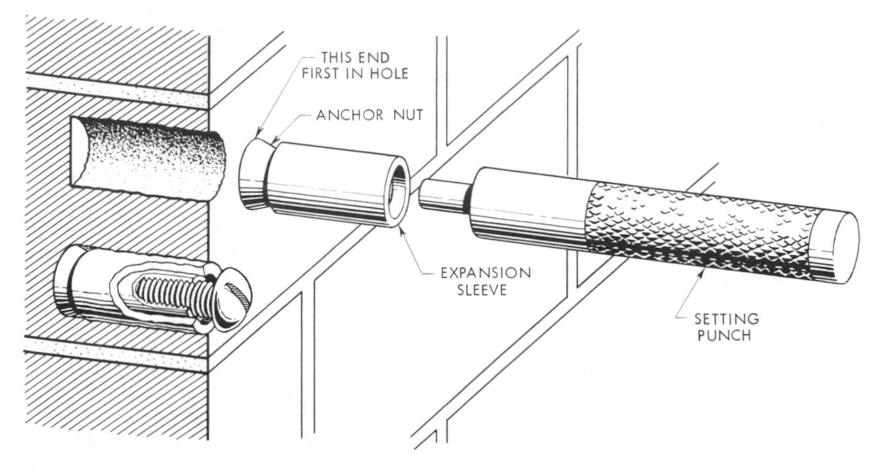

Fig. 13-59. Method of setting lead anchors.

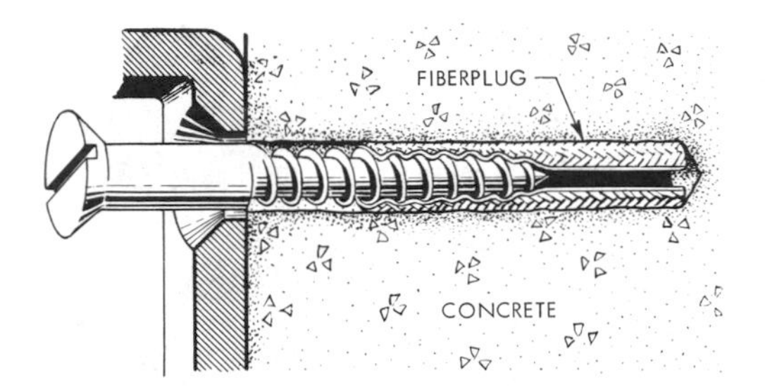

Fig. 13-60. Illustrating the use of the fiber plug anchor. (Rawlplug Co.)

wall of the hole, thereby providing a firm anchorage. The plug should fit the hole snugly. If the hole has been drilled too large, use the next larger size of screw and plug.

Plug type anchors are also made of plastic, see Fig. 13-61, and come in assorted sizes. They are unaffected by moisture, corrosion, shock or vibration. They are installed in a manner similar to the fiber plug. Sheet metal or wood screws are used

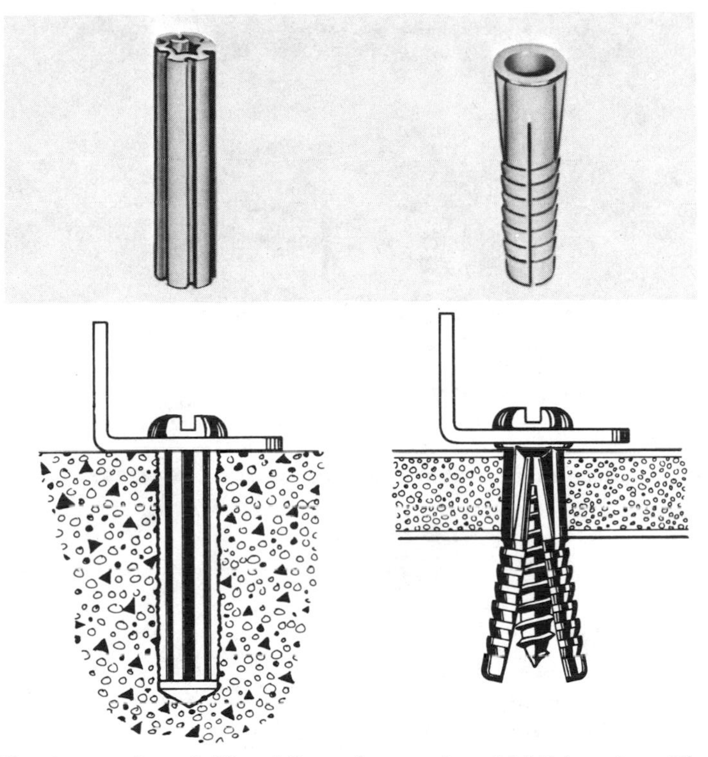

Fig. 13-61. Plug type anchors (left) and flange type anchors (right) have two different designs to handle various anchoring jobs. Illustrations show the holding action of each anchor. Anchors come in varying szes. (Ideal Industries, Inc.)

and make their own threads. Both can be used in any solid materials, such as concrete, brick, stone and tile. The plastic flange type (Fig. 13-61, right) can also be used in hollow walls and plaster.

Drive Plugs. A one-piece device for making attachments to masonry and kindred substances is shown in Fig. 13-62. This is known as a rawl drive, and it is installed by drilling a hole of the exact size of the drive, then inserting the rawl drive and

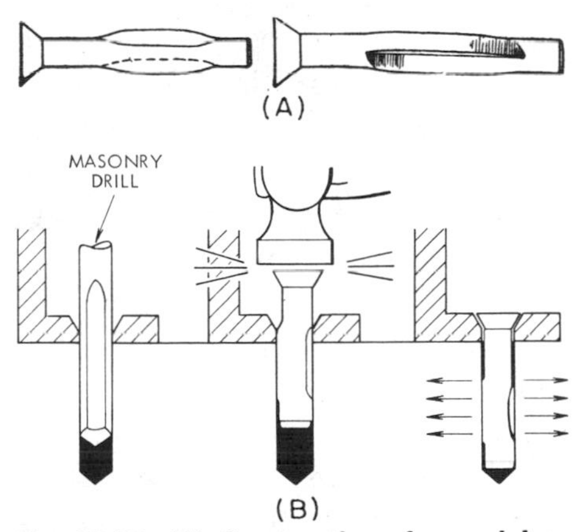

Fig. 13-62. (A) Construction of a rawlplug. (B) Installation method; first drill then drive in the rawlplug. (Rawlplug Co.)

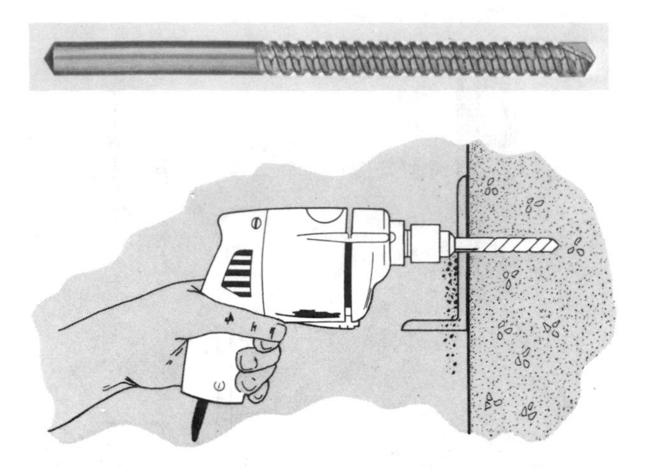

Fig. 13-63. Top: Carbide tipped masonry drill bit. Bottom: How masonry drill bit is used. (Ideal Industries, Inc.)

forcing it home with a hammer. The hole must be deep enough so the end of the drive does not reach bottom. These devices will not hold in crumbly, weak, or yielding substances and must not touch the bottom of the hole. They are available with round heads and with threaded studs. The stud type is meant for use in applications where the fixture might be moved.

Holes. Holes to accommodate anchors are commonly drilled in masonry with a carbide tipped *drill bit*, Fig. 13-63, top. This is used with a low number of revolutions per minute (RPM) electric drill motor. See Fig. 13-63, bottom. There are other types used by hand, with hammers and impact-power devices. The common *star drill* is shown in Fig. 13-64A (side view) and in Fig. 13-64B (looking at the point). These manual drills can be sharpened on a grinding wheel by following the original contour and

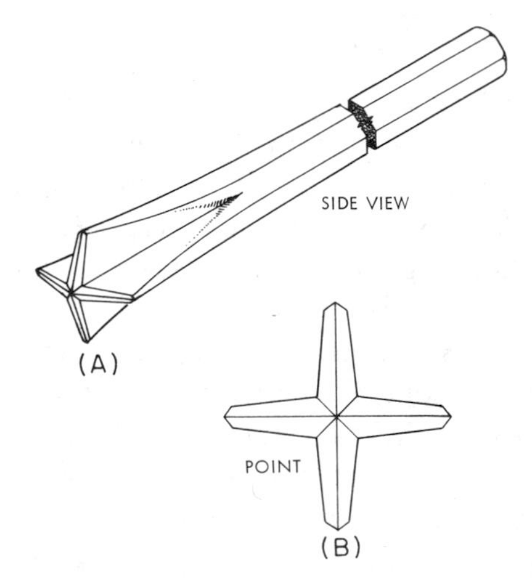

Fig. 13-64. Star 4-point masonry hand drill.

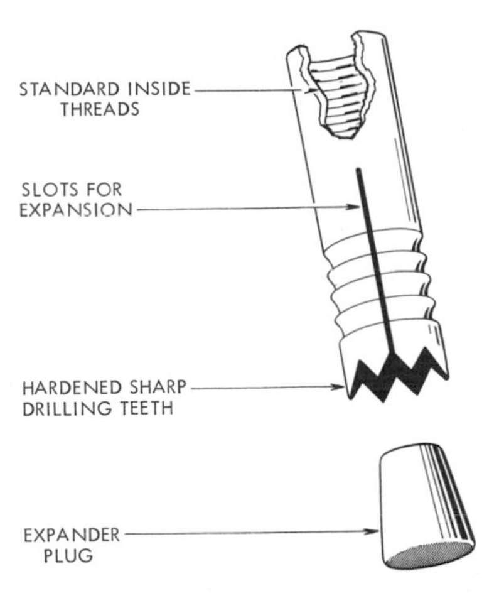

Fig. 13-65. Self drilling anchor.

the four flat sides (as shown in Fig. 13-64B). They may be used with power impact hammers or by hand with a claw hammer.

Self Drilling Anchors. Self drilling anchors for masonry are a combination drill and anchor. Each anchor bores its own hole with the use of a handle that screws into the anchor. See Fig. 13-65. The handle is then pounded with a hammer, drilling the hole with the sharp teeth of the anchor itself. Masonry dust should be removed now and then while drilling. Impact power hammers are also used with self-drilling anchors. See Fig. 13-66. When the hole is the proper depth and cleaned, the expander plug is wedged in the end of the anchor at the bottom of the hole by hammering on the driver as done in drilling. With the driver removed (either hand or impact tool) a standard cap screw or bolt may be installed to hold the desired fixture.

Homemade Anchorage. When the need arises for one or more stud anchorages in masonry and there are no suitable ready-made devices available for the purpose, they can be made as shown in Fig. 13-67. Obtain the required machine bolts. Drill holes into the masonry large enough to take the heads of the bolts and of such depth that the shoulder of the head is below the surface a distance of 1″ to 2½″ depending on the size of the bolt and load to be held. Undercut the walls of the holes and clean out the dust. Insert the head of a bolt, placing it at right angles to the surface of the masonry. Next, fill the hole with molten lead or solder and calk it well. If the masonry is cold, it is a good idea to

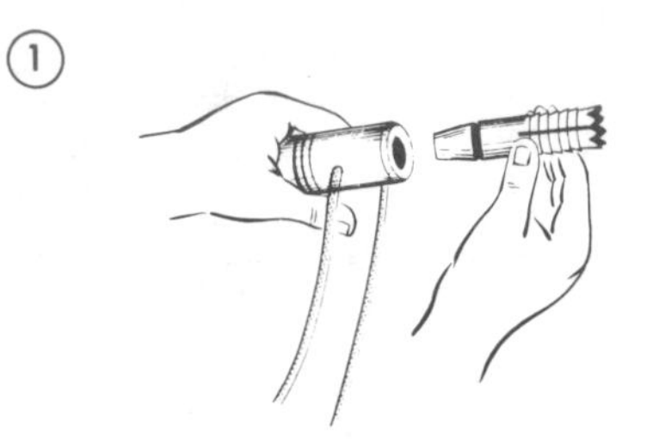

Insert tapered end of snap-off anchor into chuck head attached to any impact hammer.

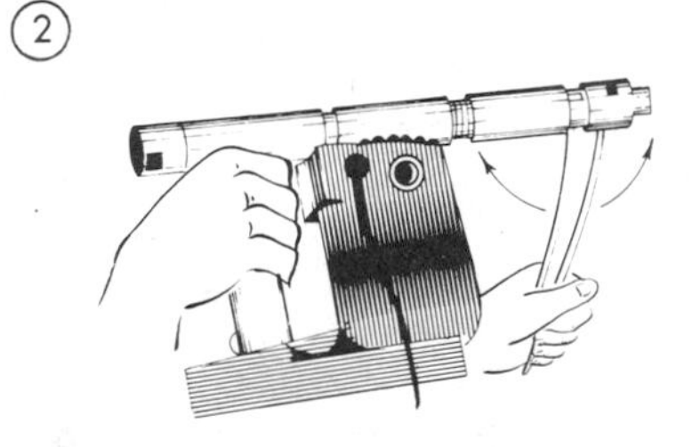

Operate impact hammer to drill into the concrete. Rotate chuck handle while drilling.

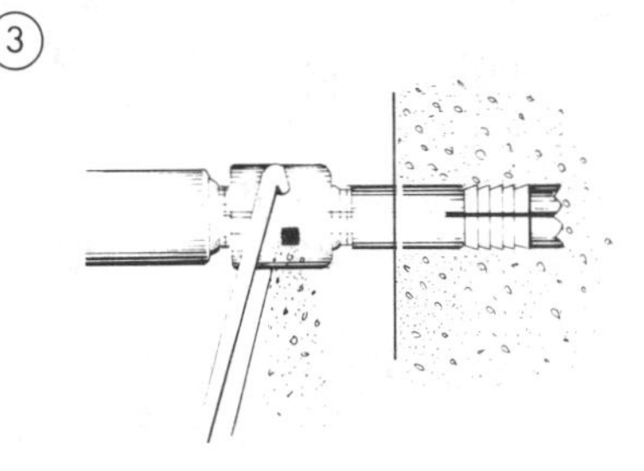

The drill is self-cleaning. Cuttings pass through the core and holes in the chuck head.

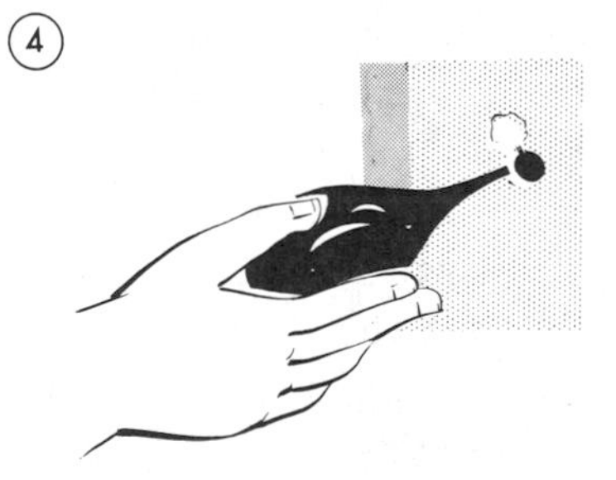

Withdraw the drill and remove grit and cuttings from the drill core and from the hole.

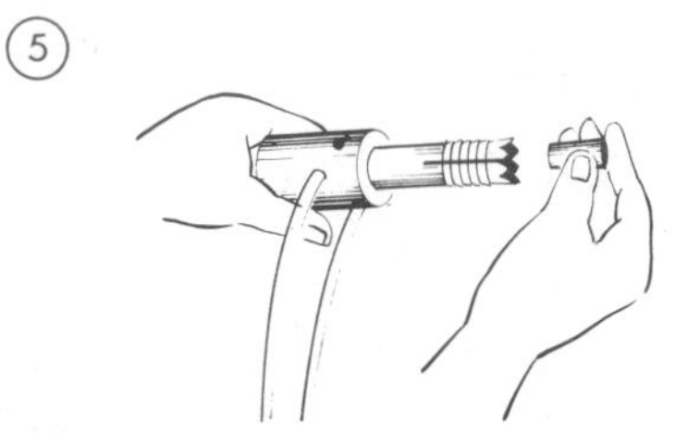

Insert hardened steel cone-shaped red expander plug in cutting end of drill.

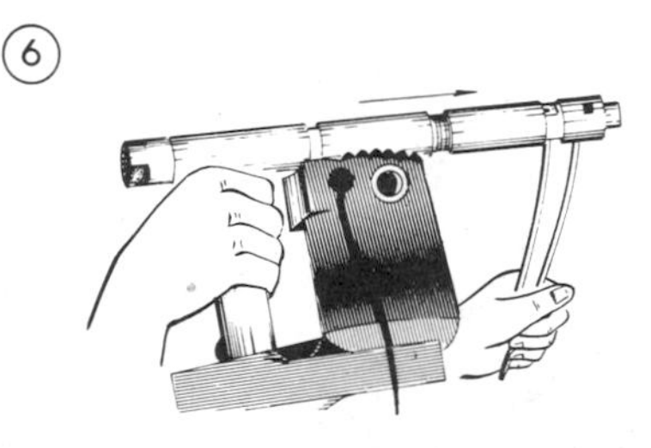

Reinsert the plugged drill in the hole and operate the hammer to expand anchor.

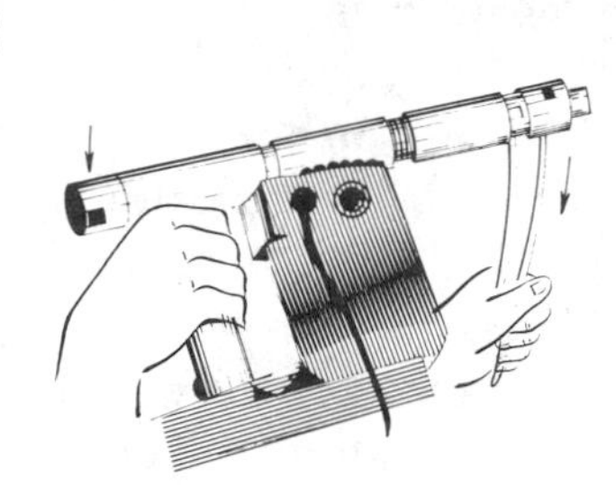

Snap off chucking end of anchor with a quick lateral strain on the hammer.

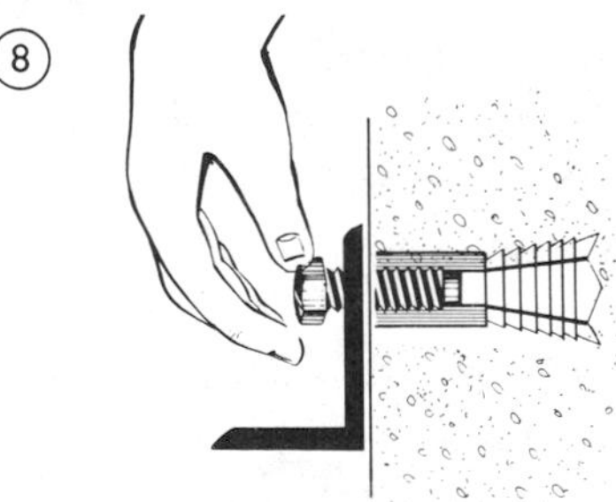

The anchor is now ready to serve as an internally threaded steel bolt hole to support any bolted object.

Fig. 13-66. Installation of snap-off type anchor. A wide selection of anchor sizes and lengths are available. (Phillips Drill Co.)

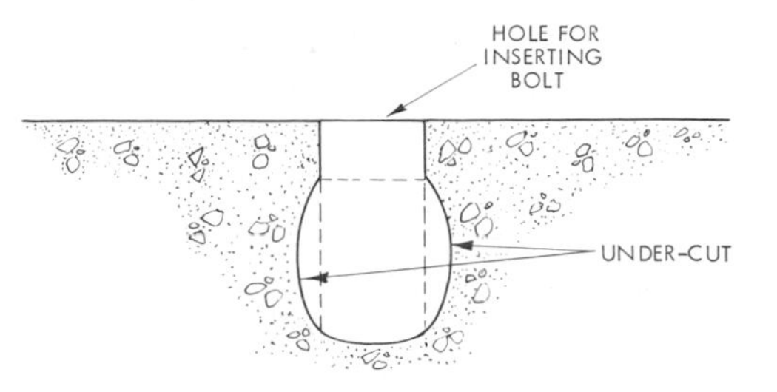

Fig. 13-67. Home-made anchorage.

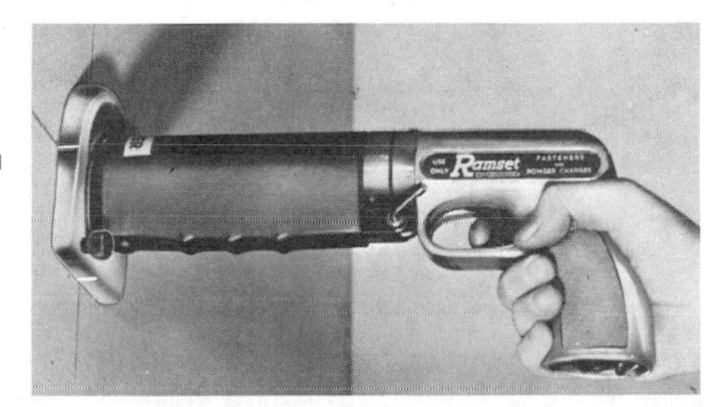

Fig. 13-68. Powder-driven fastener: position and fire. (Ramset Fastening System)

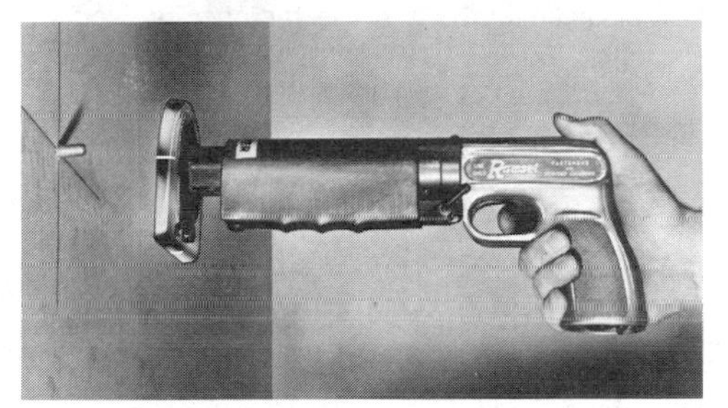

Fig. 13-69. Powder-driven fastener: pin is driven in place. (Ramset Fastening System)

warm the adjacent area so the lead does not chill before it flows into the small indentations. A washer placed against the head of the bolt before lead is poured will increase the holding power considerably.

Powder Driven Fastener

The powder driven fastener (Figs. 13-68 and 13-69) is a tool that fires a specially designed cartridge which provides the power to sink fasteners into a wide variety of construction materials. The depth of penetration can be controlled by a combination of adjustment and the use of charges of different power. Interchangeable barrels are used for different sizes of fastener. The powder actuated fastener is particularly suitable for

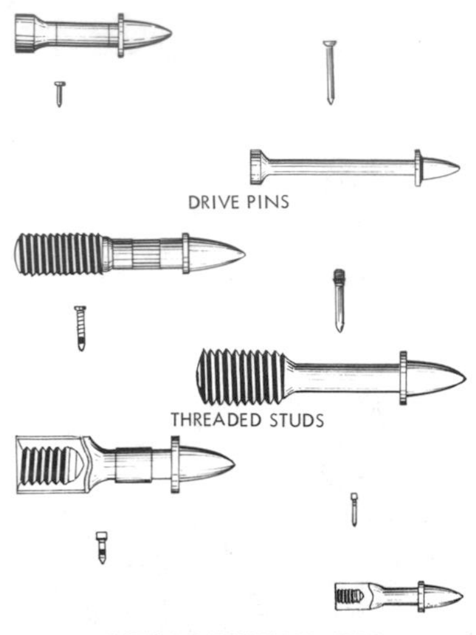

Fig. 13-70. Common type of powder driven fasteners; a large variety of sizes and lengths are available. Plastic tip keeps stud in place and is dissipated when the fastener is fired. (Practical Builder)

securing fasteners in concrete and steel, and can penetrate up to an inch of steel. Variously shaped *drive pins* or fasteners, may be used. See Fig. 13-70.

Powder actuated fasteners may be used extensively in concrete multiple-family dwellings. In using powder driven fasteners, be sure to follow all of the manufacturer's recommendations. Read and understand the instruction manual issued with each tool. Follow *all* of the safety precautions.

Because deaths have been caused by the misuse of this tool, the following lists some of the safety precautions.

SAFETY: POWDER DRIVEN FASTENERS

Have a permit and obtain authorization before using the powder driven fastener.

1. Be *thoroughly* familiar with the operating principles and instructions for the powder driven fastener. Follow *all* of the manufacturer's safety rules. (Most areas require certification for the use of this tool.)

2. Wear safety goggles and heavy gloves to protect against flying particles.

3. Do not use powder driven fasteners in an explosive or inflammable atmosphere.

4. Follow manufacturer's recommendations for firing into each different type of material. Do not fire into any material that can be nailed.

5. Determine if material is of sufficient density and thickness so that the fastener will not go completely through the material and do injury on the other side. Do *not*, for example, fire into concrete less than 2 inches thick.

6. Use the right type and size of drive pin for the job. Consult manufacturer's specifications.

7. Always use an alignment guide for firing through previously prepared holes in steel.

8. Do *not* use a fastener closer than ½″ from edge of steel or 3″ from the edge of concrete.

9. Before loading the driver, be sure the cartridge is of proper pow-

der load. (If the powder load is too great, the drive pin may go through the material and cause injury.) Select and position the power cartridge according to manufacturer's recommendations. The stronger the powder charge, the stronger the force of the explosion. Learn the color code associated with the cartridge.

10. *Always* keep the powder driven fastener directed toward the work area. *Never* point away from the work area or in general direction of anyone.

11. Keep your full attention focused on the work.

12. When loaded, position the gun and fire immediately — never leave lying around. Unload the tool if it is not possible to fire immediately.

13. When ready to fire, place the protective shield *evenly* against the work surface, press hard, and pull the trigger. (If the guard is not pressed against the work surface evenly, the driver may not fire.)

14. If the powder cartridge does not fire, keep the safety shield firmly pressed against the work surface for at least 30 seconds, then remove and safely dispose of the powder cartridge per manufacturer's recommendations.

15. Follow local and states codes for storing the powder cartridges. Store the powder driven fasteners in a designated location so that their use may be carefully controlled.

Masonry Construction Wiring

Fig. 13-71 illustrates a cut-away of residential masonry construction. The most acceptable method of wiring a building of this type is for the wireman to work closely with the mason laying the blocks.

Switches and outlets are located on the job from the scaled floor plan. When the construction blocks reach the convenience outlet elevation, boxes are made up as shown in Fig. 13-72. The figure on the left shows a raised plaster ring or box device cover, manufactured for this purpose and for tile finished surfaces. Fig. 13-72, right, requires additional mortar grouting. Galvanized EMT is used here with watertight compression box connectors. The wireman can determine from the floor plan how many of these outlet units he may assemble. All these units (and others) may be prefabricated so that they are ready when the block mason is ready.

Fig. 13-73 shows a masonry through-the-wall box, called a "through box" in the trade. They are also available for wood frame construction, with different lengths.

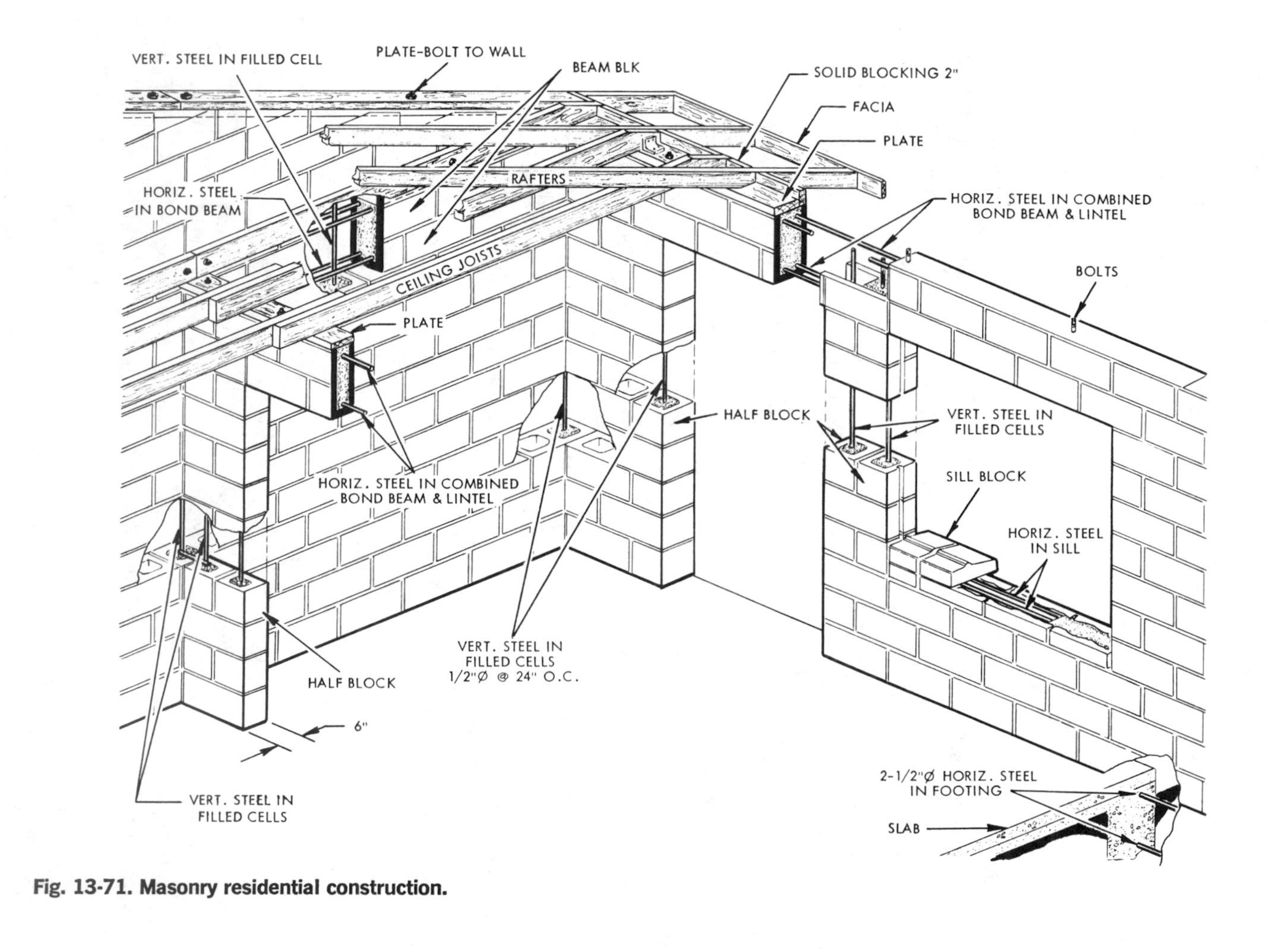

Fig. 13-71. Masonry residential construction.

Fig. 13-72. Left: Method of installing single outlet or switch box in masonry construction using masonry or square cornered plaster rings. Right: Alternate method of installing single outlet or switch box in masonry construction using general purpose plaster rings. (All Steel Equipment Inc.)

Fig. 13-73. A masonry through-the-wall box. (All Steel Equipment Inc.)

This box locates an outlet, or switch, on each side of the wall. It eliminates the need for 4″ square boxes and raised plaster rings, or device box covers, as shown in Fig. 13-72, coupled together. This box is installed with rigid galvanized conduit here. Note ½″ and ¾″ concentric knockouts available.

Fig. 13-74 shows the use of a "4-S" extension ring installed to bring the box to the masonry surface. Here a flat, blank cover will be installed as the box is being used as a wire pull box only.

Fig. 13-75 illustrates the use of a

Fig. 13-74. Use of box extension ring in masonry construction. (Minnesota Mining and Mfg. Co.)

Fig. 13-75. Three gang switch box installed with watertight EMT connector, flush mounting, in concrete block construction. (All Steel Equipment Inc.)

masonry box that needs no extension or deep plaster ring to bring it to the surface. Note the EMT installed in the rear right knock-out of this three gang, masonry switch box. The knock-outs fit the void in blocks, where conduits are installed. Full length conduits are not used during the laying of blocks. The mason would have to slip the blocks over

10 feet lengths and this is impractical for him. Lengths may be coupled together as the job progresses or full lengths and boxes may be "fished" through the block voids to precut holes.

When conduits are extended beyond the top of the finished walls, they are bent into the attic space for coupling together and completing the runs to pull boxes or to ceiling outlets. See Fig. 13-71.

Conduits with 90° bends turned up for every convenience outlet location may be installed when the floor slab is poured. This means coupling many boxes together under the structure instead of in the attic as mentioned previously. The bends are located "looking-up" to meet the outlet boxes. They may be held in place by mounting outlet boxes on top of the bends at estimated positions. Locating proper height of the boxes often presents a problem. Block mortar thicknesses vary. The method of installing conduits in the slab and up to outlets and switches does not always allow a box to be installed without cutting too much block.

EMT may be cut on the spot while blocks are going up, assuming they are too long, to lower an outlet so only one block need be cut, thus locating box elevations uniformly. It is most difficult to alter heights of rigid conduit stub-ups after the concrete is set. It can be understood that it is easier to drop a box on the end of a length of EMT down a finished block void to a precut outlet hole.

An understanding between the crafts and the department of building and safety should be sought as to questionable methods and common problems.

Wire Capacity of Conduit

The number of wires generally permitted to be pulled through conduits is the same for rigid and thin-wall conduit, size for size. It is based on the ratio of the combined cross-sectional area of the wires to the cross-sectional area of the conduit opening. Fig. 13-76 shows the actual size of thin-wall conduit (solid black circle) and of rigid conduit (outer and inner circles), and number of wires of different sizes installed.

Tables are available in code books to determine maximum numbers of conductors that may be installed in conduits. Wire insulation types used must be known in order to select the proper table column. Take care to use the correct table for new work, rewiring, and wire type.

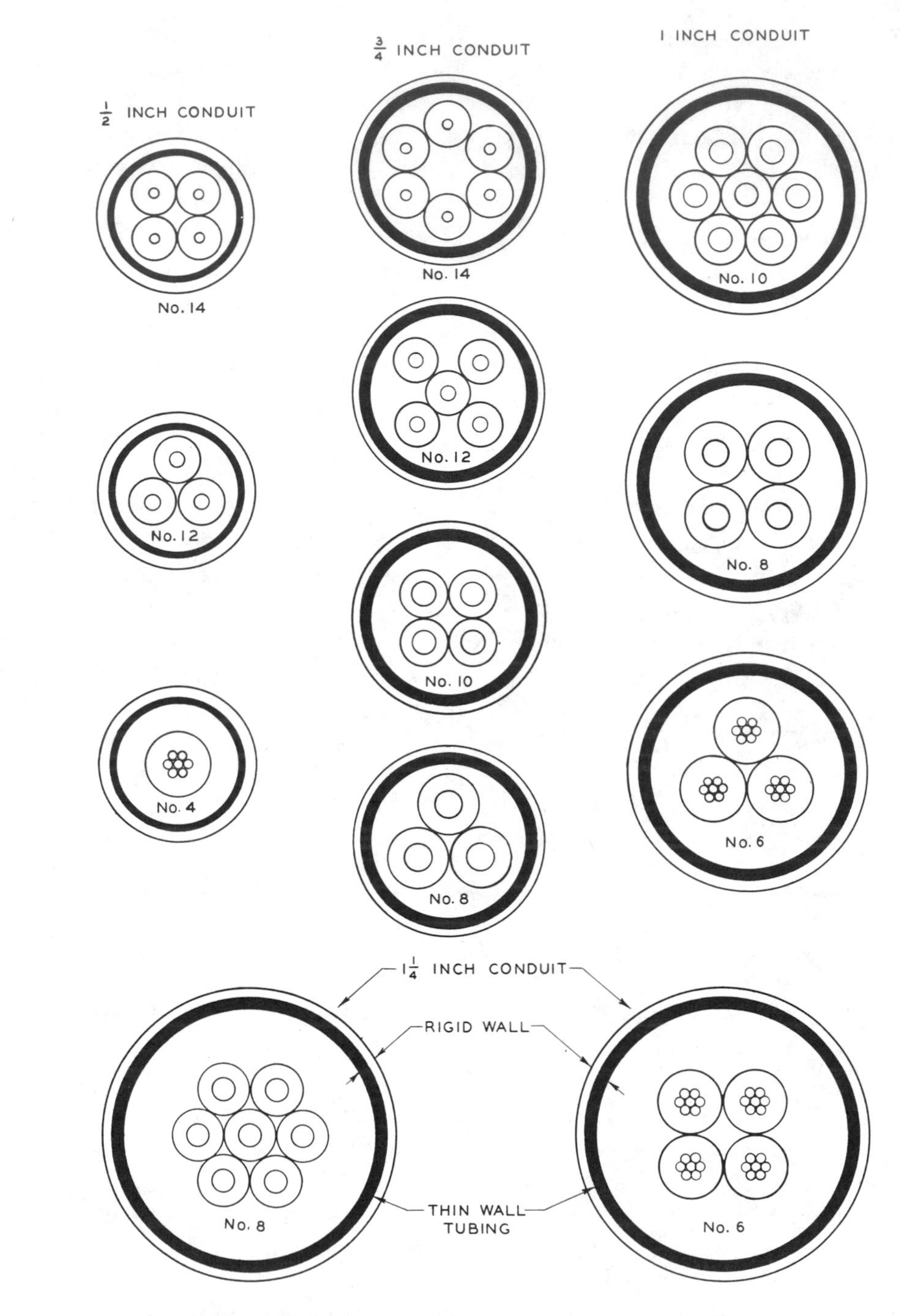

Fig. 13-76. Actural conduit sizes.

It is also important to know the maximum number of conduit bends allowed between any two boxes. Generally the maximum has been the equivalent of four 90° bends between boxes. The radius of any conduit bend should be made as large as possible but code minimums should be known.

Wire Pulling and Equipment

After conduit and outlet boxes are in place and other rough work finished, wires can be pulled in. On short runs, small conductors may be pushed through from outlet to outlet. In most cases, however, it saves time to use a fish tape.

Fish tape, illustrated in Fig. 13-77A, is made of tempered but flexible steel, and is supplied in coils of 50′, 100′ and 200′. To avoid twisting and tangling, it should be kept in a reel. The fish steel may be obtained in thicknesses of .030″, .045″ and .060″. Standard widths are: ⅛″, 3/16″ and ¼″. The .060″ x ⅛″ is the most popular size for general wiring.

The tape often comes with a hook on the end, as shown in Fig. 13-77C. If it has no hook, one can be readily made with a pair of pliers. If bent too sharply with a pair of pliers the fish tape will break because of its temper. Temper can be removed if necessary by heating the steel rod with a torch and letting it cool. Today, some form of leader is usually fastened to the end of the tape. Two such devices are shown in Fig. 13-77B and Fig. 13-77D: the smooth ball and the flexible spring. Either type makes fishing comparatively easy. They have rendered nearly obsolete the older laborious trial-and-error method for getting a fish steel through a difficult run of conduit. Journeymen wiremen when pulling wire in flexible conduit jobs generally have a leader on each end of their 50 foot fish tape (which is not in a reel holder). This saves time in searching for the proper end—either one will do for fishing conduits.

When a fish tape is inserted through a conduit and reaches an outlet, wires may be connected to the end of the hook or leader in a fashion shown in Fig. 13-78. Fig. 13-78A illustrates how wires are skinned and hooked. Wire hooks are closed tightly with pliers. In many cases the wire hooks are taped closed where the pulls are expected to be difficult. Fig. 13-77B suggests a method for pulling larger wire, such as #8. The connections are staggered so that they will pull easier through bends.

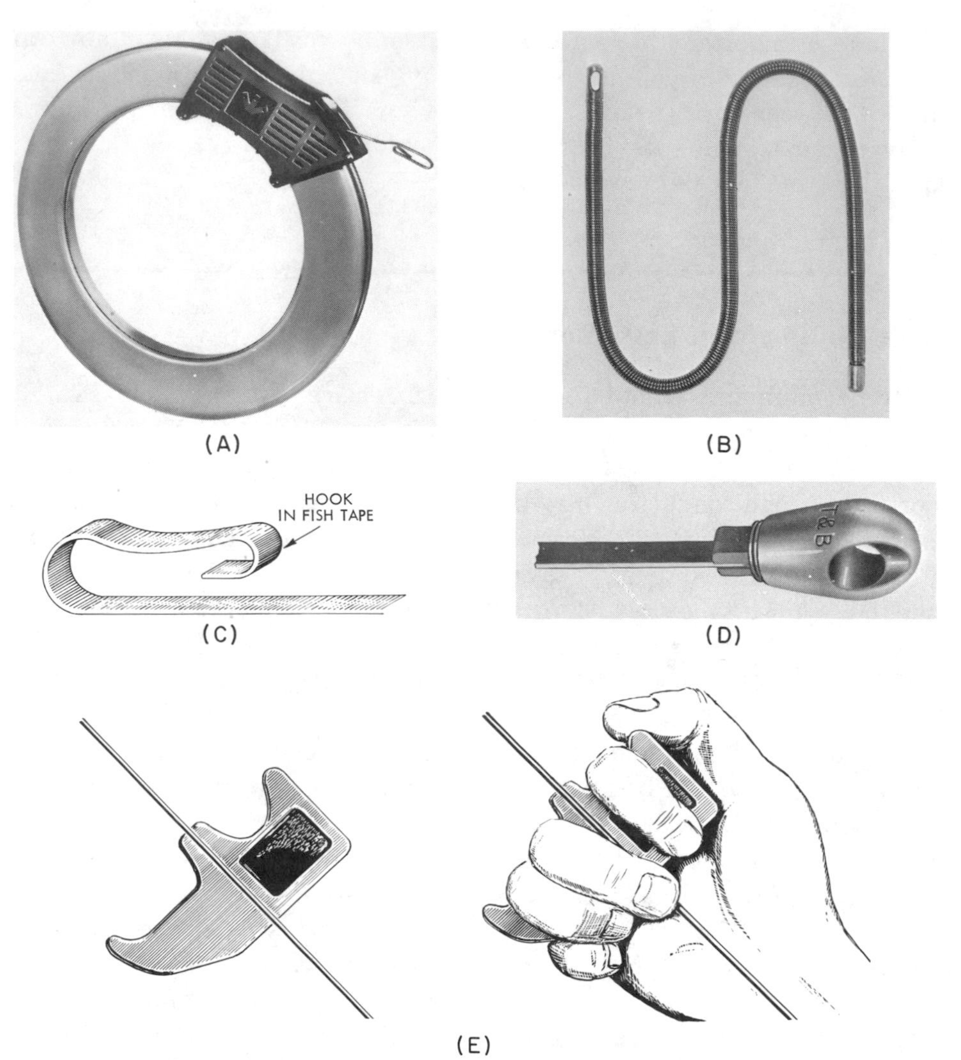

Fig. 13-77. Fish tape and accessory items. (A) Fish tape. (B) Fish tape holder. (C) Fish tape with hook. (D) Smooth ball leader. (E) Fish tape puller and method of use. (A and B: Ideal Industries, Inc. D: Thomas & Betts Co.)

Staggering is done by attaching one wire to the fish tape or "snake" (as it is often called) then attaching the second wire a short distance behind this to the bare copper conductor of the first. The third wire is, in turn, attached to the second wire, and so on. Wire ends on hooks or wraps

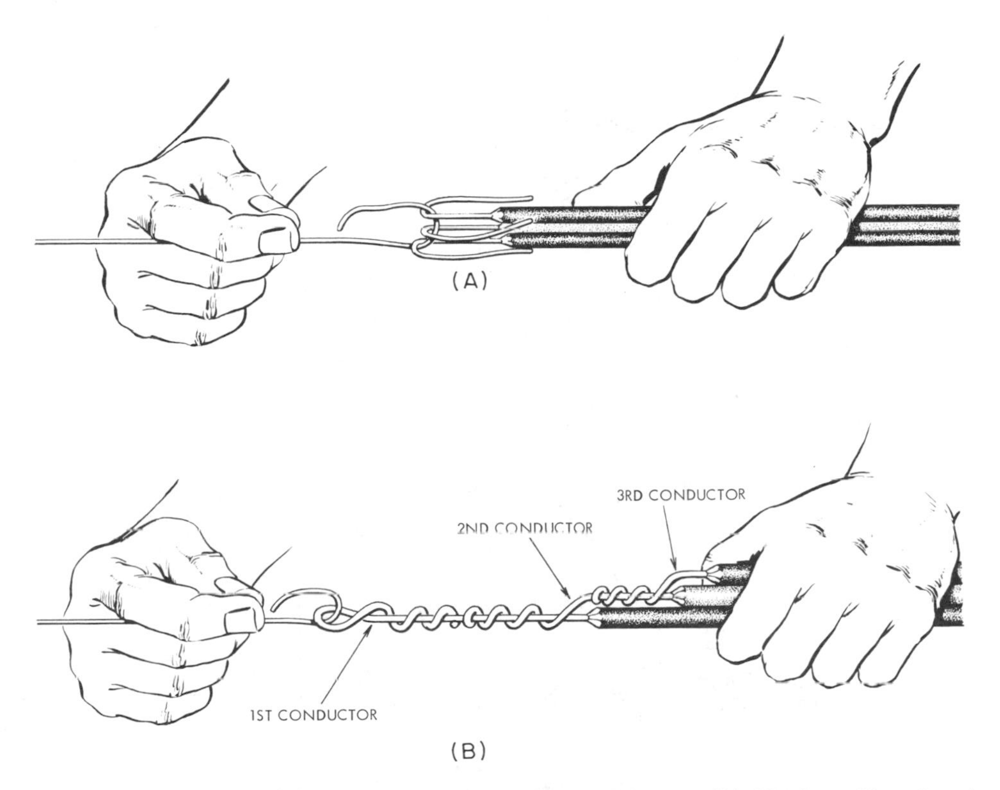

Fig. 13-78. Methods of making up wire pulling ends to fish tape. (A) All wires skinned and hooked to fish tape. (B) Larger wires are tied to each other and staggered.

should be made up so that they are not allowed to catch on couplings or connectors.

Fig. 13-79 shows a method of feeding wire directly into the conduit from the center of wire boxes. Another method is shown in Fig. 13-80. Using a manufactured reel cart the coils of wire are removed from the box and placed on reels as shown. Wire is withdrawn from the outside of the coil when using reels. Wire can be more easily moved by using the reel cart.

When several conductors must be fed into a conduit they should be kept parallel and straight, and free from kinks, bends and crossovers. Wires that are permitted to cross each other form a bulge that makes pulling more difficult, especially around bends, saddles or offsets. Whenever possible, the conductors should be fed downward, like from the second floor to first, so the weight of the wires will help rather than retard.

Wire pulling can be made easier

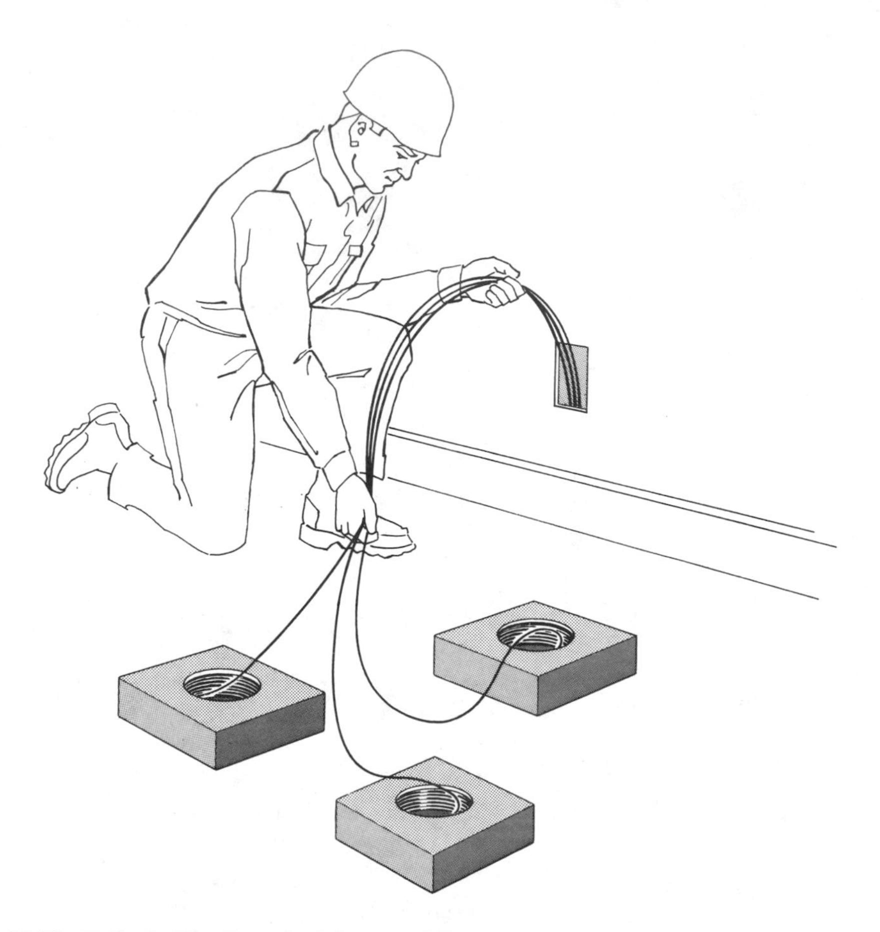

Fig. 13-79. Method of feeding wire into a conduit.

by one of the numerous pulling devices, one of which is illustrated in Fig. 13-77E. The back side, or insulation crushing point, on a pair of side cutting pliers may be used, well gripped, across the fish tape for a good pulling handle.

If the conduit is long, or complicated with bends, so that the conductors tend to stick, the wires may be coated with approved wire pulling compounds. They may be coated with talc, soapstone, soap, or other non-corrosive substance to lubricate them as they enter the conduit.

Only 6 inches of wire need be pulled out of outlet boxes for makeup purposes to avoid unnecessary splices (a decided advantage over cable wiring). Enough wire should be pulled out to reach the following outlet if near enough. See Fig. 13-81.

Fig. 13-80. Using a reel cart for feed wire. (Greenlee Tool Co.)

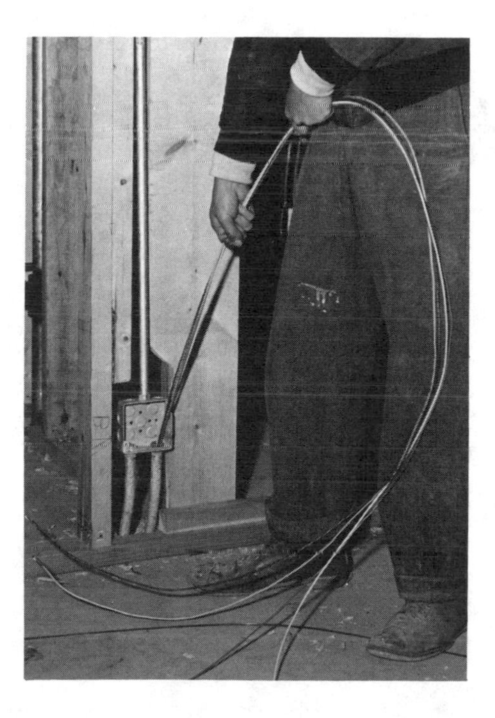

Fig. 13-81. Pulling out enough wire for next outlet to avoid a splice. (Boron Electric)

Care should be taken, however, to leave wire hooks on fish tape while doing this. If snake is removed, wire hooks should be opened to avoid injury should the hand slip.

After all wire is pulled, splices or joints are made. See Chapter 7, "How to Make Electrical Connections," for information on finishing the job. See also Chapter 6, "Switching Circuits."

QUESTIONS FOR DISCUSSION OR SELF STUDY

1. What does EMT mean?
2. What is the largest obtainable size of EMT?
3. What are the other two general types of conduit?
4. What other common name is given EMT?
5. What is the primary function of metallic conduits in the wiring system?
6. Rigid conduit is available in what lengths?
7. What is the most common finish on rigid conduit?
8. Why should the radius of bends be made as large as conditions permit?
9. A conduit hickey is used for what purpose?
10. Why must the foot be kept in close to the bend while bending rigid conduit?
11. What tool is normally used to check accuracy of a 90° bend?
12. What type of coupling is used when rigid conduit cannot be turned to tighten it?
13. How might a large, sweeping 3 inch, 90° service conduit be bent?
14. When the handle is straight up from the floor (Fig. 13-34) on an EMT bending position, what degree angle has been bent?
15. What is the total degrees bent in an offset?
16. What is the total degrees bent in a saddle?
17. How many teeth per inch on hacksaw blades are recommended for cutting EMT?
18. What does K.O. mean?
19. Why are ceiling outlet box bar hangers sometimes notched into the joists?
20. What type of coupling is used on a run of thin-wall conduit that is exposed to the weather?
21. How are indenter couplings fastened to thin-wall conduit?
22. How are self-locking connectors put onto thin-wall conduit?
23. What is the most common application for flexible metallic conduit?
24. Where are toggle bolts used?
25. What tool is used to remove burrs from the inside of freshly cut conduit?
26. In preparing to make a 10″ right-angle bend in a piece of ½″ rigid conduit, how far would you measure "backward" to determine the spot at which to place the hickey?
27. What is the most popular size of fish tape for residential wiring?
28. What device may be employed to aid a fish steel to pass around bends in conduit?
29. In general, what are the two methods for installing thin-wall conduit in wooden partitions?
30. Which end of a lead anchor must be inserted into the hole first?
31. What is a star drill?
32. Where is capacity information for conduit fill located?
33. Why should knock-out punches *not* be ground on the outside to sharpen?
34. When conduit or EMT is connected

to outlet and switch boxes individually and brought straight up the voids in concrete blocks, how and where are these conduits connected to the rest of the distribution system?

35. Why is it sometimes objectionable to lay conduit in the concrete slab pour in connecting together convenience outlet boxes?

36. What is a "through box"?

QUESTIONS ON THE NATIONAL ELECTRICAL CODE

1. Where may rigid metal conduit be used?
2. What is the general minimum electrical trade size conduit?
3. What are the code requirements for reaming rigid conduit?
4. What is the minimum radius for a 1″, 90° bend in rigid conduit for conductors without a lead sheath?
5. What size rigid conduit is required for 3 #12 AWG, TW wire for new construction?
6. What is the minimum size wire permissible for a lighting circuit on new construction?
7. How many bends are allowed in rigid conduit between outlets?
8. What are code requirements for supporting rigid conduit?
9. What shape or form box may not be used with rigid conduit?
10. How are conductors counted to determine how many are allowed in an outlet box?
11. What is the minimum free length of conductors at outlet boxes?
12. What common type wire is used for damp locations?
13. What are code requirements for boring through studs and joists?
14. What are code requirements for electrical continuity for metal raceways and enclosures?
15. Where must conductor splices be made?
16. When are conductors permitted to be installed?
17. What is the general wiring arrangement to prevent induced currents in metal enclosures?
18. What color is an "identified" conductor?
19. What color is an "unidentified" #8 conductor?
20. What is the temperature limitation of #12, TW wire?
21. What is the largest size solid conductor?
22. What is the ampacity of the following copper conductors: #14, #12, #10, #8, #6, #4, #2 and #1?
23. What size EMT is required for a 100 ampere service?
24. How shall bends be made for EMT code requirements?
25. What is the approved use of EMT?
26. What are the minimum and maximum electrical trade sizes for EMT?
27. What is the maximum number of 90° bends for EMT?
28. What are the reaming requirements for EMT?
29. What are the supporting requirements for EMT?
30. Where may flexible metal conduit *not* be used?
31. On new construction, what is the requirement for supporting flex on each side of an outlet and at intervals?
32. Generally, what is the minimum electrical trade size flexible metal conduit?
33. In remodel wiring how is flexible conduit supported in hollow fished in areas?

Multiple Family Dwellings

Basically, the problem of wiring a multiple family dwelling is not much different from that of a single family unit. Because of the greater risks involved in multiple occupancy, however, wiring details are subject to greater restrictions than with the simpler structure.

Knob-and-tube installations are probably never used for apartments and motels, even where the inspection authorities allow it in single-family residences. Nonmetallic sheathed cable is frequently employed, especially in less populous areas. One large city may permit nonmetallic sheathed cable in motels, and in apartment houses with four or less occupancies. Another city may limit this material to one- and two-family dwellings. In neither city may this form of cable be installed in a hotel. Local codes vary.

Armored cable is used rather widely in multi-family units. The use of a continuous bonding strip, which codes now require in contact with the armor, has removed the most serious objection to this type of wiring. The strip insures positive grounding continuity when the cable is properly installed.

Electrical metallic tubing (EMT) may be used in all types of dwellings. Because of high labor costs, rigid conduit is seldom found today in wood frame construction except for underground runs where permanent moisture is encountered and for service conduits. Flexible metallic conduit is used extensively, especially in large cities where electrical codes are more extensive.

Construction Procedures

Working procedures in multi-family frame buildings are similar to those for smaller units. Outlets are to be spotted, boxes set, holes drilled, cable or conduit run, and splices made. If there are a number of identical occupancies, prefabrication is often resorted to. This method, called "prefab" by electricians, is based on the principle of mass production.

For example, suppose that the occupancy consists of twelve apartments with identical physical arrangements, and that armored cable wiring is specified. First, outlet boxes are spotted and holes drilled. Then, lengths of individual runs are measured between outlet and outlet, or outlet and distribution panel. Twelve pieces of each size are prepared.

Preparation involves cutting the numerous lengths, removing enough armor at either end to allow the proper amount of exposed conductors, reaming sharp ends of armor, and skinning the wires for splicing or for connecting to switches. If holes are drilled large enough to permit connectors to slip through, insulating bushings and connectors may be installed at the same time. By lettering or numbering the various pieces according to their respective locations, twelve separate bundles can be made up, one for each apartment.

When the bundle is delivered to a particular apartment, wiring may be installed quite rapidly. Time saved through prefabrication depends upon the number of occupancies concerned. Where there are four or less units, the method offers only slight advantage. It is most valuable when there are a large number of uniform living quarters, such as in a housing project. The method can be employed also with nonmetallic sheathed cable, but the greatest possible savings are obtained where electrical metallic tubing or rigid conduit is used. This is particularly true of slab work, which is explained next.

Slab Work

Reinforced concrete floors come under the general heading of *slabs*, and the installing of conduit in such floors is known as *slab work*. Electricians also designate the conduits as *pour*, speaking of this type of work as a *pour job* when forms are used. The codes allow either rigid conduit or EMT for this application. Certain authorities, however, prohibit EMT here.

Ground Slab

The simplest type of slab is one poured directly on the earth. Figure 14-1 shows the floor of a motel room which has a plug receptacle at the middle of each wall 10″ above the floor line. The circuit drops from a panel in the north wall down to a "handy box" (outlet box). A piece of rigid conduit extends from this point to a 4″ square outlet box in the east wall, a right-angle bend being made at each location, and the conduit lying upon the earth. A second piece of conduit extends from here to another 4″ square box in the south wall. The third conduit runs from this outlet to a "handy box" (outlet box) in the west wall, the circuit terminating at this location. Wire screen is placed on top of the conduit by other crafts.

The electrician first locates exact spots for centers of outlets, taking measurements from blueprints and transferring them to the ground by means of stakes or other markers. Since rigid galvanized conduit is employed, it may rest upon or in the earth. It is a good plan to install the outlet boxes, and to fasten them to wooden stakes, short pieces of reinforcing steel, pieces of conduit, or similar materials, as in Fig. 14-1, to insure that the conduit does not move during concrete pouring operations. Conduit openings should be sealed by means of cardboard or plastic plugs under bushings so that concrete will not enter. It is then necessary to install single-receptacle plaster rings on the 4″ square boxes before the walls are erected.

When EMT is used, it should be

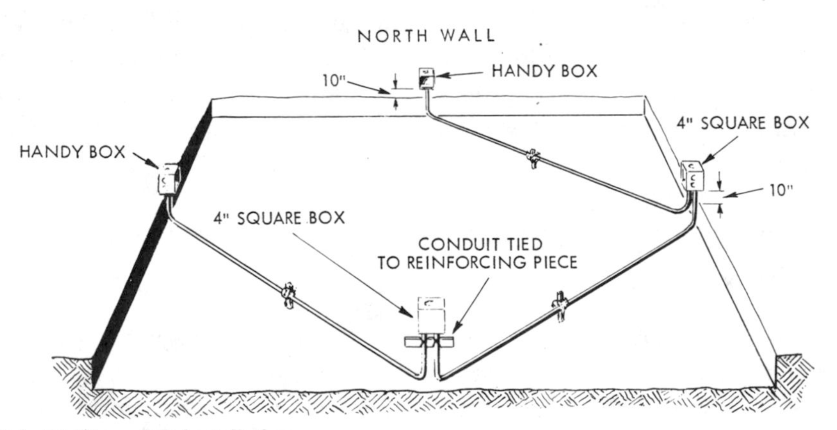

Fig. 14-1. Rigid conduit installation.

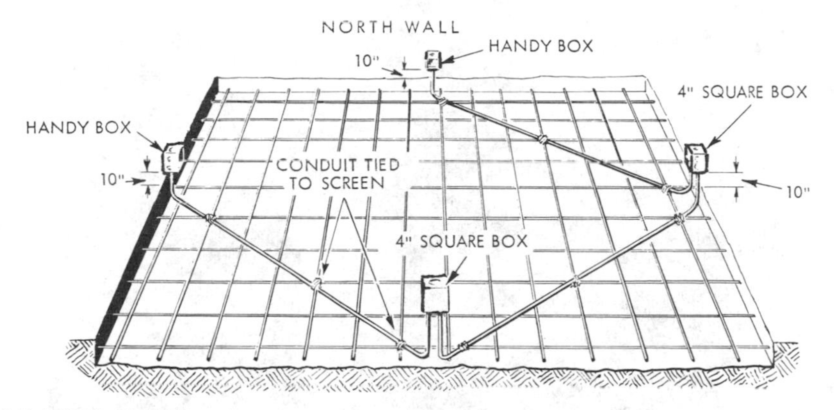

Fig. 14-2. EMT installation.

surrounded on all sides by at least 2″ of concrete to guard against damage from permanent moisture. This requirement is met usually by installing the tubing after the wire screen is laid, as shown in Fig. 14-2. The tubing is placed on top of the screen and tied to it by soft iron wire. When the screen is raised from the ground by another craft, the tubing is held up also, and a bed of concrete flows underneath. After concrete is poured it is permitted to "set" before additional work is done.

In some cases, copper heating pipes are laid in the slab. Rigid conduit should be used in this case, installing it two to four inches below the heating pipes, depending upon the maximum operating temperature.

Upper Floor Slabs

Flat Deck Slab. There are several general methods of placing concrete in buildings which have one or more floors above ground level. The flat-deck scheme makes use of a smooth surface, or deck, of plywood upon which reinforcing steel is laid before concrete is poured. The slab may be four, six, or eight inches thick, depending upon structural design, and the reinforcing steel is "woven" in two or more layers. See Fig. 14-3.

The electrician marks outlet locations on the smooth deck by means of a crayon or chalk. Ceiling outlets of rooms underneath are included, Fig. 14-4, if the concrete slab is to act as the finished ceiling of the lower story. Partition lines, too, are sometimes marked. Concrete boxes, like the one shown in Fig. 14-5, are employed for the lower floor ceiling outlets. This box consists of an octagonal ring and a cover plate. The box has projecting ears for nailing it

Fig. 14-3. Pouring concrete on a flat deck with a crane. (Republic Steel Corp.)

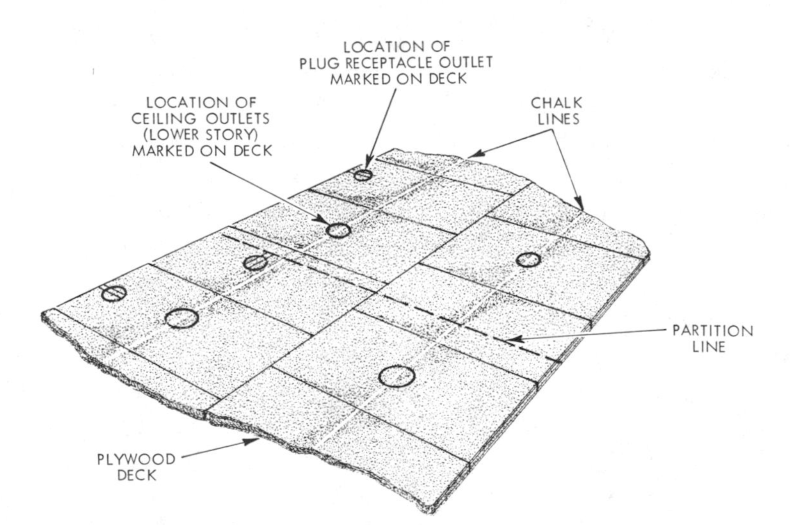

Fig. 14-4. Location of outlets marked on flat deck.

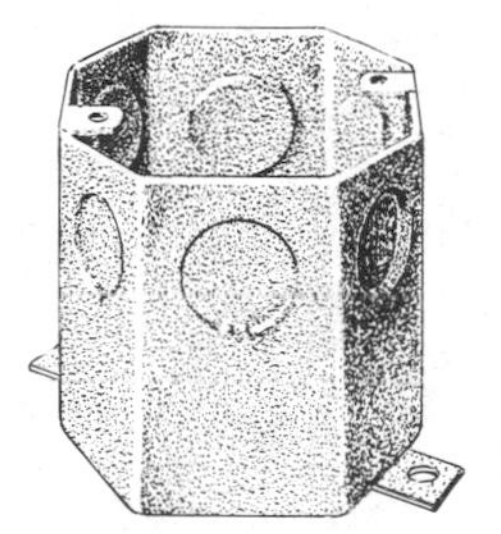

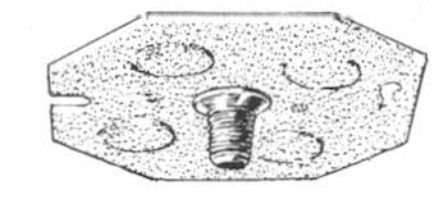

Fig. 14-5. Concentric boxes and cover plates.

Fig. 14-6. Coupling EMT on a flat deck construction with distribution panelboard below. (Republic Steel Corp.)

to the deck, and the plate may be equipped with a fixture stud if desired.

Rings are fastened to the deck, if possible, before the first layer of reinforcing steel is spread. After the steel is in place, rigid conduit or EMT is run from box to box. (The term *conduit* will include both rigid conduit and EMT throughout the remainder of this chapter, unless one or the other type is specifically mentioned.) Conduit for plug receptacles is extended from panel locations to the various outlets in much the same way as in the ground slab. Fig. 14-6 illustrates a wireman coupling EMT. Note conduits with "goosenecks" looking down. These will contain branch circuit wiring from a panel on the floor below. Plug receptacle boxes are often supported by metal brackets which are nailed to the deck.

In Fig. 14-7, conduits extend from the panel location to plug receptacle boxes. A feeder conduit is also shown. Conduits are grouped to en-

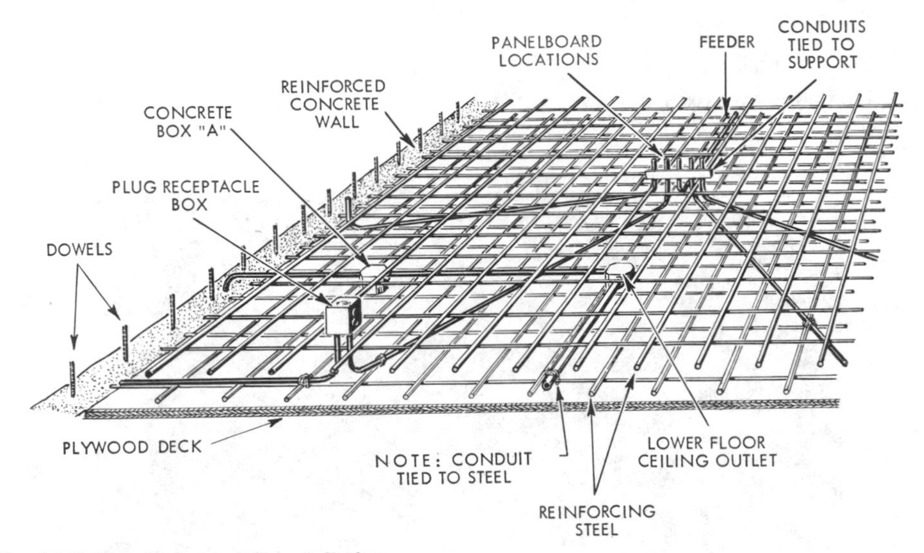

Fig. 14-7. Flat deck conduit installation.

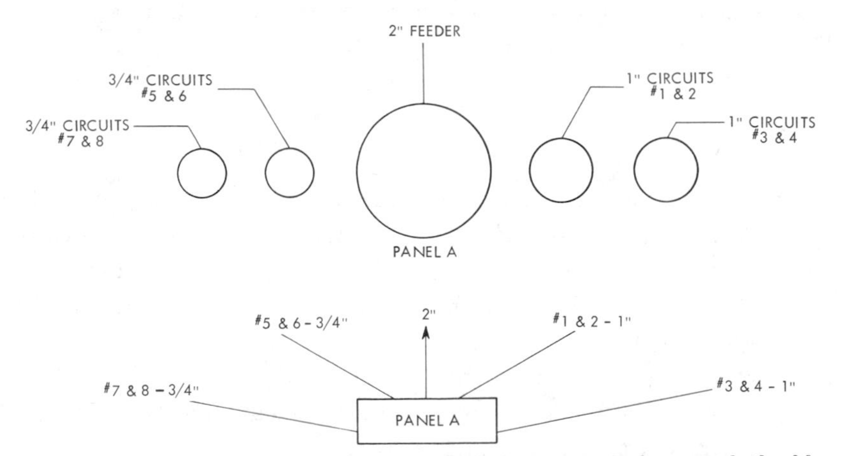

Fig. 14-8. To insure no panel conduits are omitted before concrete is poured, the blueprints should be checked with the job. Conduit sizes and circuit numbers are shown using two different methods, top and bottom.

ter the bottom of a distribution panel, which is to be installed later. The group is tied to a piece of wood or other support, such as strapping to channel iron support, to hold it firmly in place. Conduits should be checked with drawings made by the wireman from blueprints or from the prints directly to see if all have been installed. See Fig. 14-8.

Note the run in Fig. 14-7 which extends from concrete box *A*, toward the wall. It is coupled to a right-angle bend formed in a length of conduit that was placed in the wall before the decking was laid. This conduit drops to a switch outlet or a panel location on the floor below, so that ceiling light may be controlled from there.

When this portion of the work in Fig. 14-7 has been completed, steelworkers install the next layer of reinforcing steel. At this time, the electrician prepares the boxes and ties the conduits to the steel with iron wire. This procedure is necessary to prevent undue movement or damage to the conduit in the pouring operation. Fig. 14-9 shows conduits being installed to feed an upper deck. Note boxes in supporting column stuffed with paper to exclude concrete when poured. Fig. 14-10 shows electricians tie-wiring EMT to reinforcing steel on a deck to prevent movement during concrete pouring operations. (A convenience outlet box is shown mounted horizontally. This is not approved in all locations.) A ceiling outlet concrete box is also being used here as a pull box. The matter of preparing the boxes consists of taking steps to make sure

Fig. 14-9. Conduits and boxes installed in column to be formed and poured. (Republic Steel Corp.)

Fig. 14-10. Tying EMT to reinforcing steel. (Republic Steel Corp.)

that concrete will not harden inside, thereby rendering them useless.

If the builder permits, ¾″ holes are bored in the decking at the center of each box. Any wet material that enters can then drain out before it hardens. In case this method is prohibited, boxes may be filled tightly with paper, sand, or other suitable material before cover plates are installed. Paper is recommended. The filling material is readily cleaned out after the decking is stripped from the concrete slab. Where sand is used, ends of conduit projecting into the boxes must be sealed off. It is advisable to do so, regardless of the method of preparing boxes.

Fig. 14-11. Cutting an EMT stub for a convenience outlet. (Republic Steel Corp.)

Considerable saving in time is possible, where local inspection authorities allow EMT in concrete, especially if it is used with crimp or tap-on fittings. But, when other craftsmen are roughshod in treatment of EMT stubs which project above the surface after the concrete has hardened, much of the time saved will be squandered in repairing flattened or broken ends. Fig. 14-11 shows a wireman cutting an EMT stub to length. Boxes are installed to help hold conduits in place and protect stubs.

Pan Slabs. The pan type of construction is illustrated in Fig. 14-12. As shown, horizontal planking is laid upon supporting structures at regularly spaced intervals. The metal pans used here are attached to the edges of planks by nailing or by other means and are overlapped at the ends so as to provide a complete form or mold that will retain wet concrete. Molded polyester resin (fiberglass) pans are also used.

After pans are set, the electrician installs outlet boxes on the wooden form planks to support fixtures which light the story below. Conduit or tubing is run from box to box, box to switch, or box to panelboard. Conduits that lie upon the planks are raised an inch or so by patented supports.

Workmen then place reinforcing bars in the space between the pans, and lay comparatively light bars or screen on the pans to form a deck. The electrician follows, running conduit on top of the steel. Since the floor slab in pan construction is relatively thin, it is usually bad practice to have conduits cross one another. To avoid doing so, cross runs are made in the space between pans, as shown in Fig. 14-12. Conduits are tied to reinforcing bars or screen with iron wire. Outlet boxes are prepared in the same way as in flat-deck construction so they will not fill up with concrete.

Fig. 14-13 illustrates a section of completed floor that has been stripped of metal forms and their

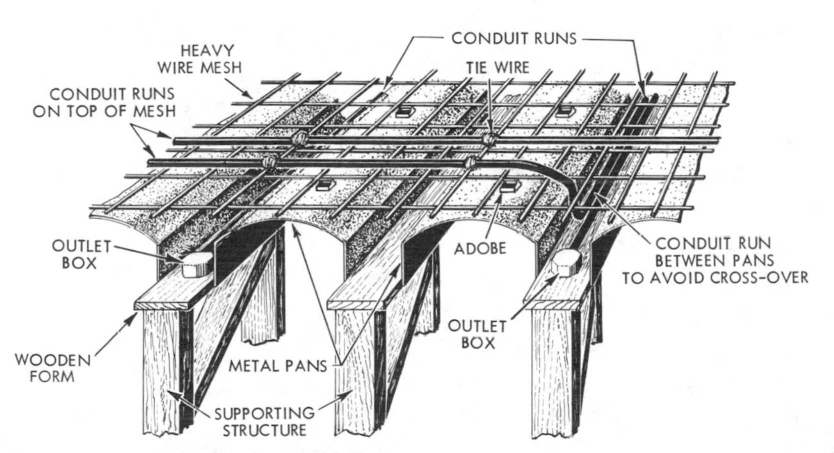

Fig. 14-12. Pan type deck showing conduit runs.

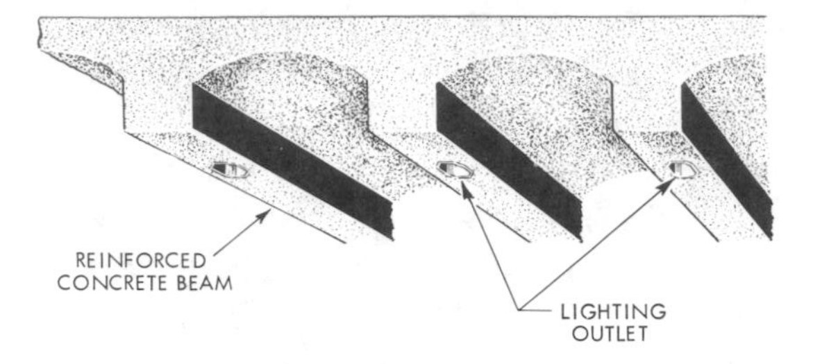

Fig. 14-13. View of pan type slab.

wooden supports. Concrete in the spaces between pans has formed reinforced beams, and a thin floor web stretches between them with lighting outlets extending along the face of a beam.

Both types of slab are in common use. In some buildings, the first floor is pan-constructed over parking garages, while upper ones are of the flat-deck type. Insofar as the electrician is concerned, the flat deck offers lesser problems.

Concrete Walls

Electrical outlets in reinforced concrete walls or partitions are handled much the same way as those in floor slabs. In Fig. 14-14A, a plug receptacle, a lighting outlet, and a switch outlet are to be installed in the wall, the receptacle facing inward, the light outward, and the switch inward. The outer half of the wooden forms has been set in place, and the reinforcing iron fabricated. The conduit has been tied to reinforcing iron exactly as in floor slabs.

Before the inner half of the

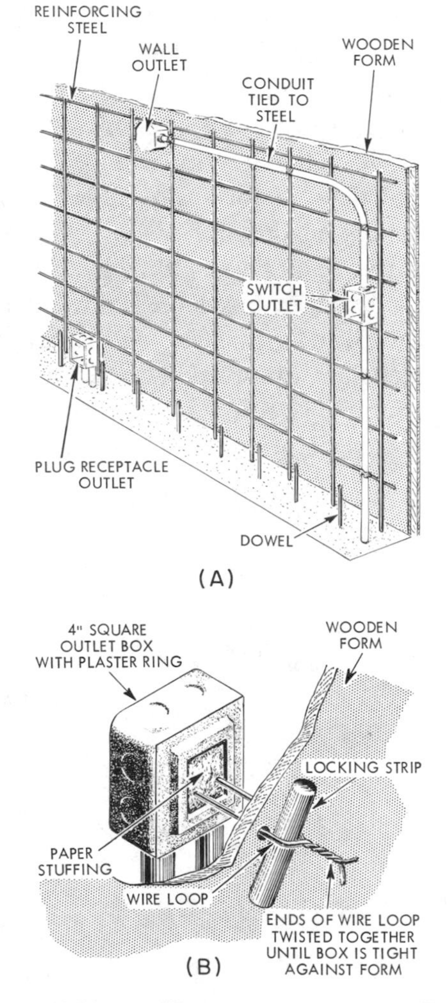

Fig. 14-14. Installing outlets in concrete walls.

wooden forms is set in place, boxes are filled with paper, and an open loop of iron wire is inserted through screw holes in the back of each box, Fig. 14-14B. The purpose of the loop is to hold the box tightly against the form, where it belongs, and to exclude wet concrete. Small holes are bored in the wood so that the iron wire loop may come through. A locking strip of reinforcing iron, conduit, or even wood, is placed between the open ends, on the outer surface of the form, before they are twisted together. Wooden wedges are also used with holes made farther apart. Wedges can be driven into a wire loop for tightening the outlet box against the concrete form. Whichever method is used they should be nailed in place so that concrete vibrators do not knock them out. Adequately sized tie wires should also be used; they should be large enough that they won't break but small enough to be able to twist easily. Even doubling it insures strength.

Chases

One additional measure connected with this type of operation is that of arranging for openings in the floor to accommodate vertical conduit risers. The builder, architect or structural engineer should be consulted regarding such holes before they are installed. Figure 14-15 shows how it is done.

If a rectangular hole is desired, so as to allow several large conduits to be placed side by side, a wooden box is constructed as in Fig. 14-14A. The box is nailed to the wooden deck and filled with sand. It is easily removed, after the concrete floor has set, leaving an opening or chase.

If a hole for a single conduit is desired, a metal sleeve or "can" is fastened to the deck, as in Fig. 14-15B.

Such openings are often made in closets or other out-of-the-way places.

After the conduits are installed, openings around them may be filled

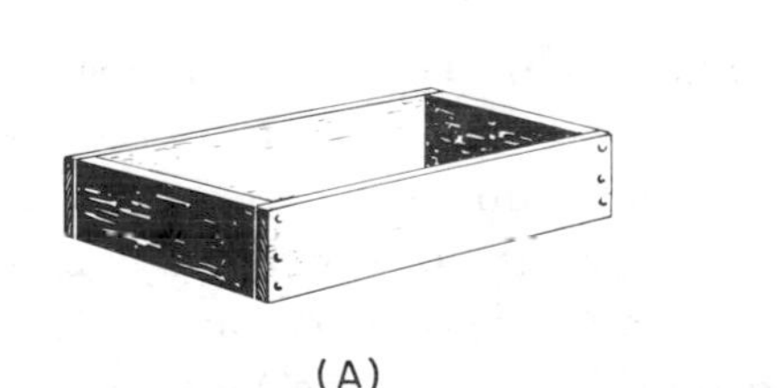

(A)

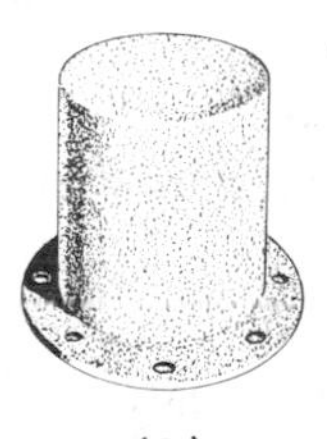

(B)

Fig. 14-15. Box and sleeve chases

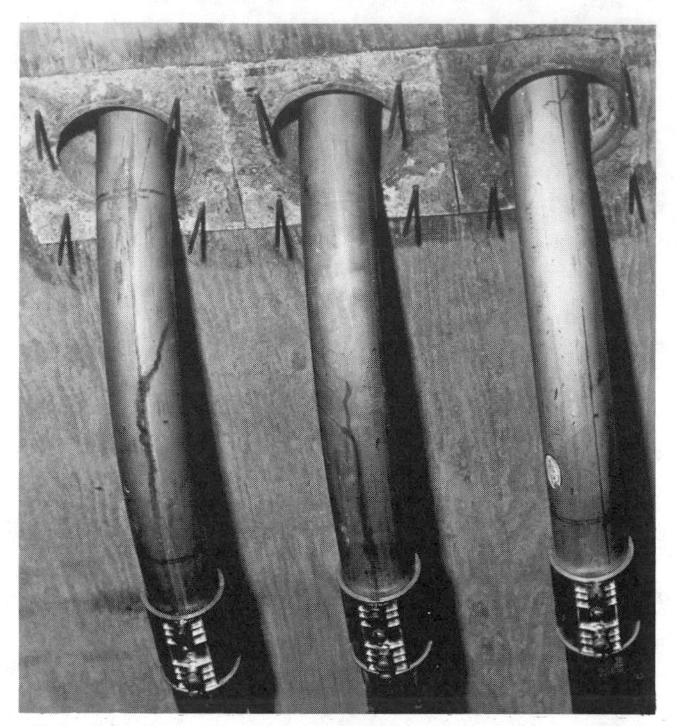

Fig. 14-16. Panelboard feeders going through sleeve chases. (Los Angeles Trade-Technical College)

in with cement or "grout," or it may remain as it is.

Fig. 14-16 shows EMT panelboard feeders installed through sleeve chases. Note nails protruding. These are sleeve mounts and will be cut off.

Service Considerations

Service requirements for multi-family units differ in some respects from those of single-family residences. They usually employ larger conductors, the location of service disconnecting means is important, and the service entrance conductors on large jobs are usually brought in underground. Methods of connecting underground service conductors to the power company mains have been considered earlier, in Chapter 11, "Installing Service Equipment."

Single-family dwellings are always

supplied by either two-wire or three-wire single-phase current. Apartment houses and hotels are often supplied by three-phase current, usually in the form of a network. The network is a four-wire arrangement that has three-phase wires and a neutral conductor. The voltage between any two of the three phase wires is 208 volts, while that between any phase wire and the neutral conductor is 120 volts. Lighting panels in the various apartments are usually supplied by a three-wire feeder consisting of two phase wires and the neutral wire. Load on the main service conductors is equalized by making sure that the same number of panelboard taps is taken from each of the three wires.

Most codes provide that each tenant of a multiple-occupancy building shall have access to his own disconnecting means. Codes also state that a multiple-unit building having individual occupancies above the second floor shall have service equipment grouped in a common accessible area. Figure 14-17A shows a multi-story apartment building. To comply with the Code, individual service switches must be grouped at a central point. Here, the selected location is on the outside, at one end of the building. It could have been placed in a more protected spot, if desired, such as in the entrance lobby or in an unlocked meter room, or underground parking garage.

When there are no apartments above the second floor, Fig. 14-17B, separate service entrance conductors may sometimes be tapped from one set of drops. Here, three separate conduits are run to the spot where service drops are attached to the building.

The presence of fire escapes on exterior walls of hotels and apartment houses requires that special attention to be given to the location of service drop conductors and the service head. Figure 14-18 illustrates a situation often met with in practice. Local requirements should be

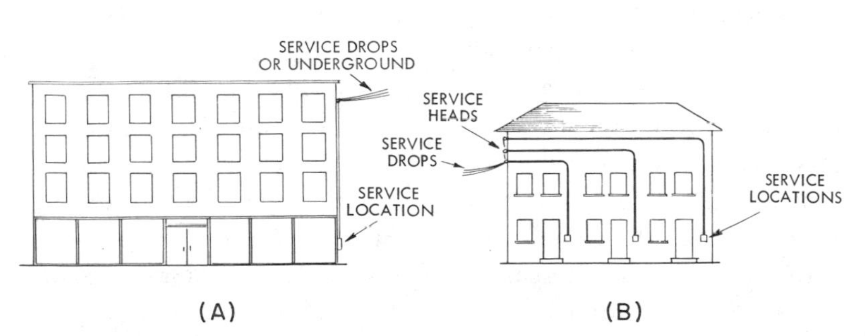

Fig. 14-17. Multi-unit services.

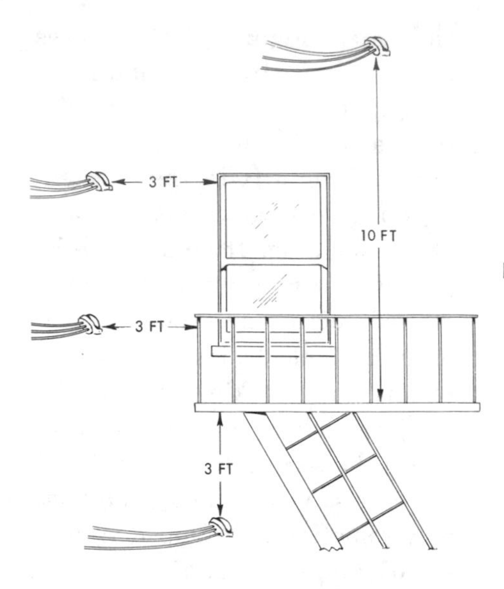

Fig. 14-18. Safe clearances of service drops.

checked. However, the following are recommendations. The service drops may not be closer than 3′ to a window unless above its top level. In such cases, the drops are considered out of reach from the window. Drops must clear the fire escape platform not less than 3′ horizontally, 3′ vertically below, or 10′ vertically above.

Switchboards and Panelboards

Most service requirements vary from city to city. Codes allow not more than six switches or circuit-breakers to act as the service disconnecting means. Figure 14-19 shows a built-up arrangement for six meters and service switches. Entrance conductors run from the service head to the "hot" gutter. Individual services are tapped to main conductors in the gutter. Apartment panels are fed from the bottoms of individual service switches.

A commercial switchboard that accomplishes the same purpose is pictured in Fig. 14-20. Meter sockets and service switches are mounted in a rectangular enclosure. The service conduit enters at the lower right side, while individual feeder conduits extend from the upper surface of the metal box. Meters may be supplied with underground services also, as shown in Fig. 14-21. The underground conduit terminates in a pull box. Horizontal conduits above are grounds. Panelboard feeder cables are ready for make-up.

With more than six individual services, a main disconnect switch is more often required. Figure 14-22

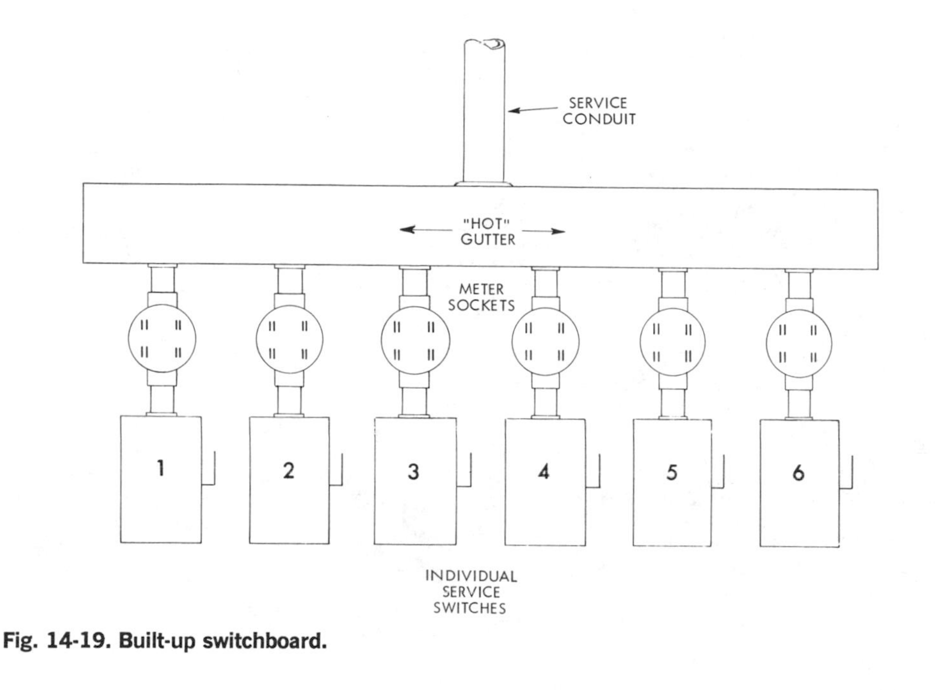

Fig. 14-19. Built-up switchboard.

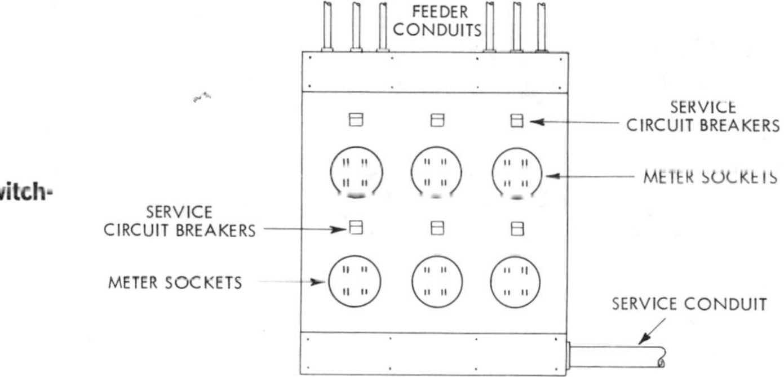

Fig. 14-20. Commercial switch-board.

shows a switchboard that has eight meter sockets, eight disconnect switches, and eight feeder conduits. This arrangement could be used for eight apartments or for seven apartments and a house feeder. The house disconnect switch may control a panel that has branch circuits for hall lights, a washing machine, and perhaps a dryer. Figure 14-23 shows a house panel with circuit breakers, and also a time switch that turns hall lights on and off at preset intervals.

The time switch has a mecha-

Fig. 14-21. Partially wired apartment meters with underground service. (Los Angeles Trade-Technical College)

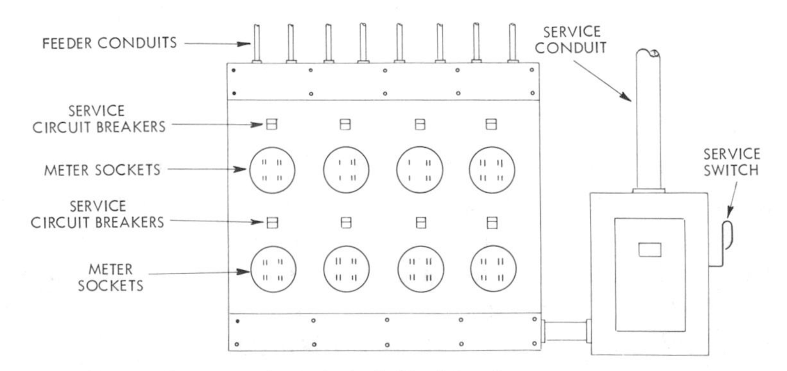

Fig. 14-22. Switchboard for more than six individual services.

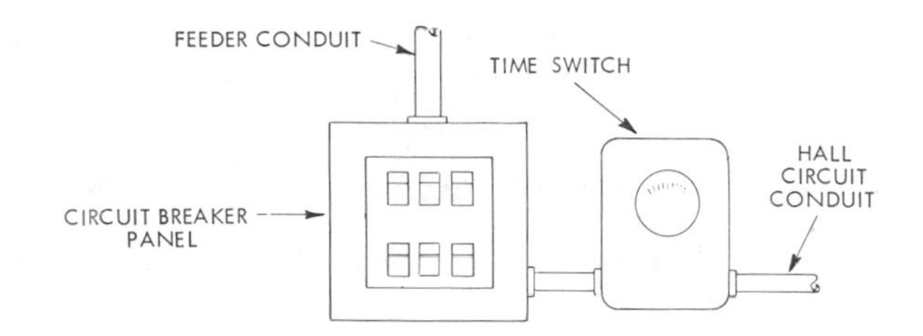

Fig. 14-23. "House" panelboard with time switch.

nism which operates a pair of switch contacts through a train of gears. A clock motor drives the gear train. An adjustable arm on the face of the dial may be set for the hour at which it is desired to close the circuit, and another arm for the hour at which the circuit is to be opened. A common practice is to set the ON arm for 6 P.M. throughout the greater part of the year, and the OFF arm for 6 A.M.

Whether or not apartments and motels are separately metered, there is usually a circuit panel inside each occupancy. Hotels often have large panels located in closets on each floor; the panel supplies a number of rooms or suites in its immediate vicinity.

Master Antenna Television

Master antenna television (MATV) systems consist of one T.V. antenna feeding many outlets. This may be compared with the old accepted idea of one antenna for every television receiver. Using one antenna also greatly improves the appearance of the building. This system is shown in Fig. 14-24, top left. A good antenna for very high frequency (VHF) and ultra high frequency (UHF) T.V. receiving frequencies is installed and connected to a solid state amplifier which in turn connects to a device called "splitter." From each of the circuits shown up to 15 T.V. outlets may be connected. This varies with different systems.

Coaxial cable, sometimes called concentric line, should be used. Coaxial cable is plastic insulated conductor, shielded with a wire mesh covering used as the second conductor at terminals. "Co-Ax" eliminates T.V. signal interferences. Better systems are installed in conduit so that they may be modified or removed in case of trouble. The amplifier and splitter are installed in a convenient location such as attic or closet space.

Telephones and Panelboards

Telephone outlets in apartment houses and hotels are usually stacked as in Fig. 14-24, top right. The conduit from the outlet in one apartment continues upward to an outlet in another apartment directly overhead. It may be noted that the tele-

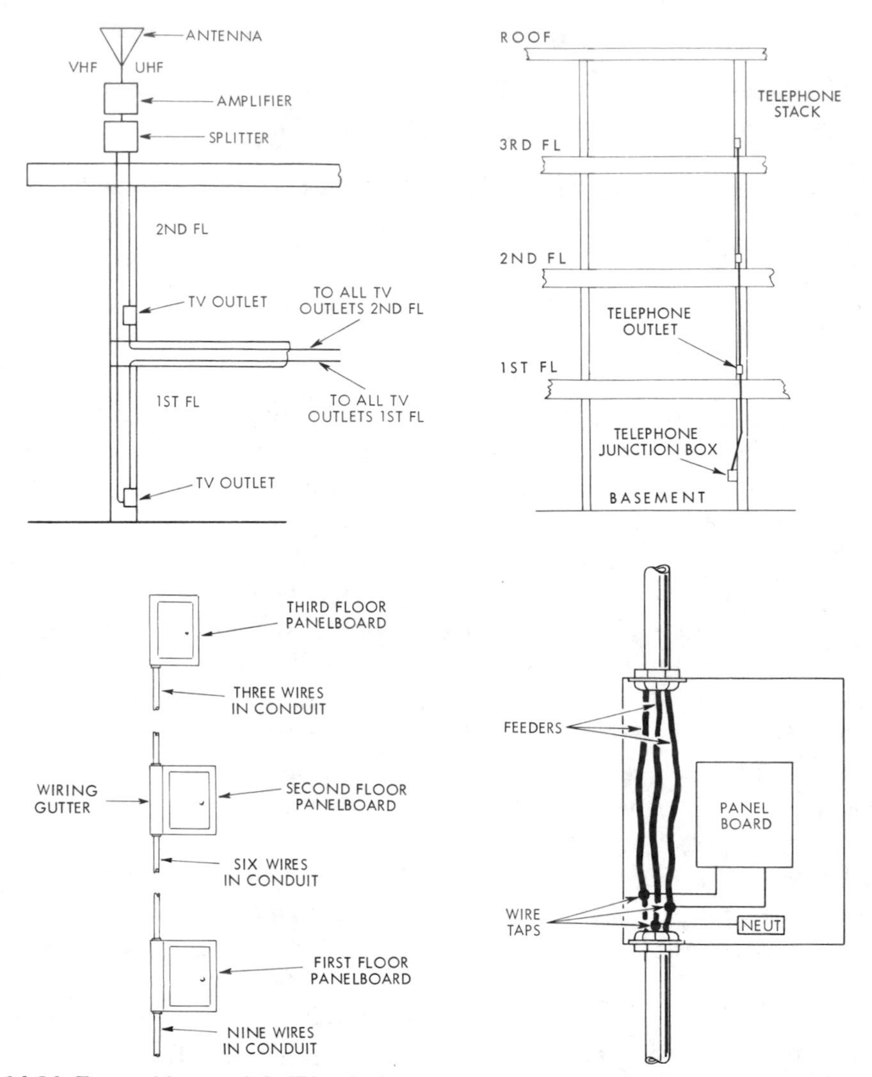

Fig. 14-24. Top and bottom left: TV, telephone and electrical distribution. Bottom right: Apartment panel connections from riser feeders.

phone stack on the top right extends upward from the basement, while the television stack on the top left descends from the roof. There will be as many separate stacks, in general, as there are apartments on a single typical floor. Telephone stacks will be run to junction boxes in the basement, and television stacks to junction boxes in the attic or on the roof. Planning departments of the telephone company should be consulted.

Lighting panelboards are sometimes fed as in Fig. 14-24, bottom left. Those on the first and second floors have wiring gutters so that leads from buses may be attached to the feeder conductors. This plan is used mainly where electricity supplied to the apartments is not separately metered. The general scheme may be employed, even where separate meters are installed at the service location, if individual feeder conductors are run to each panelboard.

Conduit entering the lower panelboard in this case (Fig. 14-24, bottom left) will have nine wires, assuming that three-wire panels are used. Six conductors will extend to the second floor panelboard, and three will continue to the one on the upper floor. It should be noted that no wiring gutter is needed for the upper panelboard. There are, of course, as many panelboards as there are apartments on a single floor. Fig. 14-24, bottom right, gives a detailed illustration of a panelboard.

Emergency Lighting

In many inspection localities, hall lighting circuits in multi-story apartment houses and hotels are classed as semi-emergency circuits. As such, they are required to be installed in conduits entirely separate from other wiring, such as that for utility plug receptacles in common hallways. In some instances, switches that control these circuits must be locked. These are not really emergency circuits within the Code definition, however, unless they are arranged for automatic power transfer supply from a special emergency service in case of necessity.

Other Installations

There are other considerations regarding electrical installations in multi-family dwellings similar to single-family dwellings. These are music and sound, radio intercommunication, alarm systems, fire and security, telephone and intercom systems to replace buzzer-pushbutton systems, and closed circuit television security systems from vestibule to apartment. These are generally specially wired systems with color coded or numbered wiring information supplied by the manufacturer for each installation.

Blower and Vacuum Fish Tape System

Fishing conduits for pulling wire was described in Chapter 13, "Methods of Conduit Wiring." The fish-steel method is applicable in all conduit systems, residential through industrial. However, with any rigid conduit or EMT system, a combination blower and vacuum fish line sys-

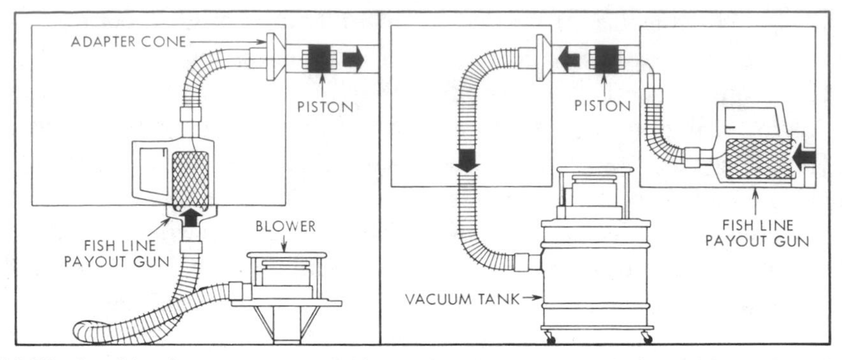

Fig. 14-25. Combination vacuum and blower fish line system. Left: It blows line; right: it vacuums line. (Greenlee Tool Co.)

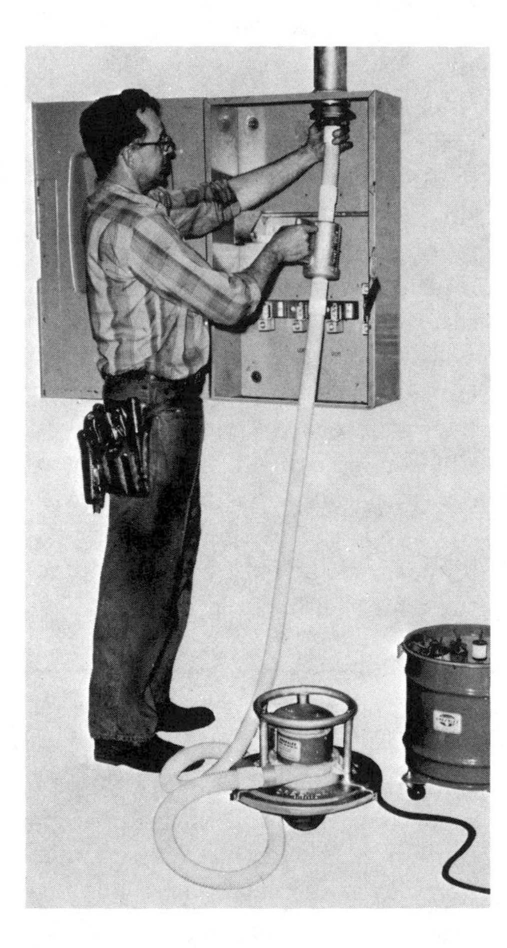

Fig. 14-26. Blowing a nylon fish line through a conduit. (Greenlee Tool Co.)

Fig. 14-27. Vacuuming a fish line through a conduit. (Greenlee Tool Co.)

tem may be used as shown in Fig. 14-25. The piston attached to the nylon fish line is lightweight plastic foam (polyurethane) and is available in different sizes matching the conduit sizes. Adapter cones for the blowing or vacuuming end vary in sizes also.

A heavier wire or cable pulling rope is pulled in with the fish line

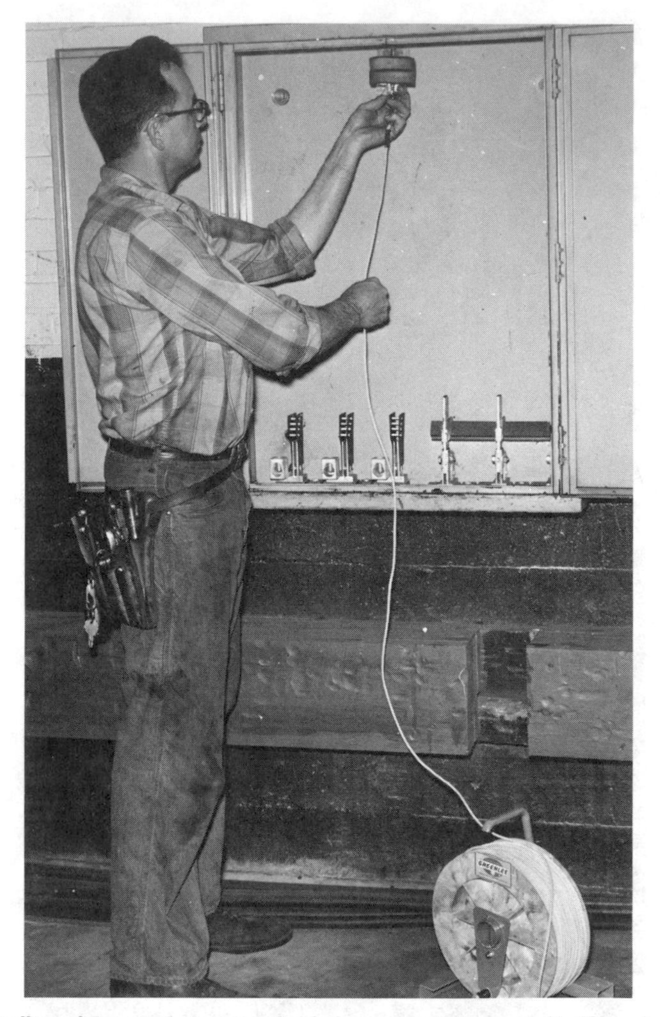

Fig. 14-28. Feeding piston and strong polyethylene rope into conduit. (Greenlee Tool Co.)

unless the runs are very short and the wire small. Fig. 14-26 shows a wireman *blowing* a fish line through a conduit. His left hand holds the adapter cone making the conduit system nearly airtight. His right hand holds the fish line payout gun. Fig. 14-27 shows a wireman *vacuuming* a fish line. Vacuuming and blowing conduits cleans conduits of water and debris.

Fig. 14-28 shows *wire pulling* rope, attached to piston, being inserted into a conduit. When pulling medium size feeder cables, half of the strands or more can be cut off when connecting to the pulling rope. This prevents a large ball-joint at the rope-cable connections. Factory-made wire and cable grips are generally used for pulling large cable. The vacuum-blower system is especially useful where there are many conduit runs to be fished such as hotels and large multi-family dwellings.

QUESTIONS FOR DISCUSSION OR SELF STUDY

1. How does wood-frame apartment house wiring compare with single-family dwelling wiring methods?
2. Upon what principle is prefabrication based?
3. How should a group of conduit stubs be treated at a panelboard location before concrete is poured?
4. How is conduit properly secured on a deck before concrete is poured?
5. What is the best method of excluding concrete from outlet boxes?
6. What are sleeve chases?
7. How is conduit supported on deck areas before concrete is poured?
8. How are concrete boxes secured before concrete is poured?
9. How are panelboard conduits supported before concrete is poured?
10. What is the advantage of installing convenience outlet boxes on EMT stubs before concrete is poured?
11. What are the advantages in using a master antenna television system?
12. Who has jurisdiction over telephone installations?
13. Why is the blower-vacuum fish line method not recommended for flexible conduit systems?
14. Name advantages of fishing conduits with the blower-vacuum method.
15. How are medium size feeder cables attached to the piston?

QUESTIONS ON THE NATIONAL ELECTRICAL CODE

1. What are service drop clearances for windows and doors?
2. How must individual services be grouped?
3. How must individual branch-circuit disconnecting means be located?
4. What is the maximum number of services without a main disconnect switch?

Forty equally sized apartments have a total floor area of 32,000 square feet. One-half of the apartments are equipped with electric ranges not over 12 K.W. each. Meters are in two equal banks with individual sub-feeders to each apartment.

5. What is the general lighting load for each apartment?
6. What size wire at 230 volts is required for 8000 watt range?
7. What size neutral for range load?
8. What is the minimum number of lighting branch circuits for each apartment?

9. What is minimum number of small appliance load branch-circuits?
10. What is minimum size feeder required for each apartment with ranges? Without ranges?
11. What size neutral is required for sub feeds with ranges? Without ranges?
12. What conduit size is required for feeders with ranges? Without ranges?
13. What size of each ungrounded feeder to each meter bank is required?
14. What size neutral feeder to each meter bank is required?
15. What size of ungrounded service conductors for the minimum main feeder is required?
16. What size neutral is required for the minimum main feeder?
17. What size conduit is required for service conductors for the main feeder for overhead service? For underground service?
18. What is the maximum distance allowed by the code for convenience outlets in multi-family dwellings?

Remodeling Wiring

Chapter 15

Remodeling wiring refers to alterations of an existing structure's wiring systems — such as adding new lighting outlets, convenience outlets or outlets for special appliances. Many times it may mean modernizing an existing electrical system in an old house. Remodeling work is sometimes called *old-house wiring* or *old work*. It covers the installation of wiring with a minimum of damage to or removal of finish. As the cables must be fished from outlet to outlet, this type of wiring is often a two-man job.

There are so many variations in reconstructing an existing electrical system that only some of the common problems are given here. Remodeling wiring requires a considerable amount of thought and ingenuity on the part of the wireman doing this kind of work. He should be thoroughly familiar with different wiring methods and, most importantly, with the building methods of the general contractor, so that he will know what's inside the wall.

He must know these things so that he may better plan an approach to a particular wiring problem without tearing out the walls.

Testing Circuits. It is a good rule to assume that all circuits are electrically alive, or "hot," until they have been checked. And the best method for checking is with a pocket device such as the voltage tester shown in Fig. 15-1.

The white wire should be checked, where present, to make sure that it is the neutral conductor. Connecting the tester between this wire and a grounded surface, such as a water pipe, should give no indication of potential. If the building has been

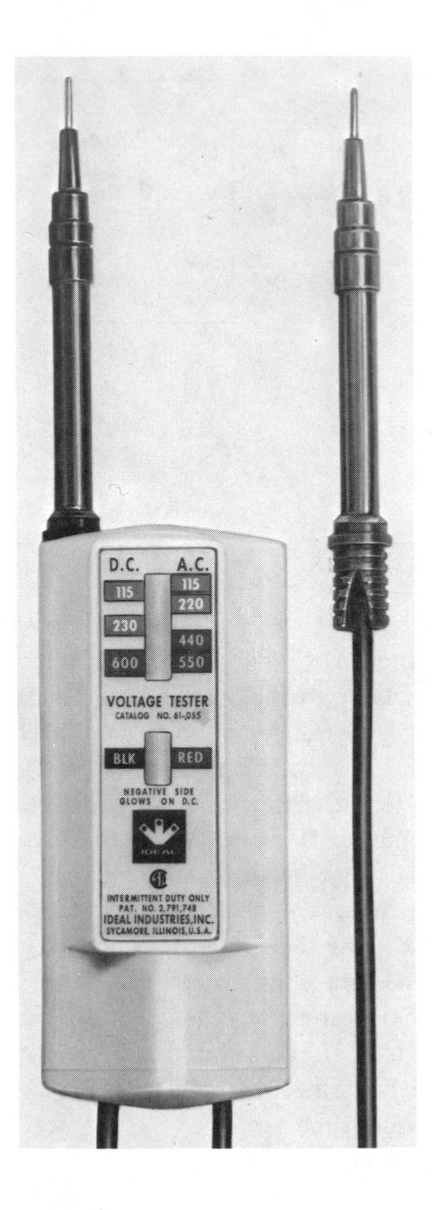

Fig. 15-1. Voltage tester. (Ideal Industries, Inc.)

wired in recent years, the NEC color code has probably been followed. For a three-wire circuit, the standard coding is: black, white, and red. If the circuit is 115-230 volts, the tester will show 115 volts between the white wire and either the black or the red wire, 230 volts between the black and the red wire, and 115 volts between either black or red wire and a grounded surface.

An inexpensive test set for check-

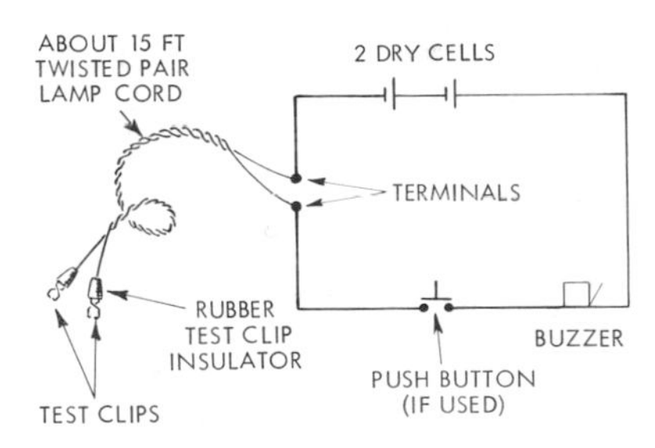

Fig. 15-2. Illustrating how to build a test set.

ing circuits for continuity or continuous circuitry, is shown in Fig. 15-2. It consists of two dry cells in series with a buzzer. The push button may or may not be included, because one of the test clips may be attached to a circuit wire, and the other clip may be used to make or break contact as desired.

Locating the Outlets

To add outlets the first step is to spot the outlets in ceilings and walls, after consulting the owner or his representative. In the choice of location for wall brackets, discretion and a knowledge of building construction are essential, because some walls offer more difficulties than others. For example, it is often impossible to fish cable in the furred space between lath on a brick wall.

The plaster-keys which overhang laths on the inner surface may be so large as to seriously impede fishing of cables. Or, the space between the top of a wall and the roof may be so small that access to a hollow space inside the wall is impossible.

The second step is to cut holes for the outlet boxes. Holes should be only large enough to accept the boxes without forcing. For ceiling outlets the boxes are round, ½″ or ¾″ in depth, depending on the thickness of the plaster. Bracket outlet boxes are either round or rectangular in accordance with the type of canopy or mounting plate used in that particular location. Switches or receptacles are 2″ x 3″, of a depth to suit the particular condition.

Installing Ceiling Outlets. Fig. 15-3 shows a handy tool which is used with a brace for cutting round holes. Cut the plaster only, the lath is left intact for the installation of shallow boxes on very old houses. A good technique for drilling the hole through the ceiling is to use a card-

Fig. 15-3. Rotary or ceiling hacksaw. (Misener Mfg. Co).

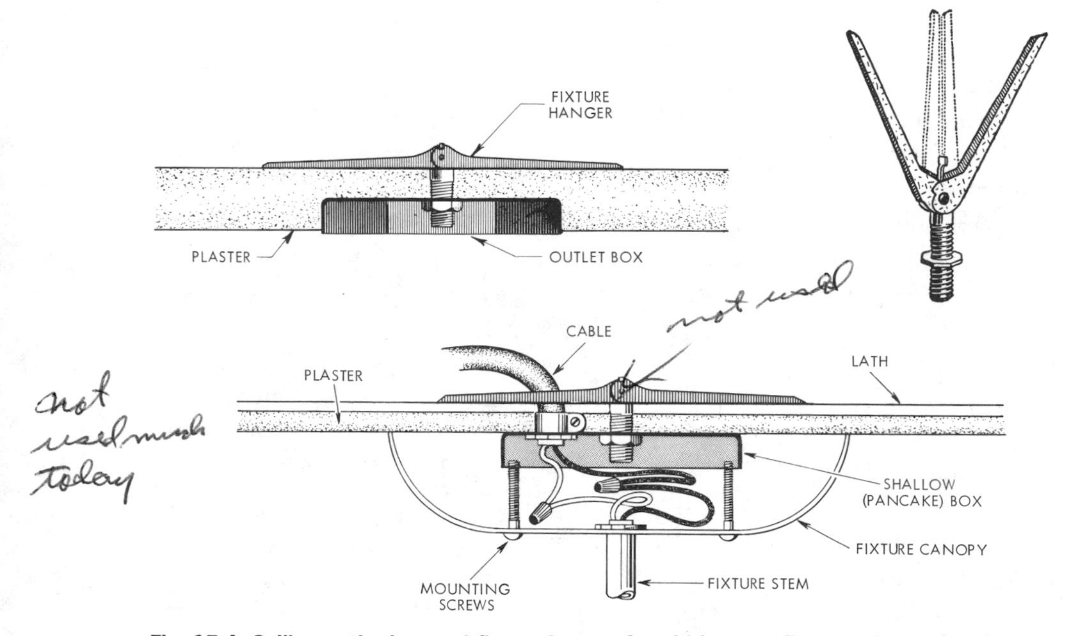

Fig. 15-4. Ceiling outlet box and fixture hanger for old houses. Bottom shows the shallow outlet box mounted on plaster surface and covered with fixture canopy.

board shoe box between the brace and the saw (shown in Fig. 15-3) to catch all loose plaster. Electricians should take great care to maintain cleanliness while working in residences. It will probably be necessary to notch the lath to permit passage of the cables and outlet-box support.

not recommended

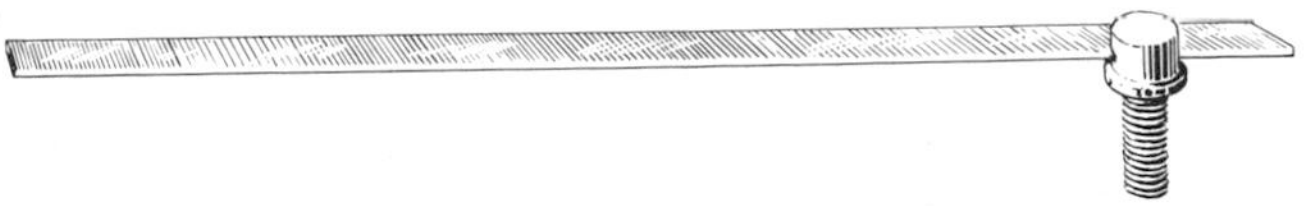

Fig. 15-5. Old-work bar hanger.

Shallow boxes can be mounted on the surface of the plaster (codes permitting) if they will be hidden by the lighting fixture or canopy. See Fig. 15-4, bottom.

Fig. 15-4 illustrates a ceiling-outlet box and fixture hanger. The view shows the hanger legs extended for supporting the box. They are held in position by the locknut on the center stem which serves also as a fixture stud. A key through the center of the nipple locks the wings securely in a horizontal position. In another type of hanger, Fig. 15-5, a stud slides freely along the bar. The whole unit is inserted through the opening, and the stud is drawn along the bar to a central position. A wire, already attached to the stud, is used for this purpose. The locknut is removed and the outlet box, with cables attached, is placed on the stud. The locknut is then set up tight, thus securing box and hanger in place. A hole about 1½″ in diameter, somewhat larger than for the other type, must be made for this hanger. With either one, the entire weight of the fixture must be carried by the lath. Hangers must be installed so as to bear on as many laths as possible. Some locations do not approve this method of installation. An alternate method is shown in Fig. 15-6. The weight of the fixture is supported by the fixture stud on the bar or box hanger which is nailed to the ceiling joists in a manner similar to that of new construction.

Installing Switch Outlets. The process of cutting a hole for a switch outlet is different from that for a lighting outlet because lath must be cut through as well as plaster. For a standard switch box, the dimensions of the hole should be approximately 3″ vertical and 2″ horizontal in older houses. Since wood lath is 1½″ wide and the spacing between laths is about ¼″, it will be necessary to cut the middle lath entirely, and to cut off ½″ from the adjacent laths.

Dig out enough plaster to locate the lath in the spot where the switch is to be installed. Outline the dimensions of a hole on the wall, using the middle of the exposed lath as a center point. Carefully remove plaster within this outline, as in Fig. 15-7. Now cut a section 2″ wide out of the middle lath, and a ½″ x 2″ strip out

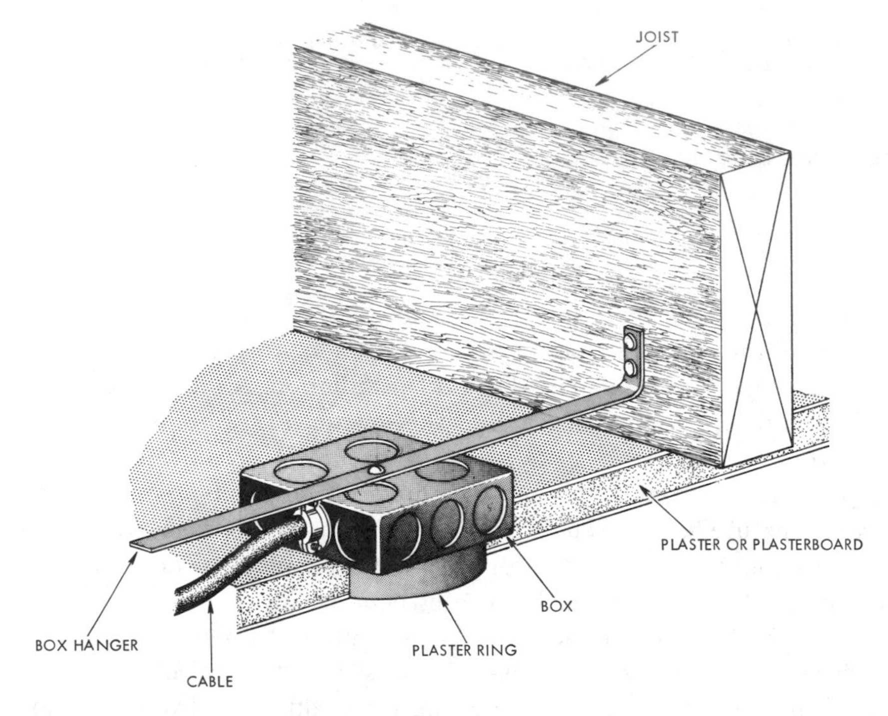

Fig. 15-6. Round hole cut through plaster and plasterboard for mounting ceiling outlet.

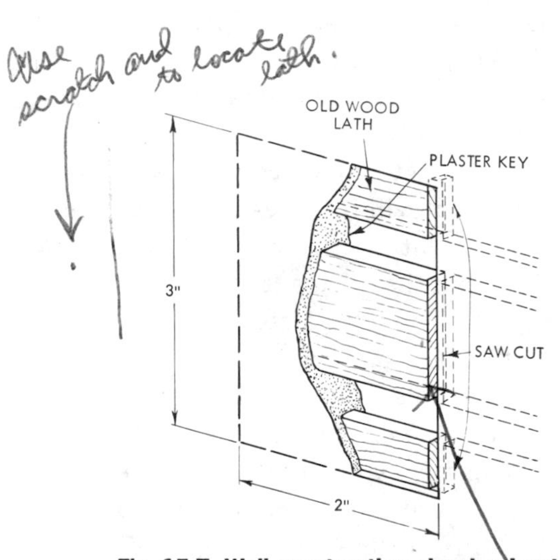

Fig. 15-7. Wall construction showing location for cutting opening in plaster for switch box in older houses.

of each adjoining lath, using a narrow, fine-toothed saw. The saw should be manipulated gently, especially after the first cut has been made, because supporting studs may be some distance away, and the spring of the lath tends to crack plaster.

Where old plaster is too brittle to sustain even the most careful sawing, a wooden anchor strip may be employed. The strip may be from 1″ to 1½″ wide, ½″ or more thick, and 8″ to 10″ long. Bore three holes through the strip large enough for

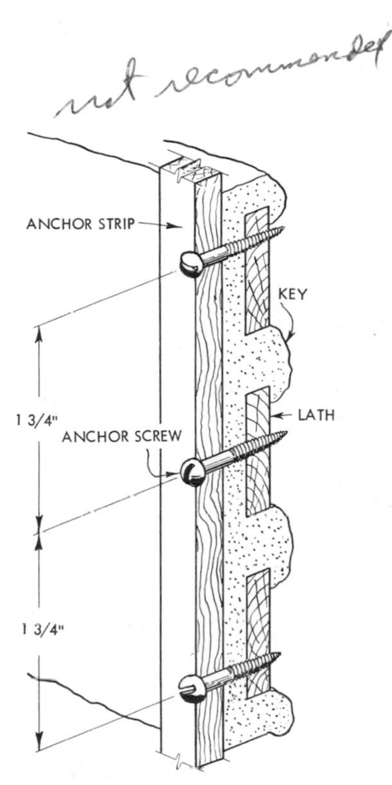

Fig. 15-8. Wall construction showing method of supporting plaster when cutting opening for outlet in very old houses.

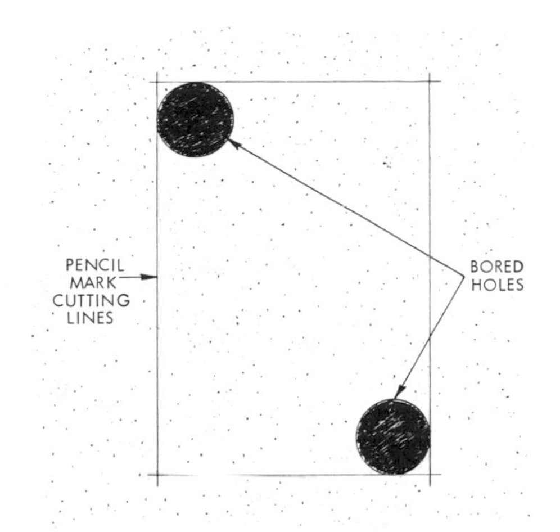

Fig. 15-9. Layout on drywall or plaster for cutting a hole to mount switch or outlet box.

the screws to pass through. The strip is installed as shown, Fig. 15-8, with No. 6 or No. 8 wood screws. Pilot holes should be drilled through the laths to prevent splitting. Saw the laths on the side where the hole is desired, and transfer the strip to the other side of the opening. If any screw holes fall outside the span of the switch plate when the job is finished, they may be filled with plaster of Paris, and tinted if necessary. But if this operation is done carefully the holes will fall inside the switch plate space.

With sheetrock or with patented lath to support plaster, such as button-board, the operation is much simpler. The hole may be cut with a thin saw, or even a sharp knife, at any desired place except directly over a wall stud. Fig. 15-9 shows a penciled layout on a wall in preparation for mounting a switch or receptacle box. Holes may be drilled with a brace and bit at the opposite corners as shown. Inserting a key hole saw (compass saw) into the holes and cutting along the penciled layout allows space for mounting a box. This can be used on dry wall or plaster construction.

Attachment of a switch box to wood lath or other wall coverings is also an easy task. Fig. 15-10A shows how a switch box is fastened to wood

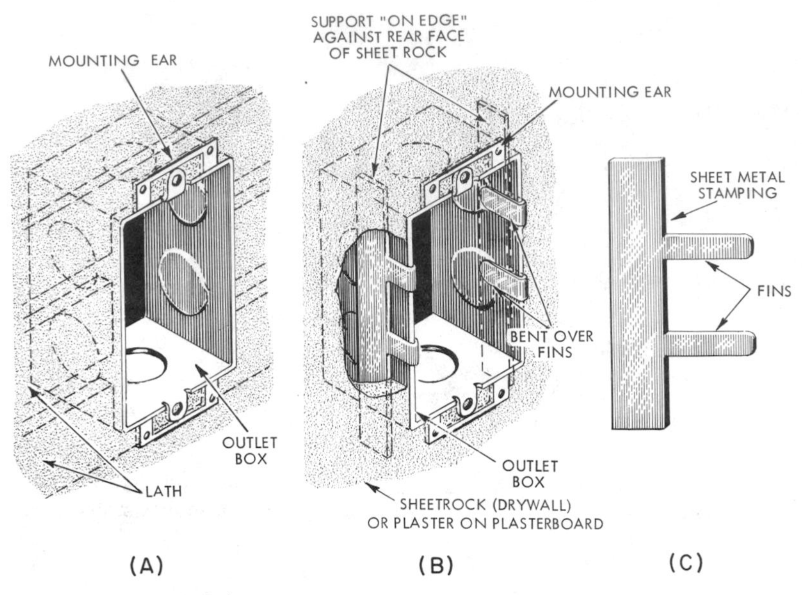

Fig. 15-10. Mounting switch boxes.

lath, a small wood screw securing the upper mounting ear to the lath against which it rests, and a second screw holding the lower ear in position.

Figure 15-10B illustrates the method of using sheet metal box supports, and Fig. 15-10C shows the part used for holding a switch box in a sheetrock wall. One of the supports is placed in the hole standing "on edge" against the back face of the sheetrock with its two projecting fins sticking out through the opening. The fins are bent over, along the wall, to hold the support temporarily while the other one is installed. The switch box is now placed in the aperture, and the fins are bent into the box and down along the inner face, as shown in Fig. 15-10B. After the second set of fins have been handled in the same way, the box is securely retained.

Another more popular type of "old work" box is shown in Fig. 15-11. It contains side cleats approximately 2½" x ½" mounted on threaded screws. These side cleats, or brackets, fold flat against the side of the box for insertion into the hole cut in the wall for the box location. When the box is in position the screws are tightened to open the cleats as shown in Fig. 15-11 and 15-12, bottom left. Fig. 15-12, top left, shows how the "L" or angle clamps are adjusted to position the front side of

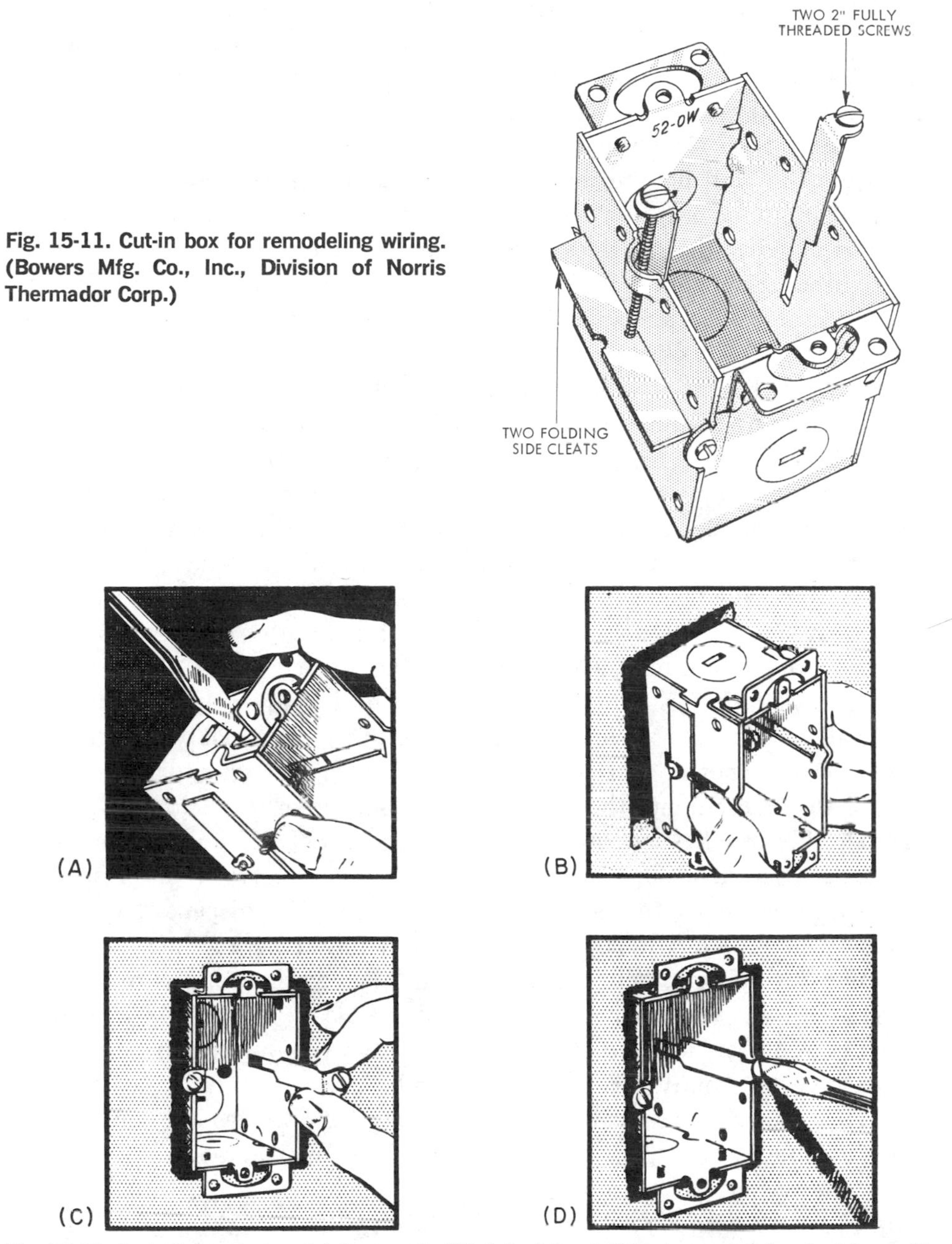

Fig. 15-11. Cut-in box for remodeling wiring. (Bowers Mfg. Co., Inc., Division of Norris Thermador Corp.)

Fig. 15-12. Installing a box cut into a wall. (A) Adjust front "L" clamps so front of box will be flush to wall; (B) fold side cleats in and insert box; (C) adjust side cleats before tightening; (D) tighten screws to secure cleats. (Bowers Mfg. Co., Inc., Division of Norris Thermador Corp.)

the box flush with the wall surface. Fig. 15-12, top right, shows how the side cleats are folded flat against the sides of the box to be inserted into a hole cut in the wall. The side cleats or inside the wall holding brackets are pulled outward until the screws are positioned on the box sides ready for tightening, as shown in Fig. 15-12, bottom left. The screws are then tightened until the cleats are pulled up tightly behind the wall surface, making a secure installation, as seen in Fig. 15-12, bottom right.

It is recommended however, that the circuit cable be fished into place first and connected to the box before the box is installed.

Plug Receptacle Outlets. The procedure for mounting plug receptacle boxes on plastered or composition walls is exactly the same as for switch boxes. Another method, however, is to cut them into the wooden baseboard. After the location is determined, and an outline drawn on the baseboard, two or four small corner holes are drilled so that a keyhole saw may cut along the outline from hole to hole. See Fig. 15-9.

A range outlet, for free standing ranges, can be mounted on the surface of the floor, thus avoiding the added problems of putting it in the wall. The outlet, however, must be located so as not to obstruct the positioning of the range. The outlet will be hidden under the range. This method requires that a hole be bored through the floor. Allowance must be made for clearance for a cable connector, conduit locknut, or flex connector.

Access or Scuttle Hole

The job of fishing cables is simple when there is enough attic space in a one-story building or, as the case may be, on the top floor of a multistory building. Usually, there is a scuttle hole for access to this space. If not, it is easy to make one in a closet or storeroom.

Locate the ceiling joists, and remove plaster between two joists over the approximate area of the desired opening. If wood lath is encountered, nail temporary wooden strips to the joists, as in Fig. 15-13, driving the nails only part way in. The strips should be about 26″ long. Nail end strips at right angles to them to mark the outline of the hole, and saw out the required amount of lath. If the ceiling is sheetrock or plaster the strips may be dispensed with; saw the required distance along the joists and then at right angles to them until the piece of material is cut free.

The strips may then be removed and a permanent frame installed, as in Fig. 15-14. Headers and a cover may be constructed as shown.

Fishing

Fishing methods are the same, regardless of whether the wiring material happens to be loomed wires in

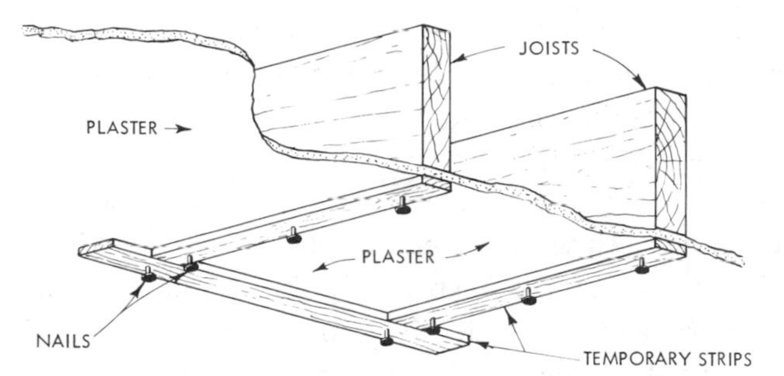

Fig. 15-13. Temporary strips for supporting plaster when cutting a scuttle hole.

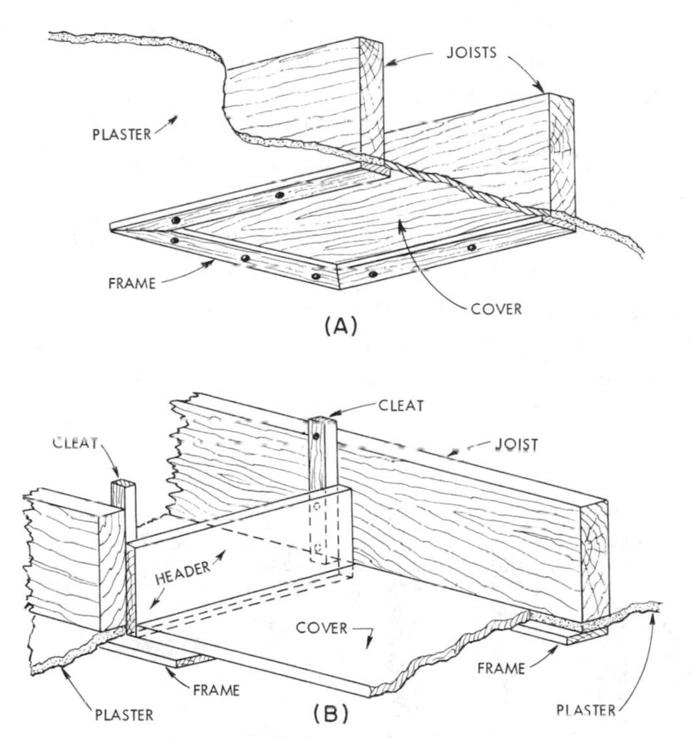

Fig. 15-14. Frame and cover for scuttle hole.

a knob-and-tube installation or one of the cables. When the term "cable" is employed in this discussion, loomed wires are included as well.

Fishing down to a wall outlet from an attic space is not a complicated procedure. The first step is to bore a hole through the partition plate directly above the point where the outlet is to be located. This point can be located through suitable measurements transferred from the lower floor. The plate is a 2 x 4 timber, often a double one, between the upper ends of the studs and the joists.

Next, drop a sufficient length of sash chain or equivalent through the hole, and fish it out from underneath with a piece of hooked wire. Attach the cable to it, and draw up enough to reach the outlet box from which the switch loop is to be supplied. Where headroom is insufficient to permit use of a drill for cutting the hole in the plate, a right angle drill motor, or chisel and hammer may serve the purpose. This difficulty arises, sometimes, when a gable roof slopes down onto the partition at a fairly small angle.

If there is an unfinished basement under the floor that is being wired, wall outlets, especially baseboard plug receptacles, may be fished from there. Bore through the floor plate or sole of the partition from underneath. If the baseboard has been removed, it may be easier to drill from above. In this case, other wall outlets around the room may be wired by running cable in bored or notched holes in the studding behind the baseboard.

Passing Obstructions

In most frame walls, braces, fire stops or bridging blocks are inserted between adjacent studs. They consist of short pieces of 2 x 4 placed at an acute angle, or else straight across. Figure 15-15 shows braces in two partitions.

Fig. 15-15. Diagonal and horizontal bridging.

Since either type of bridging hampers fishing operations, steps must be taken to pass through or around it. For example, a switch is to be installed near a door and at a point under a brace whose location has been determined by measuring the fish chain or by carefully sounding the wall with a mallet, hammer or other suitable object. First, take off the door stop, which is at least 1½″ wide. Bore a hole through the door casing and the stud, or door buck, at a point above the obstruction, and another some distance below it. Then, cut a channel in the door casing from one hole to the other, slanting off the corners where channel and hole meet, to avoid sharp bends in the cable. After the wiring is installed and the door stop replaced, there will be little evidence of work.

If the outlet is to be, say, 3′ from the door casing, the same plan may be followed, using one or, if necessary, two bit extensions to reach through the next stud. A door opening may be crossed by taking up the threshold and sawing or gouging a channel underneath it. If there is no threshold, a channel will have to be cut at the top of the door opening.

When the outlet location is too far away from a door, another method of passing the obstruction must be employed. Figure 15-16 illustrates the manner in which bridging may be passed during the process of fishing cable from the attic to a plug receptacle in the baseboard. After the bridging has been located with considerable accuracy, as shown by the dotted lines, Fig. 15-16 B, the outline of a narrow rectangle is drawn with a piece of chalk. If the wall is made of bare plaster, the material is scraped out until the lath is exposed. The lath is cut with a chisel, and a hole chiseled or gouged in the cross brace. After the cable is installed, the hole is repaired with plaster of Paris.

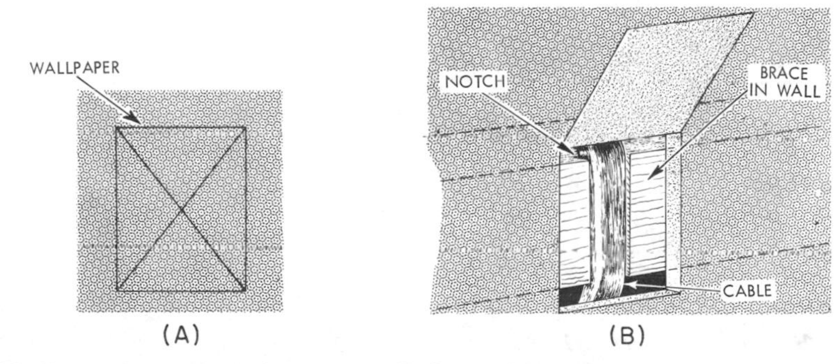

Fig. 15-16. Alternative method of passing wall obstruction.

If there is wall covering or wallpaper, it may be cut diagonally as in Fig. 15-16A, or rectangularly as in Fig. 15-16B, depending on the design of the covering. After proceeding as before, the wall covering is pasted back.

Sometimes existing single gang switch boxes may be converted to two or three gang switch locations. These switches are smaller and operate horizontally. The problem here is in fishing cable to the box. Should a single-gang box contain a neutral and a hot wire, future extension of the existing electrical system is greatly simplified.

Other methods of passing obstructions are shown in Fig. 15-17 for inside partitions. Where there is ample

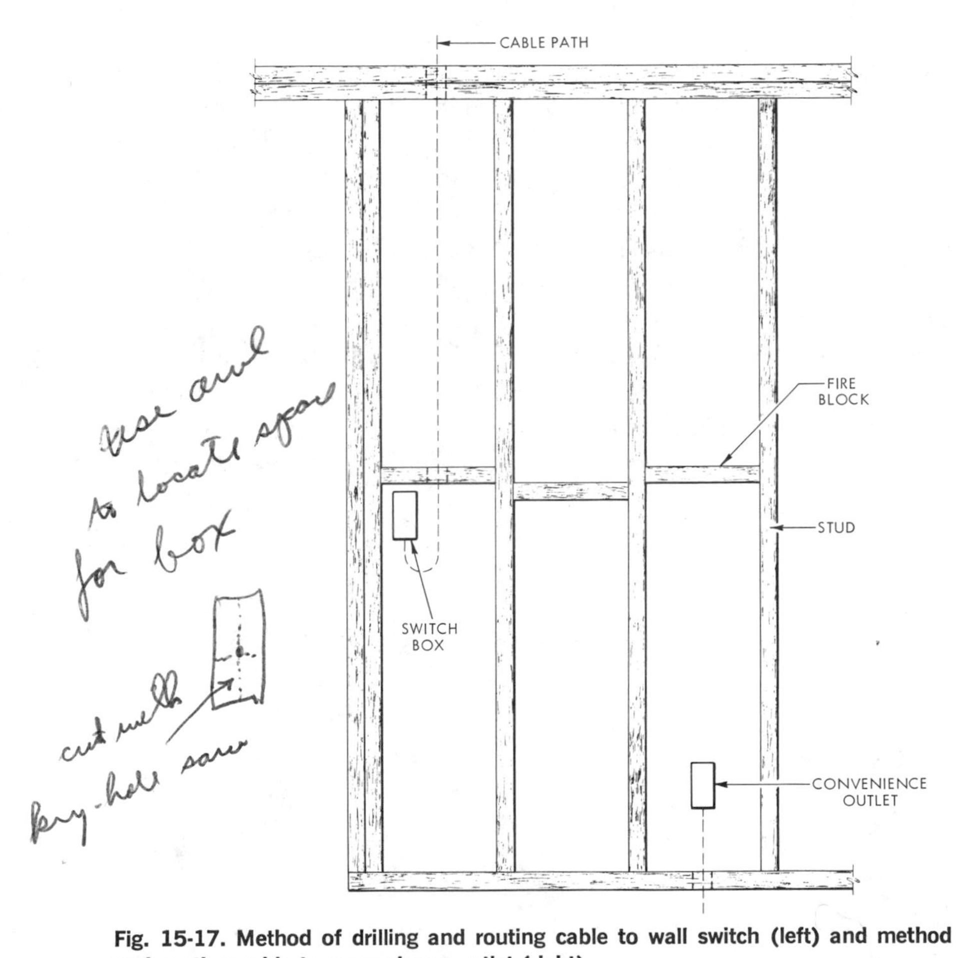

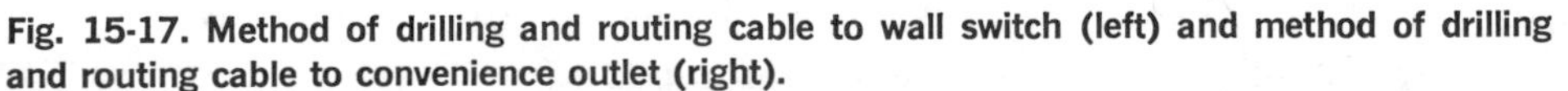

Fig. 15-17. Method of drilling and routing cable to wall switch (left) and method of drilling and routing cable to convenience outlet (right).

attic space, a hole can be drilled through the top plates for mounting a switch. With the use of a *bit extension* a hole can also be drilled through the fireblock or horizontal bridging. Diagonal bridging is most difficult to drill because of the difficulty in starting the screw of the auger bit into the wood at an angle. However, it can be done. After holes are drilled a string or cord with a screw or nut for a weight may be fished through the holes. This string and weight is sometimes called a "mouse." A fine sash chain may also be effectively used. The string may be tied to a cable and pulled up or down, utilizing the hole space cut into the plaster for the switch box. The cable can be connected to the box outside the wall before it is mounted.

A convenience outlet may be installed the same way if the house is of slab floor construction.

A more convenient method is to feed it from under the house (as shown in Fig. 15-17, right) if there is crawl space. Many times with the aid of a good flashlight, the sole or bottom floor plate of a partition can be seen through cracks in the subfloor. Measuring from an outside or other wall, the desired location for drilling can be located below. Power supply for either switch or outlet may be obtained by tapping onto other outlets or from the panel.

Some wiremen prefer removing the lower wall floor molding (base shoe) and driving a nail through the floor. This nail is then located under the house and measurements are made over to reach the center of the partition. This locates the position for the hole to be drilled for flex or cable entry. Extension cords, caps and drills should be in perfect condition for this kind of work to guard against electrical accidents. Accidents are especially dangerous in the close, hot quarters in which the electrician must often work. Poorly insulated cords or poor ground connections can cause death by electrocution if accidental contact is made.

Holes should be drilled with standard ship auger bits with no spurs. Nails are often struck in this kind of work because there is little choice in the matter of hole locations. This style of bit is not damaged so easily as other types, and it is somewhat easier to put in shape again if damaged in striking a nail.

Fig. 15-18 shows an easy method of clearing a foundation wall to supply outside outlets such as post lights, landscape lighting or convenience outlets. A nail or an awl may be used to punch a hole for location through the outside plaster. A small hole is drilled through the wooden header or a nail is again driven. If upon investigation clearance is satisfactory under the house, the nail hole may be enlarged for clearance of a conduit

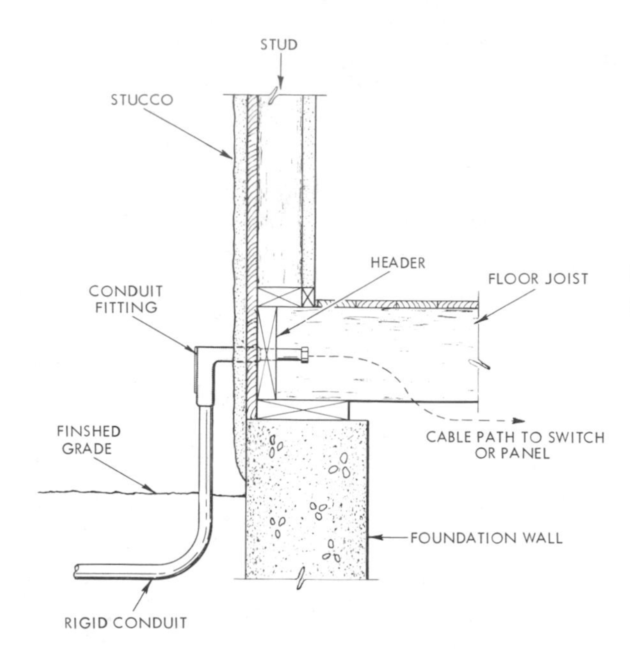

Fig. 15-18. Method of feeding landscape lighting.

nipple (a short piece of conduit threaded on both ends). The conduit fitting shown in Fig. 15-18 has an opening in the back for access of pulling wire or splicing. It is generally called an "LB" fitting: "L" means elbow, "B" means opening in the back. Here is where the cable conductors brought out from under the house are spliced to single conductors pulled into the conduit. A gasket and cover are then installed finishing the fitting. A bushing should be installed on the inside of the nipple to protect the cable insulation.

This method may be considered for some basement built homes. For slab floor construction, this method may sometimes be used in connecting to the back of a convenience outlet inside the house. A knock-out may be removed from the box from inside after measurements have decided that the approximate outside location is satisfactory. A small hole through the wall will identify this location on the outside wall and the previously described process may proceed.

Removing Trim and Floor Boards. When removing a piece of interior trim, use a thin, wide tool to distribute pressure over as large an area as possible. A narrow chisel is apt to leave a mark which is hard to remove or conceal. Start the prying effort at a point above or below the normal line of vision so that marks left by the tool will not be readily observed.

If cable must be installed at right angles to joists which are inaccessible from below, for any reason, it is necessary to take up the floor. The beginning should be made where two pieces of flooring meet, if possible, so as to have a loose end available when starting to raise the boards. First, however, the tongue must be cut on each side of this strip of flooring. An ordinary keyhole saw is likely to cut away too much stock from the edges, and cracks will be noticeable when the boards are replaced. A tool which uses a much thinner hacksaw blade, will not cause this trouble. When a hacksaw blade is used without some kind of a holder, it should be inserted so as to cut on the upstroke, because it tends to buckle if made to cut downward.

Before cutting may be started, a hole must be made through the tongue large enough to admit the saw blade. A good tool for the purpose is a painter's scraper, which resembles a putty knife, but which is wider and stiffer. Set one corner of the blade in the crack, and strike sharply with a hammer, as illustrated in Fig. 15-19.

There may be trouble with flooring nails, especially if they were not driven fully home. They are cut through easily with the hacksaw blade. While flooring is being raised watch carefully for any bulging along the edge of the adjacent strip, and apply pressure there to avoid splintering. Cross bridging runs should be located before removal of floor boards is attempted, in order to prevent the mistake to taking up a section directly above them. Cross bridging in floors consists of "herringbone" braces usually located half way between bearing partitions. They may sometimes be spotted from underneath by way of the hole

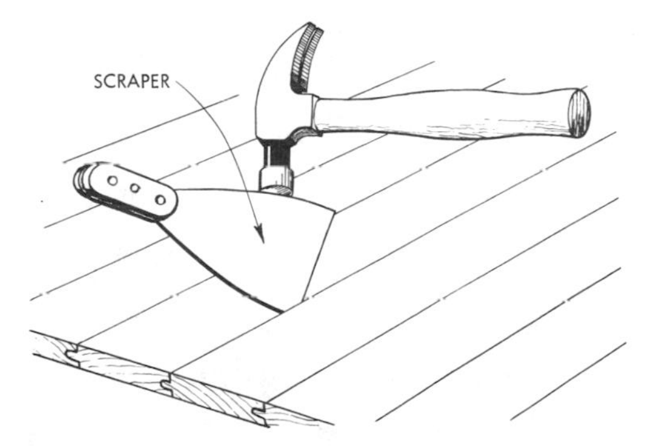

Fig. 15-19. Method of starting to cut tongue in floor board.

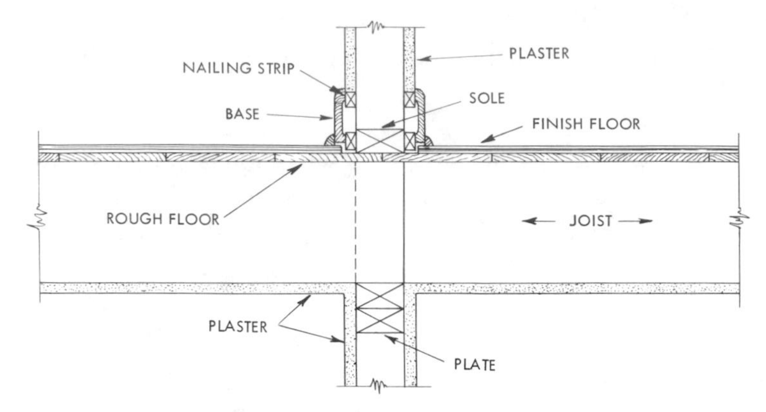

Fig. 15-20. Construction of partition wall for two story building.

cut in the ceiling for the lighting outlet below. After the boards have been raised, any rough underflooring may be cut without difficulty, using a keyhole saw or a short handsaw.

Removing Baseboard. With this work completed, a method for getting down to the wall switch location on the floor below must be considered. A common device is to remove a section of baseboard, thereby exposing the sole of the partition, usually 2 x 4 to which the studding is nailed, Fig. 15-20. Chisel a hole through this timber large enough to admit a hand. Locate the plate of the lower partition, and bore a hole large enough for the cable.

If the baseboard is flush with the plaster, and there is no other alternative, care in removing the baseboard should be exercised so as not to damage the plaster.

Cutting a Pocket. An alternative plan to removing baseboard is to cut a pocket between two joists, close to the upper wall, and then drill the plate of the lower partition. This method must be used, also, when there is no partition directly above the lower one.

If possible, locate the pocket at the end of a strip of flooring. Sometimes it pays to start at the end of a strip one or two joists beyond the desired location so that it will not be necessary to saw a floor board. Cut the tongue, and begin raising the strip from one end, going as far as possible before running the risk of splitting that portion in which the tongue is still intact.

Insert a bar, chisel, or similar object between the floor and the raised strip, as in Fig. 15-21. If there is room enough to cut the underflooring over section *A* and to perform the other work, nothing further need be done to the floor. Should more working room be needed, the strip must

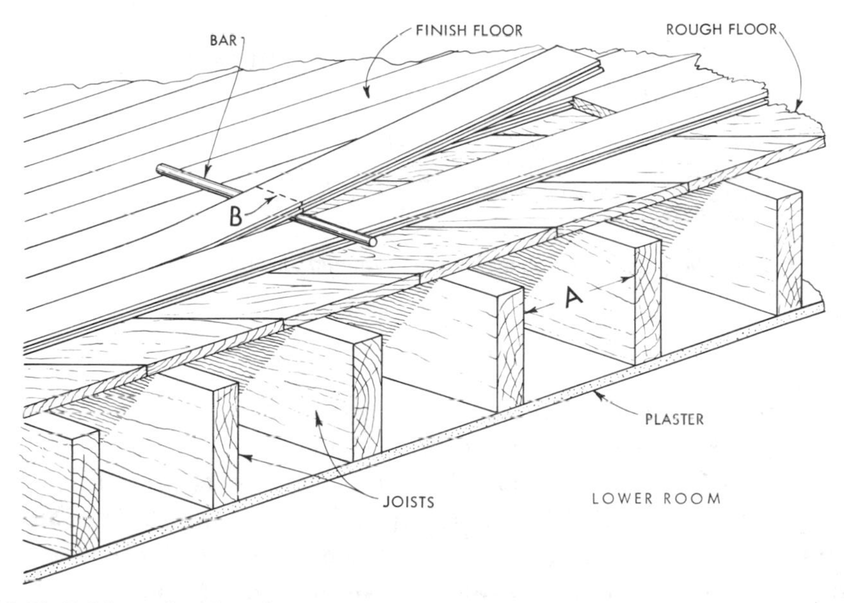

Fig. 15-21. Raising a floor board.

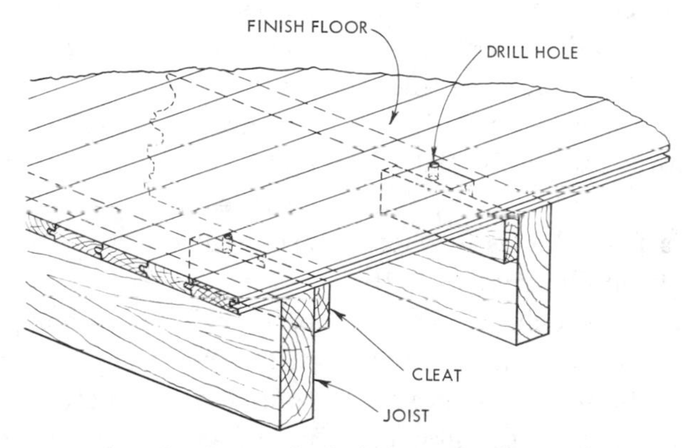

Fig. 15-22. Floor construction showing method of supporting floor strip.

be cut at point *B*, which is the middle of one of the joists, in order to get support for both ends of the strip when the board is replaced. A thin saw should be used to avoid cutting away too much stock.

When a pocket is to be cut between two floor joists, say in the middle of the room, proceed as in Fig. 15-22. Bore a 1/4″ hole through the strip that is to be cut, the hole being as close to the joist as possible. Repeat

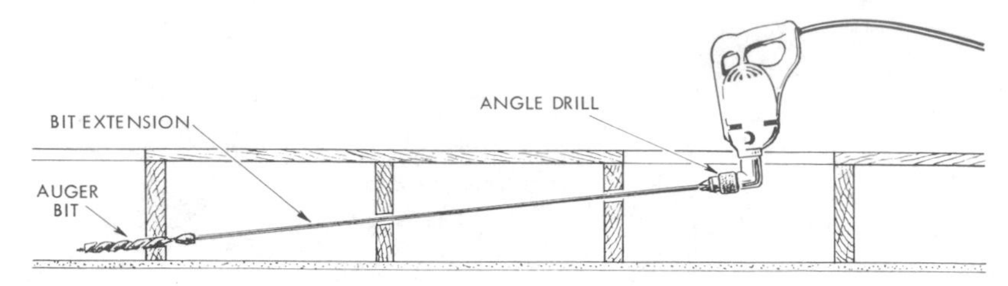

Fig. 15-23. Boring floor joists with auger bit extension and electric angle drill.

the operation at the other joist. Cut the tongue and remove the strip. Before replacing it, nail a substantial cleat against the face of either joist to support the piece of flooring.

Occasionally, time may be saved by cutting a pocket in one joist space and another pocket two or three spaces away, instead of taking up flooring all the way across. The three joists, Fig. 15-23, can be bored from one pocket and the cable fished through.

Do not cut a floor that is finished in parquetry or similar material. If cables must traverse such an area at right angles to joists, take off the baseboard along one of the sidewalls, and bore or notch the studding to take the cable.

Replacing Wooden Floor. Rough flooring must be replaced, at least over the joists, and paper or other lining that was present must be replaced in order to bring the strip to its original level. Nailing must be well done so the floor will not creak when walked on. Use flooring nails at each joist, driving them well into the strip with a nail set. The holes may be closed with putty or plastic wood which is stained to match the color of the flooring.

Installing Special Lighting

Recessed Incandescent Fixtures. Recessed fixtures are quite popular on modernization projects. The codes require that such units shall be mounted so that clearance of the metal enclosure from combustible surfaces shall be not less than ½″, except at the supporting lip.

One type of fixture, which has a built-in junction box protected from direct heat of the lamps, may be connected directly to the circuit wires. With other types, the circuit wires should terminate in a junction box not less than 12″ from the fixture, Fig. 15-24. Wires from the junction box to the unit must be approved for high temperature (usually some form of asbestos-insulated conductor) and should be enclosed in a suitable raceway not less than 4′ and not greater than 6′ long.

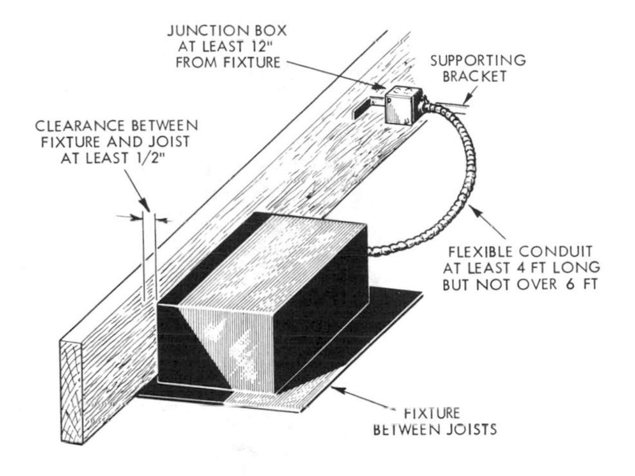

Fig. 15-24. Recessed lighting fixture without built-in junction box.

This type of fixture gives good lighting in the immediate area under the fixture.

Recessed Fluoresent Area Lighting. When lighting intensity is to be increased above that previously described without adding more surface or conventional lighting fixtures, fluorescent lighting which is recessed into the ceiling is often installed. This means cutting and removing a section of plaster and one ceiling joist. This recess in the ceiling is then boxed in, painted white, and rapid-start fluorescent fixtures are installed. Light diffusing plastic sheets cover the fixtures, hung on aluminum "T" (tee) bars. An installation is shown (looking up) in Fig. 15-25. The dimensions shown are approximate. The installation shows four, four foot fluorescent, rapid-start fixtures. The lamp will light instantaneously when energized by a wall switch.

The precise side width of the recess in the ceiling is found when one or more joists are cut out. The normal spacing of 14½" between joists (assuming 16" from center to center) is not considered enough additional lighting space for most applications. When one or more joists are cut the ceiling must then be reinforced with headers the same size as the joists, nailed at each end of the opening. No reinforcement is necessary if no one is walking above and headers are nailed in immediately.

The distance between the end headers should allow clearance for mounting fixtures. Greater width openings will require possibly two headers laminated with nails at each end. With recess dimensions estab-

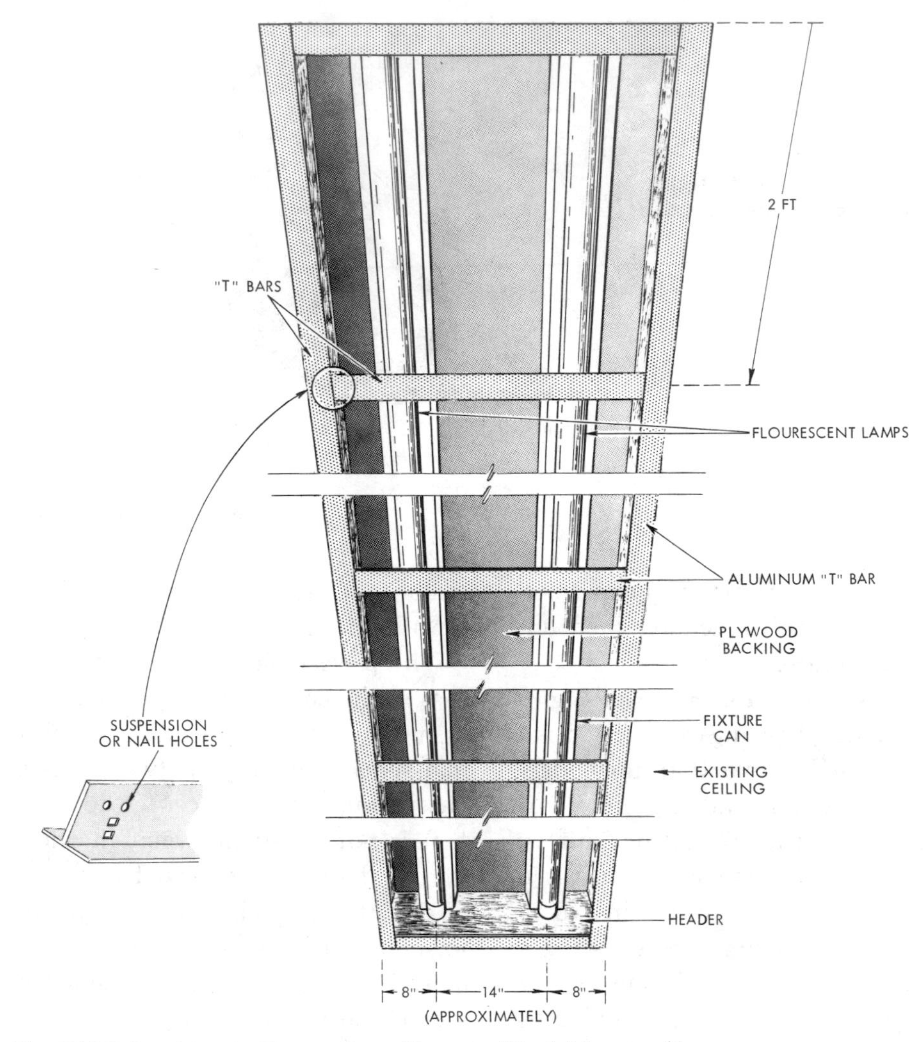

Fig. 15-25. Luminous ceiling section with one ceiling joist removed.

lished and laid out on the ceiling, the plaster can be cut with an electric power saw using a special blade for this purpose, or the same procedure may be used as has been already described for installing a scuttle hole. Should the house be very old, openings are located where they are desired but they may have to be adjusted slightly one direction or the other depending upon exact joist location. Header and ceiling joists may have to be heightened to gain 4 to 6 inch clearance under the lamps when mounted. Added 2 x 4s to the top of existing joists may be sufficient. Plywood backing is installed on top of the recess to enclose the opening. Before lighting fixtures are installed the inside of the recess should be painted white to reflect light downward.

Aluminum is rust resistant and is recommended for use in border "T" (tee) bars. Tee bars are nailed in place as shown in Fig. 15-25. One side of a "tee" border covers a plaster cut, the other side supports one side of the plastic light diffusing panels. White, translucent plastic sheets are cut to fit openings and are installed to rest on tee bars. The plastic sheets hide the fixtures. A white ceiling is the effect during daylight hours and a luminous ceiling when lamps are energized at nighttime.

This lighting is recommended for kitchen areas, bathrooms, game rooms, studies or home offices. Existing lighting intensity, location, room shape and size dictates amount of new light to be installed. The example shown in Fig. 15-25, or 1½ times this size or twice this size, will light an average kitchen satisfactorily. However, a minimum of 80 to 100 watts of fluorescent light for 50 square feet is recommended. Warm-white fluorescent tubes should be used for natural color. These installations are recommended for new construction also.

Wall to Wall Luminous Ceilings. Unfinished basement ceilings with exposed pipes, drains, air ducts, etc., can be neatly and conveniently covered, but remain accessible, with a suspended wall to wall luminous ceiling. Other rooms such as bathrooms, kitchens, dens, family rooms or home studies are also brightened with indirect, glarefree lighting with suspended luminous ceilings. These are especially adapted to homes with nine foot or higher ceilings in older homes but can be installed in eight foot high, standard height ceilings. Wall to wall luminous ceilings are desirable in new construction also.

To achieve a blend of non-glare fluorescent light and sound absorption for large areas, a combination plastic light-diffusing and acoustical panel can be installed on a suspended grid of angle and tee bars. This is similar to the recessed fluorescent area lighting with the exception that wall to wall lighting is spended below an existing ceiling.

Lighting Fixtures. Fig. 15-26 illustrates fluorescent rapid-start, strip fixtures mounted on a ceiling. Small dimensions are used for simplicity in this example. The fixtures are mounted one half the distance to the wall as they are mounted apart from each other, as seen in Fig. 15-26. This should help even light distribution assuming intensity is equal one foot on each side of each fixture. This principle is also applied to placement

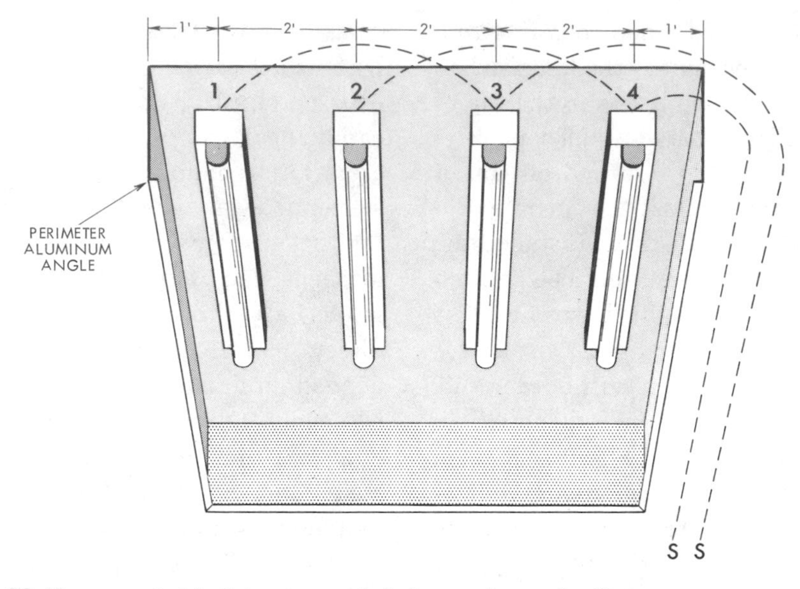

Fig. 15-26. Fluorescent strip fixtures mounted above a dropped ceiling.

of incandescent lighting fixtures. However, inefficient incandescent bulbs are avoided with luminous ceilings. This is due to excessive heat, much higher power requirement for light output, and a need for frequent replacement.

Plain white plaster above translucent panels provides excellent reflection. Gypsum board should be painted white. White enamel reflectors above fluorescent tubes are helpful when there is no white structural ceiling close by.

Noisy rated fluorescent lighting fixtures should be avoided in discriminating residential room areas. Fixture selection in this respect should be discussed with the experts or manufacturers' representatives. Economical and convenient control of light intensity is achieved by wiring fixtures on separate switches. The fixtures should be alternately connected to two switches as shown in Fig. 15-26. Greater light intensity is achieved by using double tube fixtures (fluorescent fixtures with two lamps).

Mounted fixtures as shown may be wired by cutting one hole in the ceiling for an outlet box. After mounting one fixture over the outlet box, the other fixtures may be wired together on the surface. Non-metallic sheathed cable should be avoided here.

The fixtures may also be wired independently with an outlet box for each fixture. Local code requirements should be observed.

Mounting the Suspension Grid

Wall Angles. The suspended grid is a dropped ceiling. The "T" members, Fig. 15-27, right, rest on the wall "L" angle channels, Fig. 15-27, left. Cross T's are assembled to the main T's.

To install the wall L's, the perimeter of wall should be measured and the L's cut to fit with a fine toothed hacksaw. Holes are made with a punch or drill at stud locations in the wall. A level chalk line may be snapped around the room at height above the floor that the drop ceiling is to be installed. To prevent shadows and bright spots on a fully luminous ceiling, the ceiling should be dropped one half the distance between fluorescent tubes.

A level can aid the installation of wall angles as shown in Fig. 15-27, top left. Six screws or nails per 12 foot of angle are sufficient but more may be used if desired. The corners may be lapped or cut to fit.

Main T's. The main T's, Fig. 15-27, right, should be run at right angles to the ceiling joists, two or four feet apart depending upon the size of translucent plastic to be used. Two feet by two, or two feet by four are common sizes. When room dimensions do not allow complete, even spacing, the remaining uneven spacing should be divided equally on the border for better appearance. A simple grid layout on paper will assist here and will aid in material estimating. Working from a center line to the walls will insure equal desirable borders. The ends of the main T's rest on the wall L's. The T's are suspended from above by hanger

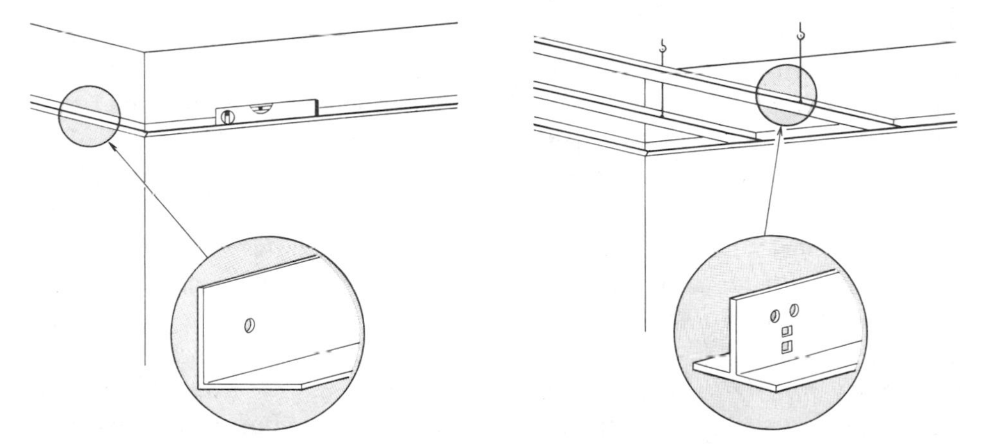

Fig. 15-27. Installing wall angles (left) and main suspended tees (right).

wire approximately every four feet. This is done by screwing hanger hooks through the plaster into the ceiling joists crossing above, see Fig. 15-27, top right.

For extended lengths of T's, *snap on couplings* are generally used for *splicing ends*. T's are available in different lengths.

Cross-T's. Cross T's are installed between the long main T's and between main T's and the wall angles. Generally there are notched holes spaced two feet apart on the main tees that cross tees clip into. The wall ends of the cross T's simply rest on the wall angle and are held in place by gravity and the weight of the plastic panels.

Panels. The panels are simply dropped into the grid. At borders and around obstructions the panels can be trimmed as needed. Flat plastic luminuous panels can be scored with a sharp knife and snapped apart or they may be cut with tin snips. Plastic louvers and flat plastic panels may be cut with a cross-cut hand saw or cut with a power saw. Acoustical panels, that may be used with translucent panels, may be cut with a long, thin, sharp knife. These acoustical panels may be used in large areas to deaden sound. They may be used in alternate or random patterns but not below lighting fixtures. They may be sized the same as the plastic ceiling pieces and dropped in places instead of the plastic. Acoustical panels are about ½″ thick and lightweight.

Surface Metal Raceways

Surface metal raceways such as the samples shown in Fig. 15-28 are channels for holding wires on interior surfaces and are designed expressly for this purpose. Specific sizes and fittings are available for nearly every code approved purpose and for different numbers and sizes of conductors. This material is most convenient where concealed wiring is impractical. The letters, "A" to "F" in the upper part of Fig. 15-28 match locations of a simple but typical installation in lower part of the figure. The convenience outlet is fished down the wall to "B", a junction box.

Surface metal raceways and their elbows, couplings, and similar fittings must be installed so that they are electrically and mechanically coupled together, while at the same time protecting the wires from abrasion.

Holes for screws or bolts inside the raceway should be so designed that when screws and bolts are in place their heads will be flush with the metal surface. Raceway and fitting bases are mounted and assembled before wire is pulled. The raceway length is cut to fit with a fine toothed hacksaw. Raceway ends are supported by mounted fitting bases with points of attachment for this purpose. Similar to these points of at-

(A) SINGLE-POLE SWITCH

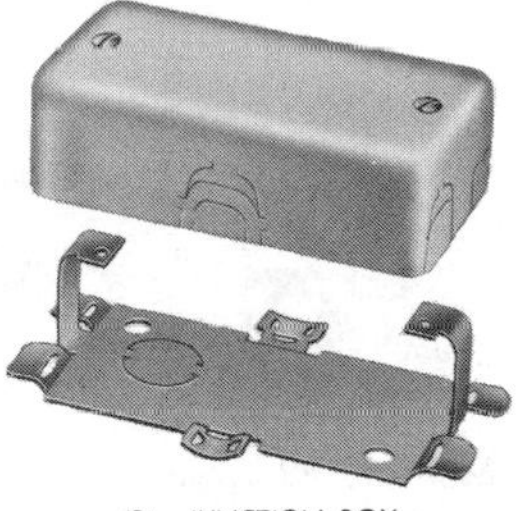
(B) JUNCTION BOX

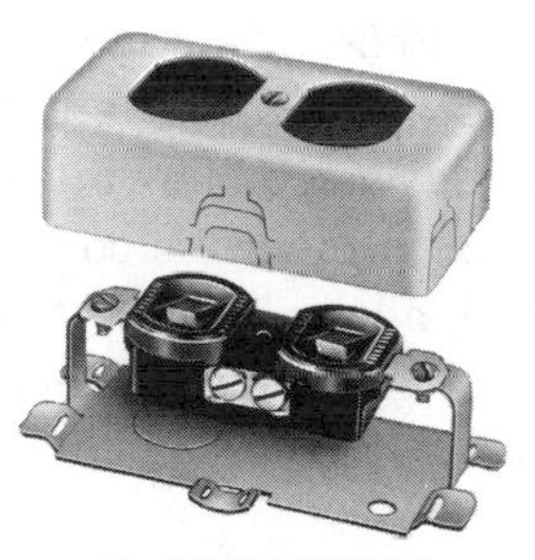
(C) DUPLEX RECEPTACLE

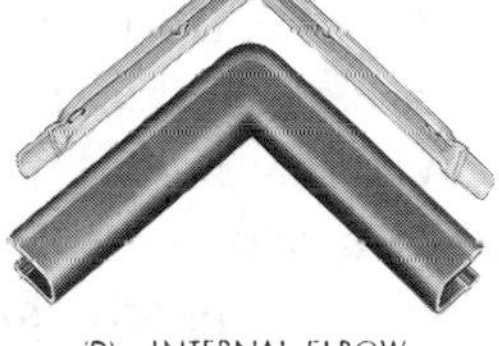
(D) INTERNAL ELBOW

(E) RACEWAY

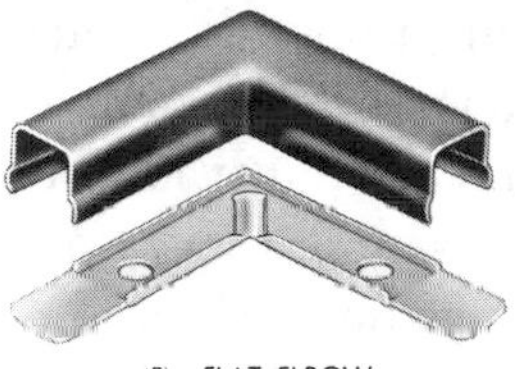
(F) FLAT ELBOW

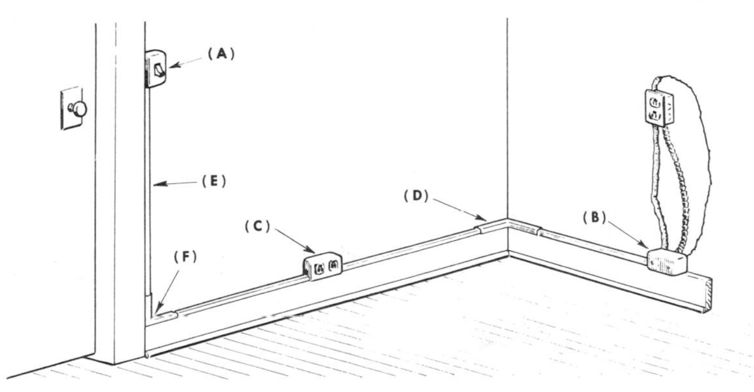

Fig. 15-28. Wiremold installation and fittings. (Wiremold Co.)

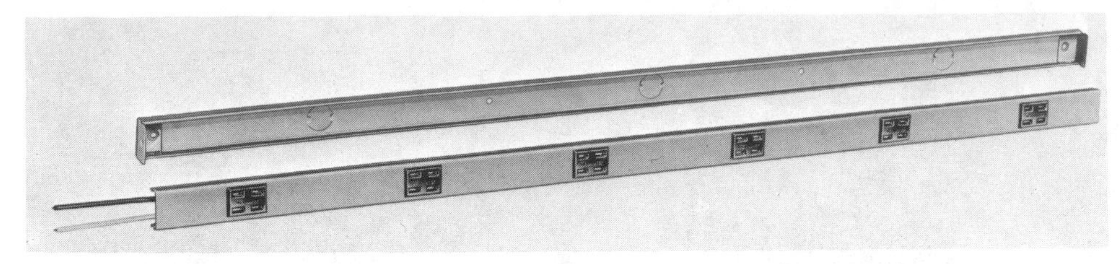

Fig. 15-29. Multi-outlet assembly. (Wiremold Co.)

tachments are double ended couplings for extending lengths of raceway. A screw mounts the coupling and also serves as a support. Other raceway supports are mere clips that are mounted first. Two hole straps are available too.

Conductors are pulled and joints and connections are made. After wiring devices are mounted, proper knockouts are removed from the device covers and they are installed completing the job.

Multi-Outlet Assemblies. Fig. 15-29 shows a multi-convenience outlet assembly applicable to work areas in kitchens, garages or areas where multiple outlets are desired. Multi-outlet assemblies are subject, in general, to the same code restrictions and allowances as those applicable to standard surface metal raceways.

Nonmetallic Surface Extensions

Nonmetallic extensions are an assembly of two insulated conductors within a nonmetallic jacket or an extruded thermoplastic covering. Surface extensions are intended for mounting directly on the surface of walls. An installation and applica-

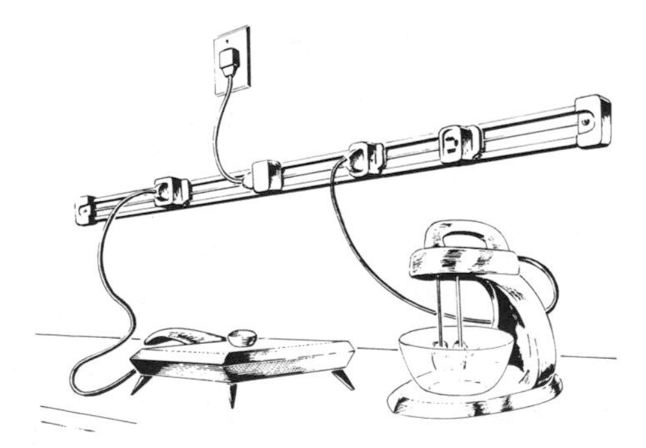

Fig. 15-30. One form of nonmetallic surface extension. (I-T-E Circuit Breaker Co.)

tion is shown in Fig. 15-30. Note that the connection to a supplying outlet is made with a cord and cap (plug). Allowances and restrictions should be checked with the Code for proper installation and locations.

Underplaster Extensions

An underplaster extension is used for extending an existing branch circuit in a building of fire-resistive construction. It is laid on the face of masonry or other material and buried in the plaster finish of ceilings or walls. The type and methods of installation of the raceway or cable for such extensions should be checked with the local code.

Remote Control Wiring

Remote control wiring, with small low voltage wires and cable, is often used in extending existing wiring systems. The system and methods of installation for new construction is given in Chapter 8, "Remote Control Wiring." Fig. 15-31 shows a method of adding more switches to an existing lighting outlet. Surface-mounted switches make the low-voltage system advantageous for rewiring, because they can be

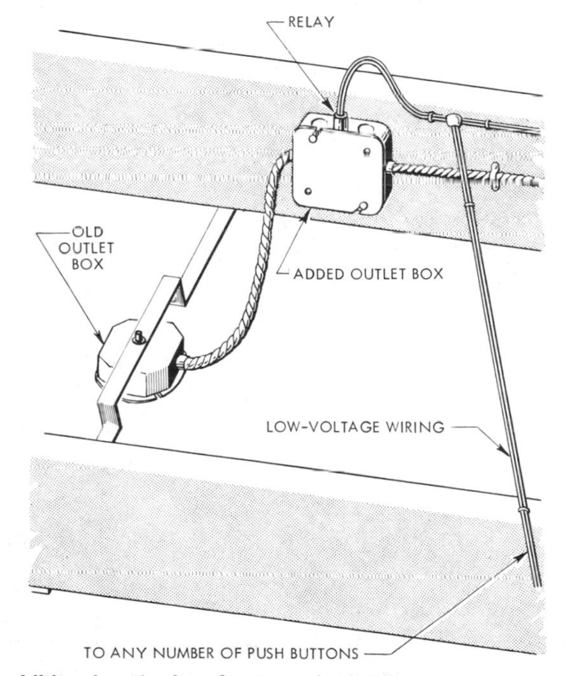

Fig. 15-31. Use of additional outlet box for mounting relay.

mounted so easily. Furthermore, the small three wire cable can be run behind moldings or along baseboards, and through areas that would not accept larger cable. If it is necessary to run low voltage cable across a plastered wall, it can be laid in a shallow groove and plastered over.

Each remodeling wiring job offers individual problems, and no single plan will satisfy all conditions. Close examination of remote-control wiring will show many ways it can be used to advantage.

Increasing Service Capacity

Electrical additions to existing branch circuits and the increase of new branch circuits to accommodate new room additions create problems in the service panel and wiring. Additions or changes are necessary to safely pass, protect and distribute conveniently the additional electrical current requirements. The more desirable large appliances, such as electric ranges and built-in cooking units, electric clothes dryers, air conditioners, electric water heaters, electric heating systems and others, all add to overloading an existing electric service center.

With this condition, 115 volt, two wire old services must be changed to three-wire 115-230 volts with increased wire size and additional fuses or circuit breakers. Existing three-wire, 115-230 volt services usually must be increased in wire size and fuses or circuit breakers to accommodate the additional current and branch circuits. This means a new main service panel, with increased wire size and more wire protective devices, larger conduit for service wires, and a larger grounding wire. Chapter 11, "Installing Service Equipment," deals with constructing new services and should be referred to because a new service must be installed in most of these cases.

Expanding Existing Wiring

Installing new services with existing wiring systems generally presents two basic problems: where to install it and how to connect it to existing branch circuits.

When new outside walls are being constructed with room additions the first problem is solved. The service may be installed as an original service flush into a plastered wall or on the surface of masonry construction. On existing walls the new service equipment may have to be mounted on the surface as shown in Chapter 11 (see Fig. 11-41). The meter may be installed outside, and the distribution panel may be located some convenient place inside. The new panel may be installed in the same location as the old.

Service heads can be located in many different locations. The power company may have to be consulted with a relocation of a service head and metering equipment. The elec-

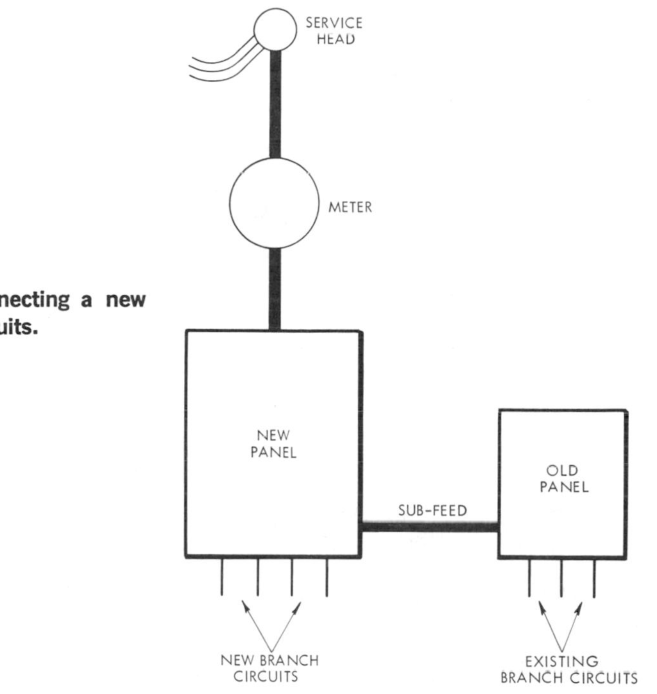

Fig. 15-32. A method of connecting a new service to existing branch circuits.

trical inspector may have to be consulted regarding a change in the service main switch and panel. Mounting of metering equipment on bedroom or sleeping quarter walls is most often discouraged.

When new service panels replace the old in the same location, the old branch circuit wiring may simply connect to the new protective devices. When conditions do not permit this installation, a sub-feeder may supply the old panel from the new. This may be seen in Fig. 15-32. The sub-feeder protective device installed in the new panel should be adequate in size to supply and protect the old panel. This circuit breaker may also serve as the disconnecting means of subfeeder and old panel.

New branch circuits may be connected and protected in the newly installed panel or the old one. This can be done with conduit or cable. A single conduit may supply another strategically located sub-panel. Several combinations are possible.

The greater the knowledge of building construction and electrical wiring systems, the easier it is for wiremen to remodel. A good remodeling wireman is a good all-around mechanic.

QUESTIONS FOR DISCUSSION OR SELF STUDY.

1. Why should all circuits be treated as hot?
2. Why is it somewhat objectionable to use a power hand saw for cutting the tongue of a floor board?
3. Why should extra precautions be taken in cutting a hole in a woodlath plastered wall?
4. What problem occurs if the nailing of floor boards is insufficient?
5. When cutting a round hole in plaster for a ceiling outlet, what step can be taken to keep the area clean?
6. In preparation of cutting a hole for a switch box in a wall, how should holes be drilled to allow space for a keyhole saw?
7. In supplying outside landscape lighting why is it not advisable to go under the foundation?
8. How are fireblocks drilled from the attic for fishing a cable through?
9. How are partitions located for drilling under the house?
10. In installing a luminous ceiling section over a kitchen work area, why must a header be nailed across ceiling joists when one is cut out?
11. Why are regular switch boxes not used when putting in a new switch?
12. What type of original plaster support was used on very old houses?
13. Generally, how are heavier lighting fixtures hung?
14. Other than on walls, where are range outlets installed?
15. What trade term is applied to passing a cable through holes in a wall from an attic to a hole cut for an outlet?
16. What does the term "mouse" mean?
17. Why are ship auger bits with no spurs recommended for old work?
18. Why are aluminum angles and tees recommended in kitchen and bathroom areas?
19. Why are not incandescent lamps used in luminous ceiling area lighting?
20. What are the main deficiencies of a very old service with increased branch circuits and power loads?

QUESTIONS ON THE NATIONAL ELECTRICAL CODE

1. What is the conductor color code for a three wire branch circuit, as specified in the NEC?
2. What is the maximum number of conductors in conduit or tubing for the following:

    New Work:
    ½", #14, type TW
    ½", #14, type THHN
    ¾", #10, type TW
    ¾", #10, type THWN

    Rewiring Existing Conduits:
    ½", #14, type TW
    ½", #14, type THHN
    ¾", #10, type TW
    ¾", #10, type THWN

3. What are the rated maximum operating temperatures of the following types of conductors: TW? THWN? THHN? RHW?

4. What type of wire is suitable for use in a damp or moist location?
5. What type of cable is approved for direct burial in earth for outdoor lighting?
6. What is the minimum depth required for direct burial cable?
7. What are the code requirements for supporting nonmetallic surface extensions?
8. What are limitations regarding the mounting of nonmetallic surface extensions on the floor?
9. In what type of building construction is underplaster extension of an existing branch circuit permitted?
10. What type of materials are permitted for use in underplaster extensions?
11. What are limitations of use of surface metal raceways?
12. What are limitations of use of a multi-outlet assembly?
13. What are limitations of boring holes in joists and rafters for extending a branch-circuit?
14. What is the requirement for free length of conductors at outlets or switch boxes for remodeling wiring?
15. What is the code restriction regarding splices in a flexible, portable cord used with a drill motor in boring holes?
16. What determines whether an increase in service size is necessary or not?
17. How is a new location of service equipment determined?
18. Who determines the location of a new service weather head?
19. How is it ultimately decided that old wiring should be replaced with new?
20. What are branch circuit electrical protection requirements for additions to a service panel?

Chapter **16**

Methods of Wiring Knob and Tube Systems

Today this method is not approved by many local codes and is rarely installed due mainly to high labor costs, as compared to other methods. It is given here because it exists in the NEC.

Open wiring, and concealed knob-and-tube wiring, is a wiring method which uses insulating knobs, tubes, and flexible nonmetallic tubing for the protection and support of insulated conductors. This wiring is run in or on buildings and concealed in hollow spaces of walls and ceilings.

Knob-and-tube wiring is the oldest type of electrical installation which meets requirements of the National Electrical Code.

Knob-and-tube wiring has a definite advantage for installations in damp or wet locations and also in buildings where certain corrosive vapors exist. Temporary installations —for example, fair grounds and construction jobs—can be more readily served by knob-and-tube wiring. It should be noted that the operating temperature will be lower than that of other wiring systems because of better ventilation afforded the conductors.

This method of wiring is divided into two types, namely, *open* and *concealed*. It is said to be open or exposed where the wires are run upon surfaces of walls and ceilings. It is concealed when hidden between ceiling and floor, or run on joists in an inaccessible attic, Fig. 16-1.

The exposed type of wiring is suitable for barns, sheds, and factories where appearance is not an important factor but where ease of servicing or altering is important.

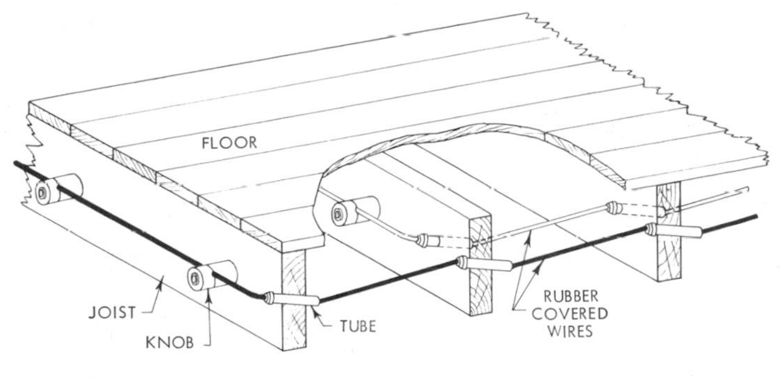

Fig. 16-1. Concealed knob and tube wiring.

Open Knob-and-Tube Wiring

Insulators

The rules for open work are simple and easy to follow. Insulators should be free from checks, rough projections, or sharp edges which might damage insulation. All knob-and-tube wiring must be done with single conductors. Wires size No. 14 AWG to No. 10 AWG are supported usually on 2- or 3-wire cleats, Fig. 16-2, and secured by screws or nails. Leather-protecting washers are used with nails. The nails or screws should extend into the wood at least as far as the thickness of the cleat. Wires size No. 8 AWG and larger are best supported on single-wire cleats, Fig. 16-3, and secured by two screws. Direction of joists must be considered when planning the wiring installation, because lath and plaster do not provide sufficient support for insulators. Supports should be every 4½′ with a joist spacing of 12″ or 16″. An insulator is required on every third or fourth joist.

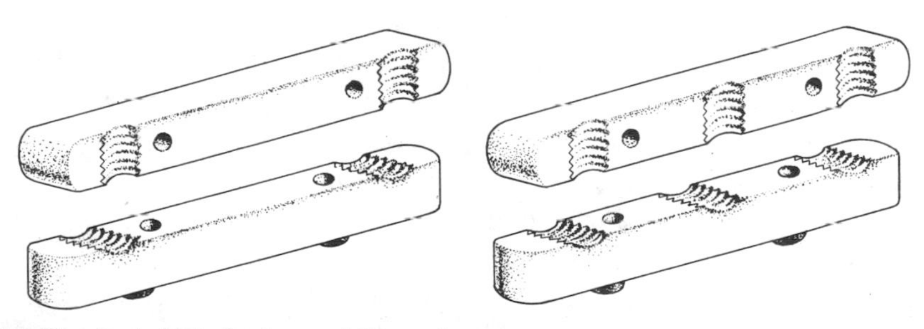

Fig. 16-2. Standard cleats for two and three wires.

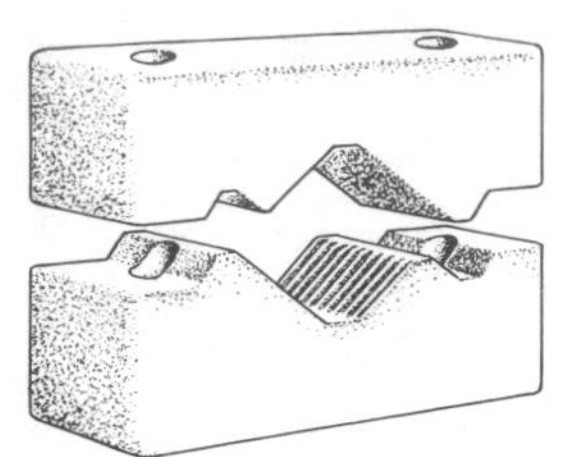

Fig. 16-3. Standard one wire cleat.

Fig. 16-4. Solid porcelain knob.

Surface Clearance. Cleat insulators should be of sufficient height to provide ½″ clearance for the surface wired over in dry locations. In damp or wet locations, a clearance of 1″ is required so that insulating knobs must be used instead of cleats.

Separation of Conductors. A separation of at least ¼″ should be maintained between the supporting screws or nails and the conductor. Cleats for voltage up to 300 volts must separate the conductors 2½″ from each other and ½″ from the dry surface wired over. Conductors that are No. 8 AWG or larger, when supported on solid knobs, Fig. 16-4, should be secured by wires having the same type of insulation as the conductors.

A split porcelain knob, Fig. 16-5 is used for conductors of No. 10 AWG gage or smaller. It is provided with a groove on each side of the nail so two wires of the same polarity and circuit can be attached. Where a tap is made from the main wire of a circuit, both conductors must be supported within 6″ of the tap, Fig. 16-5.

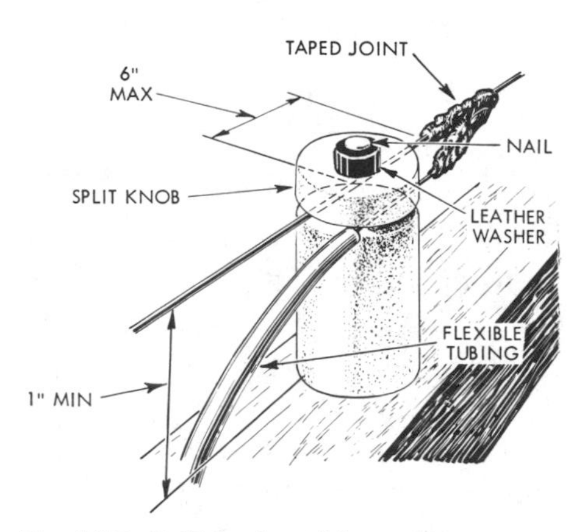

Fig. 16-5. Split knob and tap splice.

Nails for mounting knobs should be at least the 10-penny size. Screws should be long enough to extend into the wood at least one-half the height of the knob.

The correct method of making a right-angle turn with two conductors is shown in Fig. 16-6. On the left,

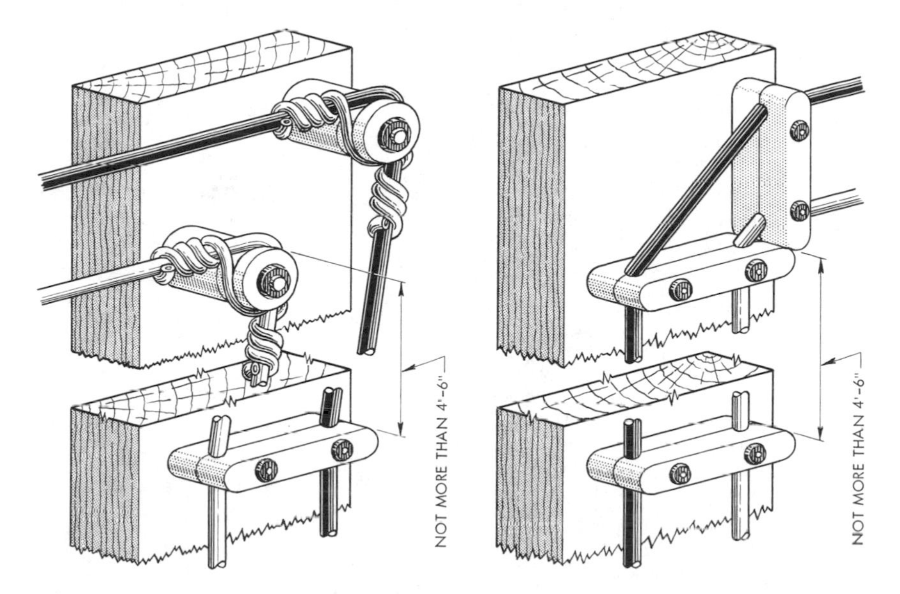

Fig. 16-6. Methods of making right angle turn with two conductors.

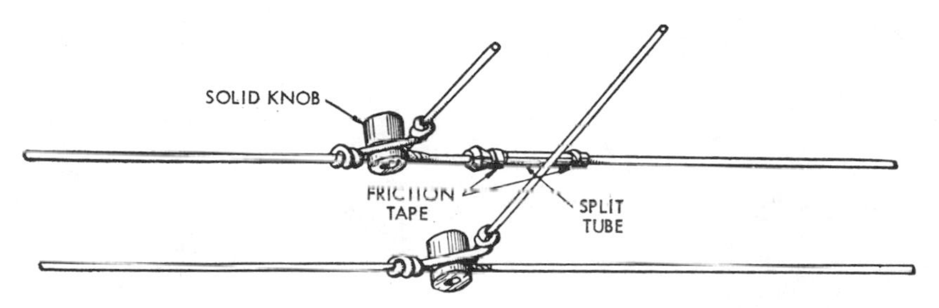

Fig. 16-7. Open wiring on porcelain knobs, and protection where wires cross.

solid porcelain knobs are used, and on the right, cleats are used. It is important that the 2½″ clearance between conductors be observed on turns as well as on straight runs. Where this is difficult the alternate method shown in Fig. 16-7 may be used.

Protection on Ceiling, Side Walls and Floors

On low ceilings, guard strips should be installed, Fig. 16-8. These must be not less than ⅞″ thickness and at least as high as the insulators.

Protection on side walls must extend not less than 7′ from the floor.

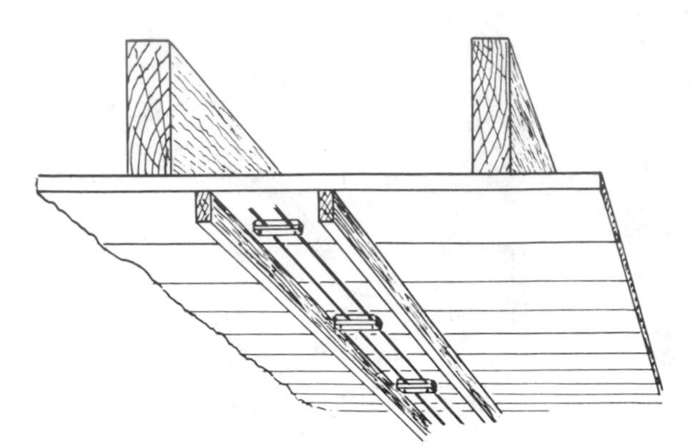

Fig. 16-8. Guard strips on low ceiling.

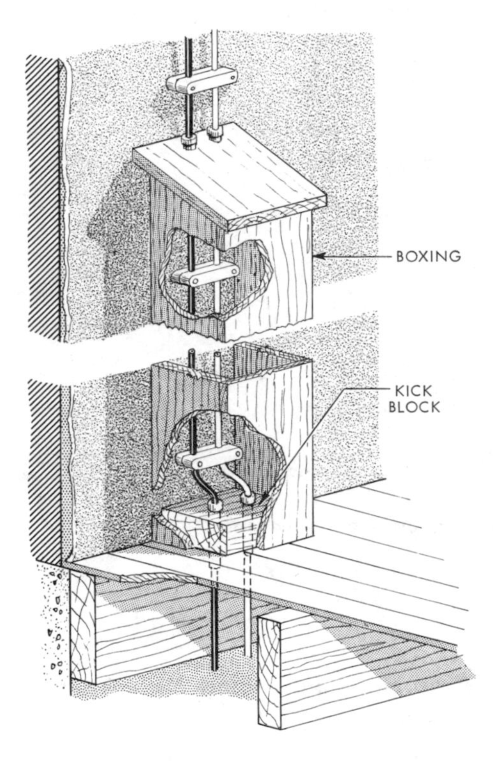

Fig. 16-9. Wood boxing for side wall protection.

It must consist of a substantial boxing closed at the top, Fig. 16-9, and must provide an air space of 1″ around the conductors. The wires, it should be noted, pass through bushed holes. When porcelain knobs are used instead of cleats, the wires must be at least 3″ apart. Wires passing through floors should be properly bushed, and a kick block should be used to guard tubes at the floor line.

Protection of Wires. Conductors may be protected from injury by the use of pipe or conduit. When wires pass inside the pipe, each wire should be inclosed in a separate nonmetallic flexible tubing, called *loom.* The loom must be in one piece and must extend from the porcelain support of the wires at the bottom to their porcelain support at the top.

When conduit is used for protection of the wires, a terminal fitting having a separate bushed hole for each conductor, Fig. 16-10A, is used. The same fitting is attached to the bottom end of the conduit (not shown). This method is also used when the open-wiring system is changed to conduit. Wires in any pipe or conduit must not contain splices.

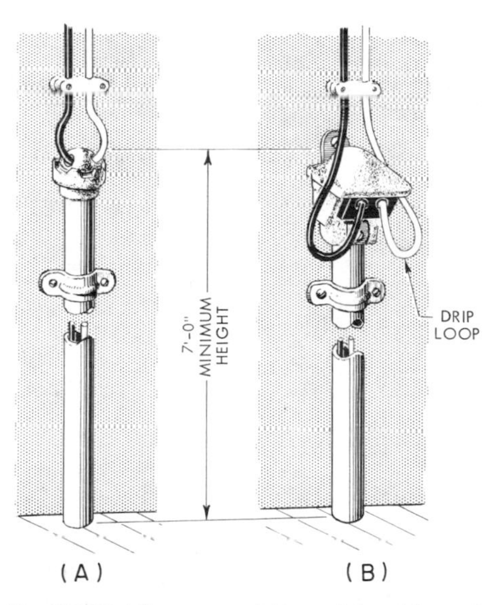

Fig. 16-10. Pipe or conduit used for side wall protection.

In wet locations a service head is attached to the upper end of the conduit, Fig. 16-10B. The wires should be arranged to form a drip loop so that water cannot enter the conduit. Conductors that form the drip loop should be at least 2″ away from iron parts of the conduit and service head so that water will drip free of the pipe.

Support for Wires. Wires require rigid support even under ordinary conditions. On a flat surface, supports should be provided at least every 4½′. If wires are likely to be disturbed, the distance between supports must be shortened.

A long board is nailed between widely-spaced beams, and the insulators are mounted on the board. The method of breaking around beams with split knobs is illustrated in Fig. 16-11. In buildings for commercial or industrial construction, cables of not less than No. 8 AWG (where not likely to be disturbed) may be separated about 6″ and run from timber to timber without breaking around beams. Drop cords should be attached to the circuit wires through a cleat, Fig. 16-12. The best installation of wall sockets and switches is with special cleat-work

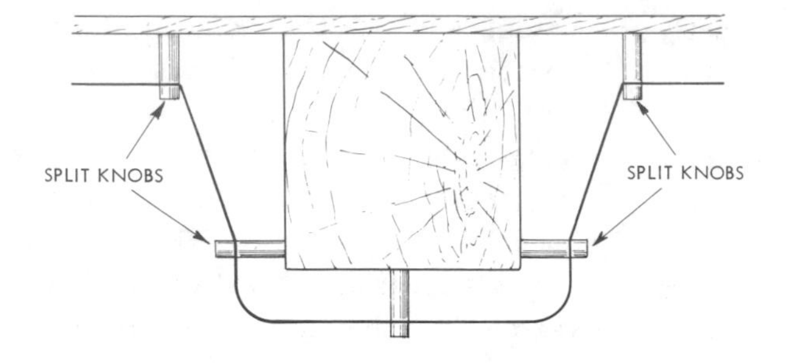

Fig. 16-11. Method of breaking around beams with split knobs.

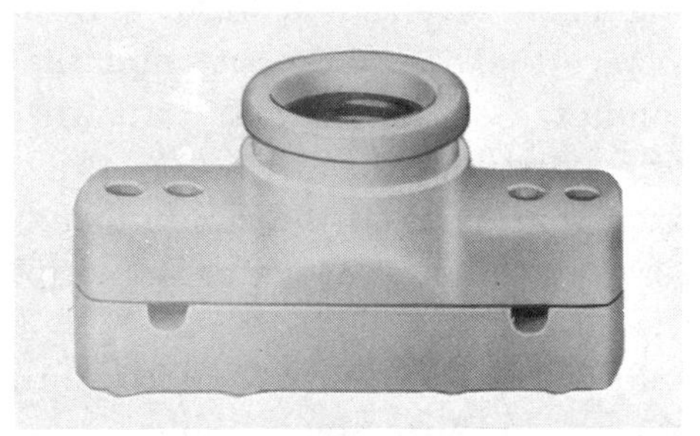

Fig. 16-12. Cleat type lampholder. (Bryant Electric Co.)

Fig. 16-13. Cleat type lampholder. (Pass and Seymour)

fittings, Figs. 16-13, 16-14 and 16-15. Other fittings, when used, should be mounted on small porcelain knobs. Conductors should be deadened on a cleat or knob, Fig. 16-16.

Precautions Against Dampness and Acid Fumes

The rules for open work in buildings subject to moisture or acid fumes are somewhat different as re-

Fig. 16-14. Surface type snap switch adaptable to open knob and tube wiring. (Bryant Electric Co.)

Fig. 16-15. Ceiling pull-switch adaptable to knob-and-tube wiring. (Bryant Electric Co.)

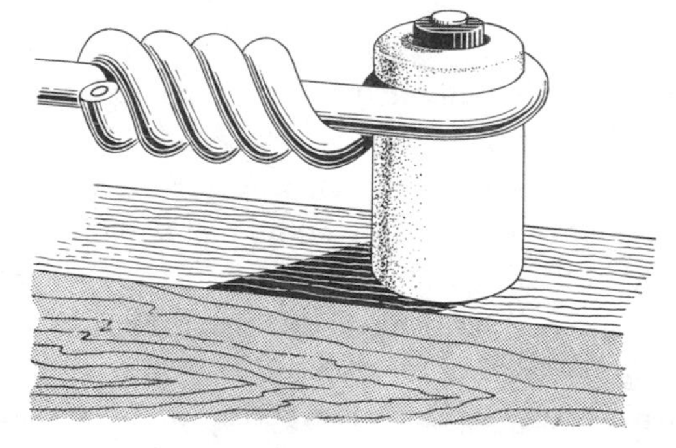

Fig. 16-16. Dead-ending a wire at a knob.

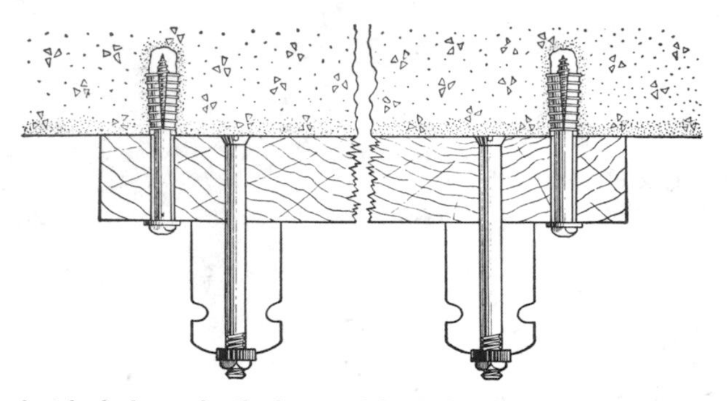

Fig. 16-17. Wood strip independently fastened by bolts and expansion shells for supporting porcelain insulators in damp places.

Fig. 16-18. Weatherproof socket. (Pass and Seymour)

gards insulator supports. Porcelain or glass knobs must be used. When installed on brick, concrete, tile, or plastered walls or ceilings, these insulators should be attached to wooden or metal strips or blocks, fastened independently by means of expansion or toggle bolts, Fig. 16-17. Nails or screws driven into wooden plugs are not permitted by electrical codes.

Wires should have standard insulating covering for protection against water and against corrosive vapors. Conductors should be separated at least an inch from the surface wired over. Sockets should be of weatherproof type, Fig. 16-18, and drop lights should be hung on No. 14 AWG standard rubber covered wire.

Cutouts and switches should be mounted on small knobs in iron cabinets.

Stringing Wires

One method of stringing wires is to fasten each conductor to a double support at one end, then pull it

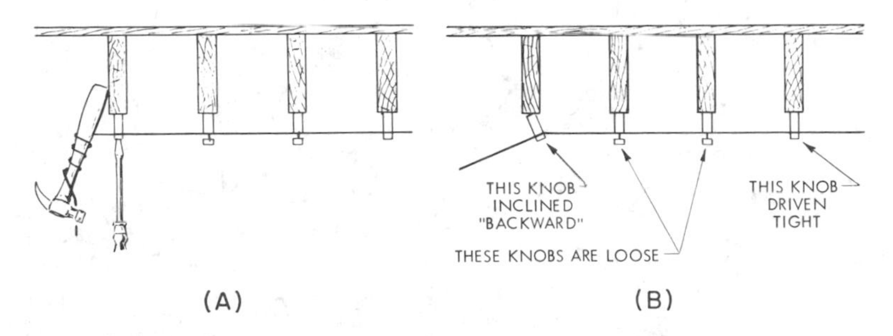

Fig. 16-19. Method of taking up slack in small conductors.

across joists to the other end of the run. After a terminal cleat or knob is installed, required insulators are attached to joists along the way. If the wires are small, one or two turns around the handle of a hammer or bar, as shown in Fig. 16-19A, will provide good leverage. An easy method for tightening small wires is to incline the end knob "backward," as in Fig. 16-19B. When the split knob is driven up, with the wire in the knob, the conductor is tightened automatically as the knob comes to an upright position.

Concealed Knob-and-Tube Wiring

Concealed knob-and-tube work, like wiring on insulators, is not permitted in congested districts of large cities. It is sometimes employed, however, in residential areas, and in small towns. The electrician should learn fundamental principles underlying this method, because there are thousands of knob-and-tube installations already existing which must be repaired and extended from time to time.

Essential Requirements. Codes prohibit the use of this type of wiring in commercial garages and other hazardous areas. Only single conductors may be run on knobs. Conductors must be supported at intervals not exceeding 4½′, separated at least 3″ from each other, and maintained at a distance not less than 1″ from the surface wired over. At distribution centers and other points where a 3″ separation cannot be maintained, each conductor must be encased in a continuous length of flexible insulating tubing (loom).

Where practicable, conductors should be run on separate timbers or studding. They should be kept at

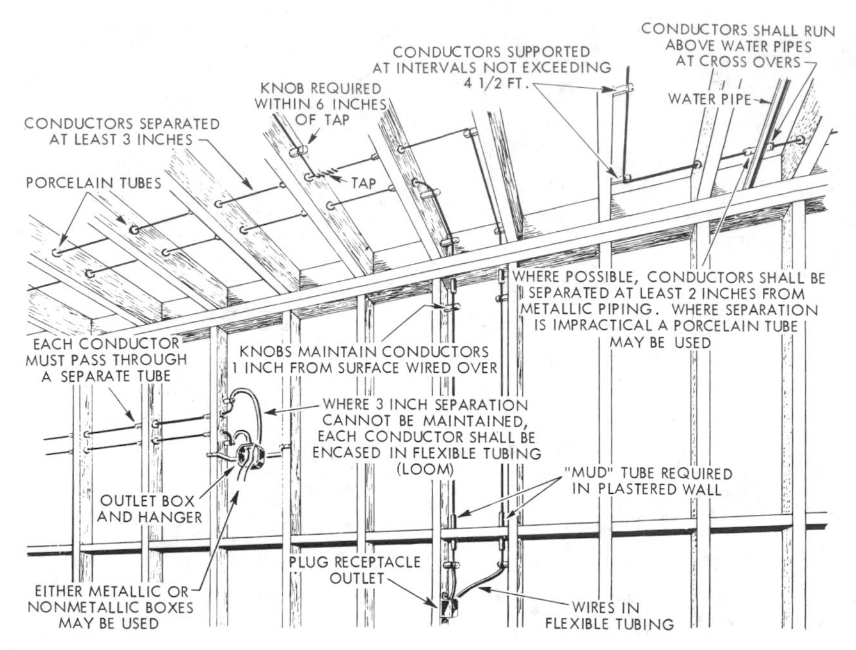

Fig. 16-20. Typical knob and tube installation.

least 2″ from piping or other conducting material, or separated therefrom by a suitable non-conductor such as a porcelain tube. In accessible attics, conductors must be protected by running boards or guard strips if within 7′ of the floor, unless run along sides of joists or rafters. Conductors passing through cross timbers in plastered partitions must be protected by an additional "mud" tube that extends at least 3″ above the timber. The more important points in knob-and-tube installation are illustrated in Fig. 16-20.

Outlet Boxes. The standard metal outlet and switch boxes may be used with knob-and-tube wiring. It is permissible, however, to use a porcelain, Bakelite, or composition plastic box for this purpose. A number of these wiring accessories are available, and give service comparable to the metal outlet box. For safety reasons, composition outlet, switch, and receptacle boxes may be desirable.

Insulated boxes, Fig. 16-21, are constructed of plastic or composition materials. Their dimensions are comparable to those of metal boxes. Insulated outlet box covers and plaster rings are also available. These boxes are exceedingly tough and resist de-

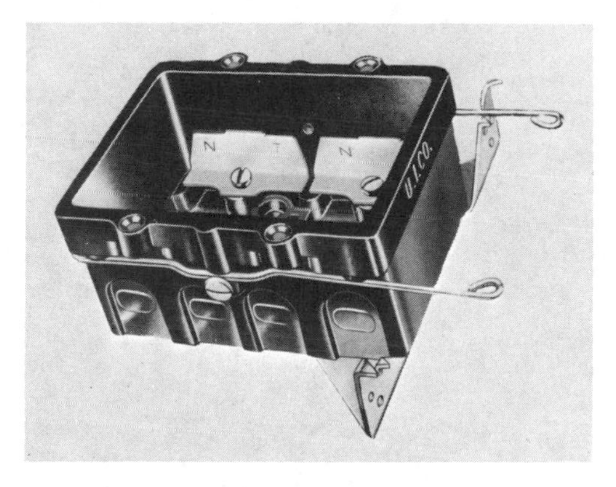

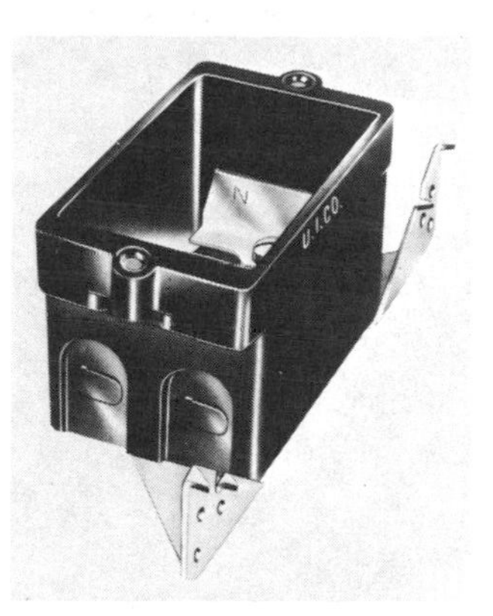

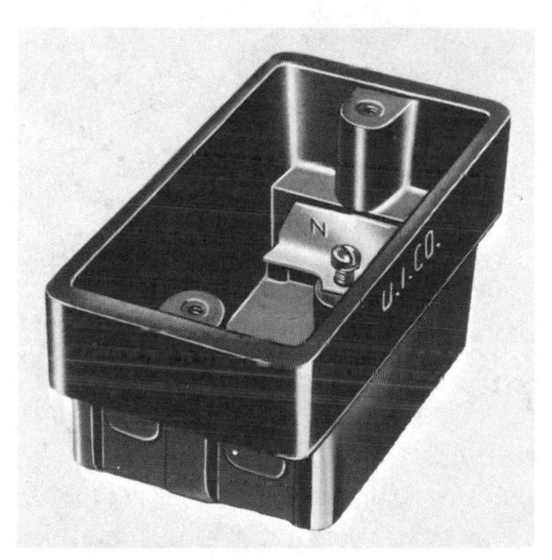

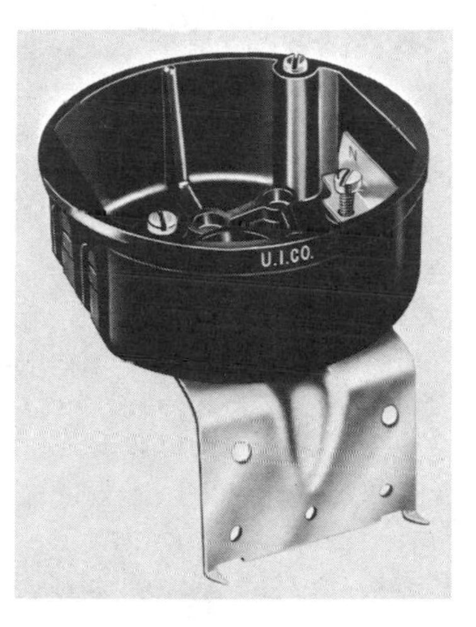

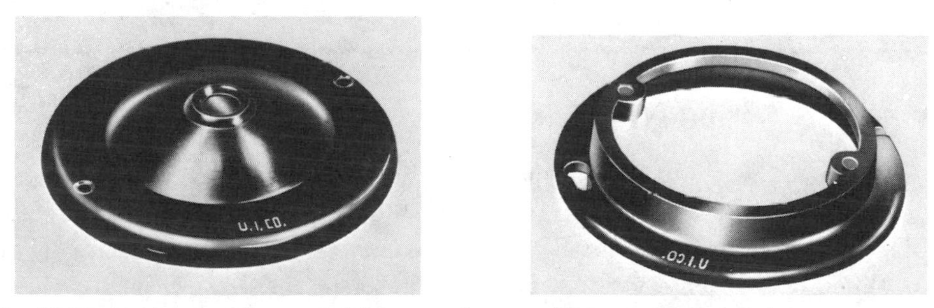

Fig. 16-21. Insulated outlets, switch boxes and covers. (Union Insulating Co.)

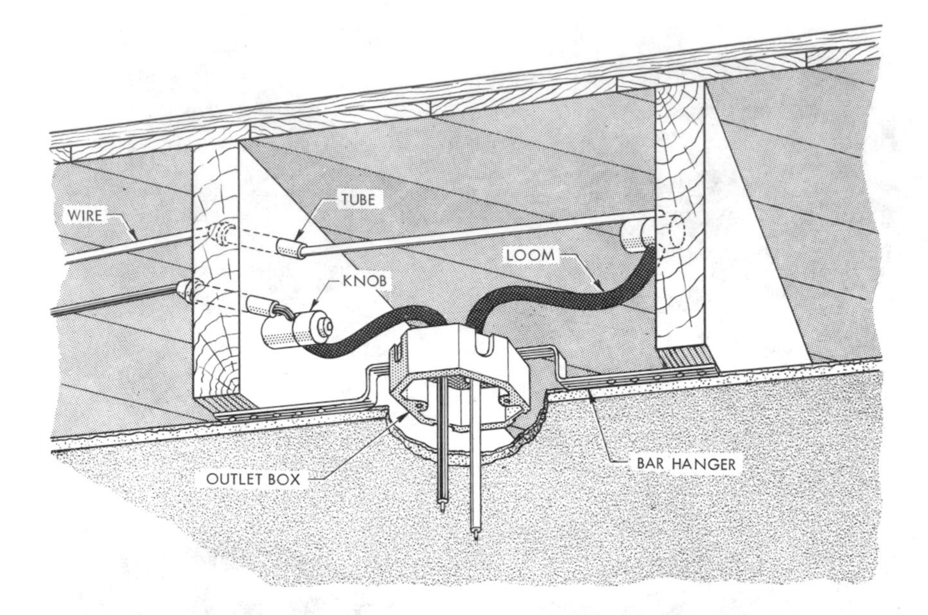

Fig. 16-22. Mounting outlet box using bar hanger.

struction due to accidental hammer blows.

When metal boxes are used, flexible tubing must extend into the box. It is secured there by clamps or other means. When insulated boxes are used, the tubing need extend only to the outer surface of the box.

The method for installing an outlet box between joists is shown in Fig. 16-22. It is important that the box extend down far enough to come flush with the ceiling finish. For lath and plaster, allow about 1″; for composition board, allow about ½″; and for masonite, allow about ½″ or ⅜″, depending on the thickness selected. Appropriate building plans should be consulted for method and type of wall finish.

Method of Wiring New Buildings

Since this form of wiring is not widely used on new construction, additional installing procedures will be outlined only briefly. Certain details which are applicable to all forms of wiring, including knob-and-tube, have been discussed in the chapters in connection with cable wiring. (See Chapters 4 and 5.)

The first step is to mark locations

of outlets with chalk or crayon, using standard symbols to indicate switches, plug receptacles, and lighting outlets. Ceiling lights are noted on the floor at the approximate centers of rooms. Locations of the fuse panel and of the service equipment are also marked. Outlet boxes are now fixed in place and the fuse panel set.

We assume that the number and distribution of circuits has already been determined. Holes are bored to accommodate the various runs, using a brace and bit, a boring machine, or, more commonly, an electric drill with a bit extension.

Wires are pulled through the holes, and porcelain tubes slipped over the conductors as they are drawn from joist to joist or stud to stud. Knobs are installed where needed, and conductors drawn tight. Wires are skinned, and joints made up, after which they are soldered and taped. Short lengths of flexible tubing are then installed, where necessary, at the various outlets and at the fusc panel. At the same time, the ends of flexible tubing are secured in the boxes. When feeder and service runs are completed, the electrician's work is done until carpenters, painters, and other mechanics have finished their jobs.

Method of Wiring Old Buildings

In wiring old buildings where it is not possible to take up flooring, nonmetallic, flexible tubing may be "fished" (shoved through) in partitions or between the floor and ceiling. Only one wire is permitted in each piece of tubing, which must be continuous from outlet to outlet or from knob to knob. This tubing must be fished through existing channels in partitions or over ceilings. By removing baseboards, fishing parallel with the floor beams, going through floor plates with porcelain tubes, and picking up the circuits in the attic and the basement, a safe and acceptable installation can be put in with flexible tubing.

QUESTIONS FOR DISCUSSION OR SELF STUDY

1. Why are knob-and-tube uncommonly used on new construction?
2. What additional step is necessary where wires pass through cross timbers in a plastered partition?
3. How much separation between wires is provided by standard cleats?
4. What type of knob-and-tube wiring would you use for a barn?
5. What is the largest size wire that may

be supported by a split knob?

6. What type of insulation is required for tie wires used with solid knobs?

7. How far should wires be separated from gas pipes?

8. What is loom?

9. What should be the clearance wired over for open wiring?

10. What should be the clearance wired over for concealed wiring?

Estimating Electrical Wiring

Chapter

17

The term *estimate* means an opinion, formed in advance, of the probable cost of a certain piece of work. The opinion is arrived at after considering each factor, including:

1. Cost of material.
2. Cost of labor.
3. Overhead cost.
4. Expected profit.

It should be emphasized that an estimate is not an exact figure, but an approximation. The precise cost may be determined only after the job has been finished.

On large construction projects, detailed drawings and specifications are made available to the estimator. There may be as high as approximately 6,500 to 7,500 different items considered on an average industrial electrical estimate. In all, there are over 72,000 electrical items which may be purchased.

On lesser operations, such as that of wiring a bungalow, drawings seldom present complete electrical layouts, while specifications are usually contained in one or two paragraphs of a letter.

Specification paragraphs in the "contract" letter set forth certain requirements which may be summarized, as an example, as follows:

All electrical material shall bear U-L approval. Workmanship shall be of high quality. The distribution panel shall be of the circuit-breaker type.

Nonmetallic cable shall be used with the minimum size being #12 AWG.

Cable shall be fastened by means of clamps in all outlet boxes. A weather-tight cover shall be installed on the plug receptacle outlet in the entry. A telephone outlet shall be installed in the hallway. The owner's approval shall be obtained in the selection of lighting fixtures which, if necessary, shall be in accordance with a list to be supplied by him upon request. Bell system chimes and the front and rear door

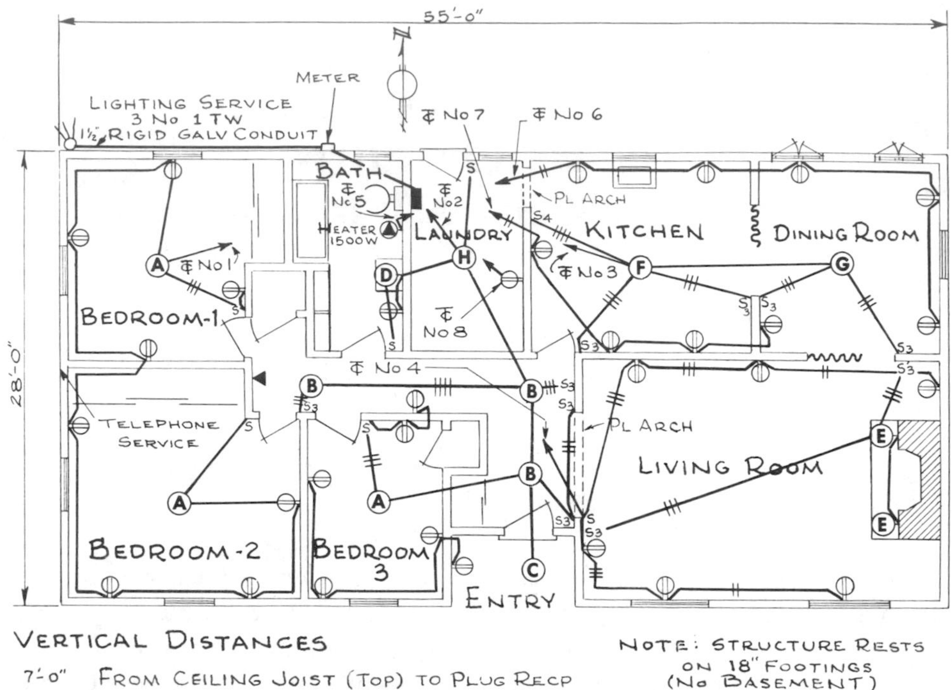

Fig. 17-1. Drawing from which estimate is prepared.

push button must have the owner's approval. Material and workmanship shall be guaranteed against inherent defects arising within a period of one year from date of completion. The completed installation shall conform to the NEC, and shall meet with the approval of the local inspection authorities.

Fig. 17-1 shows an electrical drawing which is reasonably complete, except that the conduit run to the telephone outlet in the hall and whole chime wiring system are omitted in order to simplify matters.

Preliminary Steps

The essential requirement in every type of estimating is to follow a systematic procedure. The first consideration is to become familiar with the plan, Fig. 17-1, noting the general layout of rooms, location of electrical service and telephone service. Special outlets are looked for, such as the heater in the bathroom. Location of the distribution panel is checked.

The occupancy is a one-story bungalow, having a low attic and no basement. These facts have an important bearing on the manner of running cable between certain outlets. The subfloor will be in place when the electrician arrives, so that he will not be able to run cable along sides of floor joists. At door openings, he will find it necessary to cross by way of the attic.

In the living room, for eample, it will not be possible to route cable directly across from the first plug receptacle outlet in the north wall to the switch location at the side of the archway. Instead, the cable must be carried upward to the ceiling from the receptacle outlet, across tops of ceiling joists, and down again to the switch location. Identical situations are found throughout the premises.

Branch Circuit and Fixture Schedules

Having made a general survey of the plan, it is convenient to draw up a branch-circuit schedule, Fig. 17-2. The form lists ceiling and bracket outlets, single-pole, three-way, and four-way switches, and grounding-type receptacles. The last column deals with special outlets.

Circuit *1* supplies Bedrooms 1 and 2, having a ceiling outlet, single-pole wall switch, and four convenience

BRANCH CIRCUIT SCHEDULE								
		LIGHTING OUTLETS		SWITCH OUTLETS			RECEPTACLE OUTLETS	SPECIAL
CIRCUIT	LOCATION	CEILING	BRACKET	SP	3-W	4-W	G	
1	BEDROOM-1	1		1			4	
	BEDROOM-2	1		1			4	
2	BATHROOM		1	1			1	
	LAUNDRY	1		1				
	BEDROOM-3	1		1			4	
	HALL AND ENTRY	4			4		2	
3	KITCHEN	1			2	1		
	DINING ROOM	1			2			
4	LIVING ROOM		2	1	2		6	
5	BATHROOM							1
6	KITCHEN						2	
	DINING ROOM						3	
7	KITCHEN						3	
	DINING ROOM						1	
8	LAUNDRY						1	

Fig. 17-2. Branch circuit schedule.

LIGHTING FIXTURE SCHEDULE								
CIRCUIT	TYPE							
	A	B	C	D	E	F	G	H
1	2							
2	1	3	1	1				1
3						1	1	
4					2			

Fig. 17-3. Lighting fixture schedule.

outlets in each one. Circuit *2* takes care of lighting and plug receptacle outlets in bathroom, Bedroom 3, hall and entry. The laundry ceiling light is connected to it, but the laundry plug receptacle is not. It is on a separate circuit.

In kitchen and dining rooms, lighting outlets, but not plug receptacles, are connected to Circuit *3*. Circuit *4* supplies both lighting outlets and plug receptacles in the living room. Circuit *5* has one special outlet connected to it—the bathroom heater. Circuits *6* and *7* are reserved for plug receptacles in kitchen and dining rooms. The laundry plug is connected to Circuit *8*.

The lighting fixture schedule, Fig. 17-3, is drawn up from type letters marked on the various outlets shown on the drawing (Fig. 17-1). Their exact specifications are to be determined through consultation with the owner. There are three type *A* units, three type *B* units, two type *E* units, and one unit each of types *C*, *D*, *F*, *G*, and *H*.

Branch Circuit Material Schedules

Drafting of a branch-circuit material schedule, Fig. 17-4, becomes the next order of business. It is constructed on the basis of measuring and counting. Runs of cable from outlet to outlet are measured on the plan. Other items, such as outlet boxes, plaster rings, switches, plug receptacles, plates, box hangers, ground straps, and wire nuts, are counted. The measuring can be done with the aid of a rotameter or an architect's scale, the counting by means of a tabulator or by mental arithmetic. (A *rotameter* is a calibrated meter device that is wheeled over scaled lines to mechanically tally distance on a drawing.) The mechanical aids are reserved for large commercial or industrial jobs. Here, the architect's scale and mental arithmetic will be employed.

Before starting to measure cable, it is well to note the table of vertical distances at the bottom of the Fig. 17-1 drawing. A 7′ indication marks the distance from the top of a ceil-

BRANCH CIRCUIT MATERIAL SCHEDULE																								
CIRCUIT	CABLE		BOXES				PLASTER RINGS						SWITCHES			PLUG RECEPTACLE	PLATES				WIRE NUTS	CONNECTORS	HANGERS	GRD STRAPS
	12-2	12-3	4" ○	4" □	4 11/16"	SR	4" R	4" SQ	4" SS	4" 2G	4 11/16" R	4 11/16" S	SP	3-W	4-W	GR	SS	2-G	PR	SP				
1	129-0	10-8	2			10	2						2			8	2		8		4		2	
2	152-0	35-2	4	3	1	12	4	2		1	1		3	4		7	5	1	6	1	14	2	7	2
3	53'-4"	44'-8"		1	1	5		1			1			4	1		5				4	2	2	1
4	78-8	76-10	1	3	1	5	1	1	2			1	1	2		6	1	1	6		4	2	2	
5	4'-6"																							
6	54'-0"					5										5			5					
7	54'-0"					5										5			5					
8	18'-0"																							
SUB-TOTAL	543'-6"	167'-6"																						
ADD 10% TO CABLE	54'	17'																						
TOTAL	598'	185'	7	7	3	42	7	4	2	1	2	1	6	10	1	31	13	2	30	1	26	6	13	3

ROUND FIGURES

Fig. 17-4. Branch circuit material schedule.

ing joist down to a plug receptacle. A 4′-2″ line shows the distance from top of ceiling joist down to a wall switch. The distance from a switch down to a plug receptacle is shown as 2′-10″; that from ceiling joist to top of the distribution panel is 4′; and that from ceiling joist to wall bracket is 2′. The last dimension, 2′-6″, is from panelboard down to the heater location (Circuit *5*).

Circuit *1* ends at the wall switch in Bedroom 2, passing through the various plug receptacles, switch, and ceiling outlets in both rooms, and then to the distribution panel in the laundry. The cable is No. 12-2 throughout, except for a short run of No. 12-3 in Bedroom 1, from the ceiling light to the wall switch. Distances to ceiling outlets are measured to their centers, while distances to plug receptacles and to switches are measured to the wall. Do not try to estimate fractional amounts. If the measurement is slightly beyond a given mark on the scale, call it 6″; if slightly less than a given mark, grant it the whole 1′.

Starting at the end of Circuit *1*, 4′-2″ of cable must be allowed for the distance upward from the wall switch to the top of a ceiling joist, and 7′-6″ for the run across the joists to the light in the center of the room. Another 7′-6″ to the right marks a point directly above the plug receptacle on the east wall of the room. From this point, 7′ are added to account for the vertical distance from ceiling joist down to the plug receptacle.

Continuing, the distances are 6′ from the plug receptacle to the south wall of the room; 14′-6″ to the west wall; 14′ to the north wall; 5′ to the plug receptacle in the south wall of Bedroom 1; 5′ back to the west wall; 13′ to the north wall; 6′-6″ to the plug receptacle in the north wall; 7′ vertically to the top of the ceiling joist; and 6′-6″ to the ceiling outlet.

From this point, the measurement to the distribution panel in the laundry is 17′, but 4′ must be added to provide for the run from the ceiling joist down to the panel. Another short piece of No. 12-2 cable extends from the wall switch in Bedrom 1 down to the plug receptacle, the length being equal to the sum of 2′-10″ and 1′-6″, or 4′-4″. The quantity of No. 12-3 cable in Bedroom 1 is equal to 6′-6″ from center light to the wall, plus 4′-2″ from ceiling joist down to the switch, a total of 10′-8″.

Adding these amounts, the length of No. 12-2 cable is found to be 129′, while that of No. 12-3 cable is 10′-8″. It may be observed that no allowance was made for the short pieces of cable which project from each outlet box. This matter is taken care of by the addition of 10 percent to the measured lengths, as will be shown later on. For the present, a notation of 129′ is made in the column, and 10′-8″ in the 12-3 column of the schedule, Fig. 17-4. Cable lengths for the remaining circuits are then entered in appropriate columns, including the No. 12-2 conductors of the utility circuits.

Other materials are now taken off. The method will be illustrated with respect to Circuit *1*. Center lights in the bedrooms require two 4″ octagon boxes, two box hangers, two 4″ round plaster rings, and four wire nuts that will be used for connecting the fixtures. Ten switch-receptacle boxes are needed, eight for the plug receptacles and two for the wall switches. Two single-pole switches will be required, and two single switch plates. Eight standard grounded plug receptacles will need eight receptacle plates. These amounts are noted in appropriate columns of the schedule.

Remaining circuits are examined in the same way. Circuit *2* is somewhat more complicated in its requirements than Circuit *1*. Before considering Circuit *2*, it should be explained that the NEC limits the number of conductors that may enter a given size of outlet box. Four inch boxes, octagon and square, may have to be increased in depth from 1½″ to 2⅛″, depending upon the number of wires entering and leaving and depending upon whether the box contains internal cable clamps. Codes usually require the elimination of one permissible wire should the box contain clamps. Boxes which are 4-11/16″ (5S) may be employed with the increased demands of cubic capacity.

The number of wires in the cable runs are indicated by cross marks on the cable. Where there are no cross marks, it is understood that two wires are run.

The ceiling outlet box in the laundry has four cables, none of which has cross marks. The total number of wires, then, is equal to 2 x 4, or 8. Since cable clamps are present (presumably by most codes), a 4″ square box must be used. An identical situation is found with respect to the hall lighting outlet which is just inside the front door.

A 4 11/16″ outlet box with clamps is selected for eleven No. 12 wires entering the hall ceiling outlet which is nearest the kitchen door. The other ceiling outlets are listed as 4″ octagon boxes.

A 4″ square box is needed for the two three-way switches that are ganged, or grouped, at the side of the living-room archway.

Four 4″ round plaster rings are necessary for the octagon boxes, two 4″ square plaster rings for the square boxes, one two-gang 4″ plaster ring for the ganged switches, and one 4-11/16″ square plaster ring for the hall ceiling outlet. Three single-pole and four three-way switches are needed, seven standard grounding plug receptacles, five single switch plates, one two-gang switch plate, six standard plug receptacle plates and one special weatherproof cover for the entry receptacle, fourteen wire nuts, two 1″ box connectors for the 4-11/16″ box, seven box hangers, and two ground straps.

Circuits *3*, *5*, and *7* offer no difficulties in the listing of material, but there are some points worth noting in connection with Circuit *4*. A 4-11/16″ box is used for the two-gang switch outlet just inside the living-room arch, because there are eleven wires in addition to cable clamps. Three 4″ square boxes are used on this circuit, one for the north bracket outlet at the fireplace, and the others for two plug receptacle outlets. The reason for the square box at the fireplace is readily apparent because of the number of wires. A 4″ box is substituted for the standard receptacle box in the two outlets adjacent to either side of the hall archway because six wires are present at these locations. It is good practice to use the 4″ square boxes in place of ordinary switch-receptacle boxes whenever the number of wires exceeds five when using #12 wire. Number 12 wire is more difficult to fold back into the box than #14 and requires the more cubic capacity that a larger box affords. After a ten percent margin has been added to measured wire lengths, the columns in Fig. 17-4 are totalled.

Other Methods of Take-Off. The method used here may be varied to suit individual tastes. Some estimators prefer to list wire for the whole area at one time, and to count all the various items as well. After gaining experience, one may proceed in this way. At the start, however, it is well to list materials circuit by circuit in order to recheck uncertain portions without too much loss of time.

In any event, branch circuit material schedules follow the general pattern set forth here. When the take-off is done on an area basis, separate counts are made by floors, sections, or buildings, the various locations being noted, for example, as "Floor 1," "Floor 2," and so on, instead of "Circuit *1*," "Circuit 2" and so on.

Service and Feeder Material Schedule

This schedule may be equally as involved as the branch circuit material schedule on certain types of jobs. In the present instance, however, it is much simpler. In Fig. 17-5, items are listed vertically rather than hori-

SERVICE AND FEEDER MATERIAL SCHEDULE		
ITEM	NUMBER	FEET
1½" ENTRANCE CAP	1	
1½" GALV CONDUIT		28
1½" GALV LOCKNUT	2	
1½" GRD BUSHING	1	
1½" GALV STRAPS	3	
1½" METER SOCKET	1	
SERVICE SWITCH	1	
GROUND CLAMP	1	
No. 1, TYPE TW		108 *
1½" NIPPLE	1	
SERVICE GROUND WIRE		10
DIST PANEL	1	
100 – A FUSE	2	
1¼" - 10 WOOD SCREWS	12	
* 10% NOT INCLUDED		

Fig. 17-5. Service and feeder material schedule.

zontally. Each kind of material is enumerated. A note calls attention to the fact that an allowance of ten percent has been added to both service conductors and to feeder cable.

Labor-Unit Schedule

A table of labor units is given in Fig. 17-6. It includes every electrical task which enters into the complete wiring job. The right-hand column gives the time unit for performing a particular task. The units normally employed should be largely the result of one's personal experience. Trade associations publish tables which vary in degree of accuracy. When they are used by the contractor, their suitability should be checked from time to time on the basis of actual job records. Meanwhile, the present list will serve as a guide here.

The first notation states a time of 0.4 hr (24 minutes) to install a hundred feet of No. 12-2 and 12-3 cable. This period includes boring of holes with an electric drill, pulling cable through holes or across ceiling joists, cutting the length of cable at each

LABOR UNIT SCHEDULE

OPERATION	HOURS
INSTALL 12-2, 12-3 CABLE	.4 C FT
MOUNT OCTAGONAL BOX, FASTEN CABLE	.2 EA
MOUNT SWITCH BOX, FASTEN CABLE	.2 EA
ATTACH PLASTER RING	.05 EA
CONNECT RECEPTACLE, INSTALL PLATE	.2 EA
CONNECT SP SWITCH, INSTALL PLATE	.2 EA
CONNECT 3-W SWITCH, INSTALL PLATE	.3 EA
CONNECT 4-W SWITCH, INSTALL PLATE	.4 EA
SOLDER AND TAPE SPLICE	.07 EA
MOUNT AND CONNECT PANEL	2.5 EA
MOUNT AND CONNECT SERVICE SWITCH	1.9 EA
INSTALL 1½" RIGID CONDUIT	9.0 EA
PULL IN No. 1 WIRE	2.0 C FT
CUT AND THREAD 1½" CONDUIT	.5 EA
INSTALL METER SOCKET	1.0 EA
INSTALL ENTRANCE CAP	.3 EA
INSTALL SERVICE GROUNDS	1.0 EA
INSTALL A OR F FIXTURE	.5 EA
INSTALL B, C, D OR E FIXTURE	.4 EA
INSTALL G FIXTURE	.7 EA
INSTALL H FIXTURE	.3 EA
INSTALL ¾" EMT (TELEPHONE)	4.5 C FT
INSTALL TELEPHONE OUTLET BOX	.1 EA
INSTALL BELL WIRING	1.5 EA
INSTALL "LOCAL" GROUNDS	.05 EA

Fig. 17-6. Labor unit schedule.

outlet, and stapling it where necessary.

The unit of 0.2 hr (12 minutes) for mounting an octagon box includes assembling box and hanger, spotting the outlet, nailing the hanger to the joist, and tightening cable clamps. The time for making a splice is 0.07 hr (about 4 minutes), which covers stripping of outer wrapping, skinning the conductors, twisting them together, soldering the joint, and covering it with plastic tape.

Service operations present some points worth mentioning. The operation of cutting and threading the 1½″ service conduit is given a special rating of 0.5 hr (30 minutes). If a number of pipes were to be threaded, the time for each would be somewhat less than this value. Here, it is necessary to unpack equipment and set up the pipe vise for two cuts. A truck mounted pipe vise saves time here.

The item *install service grounds* embraces connection of a bond wire to the service grounding bushing, installation of the armored ground wire, fastening the service ground wire to the ground clamp, and fastening the ground clamp to the waterpipe grounding electrode. The time also covers attachment of the feeder ground wire to the service bonding conductor.

Time for mounting the panel includes fastening it to the wall, inserting and tightening cable clamps, skinning feeder conductors and attaching them to the buses, grounding the enclosure by means of the bare ground wire in the cable, skinning circuit conductors, and securing them to the circuit breaker. Finally, the panel cover is set in place.

A special time rating is given to chime wiring, a value based upon numerous similar installations. The last item is that of installing local grounds. The term *local grounds* means those located elsewhere than at the service. They are commonly known as *equipment grounds.* (See Chapter 12, "Grounding for Safety.") In one method the wire is looped from box to box, and fastened under a convenient screw, such as that of a cable clamp. The end is attached to a copper ground strap which is wrapped around a water pipe. Three ground straps are used on this job.

Estimating Form

When the schedules are completed, information they contain is transferred to an estimating form, Fig. 17-7. Column 2, under the heading *Description,* lists every item in the schedules, as well as a number of miscellaneous components including solder, soldering paste, insulating tape, nails, low voltage wiring material, grounding wire, straps, and materials required for the telephone outlet.

From the information in the *Number* and *Feet* columns, together with unit costs given in Column 5, it is easy to determine material costs in Column 6. Labor units from Fig. 17-6 are copied, Column 7, beside appropriate items. The number of hours, Column 8, is determined by multiplying the labor unit by quantity of goods used. Finally, labor cost is found in Column 9, hours of labor being multiplied by the labor rate of $6.50 per hour as an example.

An example will show how the estimating form is used. Item 1 is given as No. 12-2 Type NM cable. Column 4 shows that 598 ft are used. Column 5 lists the cost as $5.07 per hundred feet, or 5.98 x $5.07, which equals $30.32, as per Column 6. The labor unit for installing 100 ft of this cable is given as 0.4 hr. in column 7. The time for running 598 ft of cable is equal to 5.98 x .4 hr, or 2.39 hr, Column 8. The labor cost is equal to 2.39 x $6.50, or $15.54, as recorded in Column 9.

It will be seen that labor units are not marked on every line. In case an omission is made, the labor unit which embraces this particular material is included with some other item of which it is a part. Thus, switch plates are part of the complete switch outlet, attaching of the plate being included with time for connecting the switch. In the same way, the box hanger is part of the complete outlet box assembly.

Time for running the No. 6 armored ground wire is included in the operation of installing service grounds, and the labor for this task is entered opposite *Cast Ground Clamp*. In many cases, there is a choice with respect to labor notation, particularly where the operation is special or "lumped." Thus, the labor for the chime system is entered in line with *Bell Wire*. It could just as well have been entered at *Transformer*. Bell or annunciator wire is generally used for low voltage signaling system.

When all calculations have been performed, Columns 6 and 9 are totalled to find *Material Cost and Labor Cost* for the whole job. Here, cost of material is $222.02, while cost of labor is $335.34. The total cost of material and labor is $557.36.

Final Considerations

The above amount is the estimated *prime cost* for doing the work at the fictitious costs used here. In order to remain in business, it is necessary for the contractor to add a percentage which covers *overhead expenses* and a *margin of profit*. The percentage added depends upon a number of factors which can be learned only through experience. A common value is 30 percent. In the present case this added amount is equal to: 0.3 x $557.36, or $167.21. The price which must be quoted to the customer, then, is equal to

ITEM	COLUMN 2 DESCRIPTION	3 No.	4 FEET	5 UNIT COST	6 MATERIAL COST	7 UNIT LABOR	8 HOURS	9 LABOR COST
1	No. 12-2 TYPE NM CABLE		598	5.07C	30.32	.4	2.39	15.54
2	No. 12-3 TYPE NM CABLE		185	7.68C	14.21	.4	.74	4.81
3	4" OCTAGON BOX	7		29.30C	2.05	.2	1.4	9.10
4	4" SQUARE BOX	7		43.00C	3.01	.2	1.4	9.10
5	4 11/16" SQUARE BOX	3		52.00C	1.56	.2	.6	3.90
6	SWITCH RECEP BOX	42		29.20C	12.26	.2	8.4	54.60
7	4" RD PLASTER RING	7		12.10C	.85	.05	.3	2.27
8	4" SQ PLASTER RING	4		16.60C	.66	.05	.2	1.30
9	4" SQ SINGLE SW RING	2		22.00C	.44	.05	.1	.65
10	4" SQ TWO-GANG SW RING	1		22.00C	.22	.05	.05	.33
11	4 11/16" SQ PLASTER RING	2		26.00C	.52	.05	.1	.65
12	4 11/16" SQ TWO-GANG SW RING	1		28.00C	.28	.05	.05	.33
13	SP TOGGLE SWITCH	6		.50	3.00	.2	1.2	7.80
14	3-W TOGGLE SWITCH	10		.76	7.60	.3	3.0	19.50
15	4-W TOGGLE SWITCH	1		1.45	1.45	.4	.4	2.60
16	GRD PLUG RECEP	31		.94	14.22	.2	6.2	40.30
17	SINGLE SW PLATE	13		.09	1.17			
18	TWO-GANG SW PLATE	2		.17	.34			
19	RECEP PLATE	30		.09	2.70			
20	SPECIAL RECEP PLATE (WP)	1		.64	.64			
21	WIRENUT	26		1.22C	.32			
22	CONNECTOR (1" BOX)	6		.13	.78			
23	BOX HANGER	13		25.90C	3.37			
24	GROUND STRAP	3		.30	.90			
25	1½" ENTRANCE CAP	1		1.02	1.02	.3	.3	1.95
26	1½" GALV RIGID CONDUIT		28	33.88C	9.49	.9	.25	1.62
27	1½" GALV LOCKNUT	2		.04	.08			
28	1½" GRDG BUSHING	1		.32	.32			
29	1½" GALV PIPE STRAPS	3		.02	.06			
30	1½" METER SOCKET	1		3.25	3.25	1.0	1.0	6.50
31	100-A-3PSN SERVICE SW	1		7.45	7.45	1.9	1.9	12.35
32	1" CAST GROUND CLAMP	1		.73	.73	1.0	1.0	6.50
33	No. 1 TYPE TW COPPER WIRE		108	7.97C	8.61	2.0	2.16	14.04
34	1½" CONDUIT NIPPLE	1		.45	.45			
35	ARMORED GRD WIRE		10	16.00C	1.60			
36	DISTRIBUTION PANEL (C-B)	1		15.00	15.00	2.5	2.5	16.25
37	CONNECTOR (3/4" BOX	5		.13	65			
38	100-A CT FUSES	2		.22	.44			
39	SOLDER (100 SPLICES)	2 LB		.90	1.80	.07	7.0	45.50

Fig. 17-7. Estimating form, Sheet 1.

ESTIMATING FORM									
ITEM	COLUMN 2	3	4	5	6	7	8	9
	DESCRIPTION	No.	FEET	UNIT COST	MATERIAL COST	UNIT LABOR	HOURS	LABOR COST
40	SOLDERING PASTE	1C		.20	.20			
41	INSULATING TAPE	2R		.70	1.40			
42*	No. 14 BARE COPPER WIRE		110	.01	1.10	.05	.6	3.90
43	HALL CHIMES	1		6.45	6.45			
44	REAR CHIMES	1		4.95	4.95			
45	TRANSFORMER	1		2.55	2.55			
46	FRONT PUSH BUTTON	1		1.25	1.25			
47	REAR PUSH BUTTON	1		.30	.30			
48	BELL WIRE		160	.01	1.60	1.5	1.5	9.75
49	8-D NAILS	1½ LB		.20	.30			
50	1¼" No. 10 WOOD SCREWS	12		1.20C	.10			
51	TYPE A LTG FIXTURE	3		4.15	12.45	.5	1.5	9.75
52	TYPE B LTG FIXTURE	3		2.90	8.70	.4	1.2	7.80
53	TYPE C LTG FIXTURE	1		2.90	2.90	.4	.4	2.60
54	TYPE D LTG FIXTURE	1		2.73	2.73	.4	.4	2.60
55	TYPE E LTG FIXTURE	2		2.85	5.70	.4	.8	5.20
56	TYPE F LTG FIXTURE	1		3.70	3.70	.5	.5	3.25
57	TYPE G LTG FIXTURE	1		8.90	8.90	.7	.7	4.55
58	TYPE H LTG FIXTURE	1		.70	.70	.3	.3	1.95
59	¾" EMT (TELEPHONE)		20	.80	1.60	4.5	.9	5.85
60	¾" EMT STRAPS	2		.02	.04			
61	SWITCH BOX (TEL)	1		.27	.27	.1	.1	.65
62	¾" EMT CONNECTOR	1		.12	.12			
63	¾" EMT COUPLING	1		.10	.10			
64	SW PLATE (TEL)	1		.09	.09			
					222.02			335.34
	MATERIAL COST		222.02					
	LABOR COST		335.34					
	TOTAL COST		557.36					
	LABOR RATE $6.50-HOUR							
*	4 -INCLUDES 13 LOCAL GROUNDS AT .05 HR EACH							

Fig. 17-7. Estimating form, Sheet 2.

$557.36 + $167.21, or $724.57 approximately.

Competitive bidding for jobs, or competition, has driven listed profits down to 7 to 10 percent in many cases, and overhead expenses have been cut to 5 to 10 percent. These percentages will vary depending upon such things as business conditions, competition, amount of jobs waiting or needed, workman's benefits, geographical location of jobs and others.

Although the estimate given here is based upon use of nonmetallic sheathed cable, the principles involved may be applied to other types of installation, such as knob-and-tube, armored cable, or thin-wall conduit. The take-off operation will be identifical, except for minor variation in materials, and some labor units may require adjustment for like reason. The systematic steps outlined in the example will lead to a reasonable estimate in any case.

Short Cut Method for Estimating

After completing several wiring jobs of the same general nature, a shorter method of estimating can be worked out. The procedure is based upon a knowledge of the average cost of installing individual outlets. This cost, once determined, will be found rather consistent. For example: a figure such as $6.50 for non-metallic sheathed cable, $9.00 for EMT, $8.00 for flex or whatever the average cost was found to be. This figure is then multiplied by the number of outlets. The cost of a service installation varies according to length of service run and required size of equipment. However, a flat average rate could be reached, for example, $1.50 per ampere for service estimate, or whatever average cost was used. An average service may require 4 hours labor. Lighting fixture costs are more unpredictable because of owner preferences. Chime wiring costs may be readily "lumped." Costs of the telephone outlet varies from job to job but are small. If these variable costs are subtracted from prime cost, and the remainder is divided by the total number of outlets (lighting, switches, and receptacles), a basic outlet cost is found.

Prime cost here is $557.36. The service cost is approximately $89.50, fixtures $65, chime system $18.50, and telephone $6. Subtracting these amounts from prime cost leaves a remainder of $378.36, the basic cost for outlets alone. Since there are 62 outlets, the basic cost per outlet is equal to $378.36, divided by 62, or about $6.10 per outlet. If 30 percent is added to this value, the selling cost per outlet becomes 1.3 x $6.10, or $7.93. In round figures, this amounts to, say, $8.00 per outlet. A similar job may be quickly estimated by multiplying the number of outlets by this value, and adding enough to cover service, fixtures, chime system, and telephone—after weighing these items separately.

As his operations increase in scope, the contractor may lower the unit cost through quantity buying of material and also by cutting down on the average labor charge per hour. The usual method for accomplishing the latter is to take advantage of mass production techniques and prefabrication, and of the relationship of apprentice wage scales to those of a journeyman. In the present case, if an 80 percent apprentice is employed in company with a $6.00 journeyman, the average labor rate per hour will be lowered. The apprentice will receive: 0.8 × $6.00, or $4.80 per hour. The average rate per hour will be equal to ½ × ($6.00 + $4.80), or $5.40 per hour, instead of $6.00. The electrical contractor may also be contributing to the advancement of the electrical construction industry by helping to train qualified journeyman wiremen, which is more important over a period of time.

QUESTIONS FOR DISCUSSION OR SELF STUDY

1. List three proper pricing steps for determining an electrical estimate.
2. List some good reasons why an electrical contractor may lose money or go bankrupt.
3. What is the first thing an estimator should do with plans and specifications?
4. How is a rotameter used?
5. To what point are ceiling outlets measured?
6. Upon what basis is the branch circuit schedule constructed?
7. Upon what basis is the branch circuit material schedule constructed?
8. How are free lengths of conductors projecting from outlet boxes figured into the estimate?
9. How are labor units expressed?
10. How is the boring of holes in structural members of a frame figured in an estimate?
11. What is meant by the term "prime costs"?
12. In running 500′ of cable, by what value must the labor unit be multiplied in order to determine the number of hours?
13. If it takes 0.4 of an hour to run 100 feet of 12-3 cable, how much is the labor cost at $6.50 an hour to run 550 feet?
14. How is the basic cost per outlet determined using the short-cut method of estimating?
15. How may material costs be lowered?
16. How may labor costs be reduced?

Index

Numerals in **bold type** refer to illustrations.